The Physics of Sound

The Physics of Sound

Third Edition

Richard E. Berg
University of Maryland

David G. Stork
Ricoh Innovations, Inc.
Stanford University

San Francisco Boston New York
Cape Town Hong Kong London Madrid Mexico City
Montreal Munich Paris Singapore Sydney Tokyo Toronto

Library of Congress Cataloging-in-Publication Data

Berg, Richard E.
The physics of sound / Richard E. Berg, David G. Stork.— 3rd ed.
p. cm.
Includes index.
ISBN 0-13-145789-6
1. Sound. 2. Music—Acoustics and physics. I. Stork, David G.
II. Title.

QC225.15.B47 2005
534—dc22

2004007670

Senior Editor: *Erik Fahlgren*
Editor in Chief, Science: *John Challice*
Associate Editor: *Christian Botting*
Editorial Assistant: *Andrew Sobel*
Vice President ESM Production and Manufacturing: *David W. Riccardi*
Executive Managing Editor: *Kathleen Schiaparelli*
Assistant Managing Editor: *Beth Sweeten*
Production Editors: *Mark Corsey* and *Katie O'Connell*
Director of Creative Services: *Paul Belfanti*
Art Director: *Jayne Conte*
Cover Designer: *Bruce Kensalaar*
Managing Editor of AV Management and Production: *Patty Burns*
Art Editor: *Abigail Bass*
Executive Marketing Manager: *Mark Pfaltzgraff*
Marketing Assistant: *Larry Grodsky*
Manufacturing Manager: *Trudy Pisciotti*
Manufacturing Buyer: *Lynda Castillo*
Cover Photo: Viara Gentchev, SolarHive Studios (solarhive.com); horn courtesy Michael Henry, from the estate of the late John Entwistle, bassist for *The Who*.

Printed in the United States of America
10 9 8 7

ISBN 0-13-145789-6

Pearson Education Ltd., *London*
Pearson Education Australia Pty., Limited, *Sydney*
Pearson Education Singapore, Pte. Ltd
Pearson Education North Asia Ltd., *Hong Kong*
Pearson Education, Canada, Ltd., *Toronto*
Pearson Educación de Mexico, S.A. de C.V.
Pearson Education—Japan, *Tokyo*
Pearson Education Malaysia, Pte. Ltd.

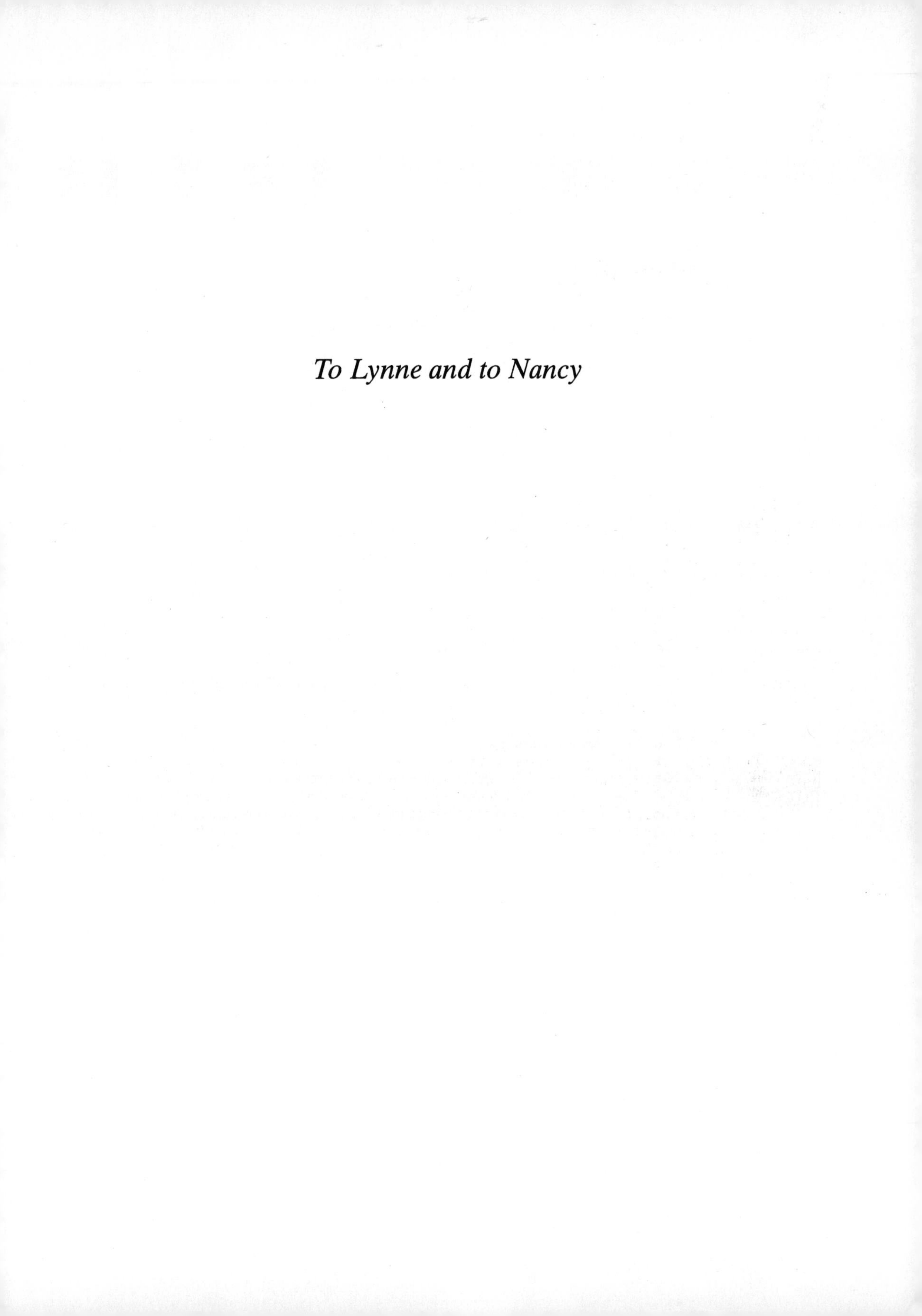

To Lynne and to Nancy

Contents

About the Authors xiii
Preface xv
Acknowledgments xvi

Chapter 1 Simple Harmonic Motion and Applications 1

1.1 Fundamental Definitions 1
1.2 Simple Harmonic Motion 4
1.3 Application to Sound 8
1.4 Damped and Driven Oscillations 11
1.5 Combinations of Simple Harmonic Oscillations 14
Summary 18
Questions 18
Problems 21
References 21

Chapter 2 Waves and Sound 23

2.1 Transverse and Longitudinal Waves 23
2.2 Fundamental Properties of Waves 29
2.3 General Behavior of Waves 34
2.4 Addition of Waves 47
2.5 Beats 50
2.6 The Doppler Effect 51
2.7 Shock Waves and Sonic Booms 58
2.8 Ultrasonics 59
2.9 Infrasonics 63
Summary 64
Questions 65
Problems 66
References 67

Chapter 3 Standing Waves and the Overtone Series 68

3.1 Transverse Standing Waves 68
3.2 Resonance and the Overtone Series 71
3.3 Mersenne's Laws 77
3.4 Longitudinal Standing Waves 79
3.5 Other Standing Waves and Applications 86
Summary 89
Questions 89
Problems 91
References 91

Chapter 4 Analysis and Synthesis of Complex Waves 92

4.1 Synthesis of Complex Waves 92
4.2 Fourier Analysis and Fourier Spectra 98
4.3 Analysis of Tone Quality 104
4.4 Resonance Curves and Musical Sound Production 107
Summary 115
Questions 116
Problems 117
References 118

Chapter 5 Electronic Music and Synthesizers 120

5.1 Combination of Waves and Modulation 120
5.2 Analog Synthesizers and Synthesis of Musical Sounds 127
5.3 Digital Synthesizers and Keyboards 133
5.4 MIDI, Computers, and Contemporary Electronic Music 136
Summary 141
Questions 141
Problems 143
References 143

Chapter 6 The Human Ear and Voice 145

6.1 Peripheral Auditory System 145
6.2 Place Theory of Hearing 148
6.3 Amplitude Response of the Ear 151
6.4 Logarithms and Sound-Intensity Scales 153
6.5 Periodicity Pitch and Fundamental Tracking 156
6.6 Aural Harmonics and Combination Tones 157
6.7 Ohm's Law of Hearing 159
6.8 Masking 160
6.9 Binaural Effects 161
6.10 Hearing Loss 163
6.11 Cochlear Implants 165

6.12 Anatomy of the Vocal Tract 166
6.13 Vocal Formants 167
6.14 Analysis of Vocal Sounds 171
Summary 177
Questions 178
Problems 180
References 180

Chapter 7 Sound Recording and Reproduction 182

7.1 Electrical Circuits and Ohm's Law 182
7.2 Reproduction and Amplification Systems 185
7.3 Microphones 188
7.4 Loudspeakers 191
7.5 Preamplifiers 196
7.6 Power Amplifiers 198
7.7 AM-FM Tuners 198
7.8 Tape Recorders 199
7.9 Digital Sound Reproduction and the Compact Disc 202
7.10 Compression and MP3 210
Summary 212
Questions 213
Problems 214
References 215

Chapter 8 Room and Auditorium Acoustics 216

8.1 Criteria in Acoustical Design 216
8.2 Problems in Acoustical Design 221
8.3 Control of Reverberation Time 225
8.4 Design of Auditoriums 227
8.5 Home Listening Rooms 231
Summary 233
Questions 234
Problems 235
References 236

Chapter 9 Musical Temperament and Pitch 237

9.1 Background and Historical Perspective 237
9.2 Historical Development of Pitch Level 242
9.3 Pythagorean Temperament 244
9.4 Just Temperament 247
9.5 Mean-Tone Temperament 249
9.6 Closed Unequal Temperaments 250
9.7 Equal Temperament 252

9.8 Recent Innovations in Tuning and Temperament 255
Summary 256
Questions 257
Problems 258
References 259

Chapter 10 Woodwind Instruments 260

10.1 History of Woodwind Instruments 260
10.2 Woodwind Tone Quality 264
10.3 Recorders 272
10.4 Flutes 276
10.5 Clarinets 280
10.6 Saxophones 282
10.7 Oboes 285
10.8 Bassoons 286
10.9 Pipe Organs 288
Summary 292
Questions 293
Problems 294
References 295

Chapter 11 Brass Instruments 297

11.1 History of Brass Instruments 297
11.2 Sound Production in Brass Instruments 303
11.3 Trumpets 308
11.4 Trombones 310
11.5 French Horns 312
11.6 Other Contemporary Brass Instruments 313
Summary 314
Questions 315
Problems 316
References 317

Chapter 12 Stringed Instruments 318

12.1 History of Stringed Instruments 318
12.2 Theory of Bowed Instruments 322
12.3 The Violin Family 325
12.4 Plucked Stringed Instruments 330
Summary 332
Questions 333
Problems 334
References 334

Chapter 13 The Piano 335

13.1 History of the Piano 335
13.2 Construction and Action of the Piano 338
13.3 Piano Strings and Sound Production 341
13.4 Recent Innovations for the Piano 345
Summary 348
Questions 348
Problems 349
References 350

Chapter 14 Percussion Instruments 351

14.1 Bar Instruments 351
14.2 Chimes and Triangles 356
14.3 Membranophones 357
14.4 Gongs, Tam-Tams, Cymbals, and Bells 362
Summary 363
Questions 364
Problems 366
References 366

Appendix A Elementary Music Theory 367
Appendix B Terms and Units 371
Appendix C Prefixes with Common Units 372

Glossary 373
Index 386

About the Authors

Professor Richard E. Berg received his B.S. degree in music from Manchester College (Indiana), with emphasis on piano and clarinet, and M.S. and Ph.D. degrees in physics from Michigan State University. After completing his Ph.D. thesis in the area of cyclotron design, he began work on the construction of the cyclotron at the University of Maryland. This work included design and construction of the external beam transport system, design of solid state radiation detectors, and support for research in nuclear physics using the cyclotron. In 1972 he became the director of the University of Maryland Physics Lecture-Demonstration Facility, which has since developed one of the largest and most diverse collections of physics demonstrations in the world. He has initiated courses in *Physics of Music*, the *Physics of Music* laboratory, and an honors course, *Nuclear Physics and Society*, involving applications of nuclear physics and radiation to contemporary society. Professor Berg has sung and played renaissance wind instruments with the University of Maryland Collegium Musicum for over 20 years. He has also played harpsichord and recorder in a smaller group known as the *Go for Baroque Ensemble*. Professor Berg has been active in physics outreach programs, annually presenting a series of public demonstration programs called *Physics is Phun*, which has been attended by more than 100,000 people since 1982. Over his career he has presented more than 500 traveling demonstration programs to area school groups and more than 300 smaller programs at the University of Maryland for visiting groups. In the photograph Professor Berg is shown demonstrating the twelve-harmonic variable frequency digital Fourier synthesizer designed and constructed at the University of Maryland.

David G. Stork is Chief Scientist of *Ricoh Innovations, Inc.,* and Consulting Professor of Electrical Engineering at Stanford University. He received his B.S. degree in physics from the Massachusetts Institute of Technology, and his M.S. and Ph.D. degrees in physics from the University of Maryland. Dr. Stork is an accomplished orchestral and chamber timpanist/percussionist, has performed in major concert halls throughout the United States, and performed on more than a dozen compact disks, including four world premier recordings. His principal research interests are in pattern classification, machine learning, and novel uses of the internet. He is an award-winning teacher (*Ralph D. Myers Teaching Award*, University of Maryland) and publishes and lectures widely on his research and scholarly topics as diverse as Renaissance painting and the relation of science fiction to science fact. His other books include *Pattern Classification* (2nd ed., Wiley 2000, W. R. Duda and P. Hart), *Speechreading by Humans and Machines* (Springer, 1996, W. M. Hennecke), *Seeing the Light* (Wiley, 1986, W. D. Falk and D. Brill), and *HAL's Legacy: 2001's Computer as Dream and Reality* (MIT 1997), the latter serving as the source for his PBS television documentary "2001: HAL's Legacy." Dr. Stork sits on the editorial boards of four international journals and is a member of IEEE (Institute of Electrical and Electronics Engineers), ACM (Association for Computing Machinery), OSA (Optical Society of America), INNS (International Neural Network Society), and the Sigma XI Honorary Research Society.

Preface

The Physics of Sound was written for an introductory course in acoustics for nonscientists. A background in neither physics nor mathematics above high school algebra is required. Traditionally, such courses have been tailored to music majors; nonmusicians either do not enroll or do not fully appreciate the physical principles because they are applied almost exclusively to musical topics. We have tried to avoid this limitation by dividing the text into three main sections.

Chapters 1–4 present the basic physics essential for virtually all topics in the text: simple harmonic motion, wave principles, resonance, standing waves, the overtone series, Fourier synthesis, and spectrum analysis. No previous musical knowledge is required to appreciate these chapters, and a brief summary of the basic musical notation used in these chapters is provided in Appendix A. Applications and illustrations come from a variety of musical and nonmusical areas. We have revised Chapter 1 with additional discussion of SHM and apply these concepts to debunking psychokinetic myths. Our discussion of wave properties of sound in Chapter 2 has been revised to include a discussion of modern applications such as noise cancellation technology, highway noise barriers, and ultramodern sonogram technology.

Chapters 5–8 illustrate the principles outlined in Chapters 1-4 and are of general interest to the musician and the nonmusician alike. The use of musical concepts and notation has been minimized so as to retain the broadest base of appeal, but the more important musical aspects of each topic are still included. For example, Chapter 7 on sound reproduction and Chapter 8 on room acoustics illustrate the principles presented in earlier chapters and are of substantial interest because of the part they play in our daily lives. We use the discussion of analog synthesizers in Chapter 5 as an opportunity to investigate the differences between sounds by studying how these sounds are created in a synthesizer. Our treatment of digital synthesizers has been updated to include use of computers as an integral part of the synthesizer system. The material on MIDI systems includes further discussion of how these systems are used to compose music and to aid musical performance using contemporary computer programs such as Sibelius. We have also added a section on physics and synthesizers in contemporary electroacoustic music composition. Chapter 6 discusses the principles of the ear and the voice at an elementary level, but has been updated with a section on cochlear implants, which have literally created a revolution in education of hearing disabled children. We have also included a section on the use of audio spectrograms in teaching language to the hearing disabled. Chapter 7 has been thoroughly revised from the second edition to include a discussion of MP3 using the concepts of spectral analysis and masking covered previously. Chapter 8 has been updated with a discussion of visual and musical features for a new auditorium at the University of Maryland. The first eight chapters contain the core material for a one-semester course in the physics of sound and music.

Chapters 9–14 are more specialized. Each of these chapters independently treats a different aspect of *musical* acoustics and is best (though not exclusively) understood by those with some musical experience. Photographs of families of contemporary instruments have been added to the chapters on woodwinds and brasses. Any of the final six chapters could be studied in class or could naturally be assigned to students on an individual basis.

Unlike most authors of elementary acoustics texts, we have treated the historical development of instruments, paying particular attention to acoustical developments. We have also tried to relate the physical principles of contemporary instruments to performance technique; the knowledge of *how* and *why* an instrument works and its limitations and problems should improve one's performance.

For the third edition, each chapter now includes a summary, with highlighted words defined in a glossary near the end of the book, and the number of questions and problems at the end of each chapter has been significantly expanded. The questions and problems for the last six chapters require greater sophistication than those for the earlier chapters and could form the basis for student projects.

A text is but one of the resources useful in helping the student to learn the material. Our course as taught at the University of Maryland uses a large number of demonstrations, audiotapes, films, and videos; information regarding these materials is available on the Web. A four-hour set of videos, "Demonstrations in Acoustics," including many of the demonstrations used in our course, is available in DVD format from the University of Maryland. A Solutions Manual (0-13-185594-8) is available from the Publisher for instructors using the text. When fully integrated, we believe that these resources will bring to the nonscience student a deeper understanding and appreciation of acoustics, the physics of sound, and of science in general.

ACKNOWLEDGMENTS

We are grateful to a number of reviewers for their detailed comments and suggestions; for the first edition: J. Gerard Anderson, University of Wisconsin, Eau Claire; Stanley H. Christensen, Kent State University; Willard Larkin, Department of Psychology, University of Maryland; Bruce Daniel, Pittsburg State University; John Guillory and Angelo Bardasis, University of Maryland; Thomas Rossing, Northern Illinois University; for the second edition, Bruce Daniel, Pittsburg State University; and Frederick Slee, University of Puget Sound; and for the third edition: David T. Bannon, Oregon State University; Clement Burns, Western Michigan University; Martin Kamela, Elon University; Luke Keller, Ithaca College; Fredrick Olness, Southern Methodist University; James McLean, SUNY at Geneseo; Aaron Lindenberg, University of California,

Berkeley; Herbert Jaeger, Miami University (Ohio); Marilyn F. Bishop, Virginia Commonwealth University, and Michael Grady, SUNY at Fredonia. Deep thanks go to our students, many of whom have provided thoughtful comments on both the book and the course. A special thanks is extended to Joan Wright Hamilton for her excellent work on the figures for the first and second editions.

We have made the best efforts to make this text as error-free as possible, but the final responsibility for all errors lies with the authors, and we welcome all comments and corrections. Please contact Professor Richard E. Berg at the University of Maryland, reberg@physics.umd.edu.

RICHARD E. BERG
DAVID G. STORK

Chapter 1

Simple Harmonic Motion and Applications

In this chapter we shall first define the physical terms that will be used throughout the book and then discuss the single concept most basic to the study of sound—simple harmonic motion—and its direct application to sound and resonance.

1.1 Fundamental Definitions

To understand physics, we must first understand its language. Its definitions are exact, formulated in the language of mathematics. For our purposes, however, more qualitative and operational definitions will suffice. The most fundamental quantities in physics are discussed next; they will be used regularly in our study of acoustics. Other important quantities will be defined throughout the book as the need arises.

Position, length, or distance (symbols x and y)

This quantity is simply a measure, in the normal everyday sense, of how far one point is from another. Units of length can be miles, feet, inches, or meters (m). We shall primarily study motion in one dimension, as shown in Figure 1-1. Position relative to the origin of coordinates ($x = 0$) can be positive (to the right) or negative (to the left); dots have been placed on the line at $x = 6$ centimeters (cm), $x = 2$ cm, and $x = -4$ cm. We shall regularly use the metric system; in this system the prefix *deci-* means 1/10, *centi-* means 1/100, *milli-* means 1/1000, *micro-* means 1/1,000,000, *kilo-* means 1000, and *mega-* means 1,000,000.

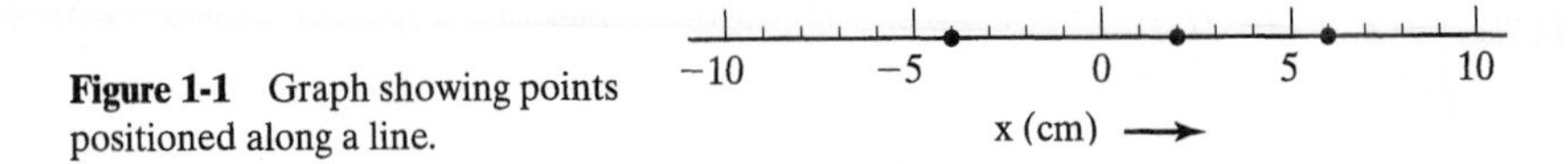

Figure 1-1 Graph showing points positioned along a line.

Some metric length units are meters (m), kilometers (km), centimeters (cm), and millimeters (mm), where 1 cm = 0.01 m, 1 mm = 0.001 m (so 1 m = 100 cm = 1000 mm), and 1 km = 1000 m. One meter is approximately 39.37 inches; 1 inch (in.) is 2.54 centimeters.

Time (symbol *t*)

For our purposes, we simply observe that time as we know it passes and is divided into seconds (s), hours, days, and so on. Units of time that will be used regularly in discussing sound include the millisecond (ms), where 1 ms = 0.001 s (so 1 s = 1000 ms), and the microsecond (μs), where 1 μs = 0.000001 s = 0.001 ms (so 1 s = 1,000,000 μs).

Velocity or speed (symbol *v*)

Velocity and speed, which should not be used synonymously, have different definitions in physics. *Speed* is distance traveled per unit of time. *Velocity*, a more general term, refers to both the speed and the direction of the motion, and is therefore a *vecto*r quantity. An object moving around a circle at a constant speed is continually changing its direction; therefore its velocity is continually changing. The examples that we use will be primarily one-dimensional; we will refer to velocity to include both the magnitude and the direction of the motion.

For example, if the dot in Figure 1-1 moves uniformly from $x = 2$ cm to $x = 6$ cm in 1 s, its velocity is +4 centimeters per second (+4 cm/s). If it moves from $x = 6$ cm to $x = -4$ cm in 2 s, $v = (-10 \text{ cm})/(2 \text{ s})$ or $v = -5$ cm/s. Convince yourself that 1 m/s = 1 mm/ms.

The speed of sound in air is approximately 345 m/s or 1100 ft/s and varies with temperature; the speed of light is approximately 300,000,000 m/s or 186,000 miles/s.

Acceleration (symbol *a*)

Acceleration is defined as change in velocity per unit of time. For example, suppose a ball rolls down a ramp starting with zero velocity (i.e., it starts at rest), and after 1 s its velocity is 1 cm/s, after 2 s its velocity is 2 cm/s, and after 3 s its velocity is 3 cm/s. The increase in velocity is 1 cm/s for each second it is rolling down the ramp, which can be written $a = (1 \text{ cm/s})/(\text{s})$ or $a = 1 \text{ cm/s}^2$. Technically, acceleration is a vector quantity too, because it points in a particular direction. Deceleration, or slowing down, is acceleration in the direction opposite to the motion. The acceleration of gravity is 9.8 m/s^2—that is, if you drop a heavy object near the surface of the earth, it will gain speed at a rate of 9.8 m/s for each second it falls.

Mass (symbol *m*)

Mass is a measure of the amount of matter in an object. A 1-gram mass of iron, for example, contains a certain number of iron atoms, which remains constant on the earth, on the moon, or in outer space. The most often used units of mass are the gram (g) and the kilogram (1 kg = 1000 g). One kilogram weighs about 2.2 pounds (lb) on the surface of the earth. A 5-cent coin has a mass of about 5 g.

Density (symbol ρ)

Density, or mass per unit volume, is represented by the symbol ρ, the Greek letter *rho*. When the pressure in the atmosphere or a wave increases, so does the density of the air through which the wave is propagating. The *mass per unit length,* or linear density, of stretched wires is important in the design of stringed instruments and the piano.

Force (symbol F)

Force can be thought of as a push or a pull. If you push or pull on a mass at rest with some net force, it will begin to move in the direction of the force. *Weight* is the *force of gravity* pulling a mass toward the center of the earth or other celestial body and will be different on the surface of the moon or some other planet; mass will not. In deep outer space, where there is almost no force of gravity, all objects are *weightless*, but they still have mass.

The metric unit of force is the newton (N), named after the physicist Isaac Newton (1642–1727). A 100-lb boy standing on the earth is held to the ground by a downward gravitational force, his weight, of about 445 N. The surface of the earth pushes up on the boy's feet with an equal force. These two forces balance, and the boy does not move up or down. On the moon, his weight would be only about 74 N, because the force of gravity is less on the surface of the moon than on the earth.

Pressure (symbol p)

Pressure is defined as force per unit area, and can be measured in units of pounds per square inch or in pascals (Pa), where a pascal is a newton per square meter. An example of pressure can be seen by considering an elephant walking on a sandy beach. Although the elephant weighs a great deal, its feet are very large, and the pressure, or force per unit area of its feet, exerted on the sand is relatively small. The weight is spread out over the large area. The elephant therefore leaves very shallow footprints in sand. On the other hand, a woman wearing high-heeled shoes exerts more pressure on the sand, because her smaller weight is exerted on the very small area of the heels. The prints of the heels in the sand would be deeper than the footprints of the elephant.

Air pressure is exerted on everything in contact with air; rapid changes in air pressure cause vibrations of the eardrum, which we hear as sound. Usually, air pressure on the outside of the eardrum is balanced by pressure in the ear, because air can flow into the mouth and then through the eustachian tube to the middle ear, behind the eardrum. Congestion of the eustachian tube restricts the airflow from the atmosphere to the space behind the eardrum. Slow changes of air pressure in the atmosphere can then create pressure differences between the atmosphere and the space behind the eardrum, causing large forces on the eardrum, which can be painful. Such an effect is often observed at the beginning and end of airplane flights, when the air pressure changes with changes in altitude. Atmospheric pressure on the surface of the earth is about 14.7 lb/in^2 or about 100,000 Pa.

Again, consider the 100-lb boy. Suppose the area on the bottom of each foot is 0.01 m^2. His weight, 445 N, is distributed over both feet, 0.02 m^2, and the pressure on the bottom of each foot is 22,250 Pa. When the boy stands on one foot, the area over which his weight is distributed is halved, so the pressure on his foot doubles to 44,500 Pa.

1.2 Simple Harmonic Motion

The single most important concept in the study of waves and sound is that of simple harmonic motion (SHM). We shall now study SHM and some of its properties, and in Chap. 2 demonstrate how SHM is basic to waves and sound.

Periodic motion is any type of motion that repeats itself after successive equal time intervals. Some examples of periodic events are twirling a rock on a string around your head, the rotation of the earth on its axis, the revolution of the moon around the earth, a mass bouncing up and down on the end of a spring, a swinging pendulum, a blinking warning light, the vibration of a tuning fork, and the vibration of the reed of a clarinet or a singer's vocal folds, producing a constant musical tone.

Simple harmonic motion is a specific type of periodic motion (for instance, a mass bouncing up and down on the end of a spring) that arises from the conditions described in the next paragraph and whose graph is a sine or cosine shape, as shown in Figure 1-2. This graph shows the position of one dot in Figure 1-1 over time. Notice that the motion repeats itself after 2 s in this example, and, as with all SHM, it has a particular characteristic smooth shape. Other names for SHM or waves of this character are simple, pure, sine, cosine, or sinusoidal motion or waves.

Two conditions must be met to produce SHM in a mechanical system. First, an *equilibrium position* must exist; that is, there must be one position at which the object executing the motion would remain at rest if placed there and to which it returns if displaced and released. Second, the force tending to pull the object back to its equilibrium position must be a *linear restoring force:* that is, the force must be linearly proportional to the distance of the object from the equilibrium position, as explained next.

As an example of these two conditions, let us consider a particular mass hanging on the end of a flexible spring attached to a fixed point, as shown in Figure 1-3. The equilibrium position is the position of the mass when it is hanging at rest directly below its suspension point. We consider only vertical motion of the mass along the direction of the extended spring, above and below the equilibrium position. If the mass is lifted above its equilibrium position, the force of gravity and perhaps also the compression of the spring will force the mass back down, toward the equilibrium position. Conversely, if the mass is pulled down, the spring tends to pull it back up, toward the equilibrium position. If it requires 1 N of force to lift the mass up 1 cm, it will require 2 N for a 2-cm

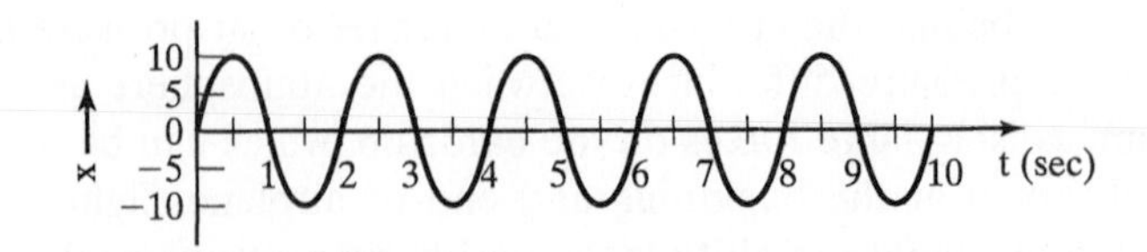

Figure 1-2 Graph of simple harmonic motion (SHM).

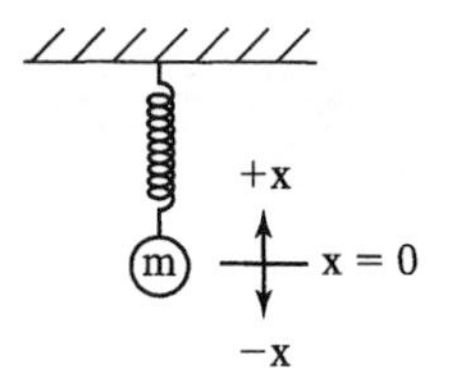

Figure 1-3 Mass hanging from a spring; equilibrium position is labeled $x = 0$.

displacement, 3 N for a 3-cm displacement, and so forth, because for this system the restoring force is linear. Likewise, if the restoring force is linear it will require 1 N of force to pull down the mass 1 cm from the equilibrium position as well. This is an example of a linear restoring force with a force constant (or force per unit of stretch) of 1 N/cm. Such a linear system is said to obey Hooke's law, named after the English physicist Robert Hooke (1635–1702).

When the mass on the spring in the example in Figure 1-3 is lifted 3 cm above its equilibrium position and released, it begins to accelerate downward, passing through its equilibrium position, and executes SHM as shown in Figure 1-4. At time $t = 0$ the mass is released from rest 3 cm above ($x = +3$ cm) the equilibrium position ($x = 0$). The mass starts moving, builds up speed as it passes through the equilibrium point, then slows down to zero velocity by the time it reaches the point $x = -3$ cm below the equilibrium point. The mass then starts upward, moving faster until it goes through the equilibrium point, then slows down to zero velocity by the time it reaches its original position, $x = 3$ cm. The motion then repeats itself every 2 s, as shown on the graph.

The maximum displacement A of the mass from the equilibrium position (in either direction) is called the *amplitude,* and the repetition time T is called the *period* of the SHM. For this example, $A = 3$ cm and $T = 2$ s. Note that the amplitude of any oscillation, including SHM, can be expressed in units appropriate to the system under consideration: for example, centimeters for a mass oscillating on a spring, pressure units like pascals for a sound wave traveling through air, and volts for an electrical signal in a stereo set.

The period T is related to the frequency f, which is the number of periods (or cycles) per second:

$$f = \frac{1}{T} \text{ or } T = \frac{1}{f}. \tag{1.1}$$

A high-frequency oscillation has many vibrations per second; to fit many vibrations into a 1-second time interval, the period must be short. Thus, a high-frequency oscillation has

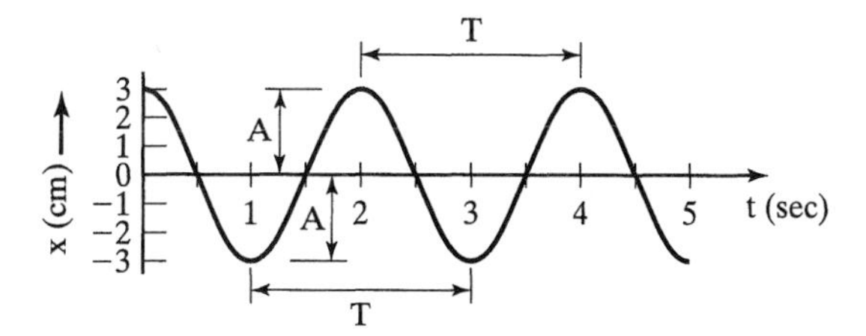

Figure 1-4 SHM with amplitude $A = 3$ cm and period $T = 2$ sec.

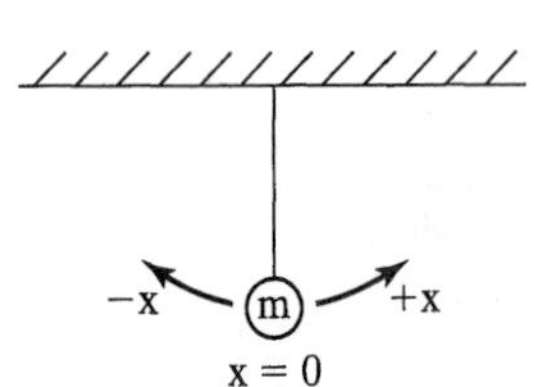

Figure 1-5 Pendulum consisting of mass hanging on end of string; mass swings left to right in the plane of the paper.

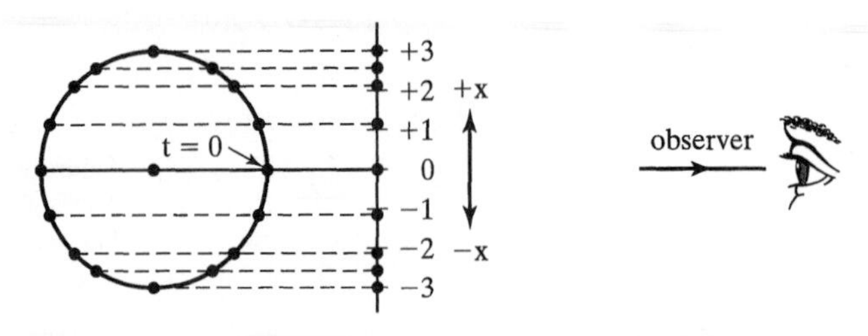

Figure 1-6 SHM as the projection of uniform circular motion along a line in the plane of the paper.

a short period; a low-frequency oscillation has a long period. If we know that the period is 2 s, the frequency f is

$$f = \frac{1}{T} = \frac{1\ \text{cycle}}{2\ \text{s}} = 1/2\ \text{cycle per second} = 0.5\ \text{Hz}. \tag{1.2}$$

The unit of frequency is the hertz (Hz), which is 1 cycle/s, named after Heinrich Hertz (1857–1894). For a 500-Hz audio-frequency oscillation, the period is

$$T = \frac{1}{f} = \frac{1\ \text{cycle}}{500\ \text{cycles per second}} = \frac{1}{500/\text{s}} = 0.002\ \text{s} = 2\ \text{ms}. \tag{1.3}$$

Thus the period of a 500-Hz oscillation is 2 ms. Check to see that if the period of an audio-frequency tone is 0.5 ms, its frequency is 2000 Hz or 2 kilohertz (kHz).

Another example of SHM is a pendulum consisting of a mass or "bob" hanging by a string from a fixed point and free to oscillate back and forth in a plane, as shown in Figure 1-5. The equilibrium position is the point directly below the suspension point, where the pendulum will hang motionless. The force required to pull the pendulum away from its equilibrium position is linearly proportional to the displacement of the bob from equilibrium, as shown in Figure 1-5, so long as that distance is small. When displaced through some small angle and released from rest, the bob executes SHM in the plane of the paper. Its motion can be described by a graph similar to the one in Figure 1-4.

Some important details of SHM can be illustrated by comparison of SHM with uniform circular motion, which is the motion of an object around a circle with constant speed. Uniform circular motion, viewed from a distant point in the plane of the motion, is SHM, as can be seen using Figure 1-6; SHM is the projection of uniform circular motion on a line in the plane of the motion. Points are shown on the circle corresponding to the position of the object at 16 equal time intervals for one period of the motion (one complete circle). The projections of these points on the line are shown and give the position of an object, such as a mass on the end of a spring, executing SHM along the line; the dots represent equal time intervals. The radius of the circle is 3 units; thus the amplitude of the projected motion is 3 units.

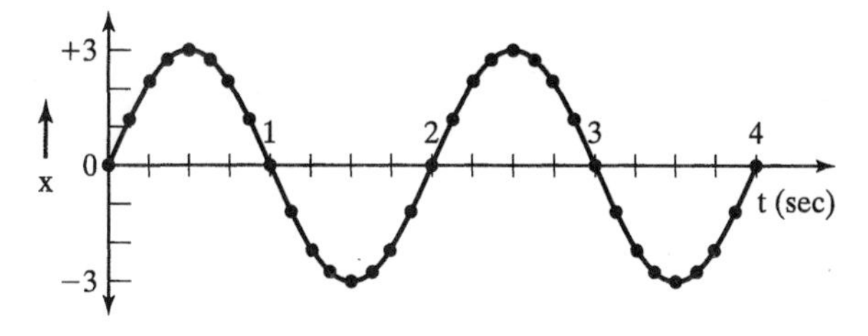

Figure 1-7 SHM position versus time at equal time intervals.

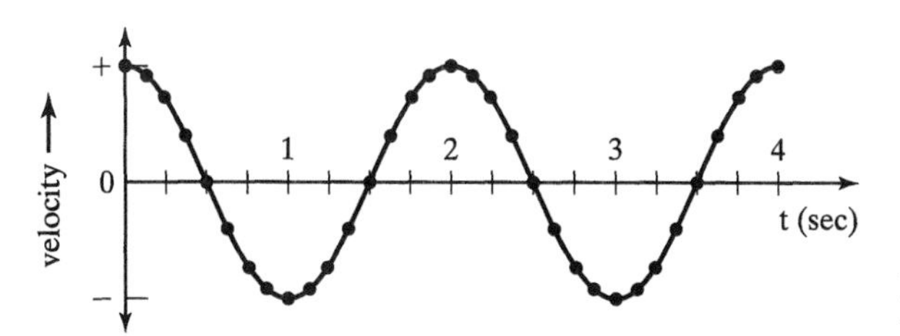

Figure 1-8 SHM velocity versus time at equal time intervals.

Notice that the projection of the distance traveled in successive equal time intervals is not the same and that therefore the velocity along the vertical line is continually changing. The velocity is least when the distance between two successive points is smallest, and, in fact, the velocity is zero when the projection is at either extreme of displacement. That is, for a brief instant of time, the point is moving neither up nor down, but is at rest. If the x position of the point in Figure 1-6 is plotted after each time interval, beginning with $x = 0$, the graph, shown in Figure 1-7, is a sinusoidal or SHM curve. Starting at $t = 0$, the point moves a large distance in the positive direction in the first time interval and less in each succeeding time interval until it reaches its maximum position. Its velocity is thus large and positive at $t = 0$ and decreases to 0 at $t = 0.5$ s, when the largest positive position is reached, as shown in Figure 1-8. The velocity of the point at any given time can be determined by examining the slope of the graph in Figure 1-7. When the slope of the graph is large, the velocity is high; a large distance is traversed in a given time. When the slope of the graph is 0, as at $t = 0.5$ s, the velocity is 0. (The slope is graphed in Figure 1-8.) The direction of the motion then changes; that is, the velocity becomes negative, and the point moves down through 0 toward its extreme negative position. Looking at the spacing of the points, one can see that at $t = 1$ s, as the point passes through $x = 0$, its velocity has the largest negative value, as shown in Figure 1-8. As time progresses, the position and velocity repeat once per period, as shown. Make sure you understand how position and velocity of SHM are correlated, as shown in the figures. Note specifically that velocity as a function of time for SHM also has the unique SHM shape.

Also by analogy between uniform circular motion and SHM, we can divide one period of SHM into 360° of phase ϕ (lowercase Greek letter *phi*), as shown in Figure 1-9. If one period is 360° of phase, the half-period is 180° and the quarter-period is 90°. Two curves are said to be *out of phase* if they differ in phase by 180° at all times, as shown in Figure 1-10. One of the curves would have to be moved forward or backward in time by one half-period, or 180° of phase, to make the two curves appear the same, or *in phase*. In

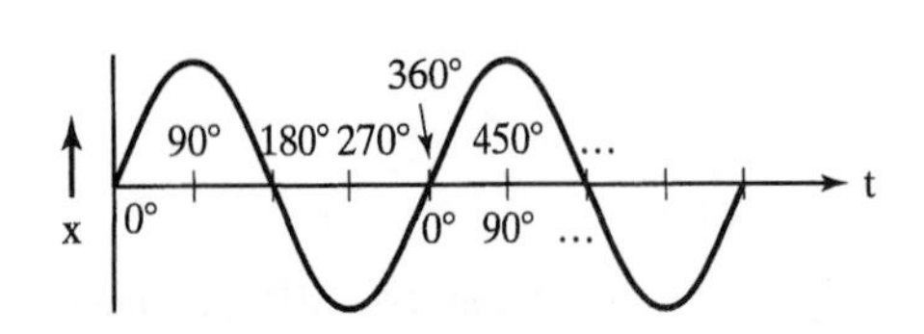

Figure 1-9 Phase of SHM curve.

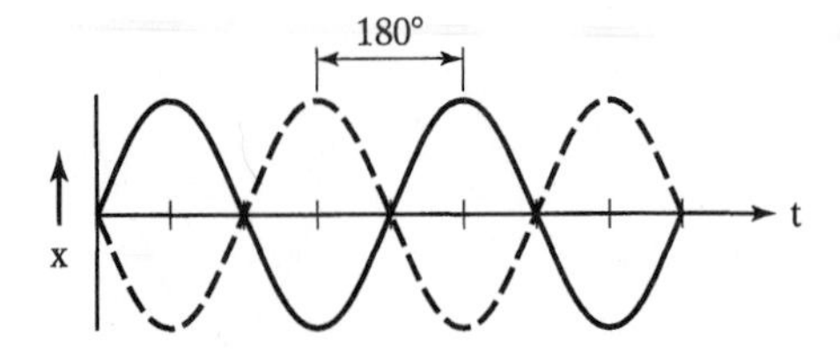

Figure 1-10 Two SHM curves differing in phase by 180° (out of phase).

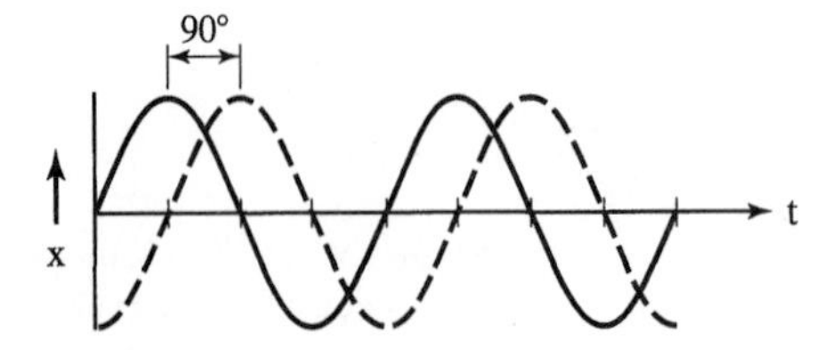

Figure 1-11 Solid curve 90° ahead of dashed curve.

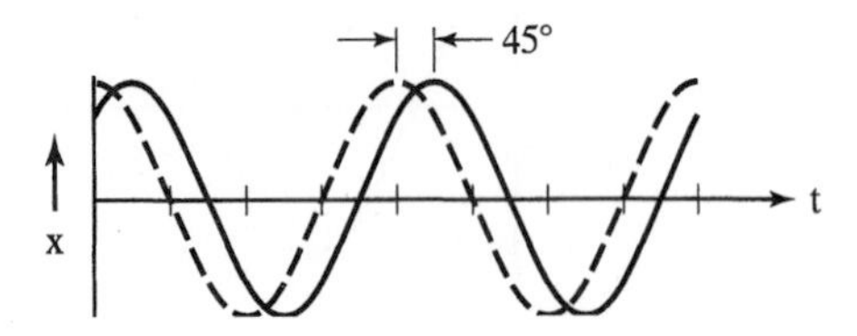

Figure 1-12 Solid curve 45° behind dashed curve.

Figure 1-11 the solid curve is 90° in phase ahead of the dashed curve; thus the peak of the solid curve occurs earlier in time by one quarter-period than the peak of the dashed curve. The solid curve in Figure 1-12 is 45° behind the dashed curve. Relative phase between two curves is only well defined when the two periods (and thus frequencies) are the same.

1.3 Application to Sound

Our knowledge of simple harmonic motion can be applied directly to sound. In our study of acoustics we shall generally deal with frequencies in the range from 20 Hz to 20 kHz, the "audible frequency range" that human ears can detect. Sound waves are changes in air pressure occurring at frequencies in the audible range. The normal variation in air pressure associated with a musical instrument played quietly is about 0.002 Pa. The smallest pressure variation that can be heard is about 0.00002 Pa, whereas the pressure variation that produces pain in the ear is about 20 Pa. Normal atmospheric pressure is about 100,000 (10^5) Pa. The maximum change in atmospheric pressure due to changes in weather is a few percent of this average value. Notice that the *variation* in pressure of a sound wave is a tiny fraction of the ambient pressure.

We can correlate the physical properties of sound waves with our perception of pitch, loudness, and tone quality; this can be readily demonstrated using an electronic wave generator and a loudspeaker. Changing the frequency of the oscillator varies the pitch; changing the amplitude of the oscillator's signal varies the loudness.

The higher the frequency of a wave, the shorter its period and the higher its pitch. For two waves of the same frequency, the one with the greater amplitude sounds louder. We shall discuss in Chap. 6 the variation in loudness with frequency for a constant-intensity wave and the slight variation in pitch with intensity. Shown in Figure 1-13 are waves that (a) remain at the same pitch but get softer, (b) get higher in pitch but remain approximately at the same volume, and (c) become simultaneously softer and lower in pitch.

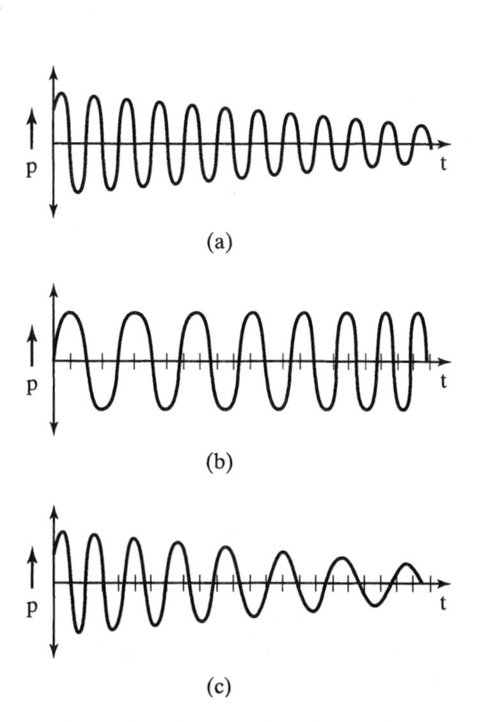

Figure 1-13 Audible waves that (a) remain constant in pitch but become softer; (b) become higher in pitch but remain at nearly the same volume; and (c) become simultaneously softer and lower in pitch.

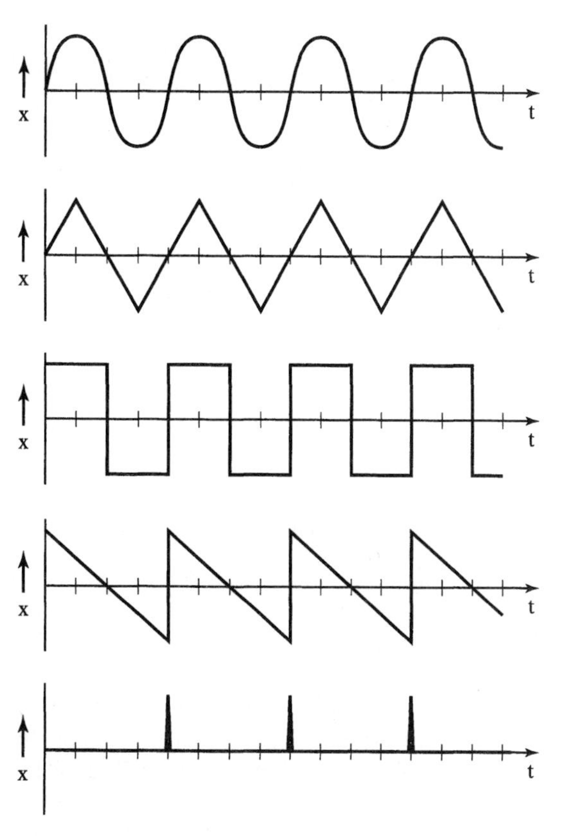

Figure 1-14 Sine wave, triangular wave, square wave, sawtooth (ramp) wave, and pulse train (series of pulses) of the same frequency.

Waveforms other than the sine wave will play an important part in our study of sound. Shown in Figure 1-14 are a sine wave, a triangular wave, a square wave, a sawtooth (or ramp) wave, and a pulse train (series of pulses). These waves all have the same period, so they all have the same frequency and pitch. The important difference between them is that they sound different; they differ in tone quality, or timbre. The variation in tone quality for different wave shapes can be demonstrated using an oscillator that can produce different wave shapes. In Chap. 4 we shall study what makes these tones sound different and under what conditions two waves with different shapes might sound very similar.

Wave shapes can be displayed using an *oscilloscope*. An oscilloscope, or "scope," is a device for displaying an electronic signal—that is, translating it into visible form on a screen. The signal from a microphone or any other electronic signal is traced onto the oscilloscope screen by a beam of electrons produced in the scope. If no signal is fed in, the beam moves at a constant velocity from left to right in a straight line, taking a specified time to cross the screen. Input of a signal causes vertical motion of the beam as it moves left to right and thus traces out the graph of the signal voltage as a function of time. Both axes can be calibrated, so we can observe the amplitude of the signal and how rapidly it varies. The graphs of sounds and other waves obtained using an oscilloscope are like the

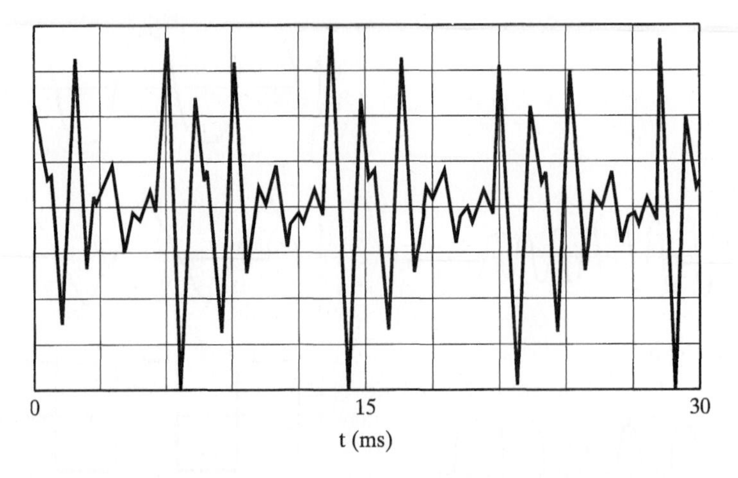

Figure 1-15 The wave shape of a male voice singing the vowel "ah" on the note C_3 = 130.81 Hz.

graphs previously shown in the various figures. Computer software has been developed as an alternative to using an oscilloscope. The signal from a microphone can be digitized using a standard computer sound card and plotted on the computer screen using readily available software.

Just as each electronic wave has a particular shape and tone, the sound wave of a note played on a musical instrument has a shape, called its *waveform*. A musical tone, which consists of periodic oscillations of air pressure, is converted into an electrical signal by a microphone. This electrical signal, or voltage, which is proportional to the pressure of the air in the sound wave, can be displayed using an oscilloscope as voltage versus time. The waveform of a male voice singing the vowel sound "ah" at C_3 = 130.81 Hz is shown in Figure 1-15. From the graph, its period is about 7.5 ms, so its frequency would be about 133 Hz.

All the preceding waveforms have one very important common feature: they are periodic. Waves do not have to be periodic; Figure 1-16 shows a wave shape of "noise,"

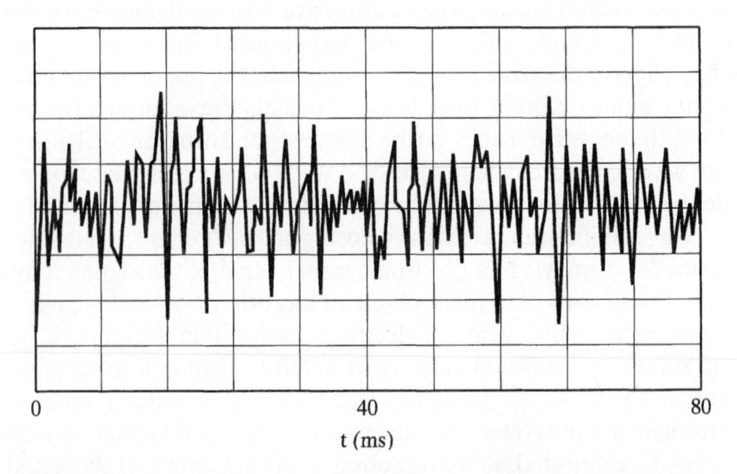

Figure 1-16 The wave shape of "noise" from a Moog Rogue analog synthesizer.

which contains no observable periodicity. This is the difference between musical sounds, which are periodic, and nonmusical sounds, or noise, which are nonperiodic. In Chaps. 4 and 5 we shall discuss the nature of several types of noise and the uses of noise in musical instruments and electronic music synthesizers.

1.4 Damped and Driven Oscillations

Consider a pendulum as in the example in Sec. 1.2. If it is started into motion by moving the bob to one side and releasing it from rest, it will not continue to oscillate forever, but will slowly decrease in amplitude, losing energy due to air resistance and friction in the suspension, until it stops. This process, called *damping,* is graphed in Figure 1-17. In this case, the damping is very slow (try a pendulum), but under certain circumstances it can be very rapid, as in the damping of a pendulum under water or a plucked guitar string. For the guitar, while the amplitude decreases, the period remains constant until the motion stops; that is, the pitch remains almost constant while the note becomes continuously softer. In damped harmonic motion the period remains constant while the amplitude decreases.

Now consider the case of driven, or forced, oscillations of a pendulum, as illustrated in Figure 1-18. If we pull the pendulum through a small displacement in the $-x$ direction and release it from rest at $t = 0$, it will undergo SHM. Suppose each time the pendulum reaches its extreme $-x$ position, we give it an additional sharp force, or push, in the $+x$ direction; this will cause the amplitude of the motion to increase slightly each cycle, as shown; this is similar to pushing a child on a swing. Arrows indicate the

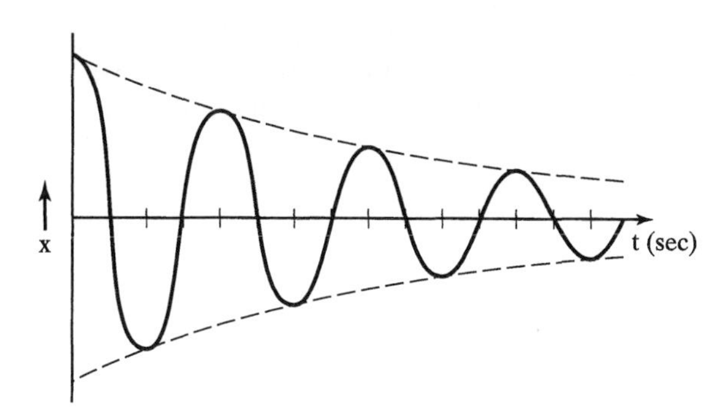

Figure 1-17 Damped harmonic motion.

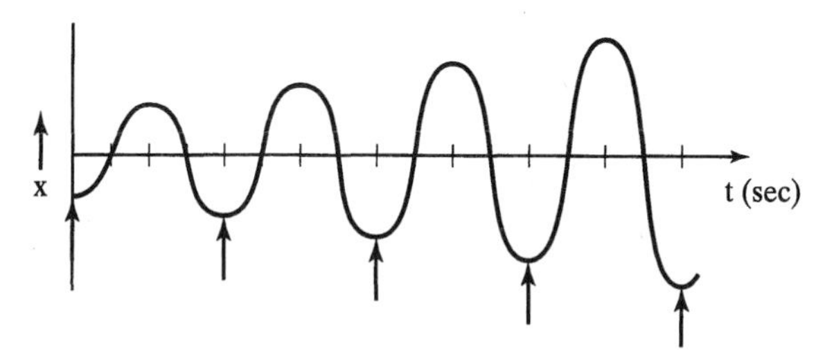

Figure 1-18 Driven harmonic motion with the frequency of applied force the same as the natural frequency of the pendulum, so that the force remains in the same phase relationship to the motion. Arrows denote times when the driving force is applied.

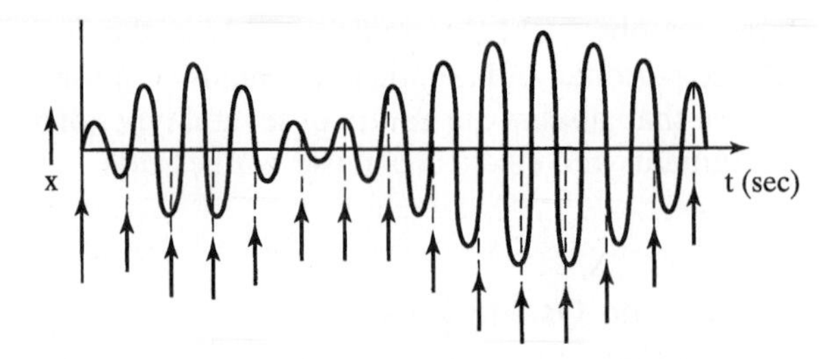

Figure 1-19 Driven harmonic motion with the frequency of applied force different from the natural frequency of the oscillator, resulting in continuous phase change between the oscillation and the force. Arrows denote times when the driving force is applied.

times at which the force is applied. This growth in amplitude resulting from application of a periodic force is called *driven* or *forced harmonic motion*. Although the amplitude increases with time, the period remains constant.

It is important to recognize that this continual increase in amplitude can occur only if the applied force continually comes at the right time, or at the proper phase, with respect to the motion of the pendulum bob; the force and the motion of the bob are then said to be *resonant* or *in resonance* with each other. In general, resonance can occur whenever the frequency of the driving force is the same as the *natural*, or normal, frequency of the oscillating system. If started at the proper phase, this phase relationship will continue. If the frequencies are the same, the system will adjust to the proper phase relationship for a resonance to occur.

Suppose now that the frequency of the driving force is slightly different from the natural frequency of oscillation of the pendulum. If the sharp force starts in phase with the swing of the pendulum, after some time the force and oscillation will become out of phase. The force, which originally was causing an increase in the amplitude of the motion, will now be opposing, or tending to decrease, the motion. This situation is shown in Figure 1-19; again, arrows indicate the force. The continual phase change between the force and the motion can easily be observed.

Another type of resonance, called a *coupled* or *coupling resonance,* occurs when two mechanical oscillatory systems with the same (or simply related) natural frequency are connected, or *coupled*, mechanically, so that vibrational energy can be transferred from one oscillator to the other.

As an example of coupled resonance, consider a set of pendula attached to a rod that can rock slightly in loosely fitting holes in the frame in which it is mounted, as shown in Figure 1-20. If the pendulum at the left is started in motion (by lifting the bob out of the plane of the paper and releasing it), its swinging will start the rod rocking back and forth at the same frequency. The rocking motion of the rod will then act as the driving force for the other five pendula. The middle pendulum of the five on the right has the same length and thus the same natural frequency as the moving pendulum on the left; it will therefore be driven into oscillation by the pendulum on the left. The other four, having different lengths, and therefore different frequencies, are not driven

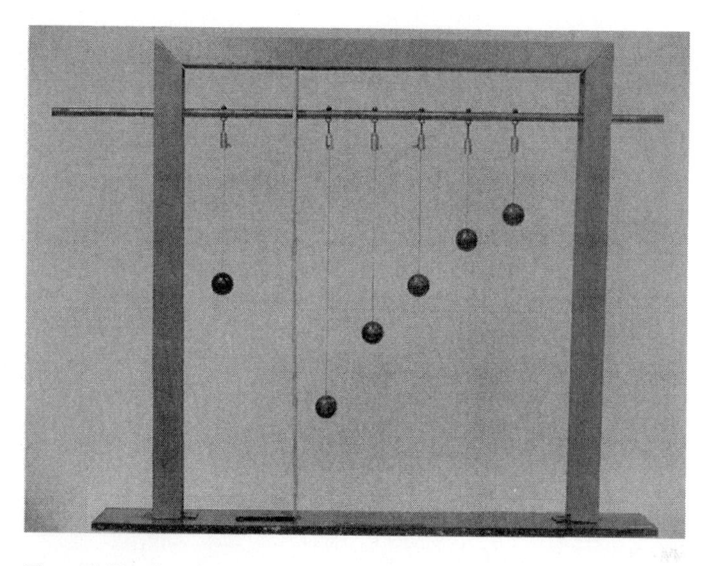

Figure 1-20 Coupled pendulum system. Suspension rod is free to rock in holes in the support plates.

to large amplitudes. In fact, in this demonstration the motion transfers almost completely back and forth between the two identical pendula, while the other four pendula, not being in resonance, execute smaller oscillations similar to those shown in Figure 1-19. Their motion goes in and out of phase with the motion of the driving pendulum at the left, so the amplitudes of their oscillations will alternately increase to a small level and then decrease to zero.

One of the most dramatic mechanical resonances arose when, owing to the action of the wind, certain oscillations built up in the Tacoma Narrows Bridge. Eventually the amplitude of the oscillations exceeded the elastic limit (breaking point) of the material in the bridge and caused its collapse on November 7, 1940.

Another example of resonance can be illustrated by using two identical tuning bars, say at 440 Hz. One is struck, and then held near the other. A resonance condition exists between the two tuning bars, with air providing the coupling, and *sympathetic vibrations* of the second tuning bar result. If either of the tuning bars is tuned slightly higher or lower, at 441 Hz for instance, the two do not have the same natural frequencies; no resonance condition exists, and therefore no significant sympathetic vibrations result.

Pendula and simple harmonic motion have found their way into magic and the occult. One of the topics for charlatans throughout the ages has been in communication using "brain waves" or other indirect forms of communication, or sending out "brain-wave vibrations" to make something move, as in the case of the Ouija® board. Mental control over physical motion, whether or not it exists, is called *psychokinesis*. Debunking this concept has been the subject of serious scientific research for over 100 years. A simple psychokinetic device, the *psychoacoustic vibration transducer*, consists of a straw with four pendula of different lengths made from thread and paper clips, as shown in

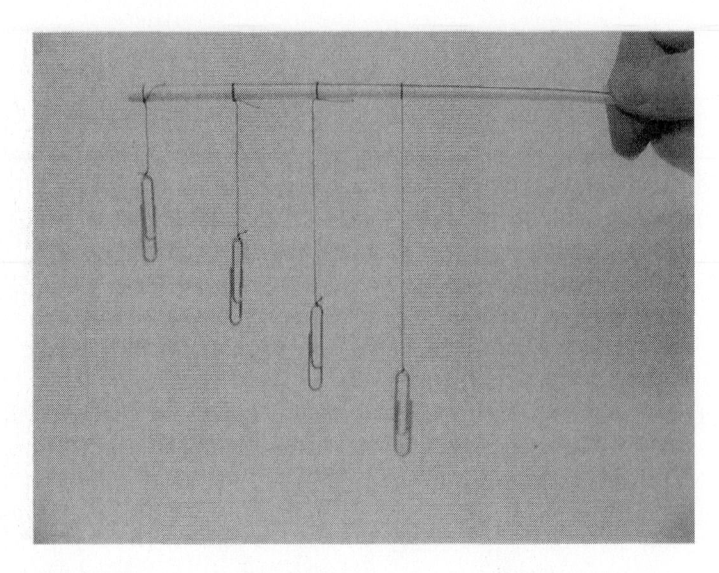

Figure 1-21 A carefully crafted "psychoacoustic vibration transducer."

Figure 1-21. It is claimed to convert low frequency "brain waves" or *psychoacoustic* vibrations from the participants into vibrations of one of the pendula. The demonstrator holds the device up in front of a group of people, stops all motion of the pendula, and asks the group to concentrate their "psychic" energy on a pre-selected pendulum known to the entire group as well as the demonstrator. Shortly after the group begins to focus on that pendulum, it begins to move, attaining a large amplitude of oscillation, while the others either remain at rest or oscillate with much less amplitude.

In fact, there is no such thing as psychoacoustic waves, and this explanation is entirely without scientific basis. Although members of the group may be inherently skeptical, unless they understand the concept of driven mechanical resonance they are compelled to accept the demonstrator's phony explanation. What is really happening is a driven mechanical resonance. The demonstrator moves the straw back and forth almost imperceptibly at the natural frequency of the selected pendulum. That pendulum experiences a *resonance* and the amplitude of its oscillation rapidly begins to increase, like the pendulum shown in Figure 1-18. The other pendula react like that shown in Figure 1-19; if they get in phase with the driving force they will begin to oscillate, but a short time later the motion becomes out of phase with the driving force and is rapidly reduced.

If you have any doubts that people actually believe this nonsense, type "psychic pendulum" or "psychokinesis pendulum" into your Web browser's search engine.

1.5 Combinations of Simple Harmonic Oscillations

Several interesting applications of simple harmonic motion involve two or more simple harmonic oscillations occurring simultaneously. One particularly beautiful example of this is the children's drawing toy called the *spirograph*®, which uses a circle rotating within a larger circle to create beautiful patterns made up of two mutually perpendicular

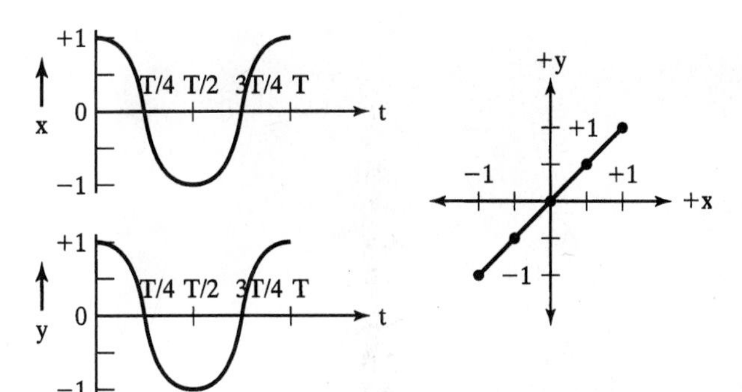

Figure 1-22 Reference graphs (left) and Lissajous figure with $A_x = A_y$ and $f_x = f_y$. The x and y oscillations are in phase.

oscillations. A number of Web sites discuss the mathematical nature of these oscillations, and the "Spirograph Nebula" is so named for its similarity to the drawings produced by this toy. Laser light shows often employ this technique to create fascinating patterns of laser light on clouds or walls. These patterns, called *Lissajous figures* after the French mathematician Jules Antoine Lissajous (1822–1880), have been used by physicists in comparing frequencies and phases of oscillations, among a host of other applications.

How these patterns are created can be understood by referring to Figures 1-22 and 1-23. Consider first the combination of x and y motion with equal amplitudes, in phase, as shown in Figure 1-22. One period T is shown in the *reference graphs* at the left; the motion repeats itself after each period. Shown at the right of the graphs of x and y is the curve traced out by a point executing these motions simultaneously. Following the motion, we notice that at $t = 0$, $x = 1$ and $y = 1$; at $t = T/4$, $x = 0$ and $y = 0$; at $t = T/2$, $x = -1$ and $y = -1$; at $t = 3T/4$, $x = 0$ and $y = 0$; and at $t = T$, $x = 1$ and $y = 1$ (the same as $t = 0$). In fact, $y = x$ for any time t, which is simply the equation of a line at an angle of 45° with respect to the axes, as seen. Now suppose that the relative phase of the oscillations is changed so that the x oscillation is 90° in phase *ahead* of the y oscillation. The oscillations, along with the resultant Lissajous figure, are shown in Figure 1-23. In this case, the point moves counterclockwise in a circle of radius 1: at $t = 0$, $x = 0$ and $y = -1$; at $t = T/4$, $x = 1$ and $y = 0$; at $t = T/2$, $x = 0$ and $y = 1$; at $t = 3T/4$, $x = -1$ and $y = 0$; and at $t = T$, $x = 0$ and $y = -1$ (the same as $t = 0$). Two other phase relationships are interesting: the x motion *out of phase* with the y motion, and the x motion 90° behind the y motion. These two cases are the subject of an exercise.

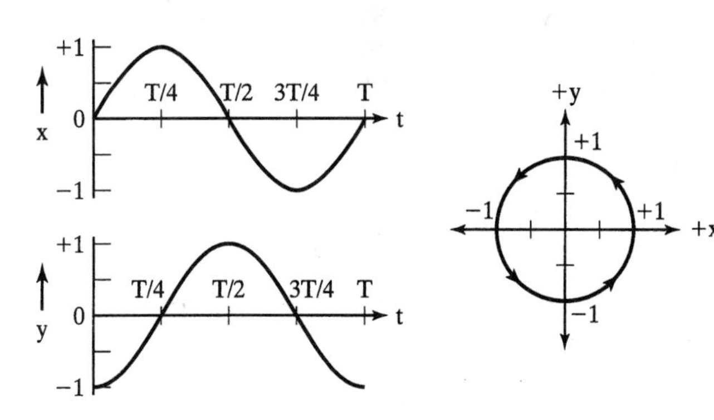

Figure 1-23 Reference graphs (left) and Lissajous figure with $A_x = A_y$ and $f_x = f_y$. The x oscillation is 90° in phase ahead of the y oscillation.

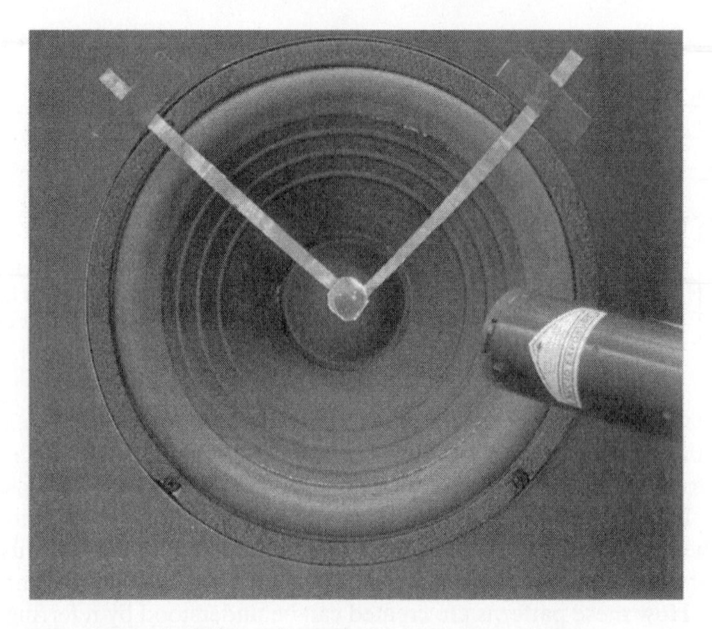

Figure 1-24 Nine-inch woofer with mirror assembly and laser.

A well-known example of Lissajous figures is the laser sound-and-light show. Music is played on a speaker that has been fitted with a mirror suspended by two perpendicular copper strips, as seen in Figure 1-24. A laser beam is reflected off the mirror onto a projection screen. The suspension creates some twisting motion that can be enhanced by displacing the mirror slightly to one side of the axis of the speaker. Because the motion of the mirror then contains two perpendicular components, the reflected light moves in two-dimensional patterns, as seen in Figure 1-25. The pattern at the left is for a simple wave, and the pattern at the right is for a more complicated musical tone.

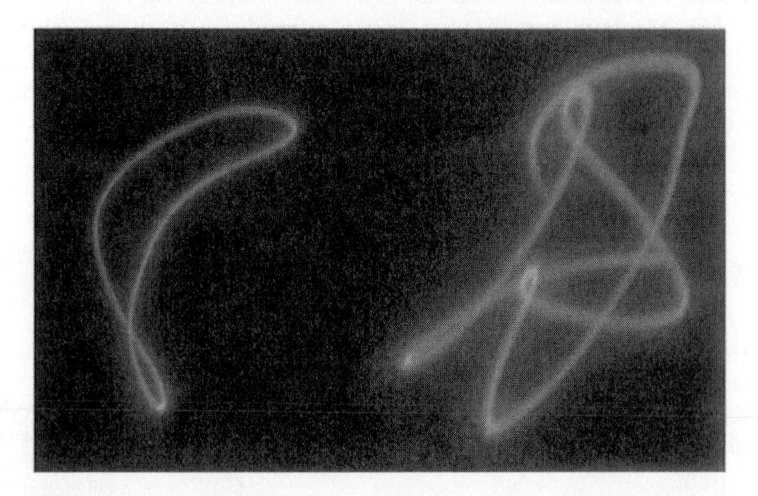

Figure 1-25 Laser light pattern from a 100 Hz tone (left) and music (right).

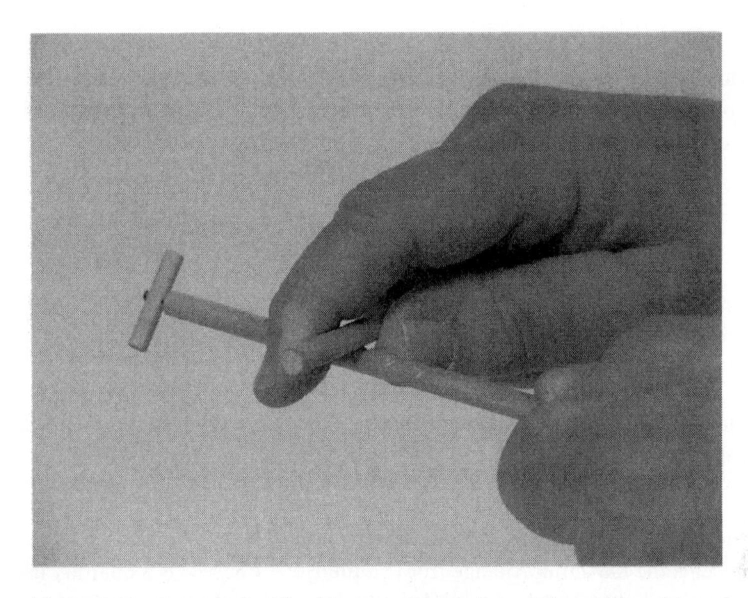

Figure 1-26 A notched stick with rotor with hand carefully positioned to produce "psychokinetic" propeller reversal when the notched stick is stroked.

An interesting toy using two mutually perpendicular oscillations, sometimes also claimed to demonstrate psychokinetic effects, is a *notched stick with rotor*, photographed in Figure 1-26. One stick has six or eight uniformly spaced notches and a well-balanced, loose-fitting propeller on the end. When you hold your hand as shown in the photograph and stroke the notches on top of the long stick with the shorter stick, the propeller rotates. If you push your thumb against the side of the stick as you stroke the rod back and forth, the propeller will rotate counterclockwise, as viewed by the operator. In this case, you are using your thumb to couple the vertical oscillations of your hand as you stroke the notches into horizontal oscillations of the notched stick. The horizontal oscillation is 90° in phase ahead of the vertical oscillation of the end of the notched stick, causing the end of the stick to rotate rapidly counterclockwise and thereby imparting a counterclockwise circulation to the propeller. If you contact the *opposite* side of the notched stick with your *forefinger*, you reverse the phase of the horizontal oscillation with respect to the vertical oscillation, so the horizontal oscillation is 90° in phase behind the vertical oscillation of the end of the notched stick. This reverses the rotation of the end of the stick and thus the circulation of the propeller. Sliding your hand back and forth a very small amount as you stroke the notched stick, so that either your thumb or your forefinger touches its side of the notched stick, the propeller can be made to reverse rotation direction apparently at will, and the reversal can (erroneously) be attributed to psychokinetic powers.

An old Appalachian folk toy called the "Gee Haw Whimmey Diddle"® uses the sound of the stick scraping over the notches as a rhythm instrument for folk music. If you use it to answer questions, you might call it a "Ouija® windmill." A "hooey stick" reverses direction when a selected observer shouts "hooey!"

The speed of sound can be determined by a very beautiful experiment using Lissajous figures; this technique is the subject of one of the questions at the end of the chapter. A more basic method of determining the speed of sound in air will be discussed in detail in Section 2.1.

SUMMARY

(***Section 1.1***) The physical description of motion in physics uses a well-defined system known as **metric units**. Fundamental metric units include the **kilogram** (kg) for **mass**, the **meter** (m) for **distance**, and the **second** (s) for **time**. Other quantities are derived from these fundamental units, such as **velocity** (m/s), **acceleration** (m/s^2), **force** (N), and **pressure** (Pa). (***Section 1.2***) **Periodic motion** is motion that repeats after some specific time interval. **Simple harmonic motion (SHM)** is a specific type of periodic motion that consists of a single vibration frequency, and is mathematically the **projection** of **uniform circular motion**. Such simple systems as the mass on a spring and the pendulum execute simple harmonic motion. The **frequency** f (in Hertz) and the **period** T (in seconds) of simple harmonic motion are related by the equation $fT = 1$. The **amplitude** A represents the maximum change from equilibrium. **Phase** ϕ is a quantity used to define the time relationship between two waves. (***Section 1.3***) The physical properties of waves are related to the psychophysical responses perceived by our brains. The **amplitude** A of a wave is related to its **intensity** or **loudness**, the **frequency** of a wave is related to its **pitch**, and the **wave shape** is related to the **timbre**, or **tone quality**, of the wave. (***Section 1.4***) **Damping** refers to the loss of energy in a vibrating system, causing a decrease in the amplitude of the oscillation. A **driven resonance** occurs when a system with some **natural frequency** is driven by an external periodic force with the same frequency. Two vibrating systems with the same natural frequency transfer energy between each other by means of a **coupling resonance**. **Psychokinesis** is the mythical claim that waves from the brain can actually make physical objects, such as pendulums, move. (***Section 1.5***) Combinations of simple harmonic motions such as **Lissajous figures** have applications in physics, music, and art.

QUESTIONS

1. Define and state the units of the following fundamental physical quantities:

a. position
b. time
c. velocity
d. acceleration
e. mass
f. force
g. weight
h. pressure
i. density

2. **a.** What is a linear restoring force?
   b. How is SHM related to a linear restoring force?

3. **a.** Give examples of motion that is periodic but not simple harmonic.
   b. Give examples of SHM.
   c. Is the brightness of a flashing automobile turn-signal light an example of simple harmonic motion? Is it periodic?

4. Draw graphs of the following:

   a. a triangular wave of period $T = 5$ ms and amplitude $A = 2$ V
   b. a square wave of period $T = 10$ ms and amplitude $A = 3$ V
   c. a sawtooth wave of period $T = 8$ ms and amplitude $A = 1$ V

   Calculate the frequencies of these waves.

5. Draw a 100 Hz sine wave with a 3 V amplitude. For each of the following parts, draw a wave that differs from this sine wave in the characteristic specified:

   a. greater in frequency
   b. lower in amplitude
   c. different in wave shape
   d. greater in period
   e. different in phase
   f. greater in intensity
   g. different in tone quality
   h. lower in pitch
   i. louder

6. **a.** Draw a graph of motion that decreases in amplitude with time but remains constant in period.
   b. Draw a graph of motion that decreases in *both* period and amplitude as time progresses. Which is damped harmonic motion? Which is not? Explain.

7. **a.** Draw a graph of two sinusoidal waves of the same frequency and amplitude that differ in phase by 180°.
   b. What term do we apply to two waves that have this phase relationship?
   c. Draw a graph of two sinusoidal waves of the same frequency but with a phase difference of 90°.
   d. Identify which wave is ahead in phase.
   e. Do the same for two waves that differ in phase by 45°.

8. Describe physically the relationship between the motions of two pendula whose oscillations

   a. are in phase
   b. are out of phase
   c. differ in phase by 90°

9. A mass is suspended on an ideal spring. It is lifted up 5 cm and released from rest at $t = 0$, executing simple harmonic motion with a period of 1 s. Draw a graph of the motion beginning at $t = 0$ and including two full periods of the oscillation. Assume that the equilibrium position for the weight on the spring at rest is $x = 0$, and that up is positive.

10. Define *resonance*. Give some examples of resonance from music and other fields.

11. A pendulum, initially at rest, can be placed into motion in a plane by a series of small pushes at equal time intervals, similar to the way you would push a small child on a swing. Describe what happens as the swing of the pendulum becomes bigger *using the vocabulary of physics*. What is the name for this phenomenon?

12. Draw graphs of a musical tone that:

    a. becomes louder with time

    b. becomes lower in pitch with time

    c. becomes simultaneously softer and higher in pitch

13. Construct a "psychoacoustic vibration transducer" as described in Section 1.4 of this book and learn how to make it work. Try it on people of various levels of physics sophistication, including children, and record their reactions. Ask people why they do or do not believe it. Explain the responses in terms of how people understand physics.

14. Figures 1-22 and 1-23 show Lissajous figures for x and y oscillations of equal amplitude with the x oscillation *in phase* with the y oscillation and with the x oscillation 90° *ahead* of the y oscillation. Draw the reference graphs and the Lissajous figures for

    a. the case in which the x oscillation is *out of phase* with the y oscillation

    b. the case in which the x oscillation is 90° *behind* the y oscillation

15. Make a "notched stick and rotor" using a $^1/_4$" wooden dowel rod, as described in Section 1-5 of this book and learn how to make it reverse when a subject says "hooey!" Try it on people of various levels of physics sophistication, including children, and record their reactions. Ask people why they do or do not believe it. Explain the responses in terms of how people understand physics.

16. Set up an oscilloscope in the xy mode to show Lissajous figures with sound waves as follows: connect an oscillator, set at about 5000 Hertz, to both a loudspeaker and the horizontal input of an oscilloscope, and connect a microphone to the vertical input of the oscilloscope. (You may need to use an amplifier to increase the sound level for the loudspeaker or to boost the microphone output for the oscilloscope.) Position the microphone in front of the loudspeaker and vary their spacing to change the shape of the pattern. Explain why the pattern is changing. Determine one wavelength λ of the sound wave by changing the Lissajous pattern through one cycle of its pattern (for example, starting in phase and moving the microphone until the pattern becomes in phase again). Use $S = f\lambda$ to determine the speed of sound S.

17. Stretch a string tightly between two points at the same height. Hang two pendula of slightly different lengths from the stretched string. Start one of the pendula in motion perpendicular to the stretched string and notice what happens. Change the length of one of the pendula and repeat the experiment. Summarize your results as to how the length of the second pendulum relative to the first pendulum affects their interaction.

PROBLEMS

1. Determine the frequencies of oscillations with the following periods: 1 s, 0.2 s, 10 ms, 0.0002 s, and 20 μs. (Use kilohertz and megahertz where appropriate to keep the numbers simple.)
2. Determine the frequencies of oscillations with the following periods: 0.5 s, 0.1 s, 20 ms, 0.001 s, and 40 μs. (Use kilohertz and megahertz where appropriate to keep the numbers simple.)
3. Determine the periods of oscillations with the following frequencies: 0.5 Hz, 2 Hz, 100 Hz, 5000 Hz, and 10,000 Hz. (Use milliseconds and microseconds where appropriate to keep the numbers simple.)
4. Determine the periods of oscillations with the following frequencies: 1 Hz, 20 Hz, 250 Hz, 10,000 Hz, and 50,000 Hz. (Use milliseconds and microseconds where appropriate to keep the numbers simple.)
5. Convert the following to Hertz: 1.5 kHz, 500 kHz, 2 MHz, and 1000 MHz. Why is it helpful to use the *kilo-* and *Mega-* prefixes?
6. The frequency range of human hearing is about 20 Hz to 20 kHz. What are the periods of these oscillations?

REFERENCES

ROSSING, THOMAS D., F. RICHARD MOORE, and PAUL A. WHEELER, *The Science of Sound*, third edition. Reading, MA: Addison-Wesley Publishing Company, 2002. Excellent book; much more detail and at a significantly higher level than this text.

HALL, DONALD E., *Musical Acoustics*, third edition. Pacific Grove, CA: Brooks/Cole Publishing Company, 2002. Excellent book; more emphasis on musical applications than this text.

PRINT DEMONSTRATION REFERENCES

FREIER, G. D., and F. J. ANDERSON, *A Demonstration Handbook for Physics*, Second Edition. College Park, MD: American Association of Physics Teachers, 1981.

CARPENTER, D. RAE, and RICHARD B. MINNIX, *The Dick and Rae Demo Notebook*. Lexington, VA: Virginia Military Institute, 1993.

EDGE, R. D., *String and Sticky Tape Experiments*. College Park, MD: American Association of Physics Teachers, 1987.

These three books contain a large number of physics demonstration experiments ideal for either use in class or construction by students, and form an excellent nucleus for developing neat, easy-to-make classroom experiments. See especially their sections on vibrations, waves, and sound.

WEB DEMONSTRATION REFERENCES

Physics Instructional Resource Association (PIRA): Physics Resources on the Web:
http://www.wfu.edu/physics/pira/Resources.htm

PIRA is an umbrella group for educators involved in developing physics classroom demonstrations, and virtually any experiment used in a physics class can be discovered on this site.

The University of Maryland Physics Lecture-Demonstration Facility Web site:
http://www.physics.umd.edu/lecdem/
contains a library of over 1500 demonstration descriptions. The page:
http://www.physics.umd.edu/lecdem/services/demouse/phys102sugg.htm
is a list of over 200 demonstrations used with the course for which this textbook was written and organized by chapter and section in this textbook, all linked to the demonstration pictures and descriptions.

AUDIOVISUAL DEMONSTRATION REFERENCES

Physics Demonstrations in Sound and Waves, Parts I, II, and III, Physics Curriculum and Instruction, three videotapes, each with eight segments about 3 minutes in length, for a total of less than 30 minutes apiece. The tapes are of excellent quality and cover several areas of vibrations and sound.

Berg, Richard E., and David G. Stork, *Demonstrations in Acoustics.* College Park, MD: University of Maryland, Department of Physics, 1980. This is a set of four 1-hour color videos that show many of the experiments described in the present text being performed and gives brief explanations of them.

The Science of Sound, Folkways Record Album No. FX6136, Descriptive Literature by Bell Telephone Laboratories. Available from Frey Scientific Co., Mansfield, OH. This is an excellent recording, covering many areas of acoustics, that can be readily used in class lectures. The records contain short segments illustrating such topics as the overtone series, tone quality, filtering, distortion, reverberation, the Doppler effect, and others.

Chapter 2

Waves and Sound

In this chapter we shall investigate the important features of waves and wave motion, and apply the concepts introduced in the first chapter to sound and ultrasound. Each musical and nonmusical topic we consider will be related to the fundamental physical law or laws that govern it. Examples and applications will be taken from familiar areas and everyday phenomena. Finally, we shall study the addition of waves and some of its important applications. Most of the concepts studied in this chapter will be relevant to the understanding of the advanced topics in later chapters.

2.1 Transverse and Longitudinal Waves

A *wave* possesses the following properties:

1. A wave is a disturbance within some medium; the disturbance propagates (or travels) with some velocity, which depends on the medium. Most of the waves that we shall study will be periodic.
2. A wave transfers energy; by virtue of the energy it carries from its source to its destination, a wave can perform work. For example, a rope wave could perhaps move a pump handle up and down to pump small amounts of water, a sound wave can make an eardrum vibrate to create the sensation of sound, waves from the sun can heat and illuminate the earth, radio waves can make electrons in a radio antenna oscillate to enable reception of the signal, and so on.

Two basic types of waves, transverse and longitudinal, are distinguished by the direction of the wave oscillation (that is, the direction of the motion of the vibration in the medium) relative to the direction of propagation of the wave.

In a *transverse wave* the oscillations forming the wave are perpendicular, or transverse, to the direction of propagation. Examples of transverse waves are rope waves, waves in slinky springs in which the wave is created by motion perpendicular to the slinky, and the pulse traveling down a whip as it is cracked. Each point of the whip (the medium) moves up and down while the wave pattern propagates, say, left to right.

A very important class of transverse waves is electromagnetic (EM) waves, which are unique, in that they require no medium in which to propagate. They are exemplified by light rays from the sun, which pass through the vacuum of outer space on the way to the earth. The various types of EM waves are similar in all their properties except frequency (or equivalently, as we shall soon see, wavelength). In order of increasing frequency, the types of EM waves are radio, television, microwaves, infrared, visible, ultraviolet, x-rays, and gamma rays.

In a *longitudinal wave* the motion of the points of the medium (forming the wave) is in the same direction as the direction of propagation of the wave pattern. Examples of longitudinal waves are slinky spring waves created by expansion and compression of the spring coils, sound waves, and ultrasonic waves (sound waves of frequencies above the audible range). Shock waves, such as sonic booms, are also longitudinal. The propagation of sound waves, and longitudinal waves in general, always requires a medium, for reasons that will soon become obvious.

A demonstration that illustrates a fundamental difference between sound and light waves is the "bell in vacuum," illustrated in Figure 2-1. A ringing bell is suspended

Figure 2-1 Bell-in-vacuum demonstration. A vacuum pump (at bottom) removes the air from the jar (upper right) in which the bell is suspended. A large gauge (upper left) indicates the air pressure in the jar.

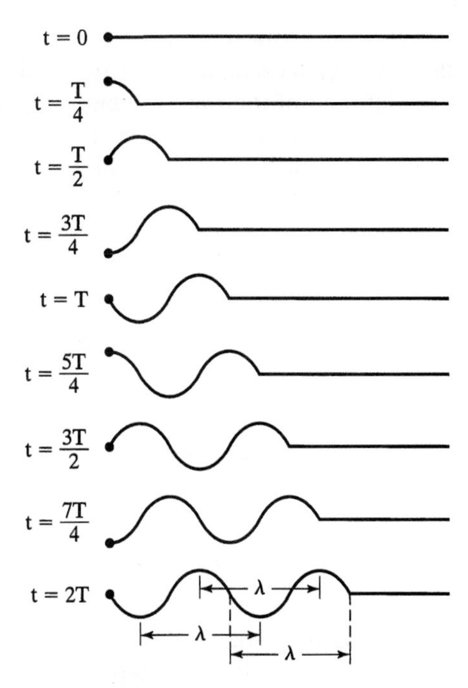

Figure 2-2 Transverse waves in a rope, created by sinusoidal oscillation of the end of the rope.

by a small spring inside a glass jar from which the air is evacuated by means of a mechanical vacuum pump. The spring suspension minimizes direct transmission of the sound to the glass jar. As the air is pumped out, the sound of the bell becomes softer; *if the jar were to be entirely evacuated*, the ringing of the bell would become inaudible. We are not surprised that the bell remains visible in the jar even when the air is entirely evacuated. Light waves (and EM waves in general) do not require air or any other medium in which to propagate, whereas sound waves (and longitudinal waves in general) do require a medium—air in particular for this experiment—in which to propagate. All transverse waves other than EM waves, and all longitudinal waves, require a medium. In fact, this experiment is more complex than this simple analysis indicates, but the conclusion is correct.

One of our goals will be to understand the low-amplitude transverse waves in a guitar or violin string. To help visualize these transverse oscillations, we shall first consider the slower-moving, larger-amplitude transverse waves in a rope.

Let us consider how a transverse rope wave is produced by rapid sinusoidal transverse oscillation of one of its ends. Consider the rope shown in Figure 2-2, in which at time $t = 0$ the left end of the rope is suddenly started in simple harmonic oscillation in the vertical direction with a period T (frequency $f = 1/T$).

In the first time interval of $T/4$, the end of the rope moves up to its maximum positive value, causing a displacement of the rope. The wave thus produced moves from left to right, the direction of propagation. After one half-period, $T/2$, the oscillating end of the rope returns to the equilibrium point, and the corresponding sinusoidal displacement of the rope moves along the rope, as shown. The figure shows the resultant wave after successive time intervals of one quarter-period, $T/4$, until two periods have passed.

Notice that the sinusoidal motion of the end of the rope produces a periodic wave that is also a sine curve; each point in the medium oscillates vertically in SHM. In fact, the time variation in the position of the left end of the rope has been converted into a variation in shape along the length of the rope. The distance along the rope after which the spatial variation repeats itself, called the *wavelength* (symbol λ, lowercase Greek letter lambda), is shown in Figure 2-2.

The figure also shows that in a time interval of one period, T, the wave has traveled a distance of one wavelength λ along the rope. The speed of propagation v of the wave along the rope is then equal to the distance traveled, λ, divided by the time interval T:

$$v = \frac{\lambda}{T}. \tag{2.1}$$

Because $T = 1/f$ *or* the frequency $f = 1/T$, it follows that the velocity obeys the relation:

$$v = \frac{\lambda}{T} = \lambda f. \tag{2.2}$$

The mathematical formula $v = \lambda f$ is true for any wave of any type and frequency. For instance, the preceding discussion also applies to both longitudinal and transverse waves in long slinky springs. The speed of the wave depends on the type of wave (e.g., rope, sound, EM) and the medium through which it is propagating.

Representation of a transverse wave in a rope is relatively easy to visualize and draw; it is like a photograph of the rope at any given time. The succession of drawings in Figure 2-2 can be thought of as photographs of the rope at the times indicated.

Now let us think about longitudinal waves. Consider a slinky spring, at rest, stretched and held so that it has some (equilibrium) length, as shown as Figures 2-3(a); shown are successive turns of the spring. If a *compression* is formed at the center of the spring by squeezing several of the turns together, they will expand when released to return to equilibrium turn spacing. Similarly, a *rarefaction*, a region of extended coil layers, will compress toward its equilibrium spacing when released. These two conditions are shown in Figures 2-3(b) and (c).

A longitudinal wave in a slinky spring is formed by a succession of such compressions and rarefactions moving along the spring. Because such a longitudinal wave is difficult and time-consuming to draw, it is usually represented by a transverse wave graph, as shown in Figure 2-3(d). The upward arc of the transverse wave graph represents a

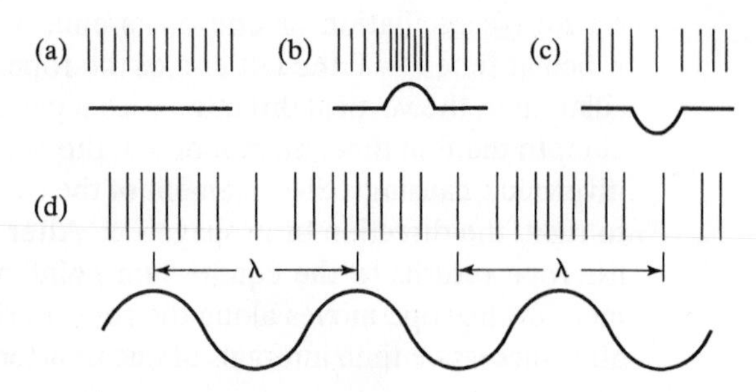

Figure 2-3 Transverse representation of (a) spring at equilibrium, (b) a compression, (c) a rarefaction, and (d) a series of compressions and rarefactions.

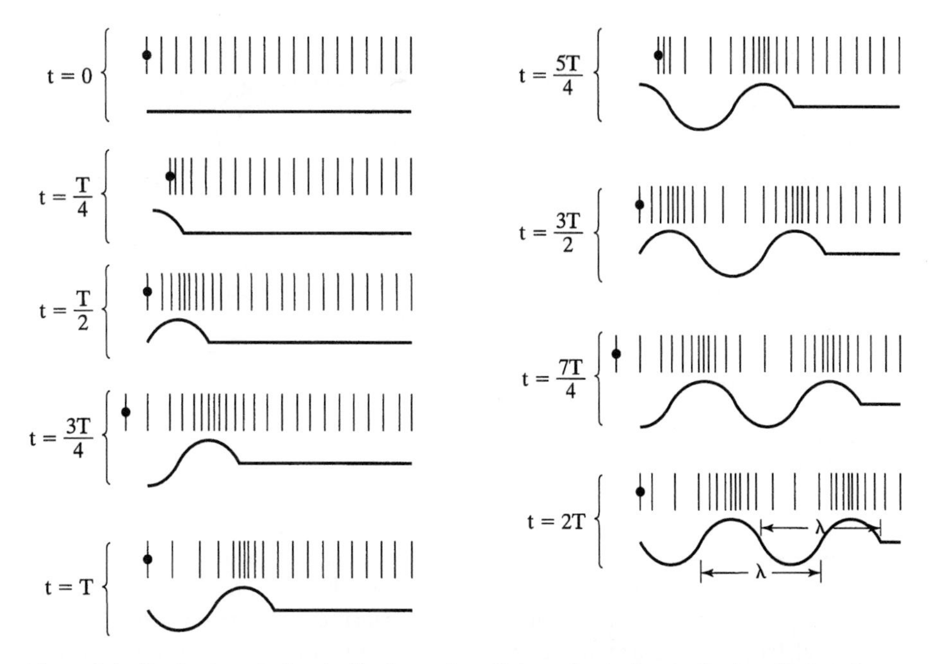

Figure 2-4 Production of a longitudinal wave in a slinky spring by longitudinal oscillation of one end of the spring, shown at time intervals of one quarter-period. Also shown is the transverse representation of the longitudinal wave.

compression in the longitudinal wave (spring spacing smaller than equilibrium spacing), and the downward arc of the transverse wave graph represents a rarefaction in the longitudinal wave (spring spacing greater than equilibrium spacing). Again, as in the case of the transverse wave, we can identify the wavelength λ, the distance after which the wave shape repeats itself. When a sinusoidal wave propagates along the spring, each turn of the spring moves in SHM along the direction of propagation.

As in the case of the transverse wave of Figure 2-2, the production of a longitudinal wave in a spring by longitudinal vibrations of the end of the spring is illustrated in Figure 2-4. The wave, consisting of sinusoidal variations in the spring coil spacing, possesses a certain wavelength, λ, and moves to the right with a wave speed v characteristic of the spring. Also shown is the transverse representation of the longitudinal spring wave.

The longitudinal waves in a slinky spring can serve as a model of a sound wave in air. In this simplified analogy, the slinky spring turns represent successive "layers" of air molecules. As in the case of the slinky, if a compression of air layers is created on the left (say, by a loudspeaker) it will tend to expand, pushing the next air layers to the right, thereby causing a compressional pulse to propagate to the right. A succession of compressions and rarefactions then forms the wave, as in the case of the slinky. Again the wave possesses the period T and frequency f of the source, but now moves with the speed of sound in air, for which we shall use the symbol S. The value of the wavelength can be calculated using $S = f\lambda$ or, equivalently, $\lambda = S/f$.

One method for determining the speed of sound is based on finding the time it takes the wave to propagate a known distance. The experiment involves an electronic

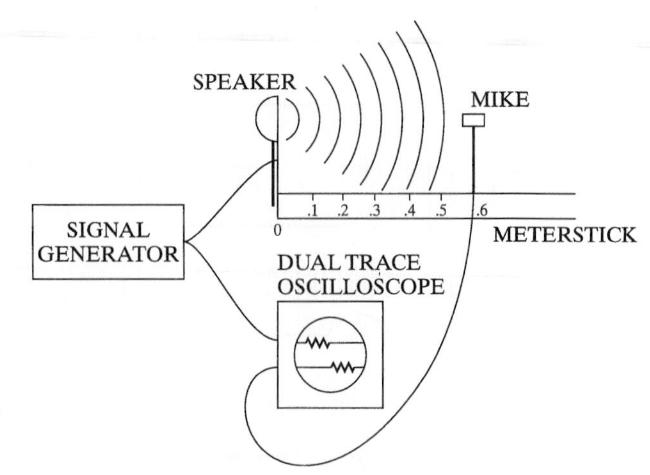

Figure 2-5 Experiment for determination of the speed of sound in air. Position of microphone along meter stick is adjustable.

signal generator, a loudspeaker, a microphone, and a dual-trace oscilloscope—that is, an oscilloscope that can simultaneously display two signals. The experimental setup shown in Figure 2-5 will now be described; the exact numerical values of the signal frequencies are not crucial but are presented as obtained in one particular experiment.

The complex signal produced by the signal generator is a 10,000 Hz sine wave that is on for 0.5 ms and then off for 4.5 ms, as shown in Figure 2-6. The effect is that short bursts of a 10,000-Hz sine wave are generated at 5-ms intervals. This signal is sent simultaneously to the top trace of the oscilloscope and the speaker. The lower trace of the oscilloscope shows the output of the microphone that picks up the sound produced by the speaker. Figure 2-7 shows the oscilloscope traces when the microphone is in its near position (labeled 1) and far position (labeled 2). In both cases the scope starts the graph each time a pulse is sent to the speaker. Note that the microphone response occurs before the next pulse to the speaker is begun 5 ms later.

As the microphone, initially placed in position 1, is moved to position 2, the response signal picked up by the microphone is consequently delayed in time, as indicated by the two curves labeled 1 and 2 on the lower scope trace. The time $t_2 - t_1$ is just the additional time the sound takes to travel from microphone position 1 to position 2, labeled $d_2 - d_1$. The speed of sound is just the extra distance divided by the time the sound takes to go that distance; that is,

$$S = \frac{d_2 - d_1}{t_2 - t_1}. \tag{2.3}$$

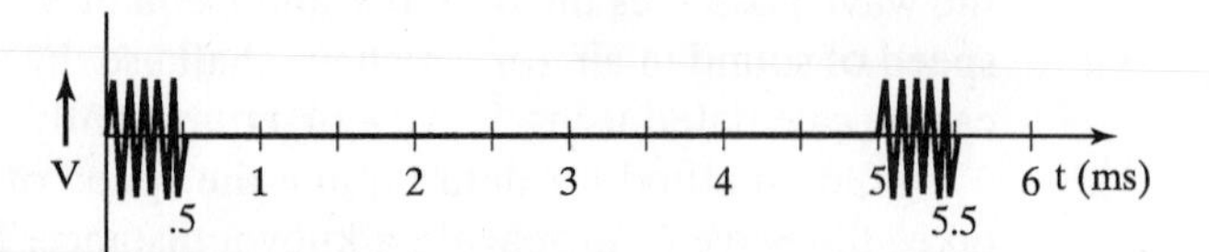

Figure 2-6 Wave-pattern input to speaker of Figure 2-5.

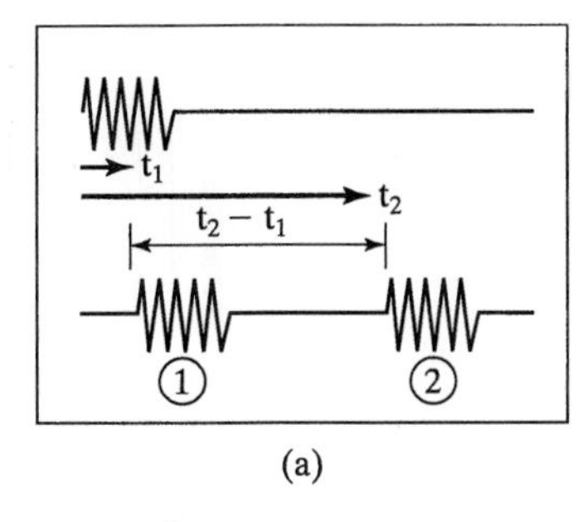

(a)

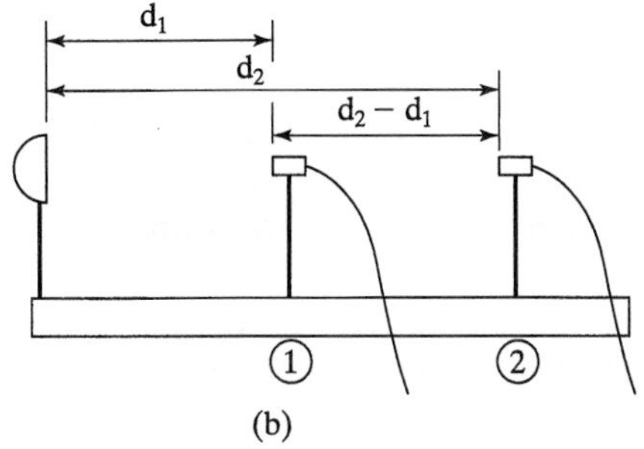

(b)

Figure 2-7 Part (a) shows the face of the oscilloscope for the speed-of-sound experiment. When the microphone is moved back a distance $(d_2 - d_1)$ as shown in (b), the sound arrives at the microphone later by an amount $(t_2 - t_1)$.

If we measure $d_1 = 10$ cm at $t_1 = 0.30$ ms and $d_2 = 60$ cm at $t_2 = 1.80$ ms, then S is calculated to be

$$S = \frac{(60 - 10)\ \text{cm}}{(1.80 - 0.30)\ \text{ms}} = \frac{50\ \text{cm}}{1.50\ \text{ms}} \approx 33{,}333\ \text{cm/sec} \approx 333\ \text{m/s}. \tag{2.4}$$

Our experimental determination gives fair agreement with the accepted value of 345 m/sec for the speed of sound near room temperature.

2.2 Fundamental Properties of Waves

We now turn to the principles that govern wave behavior, starting with four phenomena that might be considered fundamental to how waves propagate:

1. Huygens's principle
2. Superposition
3. Inverse square law
4. Polarization

These phenomena (with the exception of polarization) apply to all three-dimensional waves. We shall first take familiar examples of various types of waves to illustrate wave behavior; then, in each case, we shall apply these principles to the domain of sound and ultrasonic waves. These four principles governing waves are more fundamental than the four wave properties to be discussed in Section 2.3, which largely arise by application of Huygens's principle and the law of superposition.

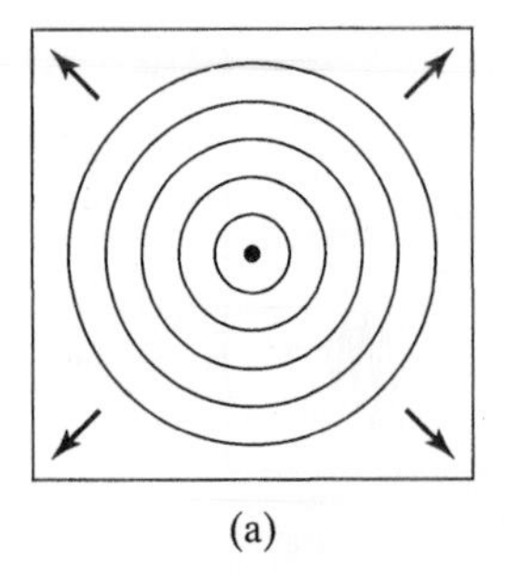

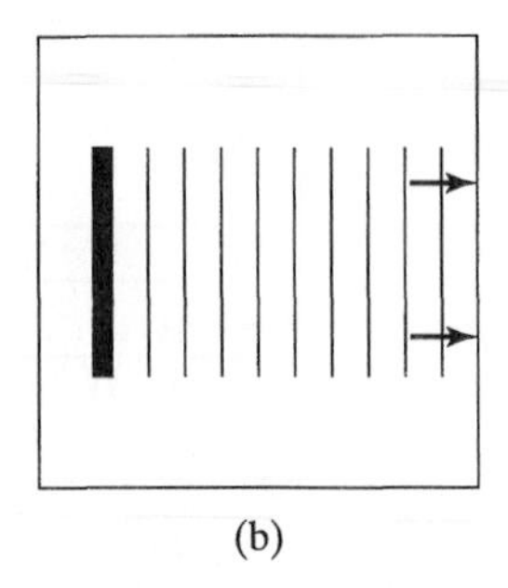

Figure 2-8 Circular waves (a) produced by a small vibrating ball, and plane waves (b) produced by a vibrating bar in a ripple tank. Periodic vibration of the source perpendicular to the surface of the water produces uniformly spaced waves. The lines in both (a) and (b) represent the crests of successive waves.

We will frequently illustrate wave behavior by means of a *ripple tank*, a shallow water tank with a small ball or straight bar that is dipped slightly beneath the surface of the water and vibrated up and down; the ball creates outgoing circular waves, and the bar creates plane waves. The water waves in a ripple tank are two-dimensional waves; that is, they propagate on the two-dimensional (plane) surface of the water. The rope waves discussed in the last section are one-dimensional, because they propagate in the one-dimensional (line) medium, the rope. Figure 2-8 shows ripple-tank representations of a circular wave and a plane wave; the lines represent successive wave crests; the spaces between the lines represent troughs. Any of the circular waves is similar to the wave obtained by dropping a pebble onto a calm water surface; the continuous succession of circular waves could be obtained by dropping additional pebbles at equal time intervals.

Huygens's principle

Let us examine in some detail the motion of waves in the presence of the barriers represented by the cross-hatched horizontal bars in Figure 2-9 (viewed from above). Here the barrier is a breakwater formed by a rectangle of material resting on the bottom of the tank and extending above the surface of the water so that the wave is only able to pass through the small opening, or slit, between the barriers. We observe that the wave emanating from the slit does not continue in a straight line, but emerges from the slit in the barrier as a circular wave. This general result, the same for all waves, constitutes the experimental basis for Huygens's principle, named for the Dutch physicist Christian Huygens (1629–1695). This principle can be stated as follows: All points on a wave act as sources of circular waves (or spherical waves in three dimensions); the total wave pattern at some later time will be the sum of all the individual *Huygens's wavelets*.

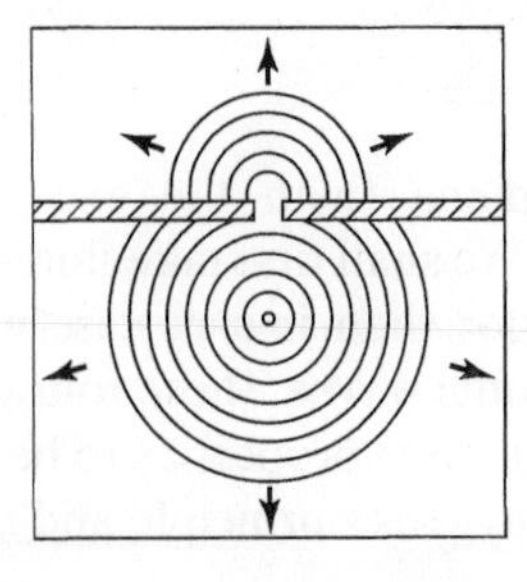

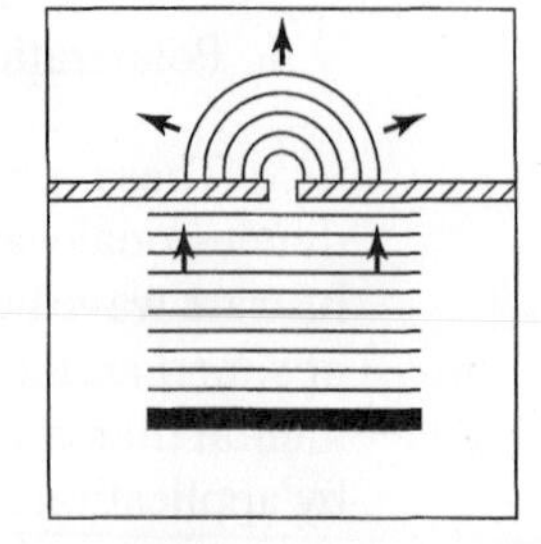

Figure 2-9 Circular and plane waves in a ripple tank. The waves are interrupted by a barrier, allowing only a small part of the wave to pass. This opening acts as a source of Huygens's wavelets.

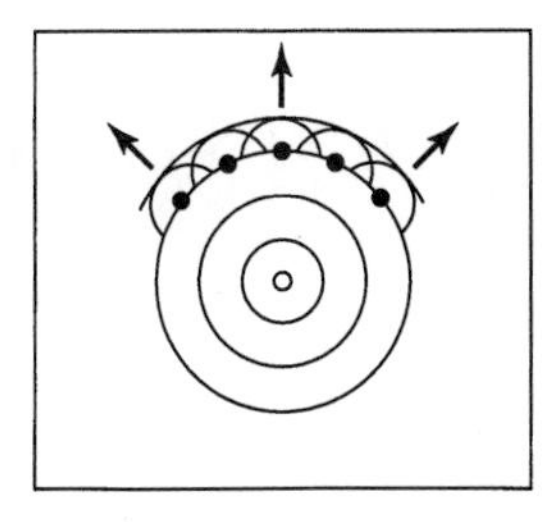

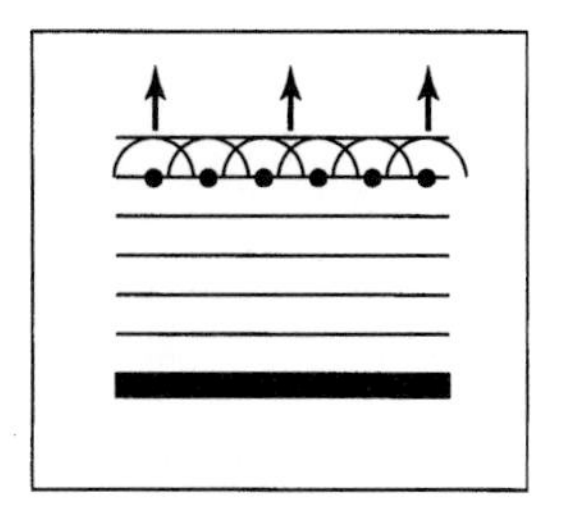

Figure 2-10 Huygens's wavelets originating from a succession of points along circular and plane wave fronts. The leading edge of all the individual wavelets forms the next circular or plane wave front.

The intensity (and thus the amplitude) of the circular wave emanating from any point (such as the point at the slit) is greatest in the direction of the original wave propagation and gradually decreases with angle, becoming zero in the opposite direction. This too can be observed in the ripple tank but is not shown in Figure 2-9.

In many circumstances, analysis using Huygens's wavelets is a convenient theoretical technique for understanding the propagation of waves. We can now see how Huygens's principle applies to the propagation of a circular or a plane wave, illustrated in Figure 2-10. If the circles represent the position of the wave crests at time intervals of one period, then one can break up any one of the circular waves into a set of points, each of which acts as a source of circular Huygens's wavelets whose sum intensity is greatest in the original direction of propagation of the wave. Part of this set of Huygens's wavelets emanating from points along the original wave front is shown in part in Figure 2-10. The crests of these wavelets from successive points on the original waves are lined up, forming a shape identical to the original wave after one period. Thus, the overall effect of the addition of all these wavelets after one period is propagation of the original wave. The wavelets propagating in directions different from the original direction of propagation consist of a wide range of phases, and will cancel each other.

We can see, however, that this argument cannot hold at the edges of the plane wave of Figure 2-10. As a result of the wavelets at the edges of the plane wave, any narrow plane wave such as the one drawn will spread sideways as it propagates. This phenomenon will be treated again in the discussion of diffraction. Another important result of this wave behavior is that the direction of propagation of a wave always remains perpendicular to the wave front. The wave front is the region defined by the farthest extension of the wave propagation and is a straight or curved line in two dimensions and a plane or curved surface in three dimensions. For instance, in the previous example of a stone dropped into water, the wave front is an expanding circle. For the sound of a firecracker in air or a flash of light from a flashbulb in three-dimensional space, the wave front is an expanding sphere. This will be important to remember in our study of diffraction, and will be dealt with in more detail in that section.

Superposition

The law of superposition can be stated as follows: The existence of one wave does not affect the existence or properties of another wave, even if they are in the same place at the same time. This is equivalent to the statement that waves add algebraically; that is, the displacement of the sum wave, $A + B$, is equal to the displacement due to wave A added to the displacement due to wave B at the same point and time. This is valid for

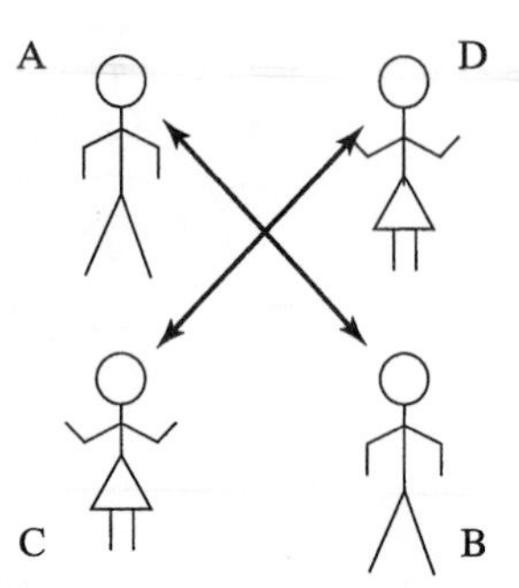

Figure 2-11 Sound waves from the conversation between persons A and B pass unaffected through the same point in space as the waves from the conversation between persons C and D. This illustrates the principle of superposition.

small-amplitude waves, which create limited stress on the medium. Waves can pass through each other without affecting each other. This clearly distinguishes waves from material things, no two of which can occupy the same place at the same time.

One can also define superposition in terms of the addition of waves, a topic we shall cover shortly. In that approach, waves are said to obey the superposition principle if the resulting displacement at a point is algebraically equal to the sum of the displacements of the individual waves at that point.

An illustration of this is shown in Figure 2-11. Persons A and B can talk with each other at the same time persons C and D talk with each other. The sounds or words from either conversation are completely unaffected by those of the other; the voice of A will not change when C produces waves that pass through the same crossing point. Not only are sound waves present in this example, but light waves of all colors (the people see each other), radio waves (a portable radio at the center of the foursome will pick up stations), radiating heat waves, and others. In fact, at any point in space there is, in general, an infinity of waves; all space is filled with a limitless number of waves of all kinds and wavelengths, none of which is affected by the existence of the others. If the persons in our example have trouble hearing each other owing to external sound waves, it results from physical or psychological processes in the auditory system rather than to any exception to the rule of superposition of waves. The superposition principle will be used to help explain several other properties of waves, such as the addition of waves and interference.

Inverse square law

Everyone is familiar with the fact that the farther one is from a source of sound, the softer it sounds, or, equivalently, the farther one goes from a sound source, the more intense the source must be made for one to perceive the same loudness. The inverse square law tells exactly how the intensity of a sound wave (or any other type of wave in three dimensions) decreases as the distance between the source and the observer increases.

Consider the two-dimensional wave created by dropping a stone into a pool of water; as time goes on, the wave expands in a circle of increasing radius, as shown in Figure 2-12(a). The energy transferred to the water by the stone when it hits the surface is contained in the wave. The wave front, the circumference of the circle, gets longer as the circle gets larger. The same energy is "spread out" over a larger circumference. In fact, if the circle doubles in radius, then the circumference, $C = 2\pi r$, is twice as large, the energy is spread out over a wave front twice as long, and the intensity (the energy per unit length) of the wave is then half as much. Stated mathematically, $I \alpha 1/r$, or the intensity is

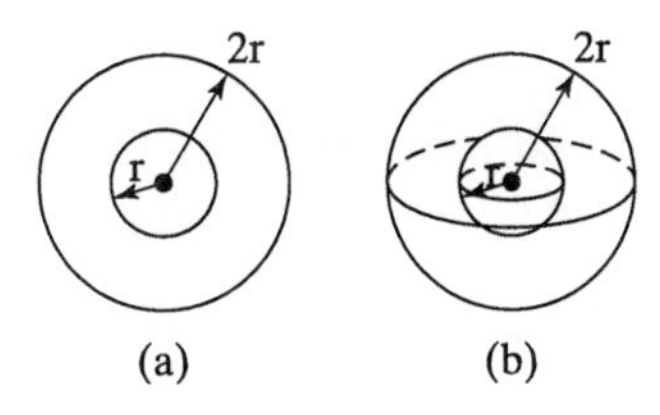

Figure 2-12 (a) Expansion of a two-dimensional water wave, illustrating the intensity law $I \alpha 1/r$ for a two-dimensional wave. (b) Expansion of a three-dimensional wave, such as a sound wave, illustrating the intensity law $I \alpha 1/r^2$ for a three-dimensional wave.

inversely proportional to the radius of the circle. Thus for two-dimensional waves the intensity is inversely proportional to the distance of the observer from the source.

Now consider the extension of this concept to three dimensions, the usual case for sound waves, light waves, and others that expand in all directions. Rather than expanding in circles, the wave will expand in spheres of increasing radius, so the energy of the wave is spread out over the spherical surface as shown in Figure 2-12(b). The area A of a sphere of radius r is given by the formula $A = 4\pi r^2$, so when the radius of the sphere is increased to twice its original radius, the area is *four* times the original area. The same amount of energy is therefore spread out over four times the original area, so in this case the intensity is only one-fourth of the original. Similarly, an increase of a factor of 3 in radius increases the area of the sphere by a factor of 9, resulting in one-ninth of the original intensity. Stated mathematically, $I \alpha 1/r^2$, or the intensity decreases in proportion to the square of the distance from the source—the inverse square law.

All types of waves—for example, an ordinary light bulb emitting light, a church bell ringing, or a radio transmitter broadcasting—obey this law when they are emitted and allowed to travel freely away from the source with no focusing or other means of confinement. One interesting light source is the laser. Because it is *coherent* light and is confined to a very narrow beam rather than spreading out in all directions, the intensity does not appear to decrease as rapidly as would be expected according to the inverse square law. These features allow a powerful laser beam to retain sufficient intensity to be observable after being reflected back to earth by a reflector on the moon.

Polarization

Whereas all the properties discussed previously apply to both longitudinal and transverse waves, polarization is a physical property unique to transverse waves. It is therefore not directly applicable to the study of sound waves in air, but we include it for completeness and because it is applicable to the study of the coupling of sound waves from the string to the sounding board of a piano, as we shall see in Chapter 13. A simple transverse wave always has a single plane in which the wave is vibrating; for example, the wave of Figure 2-2 is oscillating in the plane of the paper. This is known as its *plane of polarization*. Another possible plane of polarization for a rope wave like that shown in Figure 2-2 would be into and out of the paper. In either case, the plane of polarization is determined by the direction of the oscillations forming the wave. If we cause all the waves in a beam of light, or some other collection of waves, to vibrate in a single plane or direction, the waves are said to be *polarized*. Polarizing sunglasses filter out light polarized in the horizontal direction. Reflection off horizontal surfaces, such as water and streets, is mostly horizontally polarized, producing very bright reflection that we call "glare." Polarizing sunglasses remove most of this glare, allowing less stressful vision.

2.3 General Behavior of Waves

Now we will survey four general behaviors of waves that arise from the fundamental principles discussed earlier and that apply to all types of waves. We will illustrate these wave behaviors with sound examples.

1. Reflection
2. Refraction
3. Interference
4. Diffraction

Reflection

Everyone is familiar with an echo, a reflection of sound off some surface, the audio analog to a light wave reflecting off a mirror. The simplest type of reflection is that occurring when you stand facing the flat wall of a building and create a sound, which reflects back to you as an echo. Similar reflectors are useful if placed in back, on the sides, and on top of an orchestra in concert, because they reflect the sound waves out toward the audience.

A mathematical feature of reflection is that the angle at which an incident wave approaches a smooth reflecting surface is the same as the angle at which the reflected wave leaves the surface. This angle is generally given with respect to the *normal*, a line perpendicular to the surface at the point of reflection, as shown in Figure 2-13. We can understand this more thoroughly by drawing a sequence of Huygens's wavelets, as in the example given in Figure 2-10; such a construction is shown in Figure 2-14. In the case of the plane waves, the angles defined are between the normal and the directions

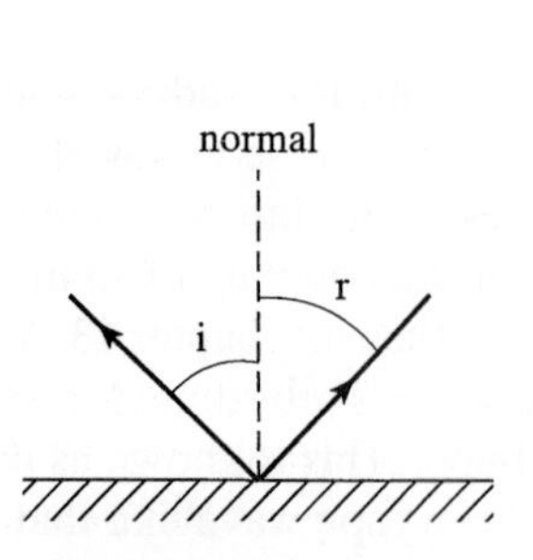

Figure 2-13 A wave incident in the direction shown reflects as indicated from a flat surface. The angle of incidence *i* and the angle of reflection *r* are equal. The normal is perpendicular to the reflecting surface at the point of reflection.

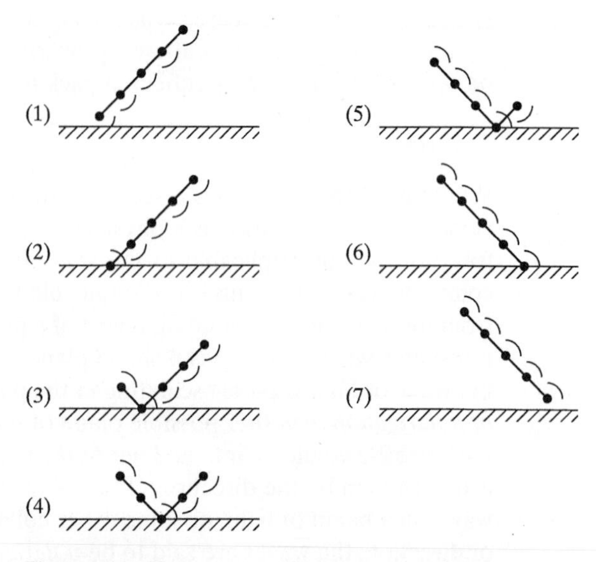

Figure 2-14 Huygens's wavelet construction showing how a wave is reflected from a smooth flat surface. The drawings show the advancing wave front and the Huygens's wavelets at a succession of times.

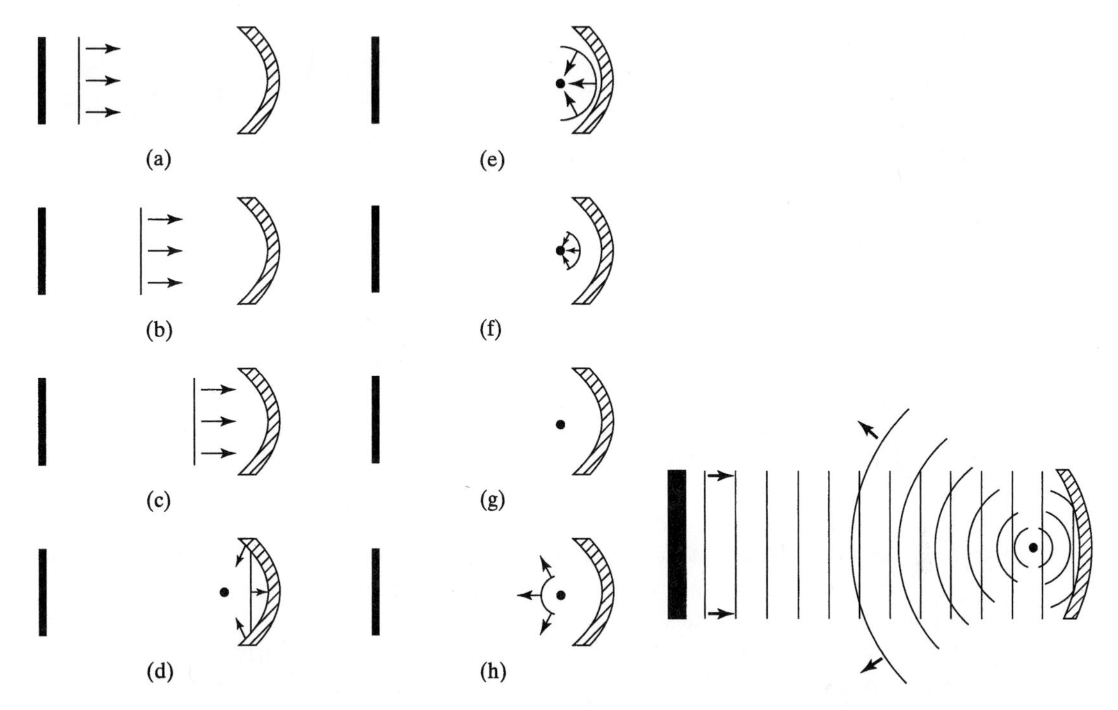

Figure 2-15 A plane wave pulse is created at the left on the ripple tank and focused to a point by a parabolic reflector. The wave is shown at a succession of unequal time intervals.

Figure 2-16 A plane wave in a ripple tank is focused by a parabolic reflector and continues to expand as a circular wave segment.

of the incident and reflected waves. Shown in Figure 2-13 are incident and reflected rays. A *ray* marks the propagation of a single point on the wave front. A ray is always perpendicular to the wave front, and for plane waves the ray is a straight line showing the direction of propagation of the wave.

If waves are reflected from irregularly shaped surfaces, the reflection must be examined in detail at the surface to determine the appropriate normals and angles. At each point along a curved surface there is a normal line. The law holds for a wave striking the different parts of such a surface. The normal is different at each point, whereas for the case illustrated by Figure 2-14, the normal at any point along the surface points in the same direction.

A more complex example of reflection is the focusing of plane waves to a point by a parabolic reflecting boundary, that is, a surface with the shape of a parabola. Figure 2-15 shows the propagation of a plane wave created by dipping the straight element at the left quickly in and out of the water in a ripple tank. The point where the entire wave comes together is called the *focus* or *focal point* of the parabola.

Figure 2-16 shows the same phenomenon, except now plane waves are being continuously created at the left and focused. If they are not stopped at the focus, they will continue on through that point and expand as a sequence of circular wave segments. It is important to point out here that to achieve a good focus the reflector must have a *parabolic*, not circular, shape. In fact, in three dimensions, the reflector must be a

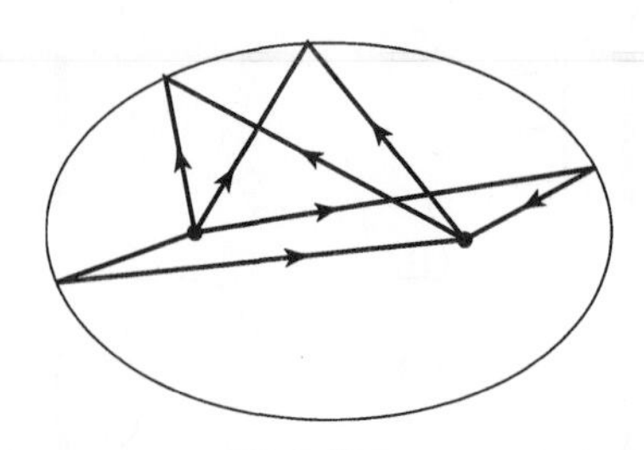

Figure 2-17 Rays emitted at the same time from one focus of an ellipse will converge simultaneously at the other focus.

paraboloid of revolution, obtained by rotating a parabola about a line through the focus and the center of the reflector. Such a mirror (in the case of light) or sound reflector is thus called a *parabolic mirror* or *parabolic reflector*.

The parabolic reflector has many important applications in the fields of both light and sound. A high-quality reflecting astronomical telescope requires a parabolic reflector to obtain the good focus necessary to resolve adjacent celestial objects. Parabolic microphones are formed by placing a microphone at the focus of a large parabolic sound reflector, which is often made of clear plastic. The microphone is placed at the focus facing the reflector *(not* the source of the sound), so that the reflected rays are collected. The signal from the microphone can be fed into an amplifier and loudspeaker to make it audible. Such microphones are regularly used at football games to capture the signals from the quarterback, and are used to record the calls of distant birds and other soft sounds.

A parabolic reflector can also be used to obtain a beam of light from a small source, as in the case of a flashlight or large searchlight, by a procedure inverse to focusing a beam of light. The source is placed at the focus and the reflector projects the beam of light. Another type of reflector is used in certain loudspeakers and some megaphones in which it is desired to project the sound in a certain direction. Some of these reflectors are hyperbolic rather than parabolic in shape, and make use of more complex reflections.

An interesting variation of this focusing property is the *whispering chamber*. In the "two-dimensional" chamber shown in Figure 2-17, an ellipse has been drawn and the positions of two points indicated. These two points are called the *foci* (plural of focus), because waves emitted simultaneously in all directions at either focus will bounce off the walls and pass through the other focus at the same time, as shown by the rays. A circle is a special case of an ellipse that has the two foci at the same point, the center. Waves emitted from the center of a circle will be focused back to the center. The three-dimensional extension of the ellipse is the ellipsoid of revolution, which also has two foci, as in the two-dimensional case. Two people, one at each focus, can whisper intelligibly to each other, while people elsewhere in the room either cannot hear or understand the words because of the low sound level at their location.

Thus far we have examined cases in which the sound wave incident to a barrier was reflected. In reality it is not quite this simple. Rather, when a sound wave is incident to a barrier between two media in which the velocity of sound is different, part of the wave will reflect off the boundary and part of the wave will be transmitted into the second medium. For example, when a sound wave reflects off a wall, part of the wave travels into the wall, possibly allowing someone on the other side to hear the sound. How much of the wave is reflected and how much passes into the second medium depends on the nature of the two media and the direction from which the wave was incident. More details on waves passing between two media will be discussed in the section on refraction.

Several interesting illustrations of reflection arise with respect to sound waves and ultrasonic waves, which in some navigational applications are referred to as *sonar* (sound navigation and ranging). Bats emit short bursts of ultrasonic waves (sound waves with frequencies above human sensitivity), listen for the reflections, and are thus able to locate objects or barriers. This procedure allows bats to fly safely in caves or dark attics and to locate insects to eat while in flight.

With the exception of very-low-frequency radio waves, radio or radar waves are not useful for underwater applications, because they are rapidly absorbed by the water. Sound and ultrasonic waves, on the other hand, can be used to determine the locations of obstacles in water and, in a more sophisticated way, the velocity of moving objects, as will be discussed in Section 2.5. Other applications of sonar include locating large schools of fish by fishing trawlers, and determining the depth of the ocean by measuring the time for the reflected wave to return to its source on a surface ship. Dolphins possess an ultrasonic sonar system to aid them in navigating and in identifying danger.

A more sophisticated application of sonar involves the location of various layers in the ground. When pressure waves from a small explosion pass downward into the earth, they will reflect off any boundary between different underground layers. By examining the amplitudes and arrival times of the returning pulses, various geological layers and their depths can be identified. The use of this technique aids in the discovery of oil, natural gas, coal, and other minerals. A similar phenomenon occurs in the atmosphere, where marked differences in the speed of propagation of sound waves exist between the various atmospheric layers.

Musical wind instruments, such as the trumpet and the clarinet, use reflections of waves inside the instrument to form standing waves, creating stable musical tones, as we shall discuss in later chapters.

Refraction

It was indicated in the previous section that, when a wave is incident upon a boundary between two different media, some of the wave is reflected and some passes into the new medium. We consider now the portion of the wave that passes into the second medium. The passing of part of the wave from one medium into another is accompanied by a change of direction of the wave front, except when the wave approaches the boundary perpendicular (along the normal) to the boundary. The bending of a wave as it passes between two different media, called *refraction*, is illustrated in Figure 2-18.

The nature of this bending can be understood by considering the following example. Suppose that a band is marching on a dry field, approaching the boundary of a muddy region at some angle of incidence (defined with respect to the normal to the boundary) as shown in Figure 2-19. The positions of any rank after successive steps are indicated by the x's in the figure. If the same tempo (time interval for each step) is maintained, the steps must get smaller when the marcher steps into the mud, owing to the increased difficulty of marching. After each successive step in the mud, that column will fall farther behind those that enter the mud later. The band must always move perpendicular to the front rank; therefore, the band will change direction at the boundary.

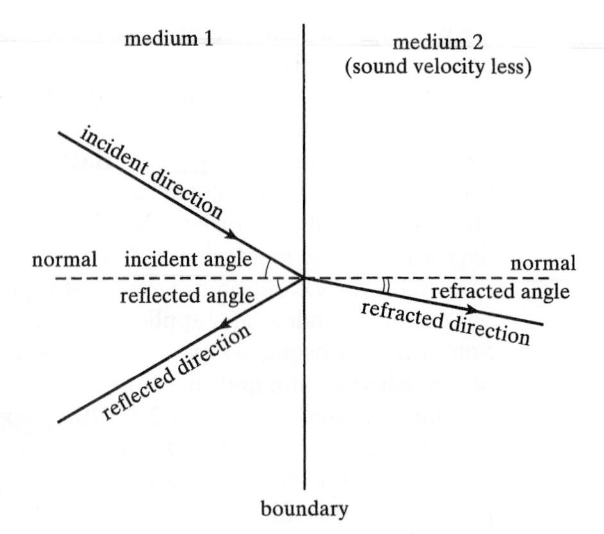

Figure 2-18 Relationship between various wave directions and the normal when a wave travels from one medium into another with a smaller wave velocity.

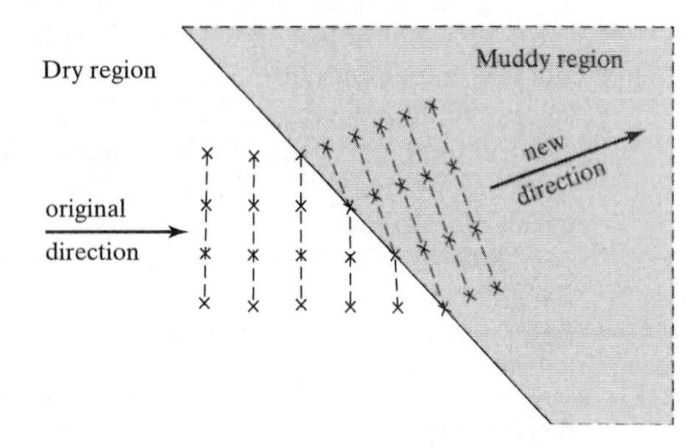

Figure 2-19 A marching band crosses the boundary from a dry region into a muddy region, maintaining the same tempo. A change in direction is necessary to keep the band marching perpendicular to the front rank.

If the reason for the changing direction of the marching band seems confusing, we can clarify this by performing an analogous experiment using waves in a ripple tank. The important concept here is that the wave does in fact move in a direction perpendicular to its wave front. This situation is illustrated by a sequence of Huygens's wavelets in Figure 2-20. After the wave hits the boundary, its velocity becomes smaller, and it goes a shorter distance in the same time interval, so the circular wavelets are smaller. As before, the line joining the extreme parts of the circular waves emitted at the same time becomes the new wave front. This process repeats itself to yield the resultant bending of the plane wave. The angle of bend is observed here to be toward the normal for the case in which the wave moves into a medium with a slower wave velocity. Conversely, the direction of propagation bends away from the normal when a wave enters a medium with a faster wave velocity. You can understand this by considering the example worked backward in time. In that case, the wave begins in the slow medium, and, upon reaching the fast medium, one side of the wave front speeds up before the other. Thus the wave bends away from the normal.

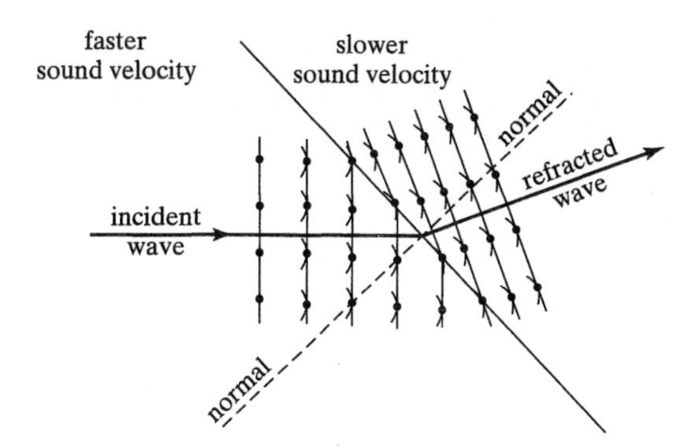

Figure 2-20 Refraction of waves, illustrated using Huygens's wavelets, analogous to the marching-band example. The arcs show the leading edges of the Huygens's wavelets.

In a ripple tank, different wave velocities are obtained by varying the depth of the water. The shallower the water, the slower the wave velocity. The same situation applies to water waves approaching the ocean shore, only in this case the depth changes continuously, creating a slow continuous change in the velocity of waves as they come in from the sea, as illustrated in Figure 2-21. As the wave approaches the shore at some angle, the part of the wave in the shallower water, closer to the shore, will move at a lower velocity. As a result, the section of the wave front more distant from the shore travels faster and will advance relative to the part of the wave closer to shore. The wave therefore bends toward the shore, as shown.

This change in wave speed with depth is also responsible for the breaking of ocean waves. As a large swell travels directly toward the shore, the front part is in water slightly shallower than the water that the rest of the wave is in. Therefore the back "catches up" to the front, until the wave shape is very steep. At a critical point, the wave breaks.

A similar effect occurs in the refraction of sound waves by the atmosphere each night and under certain weather conditions. Under normal daytime conditions, the sun heats the ground, which then radiates heat into the atmosphere; the air higher above the surface of the earth is therefore cooler. At nighttime and during the day whenever there is a thick cloud cover, the sun does not heat the ground directly, and the opposite situation occurs. Hot air rises and cold air moves to the ground, resulting in a temperature that rises with elevation above the surface of the earth, called a *temperature inversion*.

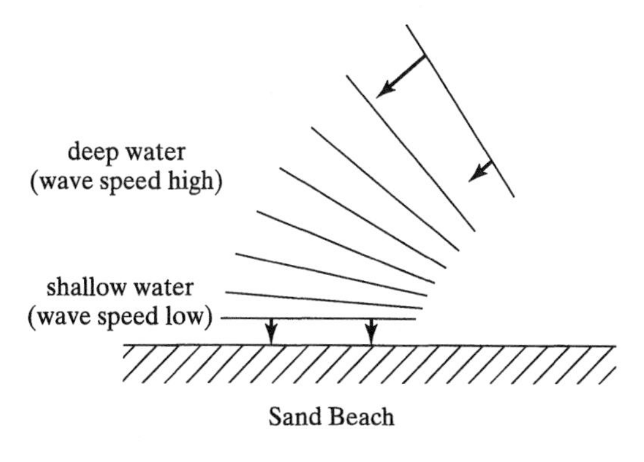

Figure 2-21 Waves approaching a beach will become parallel to the shore, because the wave velocity of the part of the wave closest to the shore is lower than that of the part farthest from the shore.

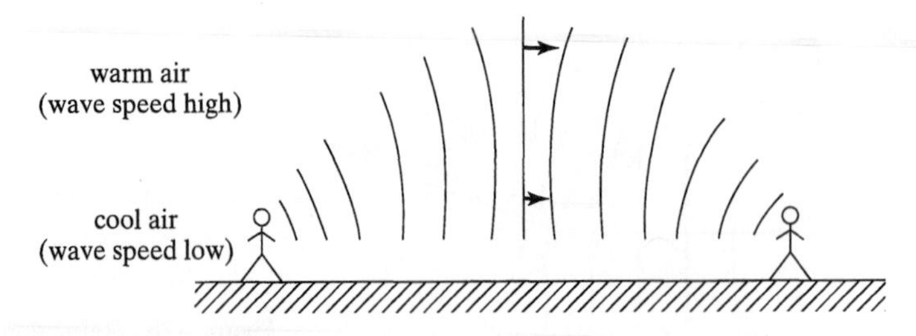

Figure 2-22 During a temperature inversion, sound waves emitted upward are refracted back toward the earth, causing the sound to travel much further along the ground.

The velocity of sound is greater in warmer air, which is higher above the ground during an atmospheric temperature inversion. A condition therefore arises that is similar to that of water waves at a beach. Sound emitted by a source at even large upward angles is bent back to earth, owing to the change in wave speed, and thus increases significantly the distance at which sounds can be heard, as illustrated in Figure 2-22. Here again, the change of medium is slow and continuous. At night, voices carry a great deal farther than during the day, particularly over a placid lake, which enhances the temperature inversion; sound from highways or outdoor concerts can often be heard for miles. This is not a psychological effect or a result of lower background noise during the night hours but rather a physical phenomenon resulting from the refraction of sound in the atmosphere.

Another refraction effect occurs when you try to speak into, or against, the wind. If a steady wind is blowing, the air will drag against the ground; this air friction causes the wind velocity to be slightly lower near the ground than higher above the ground, as illustrated in Figure 2-23. The wind changes the velocity of sound with respect to the ground: the sound wave moves through the air at the speed of sound in air, and the air itself is moving. As a result, the velocity of sound increases with elevation above the ground when the sound wave propagates in the direction in which the wind is blowing. A situation similar to that during a temperature inversion therefore occurs when you speak *with* the wind; sounds are refracted back to the earth and your voice "carries" with the wind. On the other hand, when you speak *into* the wind, the speed of sound decreases with elevation; sounds are therefore refracted upward and your voice is "lost," as illustrated in Figure 2-24.

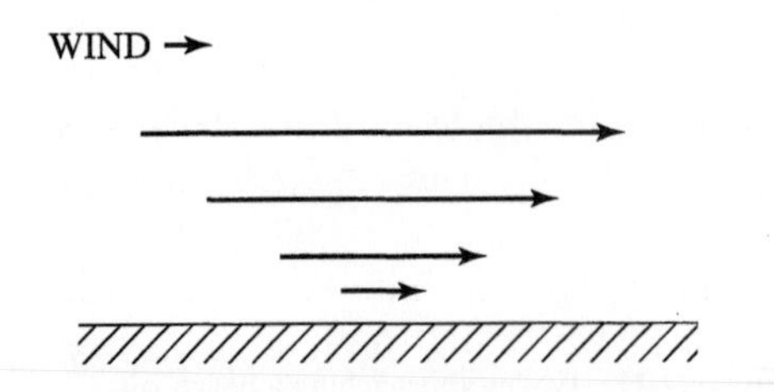

Figure 2-23 During a steady wind the air velocity is faster for points higher above the ground, due to friction between the air and the ground.

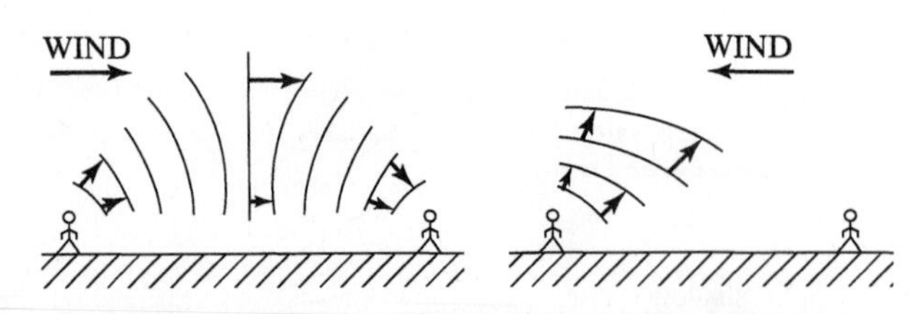

Figure 2-24 Refraction of a sound wave resulting from motion of the air. Speaking "with the wind" your voice "carries," whereas speaking "into the wind" the sound of your voice is "lost."

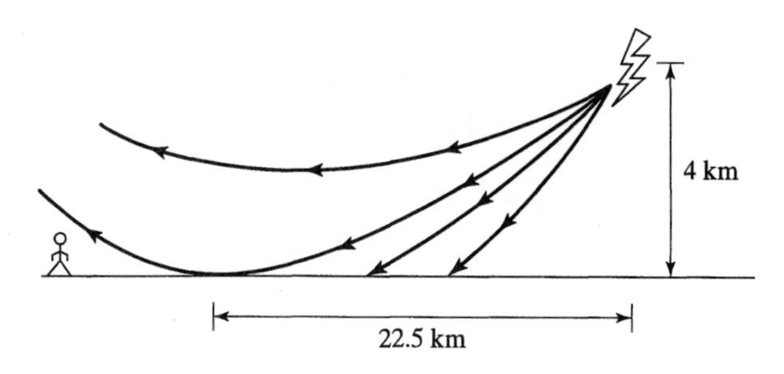

Figure 2-25 A person more than 22.5 km from a thunderhead cannot hear the thunder, because the sound is refracted back up into the atmosphere.

The refraction of sound in the atmosphere is also responsible for there being a maximum distance over which thunder can be heard. During the beginning of a thunderstorm, the usual temperature gradient in the atmosphere is found. The speed of sound is therefore greater near the surface of the earth, where the warmer air is found, and waves will bend upward, as shown in Figure 2-25. Most thunderheads are about 4 km high, and under the usual atmospheric conditions a maximum range of about 22.5 km is attained before the sound is refracted back upward into the atmosphere. In general, thunder will not be heard beyond this range.

Interference

Interference has a special meaning in the physics of waves; it refers to the combining, or addition of two similar waves. Interference can be either destructive, resulting in the effective disappearance of the waves, when they are out of phase, or it can be constructive, resulting in the enhancement of the waves, when they are in phase. Interference does not refer to one wave affecting the properties of another, which would be a violation of the principle of superposition. Rather, it refers to the addition of two waves at some point or points in space.

A demonstration of the interference of sound waves involves a Quincke's tube, which is shown in Figure 2-26. An audio-frequency sine wave is fed into the end of a tube, which branches into two parts and then recombines, with the resultant wave picked up by a microphone. If the path-length difference L_2–L_1 is one wavelength (or any integral number of wavelengths), as shown in Figure 2-27(a), the two waves will be in phase when they arrive at the microphone and the resulting sound will be loud. If, on the other hand, the path-length difference is one half-wavelength, the two signals will arrive at the microphone *exactly* out of phase, as shown in Figure 2-27(b), canceling each other, and *nothing* will be heard. This will also occur whenever the path-length difference is any odd number of half-wavelengths, i.e., $\lambda/2$, $3\lambda/2$, $5\lambda/2$, etc. The two waves are out of phase if, whenever a compression from one path arrives at the recombination point, a rarefaction from the other path arrives simultaneously, or vice versa. Two identical waves arriving at the same point at the same time combine to produce the sum of their effects; when a compression and a rarefaction of equal amplitude arrive simultaneously, their effects cancel.

A more complicated example is the moiré-pattern model of interference between two identical point sources shown in Figure 2-28. Two moiré patterns consisting of a series of concentric circles represent the crests of the two interfering waves. In the

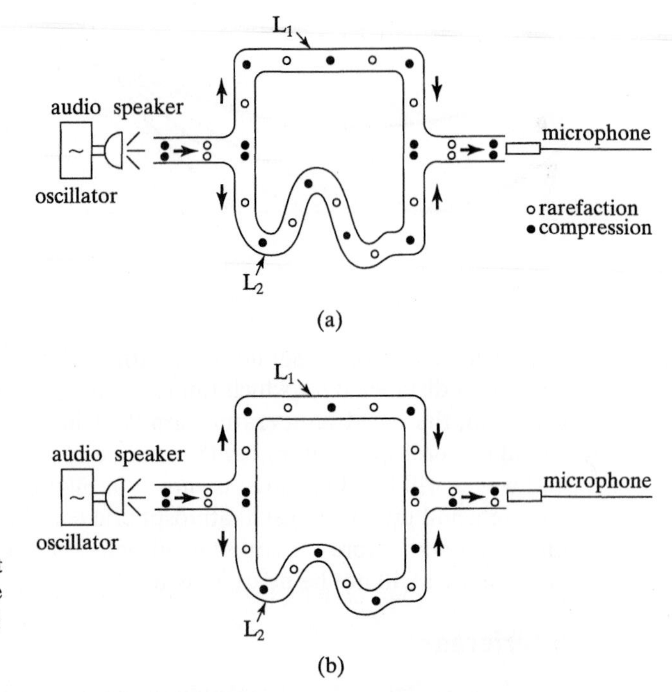

Figure 2-26 Quincke's interference tube. Part (a) shows the case in which $L_2 - L_1 = \lambda$, so the waves are in phase at the microphone. Part (b) shows the case in which $L_2 - L_1 = \lambda/2$, so the waves are out of phase at the microphone.

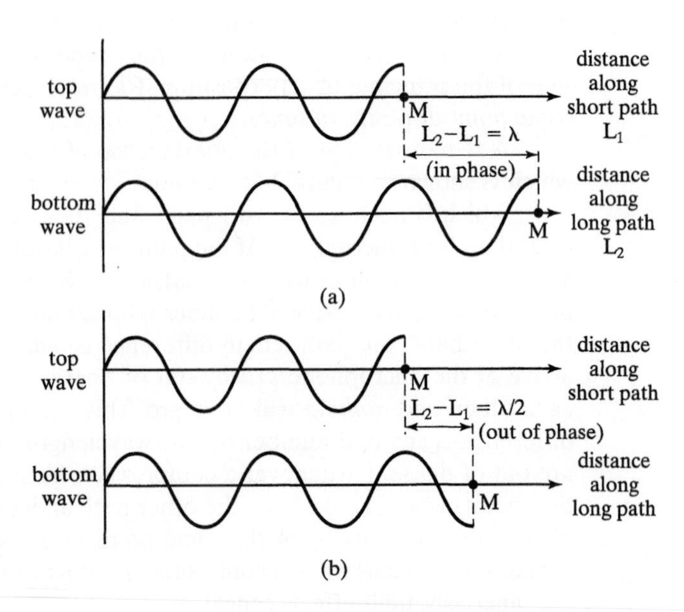

Figure 2-27 If $L_2 - L_1$ in the Quincke's tube is an integral number of wavelengths, the signals interfere constructively at the microphone to produce a sound almost as intense as the original sound. If the path length difference is an odd number of half wavelengths, the two waves interfere destructively to produce almost no sound at the microphone.

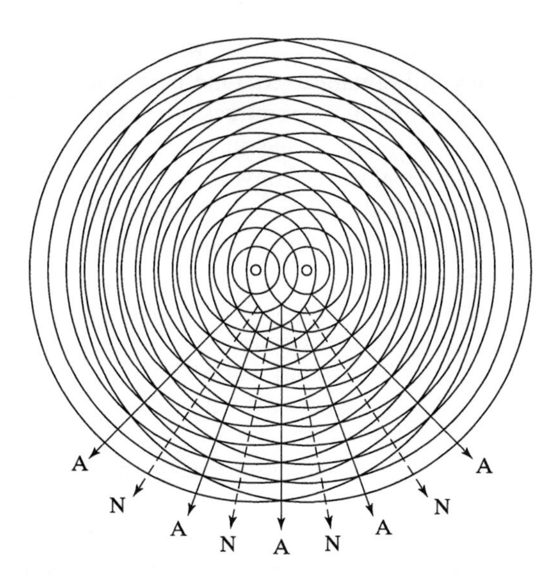

Figure 2-28 Interference between two identical point sources. A labels each antinodal line; N labels each nodal line.

sound analog, the lines represent compressions and the centers of the spaces between represent rarefactions. This could also represent ripple-tank waves photographed at an instant in the positions shown in the drawing, where the lines are the crests. Observe that there are lines called *antinodal lines*, labeled A, along which first peaks from both sources coincide, then troughs from both sources coincide, and so on, so that the motion is greatest along these lines. Between each pair of antinodal lines is a *nodal line*, labeled N, where a peak from one source always coincides with a trough from the other, and there is no motion of the surface of the water. These lines are nearly straight at large distances from the sources. The two cases are illustrated in Figure 2-29; the bottom curves show the sum of the waves along the antinodal and nodal lines at the particular instant of time chosen, beginning at the start of the antinodal and nodal lines, respectively, near the center.

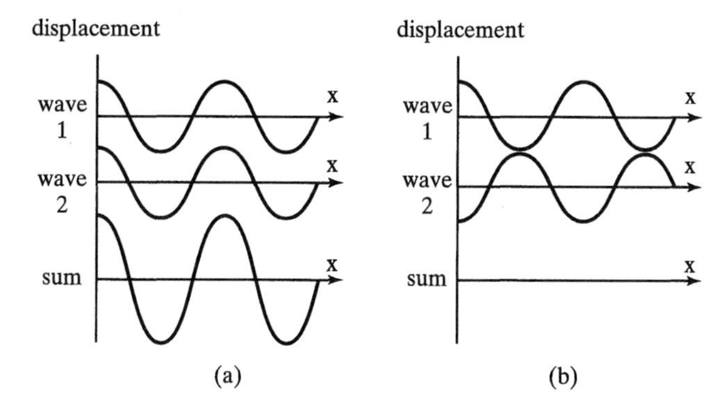

Figure 2-29 Sum of the two waves along (a) antinodal lines, and (b) nodal lines, for the interference of two point sources as shown in Figure 2-28. The coordinate x denotes the distance from the point midway between the two sources.

As these waves are moving outward, they will "interfere" with each other in such a way that the nodal and antinodal lines will remain fixed. One period later the "picture" is identical to that drawn. The waves from the two sources propagate outward, so that at any point along an antinodal line they are always in phase, producing a sum wave greater in amplitude than either individual wave. Conversely, at any point along a nodal line, the waves from the two sources will always be out of phase and cancel each other. In summary, you observe large waves moving outward along the antinodal lines and no waves at all along the nodal lines.

If, instead of the two point sources in the ripple tank, we use two speakers as sources of sinusoidal sound waves, the nodal and antinodal lines can be found by listening at various places around the speakers. The sound will be loud along antinodal lines and soft along nodal lines. This experiment works best where there are no reflections from walls and ceiling.

The source separation and wavelength of the sound affect the pattern of nodal and antinodal lines. As the centers of the moiré patterns are moved farther apart, the nodal and antinodal lines get closer together; as the centers are brought closer together, fewer nodal lines appear, and they are separated by larger angles. This is analogous to changing the distance between the two sources; the closer the sources, the greater is the angular separation between nodal and antinodal lines, and vice versa.

Keeping the source separation constant, we can investigate the effect of changing wavelength. Using concentric circles with greater radial spacing, the nodal and antinodal lines spread apart. When concentric circles with small radial spacing are used (imagine twice as many circles in the pattern), the angles between the nodal and antinodal lines decrease. Analogously, increasing the wavelength of the sound (decreasing the frequency) spreads out the pattern of nodal and antinodal lines, whereas decreasing the wavelength decreases the angles between the lines.

If the two sources are out of phase, rather than in phase, the nodal lines and antinodal lines are interchanged compared to the pattern just described. This is of considerable practical consequence in the wiring of stereo loudspeakers. With the speakers properly wired, for bass notes (long wavelength) there should be a very wide antinodal region covering the entire area in front of the speakers. If the two speakers are wired out of phase, the area in front of the speakers will become a nodal region where the low-frequency waves from the two speakers will interfere destructively.

Figure 2-30 shows an experimental method for determining whether speakers have been wired in phase or out of phase. Setting the preamplifier to the *monaural* mode, music with a loud bass sound should be played with the speakers facing each other a few inches apart, a distance much shorter than the wavelengths associated with bass frequencies. First use the normal wiring; then reverse the wires on one (only one) of the speakers, reversing its phase. If the speakers are in phase, the bass signals from

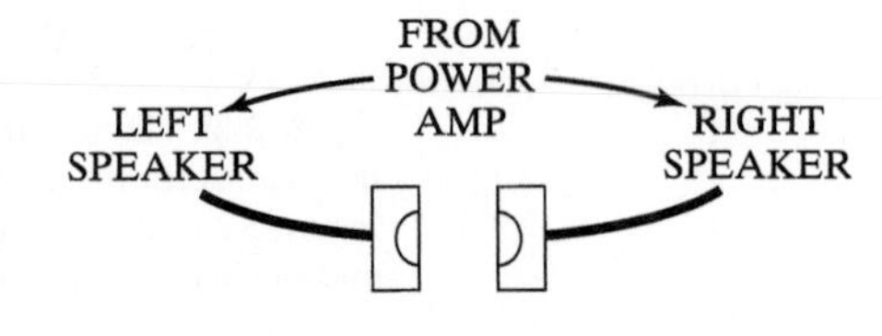

Figure 2-30 Setup for determining the proper phase relation between stereo loudspeakers. System should be set to MONO on preamp controls.

Figure 2-31 Bose QuietComfort®2 Noise-Canceling® Headphones. (Courtesy of Bose Corporation.)

both speakers add to give a loud bass sound. If they are out of phase, the bass signals will tend to cancel each other, resulting in soft bass sound. The proper phase of the two is that which gives the louder bass sound (speakers in phase).

Interference plays a crucial part in active noise-cancellation devices. One example, Bose QuietComfort® 2 headphones shown in Figure 2-31, sense ambient sound, invert its phase, and send the inverted wave to the earphones. The ambient sound and the inverted wave mix such that they interfere destructively, eliminating most of the noise. These headphones can be used to reduce noise in a work environment; music played through the same headphones can be enjoyed free of background noise.

Diffraction

We have studied refraction, the bending of waves due to a change in velocity. Waves do not need a change of medium to cause them to bend; bending within the same medium is called *diffraction*. Sounds can be heard around corners or behind barriers that cut off direct view of the source of sound. This bending of waves around corners in the same medium is an example of diffraction, and can be explained using Huygens's principle.

The drawing of the plane wave in Figure 2-8(b) is an oversimplification. Although it is accurate for the middle of the wave, near the edges the Huygens's wavelets spread out in all directions and cause the wave to develop curved edges and spread out, as shown in Figure 2-32. This is an example of diffraction. The amount of diffraction will depend on the nature of the waves and their wavelengths. For example, you can hear around corners, but you cannot see around corners, because sound waves diffract much more than light waves. In general, low-frequency sound waves will diffract more than high-frequency waves. This is one reason voices sound muffled, or reduced in high-frequency content, when heard around corners or in adjacent rooms. Also, we hear the drums and tubas of a marching band as louder than the other instruments if the

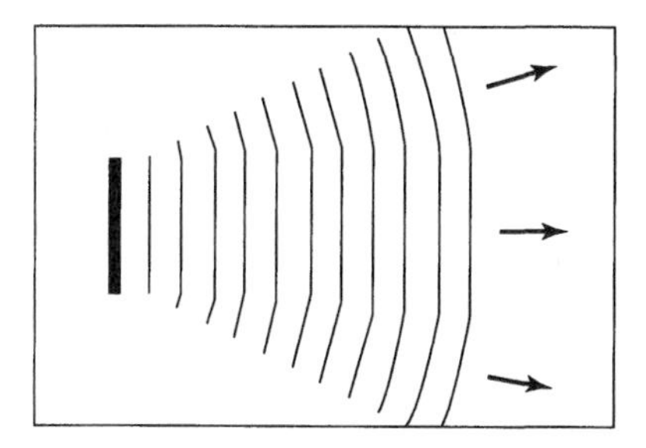

Figure 2-32 Diffraction at the edges of a plane wave.

band is around the corner of a building. This is not because the drums and tubas are playing louder than the other instruments; it is because their low-frequency waves diffract around the corner more easily than the higher-frequency waves from the flutes, for instance. Likewise, sounds from directional speakers take on a different quality when heard from other than the ideal direction.

An interesting experiment demonstrating the diffraction and interference of sound waves involves a small unmounted loudspeaker and a thin square board about 60 cm on each side, serving as a baffle. A circular hole the size of the speaker is cut in the center of the board. If music is played with the speaker alone, the sound lacks volume and has very little bass. If the speaker is placed behind the board, facing out through the hole, the sound heard by a listener in front of the board is now much louder than before the board was present, particularly in the low frequencies. Why is this so?

As the diaphragm of the speaker moves rapidly forward and backward, it creates alternating compressions and rarefactions moving in the forward direction. The signal emitted from the rear of the speaker is exactly out of phase with the signal emitted from the front, because when the speaker cone moves forward it will create a compression in front and a rarefaction behind, and vice versa. If the baffle is not present, the waves from the rear will diffract around the edge of the speaker and travel away from the front of the speaker. The total signal from the speaker is the sum of the two waves, those from the front and those from the rear. Because they are out of phase, they interfere destructively and tend to cancel; the sound heard is thus reduced in amplitude. With the board present, the cancellation process is largely eliminated for two reasons. First, owing to the geometry, it is harder for the waves emitted by the back of the speaker to diffract all the way around to the front, and, second, owing to the extra distance traveled by the wave emitted from the back, it will not be exactly out of phase with the wave emitted from the front (in general), and total destructive interference will not occur. Therefore, the sound is louder when the baffle is in place. Because the long wavelengths (low frequencies) diffract more readily than the short wavelengths, the greatest effect is in the bass range.

The board in this experiment is a type of loudspeaker baffle. We shall study more about the effects of baffles and enclosures on speakers in Chap. 7.

Diffraction introduces a limit to the effectiveness of highway noise barriers. According to the U.S. Department of Transportation Federal Highway Administration, in order to be effective a noise barrier must be "tall enough and long enough to block the view of the highway from the area that is to be protected." This typically results in a 5-dB reduction in the noise level, compared with the minimum change observable by the human ear of about 2 dB. An additional reduction of about 1.5 dB can be achieved for each meter of barrier height. Attenuation is caused by the barrier reflecting and absorbing sound. However, much of the sound energy lies at relatively low frequencies, so diffraction of waves over the barrier to the protected area places a limit on the attenuation that can be achieved. Some additional decrease in sound level may be obtained by using combinations of earthen berms and sound barriers or vegetation, and the inverse square law plays a significant role. Research into more efficient noise barriers continues, but the laws of physics clearly limit the effectiveness of these barriers, and the laws of economics limit the willingness of governments to construct them.

2.4 Addition of Waves

In our study of waves and wave motion, we shall regularly find it necessary to add two waves together graphically to obtain a sum wave. If the graphs represent the waves themselves (say displacement or velocity versus position along the wave), we say that we are adding the waves together when we add their graphs. If the graphs represent motion at a fixed position in the medium (say displacement versus time), we say that we are adding the motions that the graphs represent. Sometimes these two types of wave addition will look the same except for the labels on the two axes of the graph. We shall refer to either of these two types of graphical addition as the addition of waves.

Before discussing the addition of wave trains and waves of different periodicity, we shall examine how individual pulses add in the case of transverse rope waves. Consider two equal pulses, each consisting of one half-wave, that come together in a single rope as shown in Figure 2-33. The succession of five pictures is like a set of snapshots taken at equal time intervals of one half-period showing how the rope would appear as the pulses approach and pass each other. Here, of course, we are graphing displacement versus position. For each instant of time, the two individual pulses are shown on the first and second lines, with the sum of the two shown on the third line. The total displacement of the rope, its overall shape, is the sum of the displacements of the component waves at that point. Notice that the superposition of two identical pulses *in phase* with each other (both positive or both negative) produces a single pulse of twice the amplitude, whereas the superposition of two pulses *out of phase* with each other (one positive and one negative) produces a net effect of zero; their effects cancel.

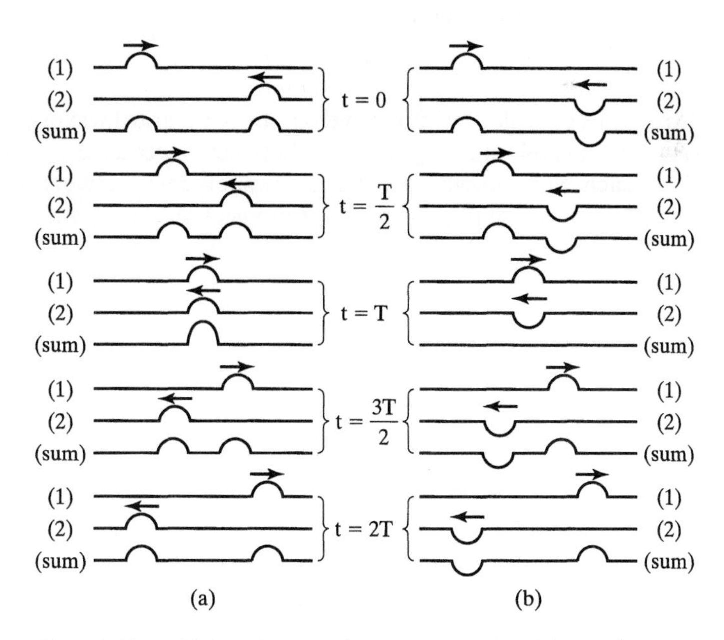

Figure 2-33 Addition of equal half-wavelength pulses (a) in phase, and (b) out of phase, as they pass each other on a rope, shown at equal time intervals of one half-period. At each point in time the individual pulses are shown, along with the overall shape of the rope because of the addition of the two pulses.

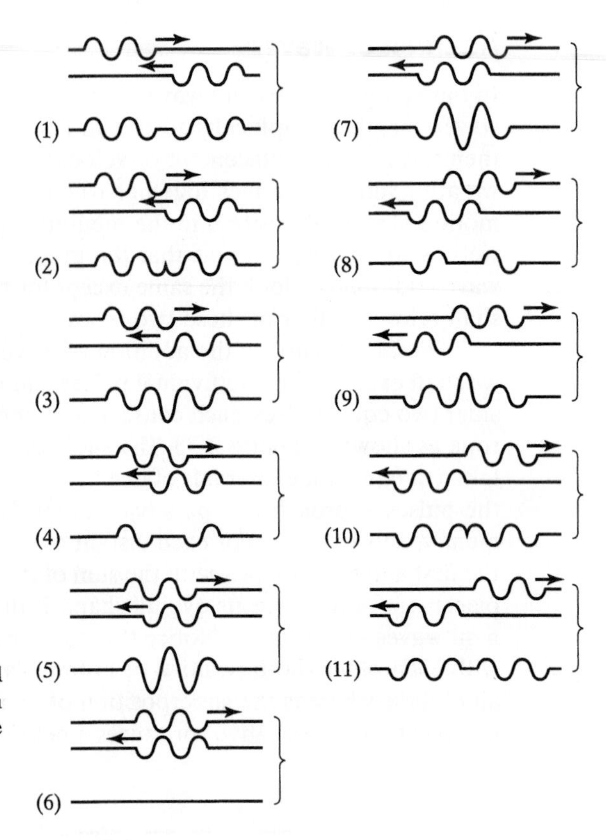

Figure 2-34 Two wavelengths of identical waves passing each other on a rope, shown at one-quarter-period intervals. Each wave is shown individually, along with the resultant sum wave, which represents the overall shape of the rope.

Now let us consider two full wavelengths of identical waves passing each other as shown in Figure 2-34. The time interval between successive pictures is one quarter-period, so each wave moves one quarter-wavelength between pictures. When two positive or two negative pulses are superimposed, the resultant is a pulse with amplitude equal to the sum of the two individual pulses; when equal but opposite pulses are superimposed, the resultant is zero.

The preceding discussion is consistent with the principle of superposition as discussed in Sec. 2.2, and the presence of one pulse does not affect the other. If the two pulses affected each other, the shape of each pulse would not be the same before and after their interaction, or they would not add in the manner described here.

Suppose that we want to add the two waves shown in Figure 2-35(a). The value of the sum wave at any given time is just the algebraic sum (positive above the horizontal axis and negative below the horizontal axis) of the two individual waves at that time. For instance, in Figure 2-35(a) at $t = 0$ ms, both waves 1 and 2 have a value of zero. The sum wave will therefore have a value of zero. Similarly, the sum wave has a value of zero at times $t = 2$ ms and $t = 4$ ms. Now consider the situation at time $t = 1$ ms. Each component wave has a value of 1 volt (V), so the sum wave will have a value of 2 V. Similarly, the value of the sum wave at time $t = 3$ ms is -2V. In fact, because the two waves have exactly the same value at all points in time (they are identical waves), the value of the sum wave is exactly twice the value of either of the component waves, as shown in

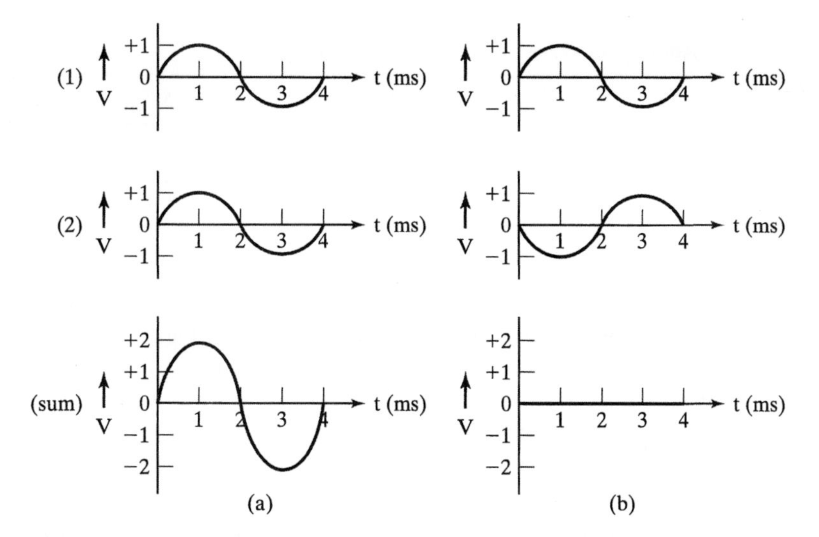

Figure 2-35 Addition of two waves of the same amplitude and frequency (a) in phase and (b) out of phase.

Figure 2-35(a). If the two waves are out of phase, as shown in Figure 2-35(b), the sum of their two values at any point in time is zero, and the sum wave is zero at all times.

A more complex situation emerges when the two component waves have different periodicity, as illustrated in Figure 2-36, although the principle and procedure by which the waves are added remain the same. As before, the procedure used in adding these waves is to add the two component waves, point by point. Because the two waves are smooth and continuous, it is only necessary to add the waves at a limited number of

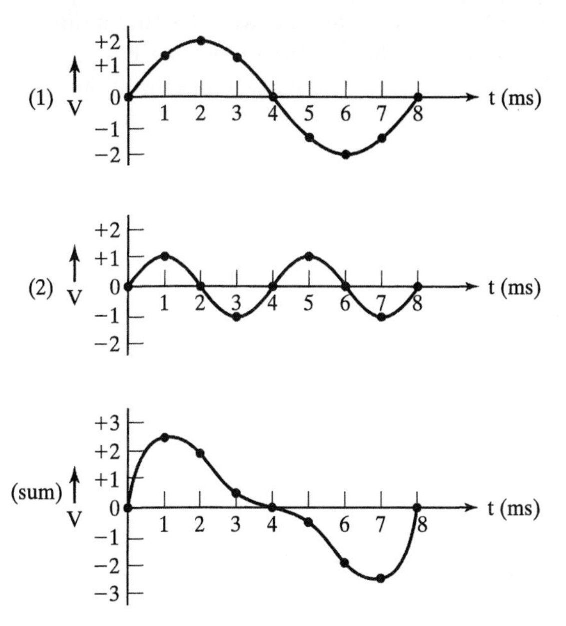

Figure 2-36 Addition of two waves of different amplitude and frequency.

Table 2-1 Addition of Waves of Figure 2-36 at 1-ms Intervals

t(ms)	0	1	2	3	4	5	6	7	8
Wave 1	0.0	1.4	2.0	1.4	0.0	−1.4	−2.0	−1.4	0.0
Wave 2	0.0	1.0	0.0	−1.0	0.0	1.0	0.0	−1.0	0.0
Sum wave	0.0	2.4	2.0	0.4	0.0	−0.4	−2.0	−2.4	0.0

points. A good general rule is to choose points spaced in time at one-quarter of one period of the wave with the shortest period, as shown in Figure 2-36; in this case that is 1 ms. This spacing ensures that all the crests and troughs of the resulting wave will be seen. After adding the two component waves at each of the chosen points, plot the sum wave points and draw a continuous smooth curve through the points. Table 2-1 shows the addition of the two component waves of Figure 2-36. For example, at time $t = 1$ ms, the value of wave 1 is +1.4 and the value of wave 2 is +1.0; therefore, the sum wave has a value of +2.4. You should check the addition and look carefully at how the sum wave curve has been drawn through the points.

The addition of waves is extremely important and will be used in our study of beats in the next section, standing waves in Chap. 3, and complex waves in Chap. 4.

2.5 Beats

We have discussed the addition of two waves with the same frequencies and amplitudes. Now we shall consider the addition of two pure waves of the same amplitude but slightly different frequencies. Those who have played instruments in a band or orchestra are probably familiar with the sound resulting when two identical instruments play the same note slightly out of tune with each other: a sum note that "wobbles" in loudness, in which the frequency of the wobbling decreases as the two notes become better in tune. This wobbling is used by musicians to tune their instruments; when the wobbling disappears, the instruments are in tune. The physical phenomenon of the change in amplitude, or loudness, of the sum of two tones with slightly different frequencies is known as *beats*.

Consider the two pure waves drawn in Figure 2-37; the first wave has a frequency of 9 Hz (a period of $^1/_9$ sec), and the second has the same amplitude but a frequency of 11 Hz (a period of $^1/_{11}$ sec). Here we are plotting the displacement of one physical point with respect to time.

The resultant wave has a frequency of exactly 10 Hz (a period of $^1/_{10}$ sec), the average value of the two component waves (9 and 11 Hz, respectively). Also notice that because their frequencies are slightly different, they go in and out of phase with each other, alternately interfering constructively (adding together) and destructively (canceling each other), with the result that the amplitude of the resultant wave varies between zero and the sum of the two components. The frequency at which the amplitude alternately rises and falls, called the *beat frequency*, is equal to the difference between the two component frequencies. In this case, the beat frequency is 11 Hz − 9 Hz = 2 Hz; that is, the amplitude of the resultant wave increases and decreases twice per second. In general, for two audio tones of equal amplitude and nearly equal frequencies of f_1 and f_2 (f_2 being greater than f_1), you will hear a tone

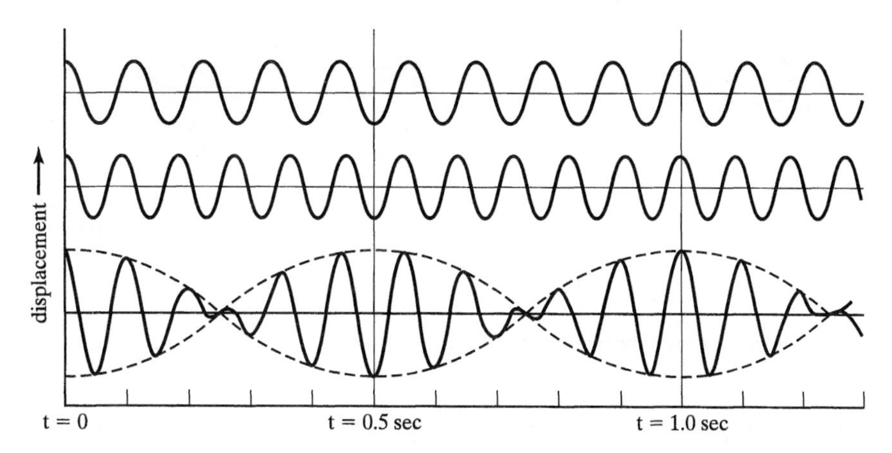

Figure 2-37 The addition of two waves of equal amplitude but slightly different frequencies results in a wave whose frequency is the average value of the two original frequencies and whose amplitude varies between zero and the sum of the amplitudes of the original waves. This phenomenon is called "beats." Dashed lines indicate the envelope.

whose frequency is the average value, $F_{tone} = (f_1 + f_2)/2$, rising and falling in intensity at the beat frequency, $f_{beats} = f_2 - f_1$. For example, if the two waves have frequencies of 439 and 441 Hz, a tone of 440 Hz will be heard, with two beats per second.

The *envelope* of the resultant wave, the curve showing the resultant amplitude as time passes, is indicated by the dashed lines in Figure 2-37. The envelope is sinusoidal in shape whenever the two original waves are sinusoidal. The full envelope consists of two sine waves of frequency $\frac{1}{2} f_{beats}$, out of phase with each other.

2.6 The Doppler Effect

Virtually everyone has experienced the rising and falling of the pitch of a train whistle as the train passes by a railroad crossing. This change of pitch, which occurs whenever there is a relative motion between the source and observer, is an example of the Doppler effect, named after the Austrian physicist Christian Johann Doppler (1803–1853). The effect becomes clearly audible when the relative velocity between source and observer is at least a few percent of the velocity of sound (1100 ft/sec or 750 miles/hour), or roughly 15 to 20 miles/hour. We do not normally observe the Doppler effect for light, which would cause everything to change color, because the velocity with which we travel never approaches a few percent of the velocity of light (186,000 miles/sec). Relative motion between our galaxy and other galaxies *is* of this magnitude, so the Doppler effect for light plays an important role in the study of the relative motions of galaxies.

As a boat moves through water, it creates waves; one can observe these waves from shore, a standing boat, or a moving boat. In any case, it seems reasonable that once the wave is created it moves over the water at a speed that is independent of any motion of the source or of the observer. That is, the wave velocity is dependent only on the medium through which it is traveling, not on the motion of either source or observer. Likewise, all sound waves travel through the air at the speed of sound in air, independent of the speed of the source or observer.

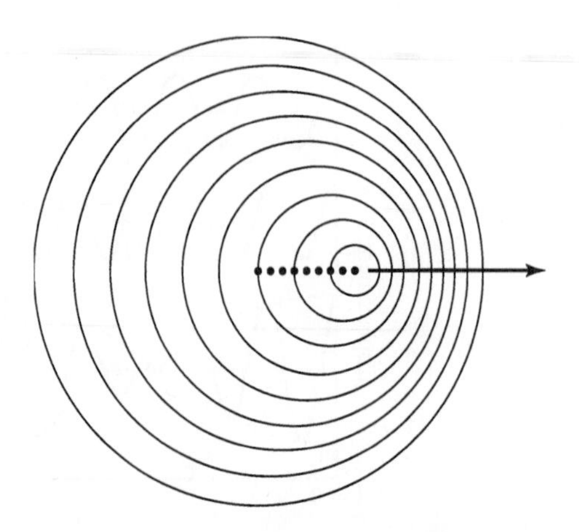

Figure 2-38 The dots show the location of a moving wave source at equal time intervals for a source moving to the right with a constant velocity. The circular wave fronts emitted by the source at each of these points are shown.

We can study the propagation of waves and the Doppler effect using a ripple tank. We have seen how a stationary point source produces circular waves in the ripple tank as in Figure 2-8(a). Now we shall study the waves emitted from such a source moving at a constant velocity. Because the source is moving at a constant velocity, it moves the same distance between the times at which it emits waves; the dots in Figure 2-38 mark the points at which the source emits each successive circular wave. Figure 2-38 also shows the circular wave fronts. The only difference between Figs. 2-38 and 2-8(a) is that, as time has passed, the center of each successive (smaller) wave has moved to the right by the amount shown in Figure 2-38, causing the circles to squeeze together on the right and spread apart on the left. The velocity of all the waves is the same with respect to the surface of the water and independent of the source velocity. A stationary observer counting passing waves in a specified time interval would count a larger number if he or she were in front of the source (at the right) and a lower number if behind the source (at the left). Thus the frequency of the waves received by the observer appears higher where the motion of the source is toward the observer and lower where the motion of the source is away from the observer. Alternatively, the wavelength is shorter in the forward direction, resulting in a higher frequency or pitch, and longer in the backward direction, resulting in a lower frequency or pitch. This is consistent with the relation $S = f\lambda$ studied in Section 2.1. All sound waves travel at the same speed within a specific medium, so a high-frequency f implies short-wavelength λ and vice versa. Notice also that in the direction perpendicular to the direction of travel, as the source passes directly by the observer, the spacing of the circular wave fronts is approximately the same as when there is no motion of the source. The frequency measured by a stationary observer as a source moves by is thus higher than normal when the source is moving toward the observer, the same as the normal frequency just as the source moves past the observer, and lower than normal when the source is moving away from the observer. By normal frequency we mean the frequency received when both source and observer are at rest.

Let us now study further why the frequency appears higher when the source is moving toward the observer. Consider first a source emitting a series of equally spaced compressional waves toward an observer, as shown in Figure 2-39. The compressions move uniformly through the air at the speed of sound, and create a constant-frequency tone for the observer; the source emits the frequency f_0, and the observer hears the

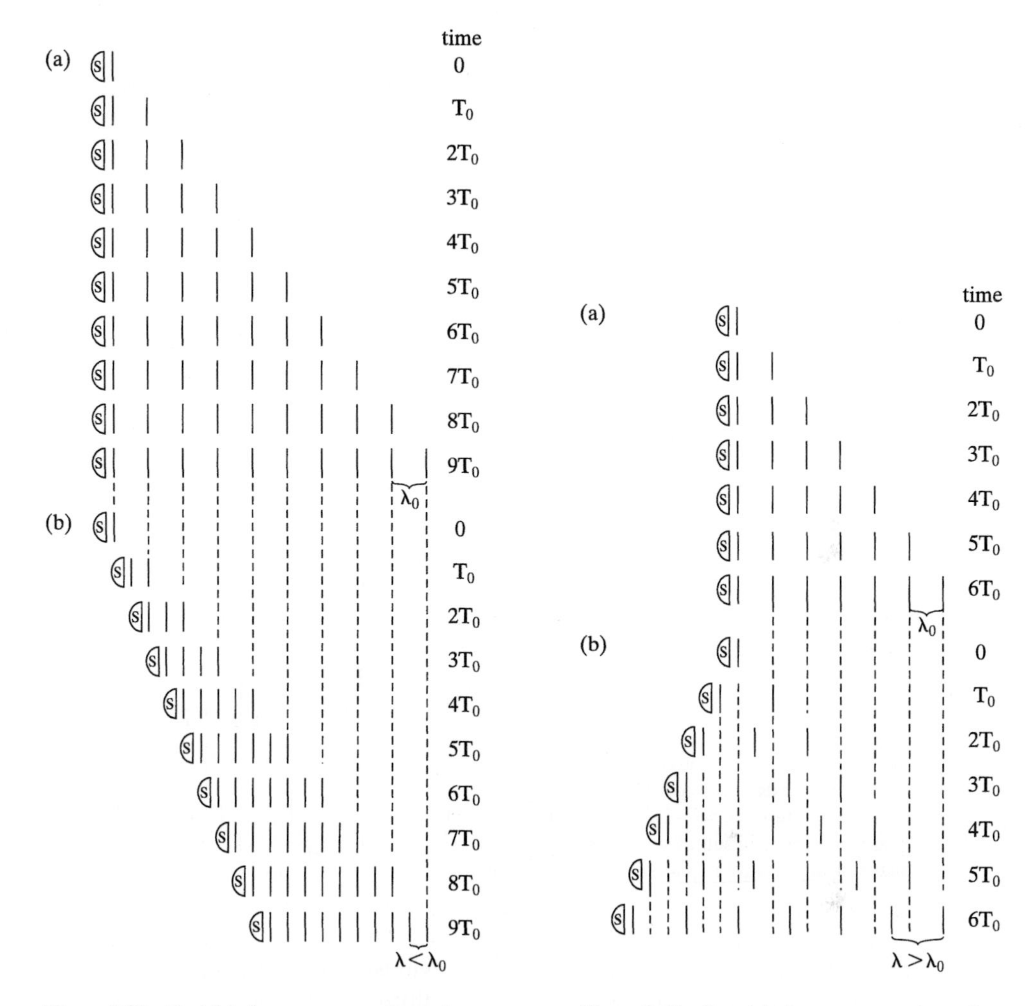

Figure 2-39 Part (a) shows compressions of frequency f_0 emitted by the source and heard by the observer. Part (b) shows the sequence of compressions emitted at the same time intervals, but with motion of the source toward the observer with velocity $v_S = S/2$, resulting in a higher observed frequency.

Figure 2-40 Part (a) shows compressions of frequency f_0 emitted by the source and heard by the observer. Part (b) shows the sequence of compressions emitted at the same time intervals, but with motion of the source away from the observer with velocity $v_S = S/2$, resulting in a lower observed frequency.

normal frequency f_0. Now suppose that the source is moving at a constant velocity of one-half the speed of sound to the right while emitting compressions at the same frequency f_0, as shown in Figure 2-39(b). Each successive compression is emitted closer to the previous one because of the motion of the source toward the observer, and the wave propagating through the air has a shorter wavelength and thus a higher frequency (twice the normal frequency), when heard by the observer.

Now consider the case in which the source is moving at a constant velocity of one-half the speed of sound *away* from the observer, as illustrated in Figure 2-40. We

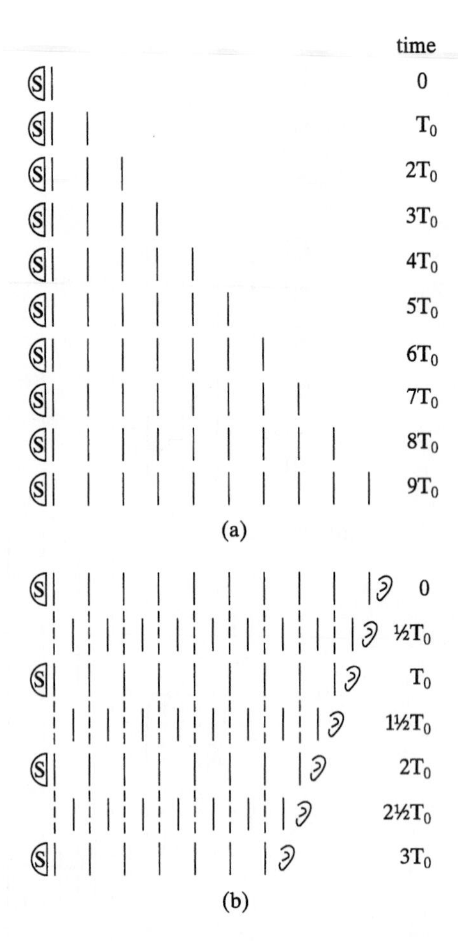

Figure 2-41 Motion of the observer toward the source increases the observed frequency above the normal frequency f_0. Here the velocity of the observer (ear) is equal to the speed of sound, and the sequence of "pictures" of the wave in part (b) is taken at a rate of two per period.

can work out the effect of the motion, as previously, to find that in this case the observer measures a frequency lower than f_0 (two-thirds of the normal frequency).

Next consider a stationary source and a moving observer as shown in Figure 2-41. The velocity of the observer for this example is equal to the speed of sound, so that, after one compression reaches the observer, it will take exactly one half-period for the next compression to arrive. During this time interval, both the observer and the next wave front will have moved one half-wavelength. Therefore, this particular velocity for the observer doubles the observed frequency, as can be read from Figure 2-42. Motion of the observer away from the source can be analyzed similarly.

The frequency f heard by the observer, as a function of source or observer velocity, is shown in Figure 2-42 for a moving observer and in Figure 2-43 for a moving source. In both figures, v_o and v_s, the observer and source velocities, respectively, are positive if the motion of the observer or source is toward the other and negative if the relative motion separates the source and observer. Figure 2-42 tells us that the faster the observer moves toward the source, the higher the pitch heard, and the faster the observer moves away from the source, the lower the pitch heard. Notice that if the observer moves away from the source faster than the speed of sound ($v_o = -S$) he or she

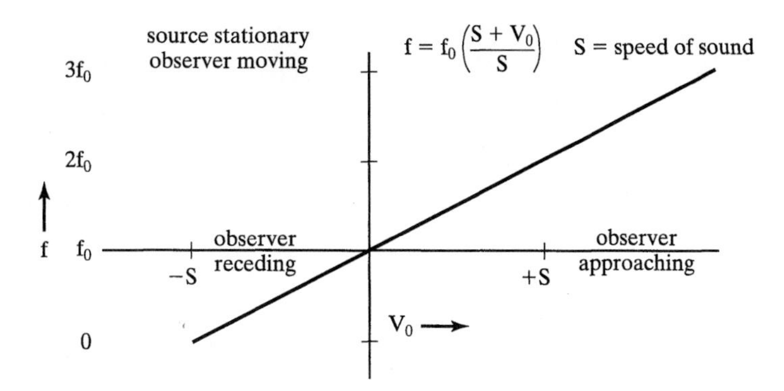

Figure 2-42 Frequency f heard by an observer moving with respect to the source with a velocity v_0 listening to a tone of normal frequency f_0 produced by a stationary source.

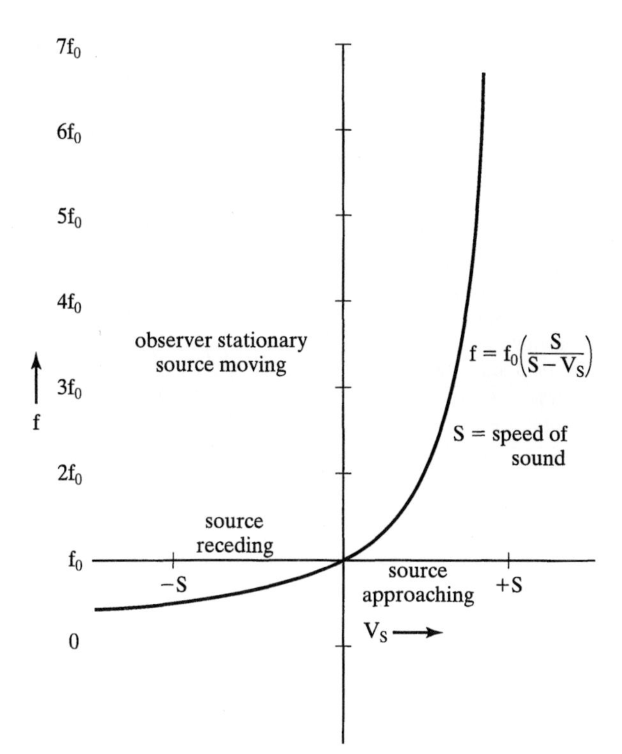

Figure 2-43 Frequency f heard by a stationary observer listening to a tone of frequency f_0 produced by a source moving with a velocity v_s.

outdistances the oncoming wave and therefore hears nothing ($f = 0$). As in the case of motion of the observer, relative motion of the source *toward* the observer *increases* the observed frequency, whereas relative motion *away* from the observer *decreases* the observed frequency, as shown in Figure 2-43. Notice that no matter how fast the source recedes from the observer, the observer will still hear something (perhaps a very low frequency), but when the source approaches the observer at the speed of sound, *all* the

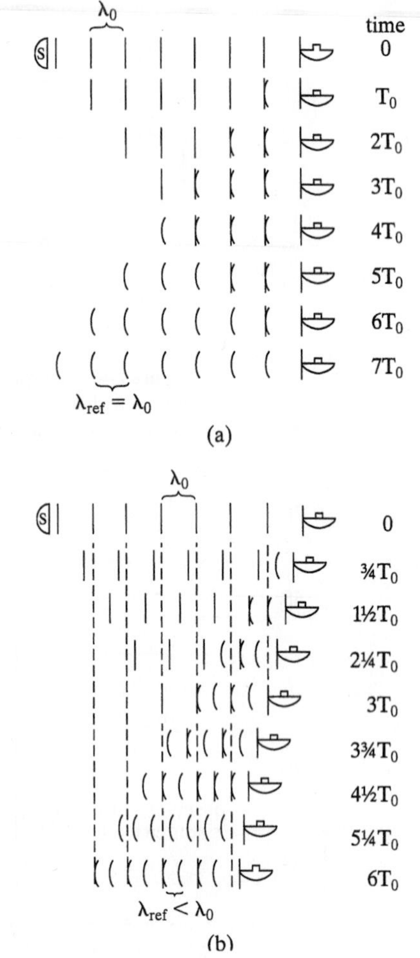

Figure 2-44 Underwater sound waves reflecting off a submarine at rest, as in part (a), return to the source with the same frequency. When the submarine is moving toward the source (here the submarine has a velocity of $S/3$), the frequency of the reflected wave is increased, as shown in part (b). Note that straight line wave fronts are moving toward the right; curved wave fronts are moving to the left after having been reflected.

waves produced by the source arrive at the observer at nearly the same time. This creates an extremely loud sound, because a large number of wave fronts are adding together at the same time and place. This phenomenon, called a *sonic boom*, will be discussed in the next section.

The Doppler effect can be used to determine the speed of underwater objects using sonar, much as the same effect is used to determine the speed of cars on a highway with radar waves. A tone of frequency f_0 is emitted in the direction of an underwater object; the waves bounce off the object as shown in Figure 2-44(a) if the object is stationary and as shown in Figure 2-44(b) if the object is moving toward the source with a velocity of one-third of the velocity of sound in water. For both the source and the object at rest, if the successive wave compressions are shown every full period, it is clear that they will bounce off the object and return to the source with the same wavelength and, therefore, with the same frequency with which they started. For a moving object, the successive wave fronts are shown at time intervals

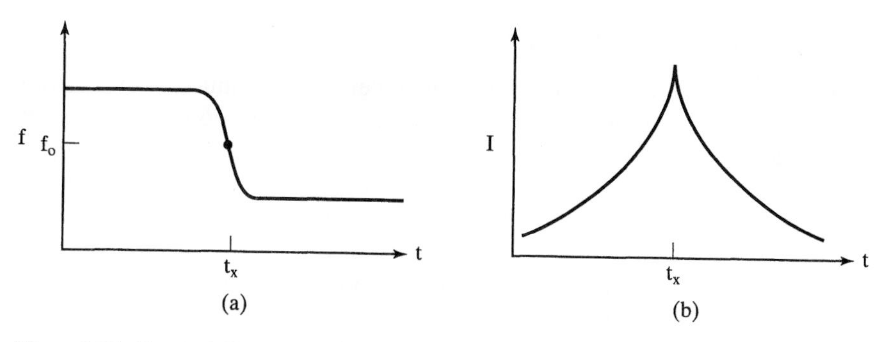

Figure 2-45 Frequency and intensity of the whistle of a train as the train passes a stationary observer; at time t_x the train is closest to the observer.

of three-quarters of the initial period to simplify the geometry. If the object is moving toward the source with one-third of the velocity of sound in water, then in three-quarters of the initial period T_0, the wave will have gone three-fourths of one wavelength and the object one-quarter of one wavelength. At this time, the next wave compression will be reflected off the object. This sequence repeats itself as shown in the figure. Thus the reflected wavelength λ_{ref} is less than the original wavelength, or, equivalently, the frequency of the reflected signal is higher than that of the original signal. In the particular case shown, the wavelength is one-half of its initial value, which means that the frequency of the reflected wave is 2 times the initial frequency. If the object is moving away from the source, a similar geometrical construction shows that the frequency of the reflected wave is decreased (or its wavelength increased).

It is important to draw specific attention to the difference between intensity (or amplitude) and frequency when discussing the Doppler effect. The Doppler effect deals only with changes in frequency due to relative motion of source and observer. Changes in intensity or loudness result from changes in the distance between source and observer (which clearly occur whenever there is relative motion between them) and are governed by the inverse square law. As an example to help distinguish between frequency and intensity effects, both the frequency and intensity of a train whistle that we hear as the train passes us at a crossing are plotted in Figure 2-45. Figure 2-45(a) tells us that the frequency heard as the train approaches is higher than the normal frequency of the whistle. Just as the train passes by, at time t_x, the observed frequency is f_0. After time t_x, the observed frequency is *lower* than the normal frequency. While the train is directly approaching or moving away from us, its frequency is constant and either higher or lower, respectively, than the normal frequency. This is because the Doppler effect is dependent only on the relative velocity; a constant relative velocity implies a constant frequency shift. Figure 2-45(b) tells us that as the train approaches its sound gets louder (according to the inverse square law), until it reaches its loudest level when it is crossing right in front of us. It then becomes softer as the distance between us and the train becomes greater. It is important to distinguish carefully frequency and intensity effects and to be aware of their origins.

2.7 Shock Waves and Sonic Booms

In this book we deal almost entirely with periodic continuous waves; however, for completeness we shall now discuss an example of nonperiodic waves that propagate like sound waves. A sound wave, as we have seen, is a disturbance that propagates through the air as its molecules effectively "bump" each other in a series of molecular collisions. In an explosion, say of a bomb or a firecracker, the air around the source is pushed out very rapidly by the force of the explosion. The *shock wave* produced is a single large compression that moves away from the explosion at the speed of sound. It is like a compression in a sound wave except that it is not periodic.

Another example of such a shock wave is the *sonic boom*. Figure 2-46 shows the sequence of circular waves produced in a ripple tank by a source moving faster than the wave speed. The result of this high source speed is that the source outdistances all the circular waves it produced earlier. If the sequence of circles representing outgoing waves from the source is made very close together, the crests from adjacent groups of circular waves join together along lines on either side of the source, forming a V-shape, as shown in Figure 2-47. A common example of this is the wake of a motorboat. If the boat were to move very slowly through the water, it would produce a series of circular waves, as in Figure 2-38. But because most boats move faster than the velocity of water waves, the characteristic V-shape results.

From a jet plane moving at a speed greater than the speed of sound, all the sound at a given instant (e.g., the roar of the motors, swish of air) is concentrated in a spherical wave that emanates from the plane at that time. The spherical wave fronts add together to form a cone-shaped surface called the *Mach wedge*, in which the sounds originating during a longer time interval come together, creating a shock wave of large amplitude. The same effect is familiar with the water waves in the wake of a motorboat. This conical wave is a nonperiodic shock wave that can be very intense. In the case of a plane moving at a speed greater than the speed of sound, the effect is known as a *sonic boom*. The sonic boom follows the plane around, and does not occur only at the moment when the plane's speed becomes greater than the speed of sound, as is often believed. For this

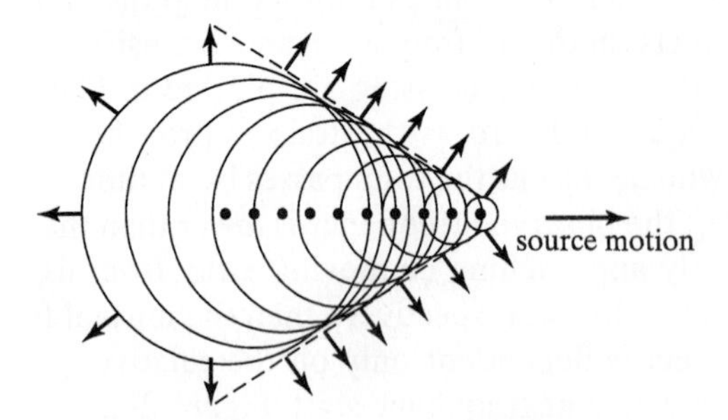

Figure 2-46 Circular waves from the sequence of points shown, in which the source velocity exceeds that of the emitted wave. The dashed lines indicate the leading edge of the combined wave front.

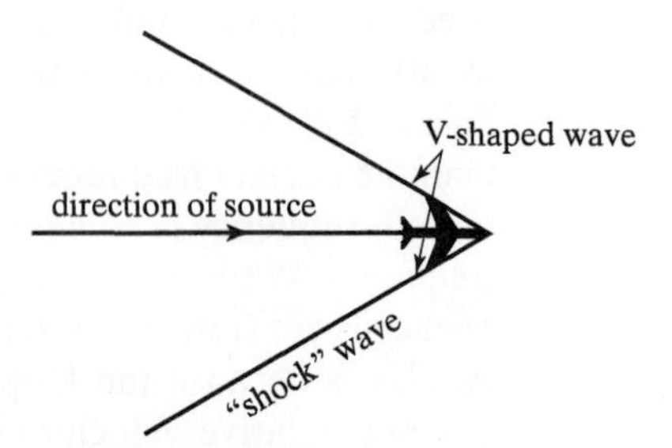

Figure 2-47 When the source velocity exceeds that of the speed of sound, the leading edges of the circular waves from the source combine to form a shock wave. The angle of the V decreases as the source velocity increases.

reason, it can be annoying or even dangerous to have aircraft continually flying at supersonic speeds over densely populated areas. By observing the circles in Figure 2-46, we can see that, although the sound is "concentrated" along the edges of the cone creating an especially loud boom, sound from the plane continues to be heard at all points inside the cone. When a supersonic plane passes overhead, you therefore hear a large boom, when the cone passes you, followed by a low-pitched Doppler-shifted roar.

In addition to explosions (such as bombs and firecrackers) and the sound from supersonic aircraft, nature has provided us with an excellent example of the sonic boom—thunder. When a lightning discharge occurs, the air is heated to extremely high temperatures and expands rapidly. In fact, this expansion may occur at a speed greater than the speed of sound in air and can therefore produce a shock wave or sonic boom. An electrical spark heats the air in the same way. The rumbling of thunder results when the sound from a long lightning bolt (sometimes over 10 miles long) arrives at the observer over a long time interval, owing to the varying distance between the lightning bolt and observer. The duration of the thunder can be related to the length of the lightning path when the path is directed radially away from the observer, whereas a bolt that goes perpendicular to the line of sight produces a sharp shock.

The crack of a whip is generally believed to be a small sonic boom caused by the tip as it travels faster than the speed of sound.

2.8 Ultrasonics

Ultrasonics refers to waves whose frequencies lie above the audible range. Ultrasonic frequencies range from about 20 kHz, the upper frequency for human hearing, to about 1.25×10^{13} Hz, the maximum frequency of sound waves, at which point the molecules in even a liquid or a solid cannot interact rapidly enough to propagate a longitudinal wave. Table 2-2 gives the range of hearing for humans and selected animals. It becomes clear that, despite the relatively large range of audibility of people, the hearing range of many other species extends beyond ours on both the upper and lower ends. Many of us have used a "dog whistle," the frequency of which is around 24 kHz, too high for us to hear but still well within the range of hearing of dogs. Bats navigate and catch flying insects for food by *echolocation*, using ultrasonic frequencies. Although

Table 2-2 Range of Hearing for a Variety of Species

Species	Range
Humans	20–20,000 Hz
Cats	100–32,000 Hz
Dogs	40–46,000 Hz
Horses	31–40,000 Hz
Elephants	16–12,000 Hz
Cattle	16–40,000 Hz
Bats	1000–150,000 Hz
Grasshoppers	100–50,000 Hz
Rodents	1000–100,000 Hz
Whales, Dolphins	70–150,000 Hz

much is known about the ears of many animals, and aspects of animal hearing have been studied to learn more about human hearing, much remains to be discovered about how certain animals communicate using the exceptional range of frequencies to which they are sensitive. Ultrasonics is one of the most rapidly growing areas of science and technology, with a large number of very important recent innovations in such diverse fields as medicine and high-tech metalworking. In this section we will review some of the more important applications and recent innovations in the field of ultrasonics.

The device most often used to both produce and detect ultrasonic waves is the *piezoelectric transducer*, a crystal, such as quartz, that can convert electrical oscillations into mechanical vibrations and mechanical vibrations into electrical oscillations. Piezoelectric crystals can be formed into a shape that provides the large variety of ultrasonic beam properties necessary for various applications; for example, a concave spherical surface will create a beam that will focus to a point.

One of the simplest applications of the use of ultrasound is in *sonar*, which can involve either sonic or ultrasonic frequencies. By sending out small bursts of ultrasound and measuring the time it takes to reflect off some object, the distance to the object can be determined if the velocity of sound in the medium is known, as we have previously seen. This procedure can be used to map the depth of lakes or oceans, to locate and track submarines under water, or to locate mines below the surface of water. Many small boats have sonic ranging devices to warn them of shallow water and thus prevent them from running aground. Many large fishing trawlers are equipped with sonic ranging devices that enable them to locate schools of fish with great efficiency.

Perhaps the most significant contemporary applications of ultrasound lie in the field of medicine. Medical uses are divided into two primary applications: diagnostic and surgical. Not only does ultrasound have many technical advantages over other medical techniques, but also it often turns out to be one of the most economical techniques, due to the simplicity of the equipment required and its ease of use.

Ultrasound reflects from any surface between two different media, as does light or audible sound. However, the sensitivity of ultrasound to differences between various internal organs and tissues allows it to "see" in much greater detail than can x-rays. In particular, ultrasound can be used to identify a wide range of soft tissues, whereas x-rays are useful primarily in seeing hard tissue, such as bones. In addition, ultrasound possesses an enormous advantage over x-rays in that it is a *nonionizing* form of radiation—that is, ultrasound in even moderate intensities does not ionize atoms or break up molecular bonds, which would damage the body tissue. This is of critical importance in obstetrics.

Several scanning techniques are used in medical diagnostics; all use the same basic idea as sonar in locating some object that is then imaged using that technique. One mode, called the *A-scan*, uses a single transducer to scan along a line in the body, with the return time of the reflected signals providing information on the location or the size of internal organs. Another mode, called the *B-scan*, uses a linear array of transducers to scan a plane in the body, with the resultant data displayed on a video screen as a two-dimensional plot. Due to diffraction effects, better resolution can be obtained by use of higher frequencies. However, the higher frequencies are more strongly absorbed, leading to practical limits on the resolution that can be obtained with ultrasonic scans of internal organs.

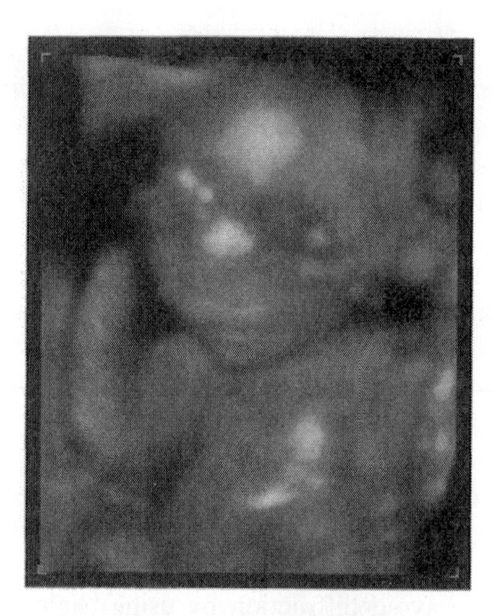

Figure 2-48 Sonogram of a fetus after about 19 weeks of pregnancy. (Scan courtesy of Dan and Beth Margulies)

Because ultrasound readily reflects off cells, it can be used to measure blood flow using Doppler reflection, for which there is no other simple procedure. Ultrasonic Doppler-effect devices exist that can be readily aimed along an artery under investigation and make a nonintrusive measurement of blood flow, allowing immediate determination of the effectiveness of the artery. One version of this device uses a 3-megahertz (MHz) beam that reflects off arterial blood with a Doppler shift of a few kilohertz, which is heard directly by the physician. Using this technique, arteriosclerosis can be readily diagnosed, and the healing of arteries can be monitored following treatment. This procedure is sensitive enough to pick up a fetal heartbeat long before a stethoscope can. B-scan imaging and the Doppler effect are used in a combination known as duplex scanning, which can be used to identify arteries and immediately measure their blood flow. These procedures have been used extensively to diagnose heart-valve defects and to identify cases of insufficient blood flow in the penis, one of the leading causes of male impotence.

Perhaps the most well-known application of ultrasound is in making *sonograms*, or ultrasonic pictures, of the fetus at various stages in its development, to ascertain that growth is proceeding normally and that no problems are occurring. The sonogram of the living fetus during pregnancy has in fact become virtually synonymous with the term ultrasound for many new parents. At this time virtually all pregnant women in the United States obtain sonograms at some time during their pregnancy. Figure 2-48 is an ultrasound image of a fetus during pregnancy. By using higher-frequency ultrasound and more sophisticated computer algorithms, images can now be obtained that have a three-dimensional appearance. Detailed features of the head, body and both arms can be seen, with the left hand covering part of the head.

Another important use for the ultrasonic imaging during pregnancy is to guide the needle when drawing amniotic fluid to be used in testing for birth defects, thus avoiding damage to the fetus or other critical internal organs. Typically these applications use ultrasound frequencies of less than 320 MHz.

Although these diagnostic applications of ultrasound involve relatively low intensity levels, surgical applications generally involve substantially higher levels. Perhaps the most well known high-intensity application involves the break-up of kidney stones. The most common ultrasonic treatment for kidney stones, *endoscopic ultrasonic disintegration*, uses a small metal rod inserted through the skin to deliver blasts of ultrasound, which cause the stones to shatter. A less-used technique, the *ultrasonic lithotripter*, focuses the ultrasound onto the kidney stones from outside the body. This is one of several recent examples of *trackless surgery*—that is, surgery that does not require an incision, or track, through the skin to the affected area. This technique has also been used to treat Parkinson's disease, by creating brain lesions in areas inaccessible by traditional surgery, and holds promise for the treatment of certain diseases of the inner ear and the semicircular canals in which blockage inhibits normal fluid flow.

Focusing intense ultrasonic beams onto small internal regions can also be used to create local heating. This technique has been used to relieve pains in joints, especially in the back and shoulder. By focusing ultrasound onto a cancerous region, creating local heating, it is believed that the cancer can be inhibited or destroyed while leaving the surrounding region unaffected.

To "see" some object, it is necessary to avoid diffraction by using "light" with a wavelength smaller than the object being viewed. Ultrasound with a frequency of about 2000 MHz (2 GHz) has a wavelength of about 0.75 micrometers in soft tissue, compared with a wavelength of about 0.55 micrometers for light. Ultrasonic microscopes have been developed which rival light microscopes in their resolution and ability to view details within a single human cell. An important advantage of ultrasonic microscopes lies in their ability to distinguish between the various parts of the cell by their differences in *viscosity*, rather than by their coloration or other features. In the case of either optical or electron microscopes, artificial contrast media such as dyes are required to distinguish the parts of the cell, killing the cells. An ultrasonic microscope can "see" within a *living* cell.

A very small ultrasonic microscope, using frequencies of about 1 GHz, can be inserted into blood vessels. A new area of medicine called intravascular ultrasound (IVUS), based on this device, is rapidly developing as a useful tool in detecting arterial plaque that may kill up to 250,000 Americans each year. This microscope seems to be useful in identifying cancerous cells in the body without the need to remove and perform a biopsy on them using standard microscopy.

Ultrasound has also become an important tool in dentistry. Ultrasonic vibrations are used with a fine grit to loosen and remove plaque during tooth cleaning, much like sandblasting a building. Ultrasonic drills can be used to remove a cavity without use of anesthesia, thus eliminating the possibility of allergic reaction to a narcotic and reducing the cost and time for the procedure.

Ultrasonics is now the object of much research work; perhaps the most fascinating is *cavitation*. When water boils, bubbles are formed and rise to the surface, leading to the sound of boiling. Several nineteenth-century physicists, who called the formation of bubbles "cavitation," studied the boiling process and its concomitant sounds. Other important sources of cavitation include the action of a ship's propeller in the water and local heating of water by intense beams of ultrasound. Because intense beams of ultrasound can be used to control cavitation, ultrasound has become a useful tool in the study of the process of cavitation, and research in cavitation using ultrasound remains important. One contemporary research subject, called *sonoluminescence*, involves study of light emission as the cavity created by high-intensity ultrasound collapses. It is believed that this process can result in temperatures hotter than that of the surface of the sun.

Just as in medical diagnostics, ultrasonic echolocation has several nondestructive industrial testing applications. Because sound waves propagate more efficiently through metals, they can be used to probe more deeply than x-rays. Because they reflect more readily off small faults or other changes in the material than do x-rays, they are useful in identifying relatively small structural problems such as holes, cracks, or corrosion in materials. Ultrasound is used to inspect welds, establish the uniformity and quality of poured concrete, and monitor metal fatigue. Partially as a result of the Three Mile Island nuclear reactor accident in 1979, an increased number of ultrasonic inspection procedures are now performed on the structural components in nuclear reactors.

Perhaps the most ubiquitous use of high-energy ultrasound is in ultrasonic cleaning. Ultrasonic waves are introduced into small tanks of liquid into which objects, such as jewelry, surgical instruments, or small machinery are placed for cleaning. The resulting vibrations and cavitation of the liquid result in turbulent cleaning action.

Ultrasonic machining uses ultrasonic vibrations to activate diamond tools or a slurry of carborundum grit to carry out the machining process. A variation of this process, known as ultrasonic drilling, can be used to form holes of virtually any shape in materials such as glass or ceramics, which are very difficult to machine or drill using standard techniques.

Ultrasonic soldering has become very important for difficult soldering procedures or where cleanliness is of critical importance. The ultrasonic vibrations clean the surface so well that the normal soldering flux becomes unnecessary, and can even prepare the surface of aluminum to take solder.

Ultrasonic standing waves with very high intensities are now being used for thermoacoustic cooling. This is advantageous, both because the amount of machinery in the refrigerator is reduced and because it does not use potentially dangerous greenhouse gases as a coolant. Temperatures in these devices have reached well below the freezing point of water.

2.9 Infrasonics

Infrasonics refers to waves whose frequencies lie below the audible frequency range of humans. Infrasonic waves occur often in nature in such phenomena as earthquakes and weather.

Perhaps the most well known infrasonic waves arise in earthquakes; there are three principal types of earthquake waves. The *S-wave*, a transverse body wave, can only propagate in solid rock. The *P-wave*, a longitudinal body wave similar to a sound wave, propagates at the speed of sound and has a very large range. The *L-wave* is the wave that forms along a boundary between two mediums such as air and the ground and is responsible for the immense damage that can occur in large earthquakes. Earthquakes that originate as P-waves within the earth are converted into L-waves when they reach the surface of the earth. The long range of the P-wave makes it useful in identifying the location of the source of the earthquake by comparing its time of arrival at three or more monitoring stations.

Underground nuclear bomb tests also create P-waves. The long range of the waves and the fact that they propagate with the speed of sound allow nuclear bomb tests to be identified and located if the blast is of sufficient strength.

Because such waves partially reflect when they pass a boundary between different media, just as do sound waves, seismic shocks can be used to locate and identify various distinctive rock formations within the earth, including those in which oil or natural gas might be found. Such seismic exploration is usually carried out prior to actually digging an oil well. This type of exploration has also helped geologists identify the various layers within the earth's structure, such as the mantle, the molten outer core, and the solid inner core.

The sound created by rapidly moving vehicles, such as airplanes, trains, or cars, as well as the sound from building air handlers or blowers also contains a significant amount of infrasound. It is believed that some people may be particularly sensitive to these infrasonic frequencies, leading to symptoms of motion sickness.

A variety of animals and birds are sensitive to infrasonic waves. Many zoologists believe that the sensitivity of such animals as elephants or cattle could provide them with an early warning of earthquakes and major weather disturbances. It has been suggested that the sensitivity of certain birds to infrasonic waves might aid their navigation or even affect their migration.

Much remains to be discovered about the effect of infrasonic frequencies on both humans and animals.

SUMMARY

(***Section 2.1***) A **wave** is a disturbance that moves with some particular wave speed and carries energy. Electromagnetic waves such as light are **transverse waves**. Sound waves are **longitudinal waves**. The **bell-in-vacuum experiment** illustrates the different behavior of sound waves and light waves in a vacuum. As a **periodic wave** of frequency f propagates, it repeats spatially in one **wavelength** $\lambda = v/f$, where v is the wave speed. The speed of sound in air at room temperature, S, is roughly 345 m/s. (***Section 2.2***) Four properties of waves can be considered fundamental or specialized, in that they are the basis of general properties that apply to all waves: **Huygens's principle**, **superposition**, the **inverse square law**, and **polarization**. (***Section 2.3***) General properties of waves arise from application of the fundamental wave properties, are common to all waves, and are directly responsible for most of the sound phenomena that we commonly observe. **Reflection** is important in echoes, focusing of sound by parabolic surfaces, and in the production of standing waves in wind instruments. **Refraction** is largely responsible for the way in which sound propagates in the atmosphere. **Interference**, which can be either **constructive** or **destructive**, includes applications involving creating standing waves in musical instruments and active noise reduction. **Diffraction** is important in how sound from a voice or other source spreads out so that it can be heard by listeners. (***Section 2.4***) The **addition of waves**, referring to similar waves closely related in frequency and amplitude, has important application in beats as well as standing waves and production of complex waves by adding together their harmonic components. (***Section 2.5***) **Beats** result from the addition of tones of closely spaced frequencies with the same amplitude. (***Section 2.6***) The **Doppler effect** explains how motion between the source and observer of a wave changes its frequency. Reflection of sound waves off moving targets, such as used in sonar, also results in frequency shifts. (***Section 2.7***) When a sound source, such as a supersonic plane, moves faster than the speed of sound, a cone-shaped **shock wave** is produced, creating a **sonic boom** as it passes stationary observers. (***Section 2.8***) **Ultrasound** has a large number of medical applications, both in diagnosis and treatment, and ultrasonic devices are common in industry. (***Section 2.9***) **Infrasound** is the basis for transmission of earthquake waves, and geological exploration commonly employs infrasonic waves.

QUESTIONS

1. A violin and a flute each play 256-Hz tones.

   a. Which wave has the longer wavelength?
   b. How might the waves differ?

2. In general, which would have a higher frequency:

   a. a violin note or a cello note
   b. a trumpet note or a trombone note

3. Which would have a longer period:

   a. a bassoon note or a piccolo note
   b. a note sung by a woman or a note sung by a man
   c. a note played by a trumpet or one played by a tuba

4. a. Compare and contrast transverse and longitudinal waves.
   b. Give two examples of each.
   c. What particular type of wave does not need a medium?

5. Describe the bell-in-vacuum experiment. What conclusions can be drawn from it?

6. To estimate how far a lightning strike is from you, count seconds between the lightning flash and the thunder that arrives shortly thereafter and divide by five to get the distance in miles. Explain why this formula works and determine the limit of its accuracy.

7. Describe the Quincke's interference-tube experiment. What principle is demonstrated?

8. Put a piece of paper, about one meter square, on a table top and position two small loudspeakers 50 cm apart centered on one side of the paper. Attach the speakers to a 2-kHz sound source with the speakers in the same phase. Use a microphone connected to an oscilloscope to locate the nodal and antinodal lines, and map out the two-source interference pattern. Change the frequency to 4 kHz and repeat the process. Explain your results.

9. a. Why can you hear around corners?
   b. All other things being equal, which can we hear more readily around corners, tubas or flutes? Why?

10. Connect a loudspeaker to a variable-frequency audio oscillator, and mount a long tube of the same diameter as the loudspeaker onto the speaker, forming a "sound collimator." Sound coming out of the tube will initially propagate along the direction of the tube. Starting at about 100 Hz, use a microphone and oscilloscope to pick up the sound emanating from the tube in the forward direction and bending at a 90° angle. Repeat these measurements for frequencies of 500 Hz, 1000 Hz, 2000 Hz, 4000 Hz, and 10,000 Hz. Describe the results of your experiment and explain. What conclusion can you reach?

11. Stretch a slinky along the floor with one person holding each end. Place several empty cola cans in a line about 30 cm from the slinky. Practice making a pulse travel down the slinky such that it just misses the cans. Then simultaneously start identical pulses from

both ends of the slinky. Can you create two pulses that will knock over only one can as they pass each other? How can this happen?

12. Draw and add together the following pairs of waves:

  a. square waves of equal amplitude with frequencies f and $2f$, beginning in phase
  b. similar waves beginning out of phase

13. Draw and add together the following pairs of waves:

  a. a square wave of frequency f and amplitude 1 with a square wave of frequency $3f$ and amplitude 1/3, beginning in phase
  b. waves similar to those except that they start out of phase

14. Draw and add together the following pairs of waves:

  a. a square wave with amplitude 1 and frequency f, and a square wave of amplitude 1/2 and frequency $2f$, beginning in phase
  b. similar waves beginning out of phase

15. Two sinusoidal tones of equal amplitude and with frequencies of 439 and 443 Hz are sounded simultaneously.

  a. Name and describe in detail (qualitatively and quantitatively) what you hear.
  b. Why is it important that the two waves have equal amplitudes for this effect to occur?

16. a. Describe qualitatively what happens to the sound heard by an observer moving with a constant speed past a stationary, steady source of sound, such as a car horn.
  b. What is the name of this effect?

17. a. A jet plane flies overhead at twice the speed of sound. Describe, with the aid of a drawing, what you hear at the ground level.
  b. Another jet flies overhead at four times the speed of sound. Draw its shock wave. Be sure to make clear any distinction between this shock wave and the one drawn in part (a). In which case does the shock wave travel faster?

18. If a jet flies directly away from you faster than the speed of sound, you can still hear the roar of its engines. But if you move away from a loud stationary jet at a speed greater than the speed of sound, you *cannot* hear the engines. Why?

PROBLEMS

1. You see lightning flash, and hear the crack of the thunder six seconds later. How far are you from the location of the lightning strike?

2. For the following frequencies, calculate their periods and their wavelengths in air (using appropriate units and prefixes): 50 Hz, 2000 Hz, 25 kHz, 1 MHz, 30 MHz, 1 GHz.

3. For the following periods, calculate their frequencies and their wavelengths in air (using appropriate units and prefixes): 50 ms, 2 ms, 0.01 ms, 5 μs, 1 μs, 0.02 μs.

4. For the following wavelengths in air, calculate their frequencies and wavelengths (using appropriate units and prefixes): 10 m, 34.5 cm, 3.45 mm, 0.02 mm, 50 μ (millionths of a meter), 1 μ.

5. The frequency limits of the range of human hearing are from about 20 Hz to 20 kHz. The speed of sound is about 34,500 cm/sec. What are the wavelengths of these waves in cm? In meters?

6. The frequency range of human singing extends from about C = 65.406 Hz for a low bass to about C = 1046.50 Hz for a high soprano. What is the range of wavelengths in air for these notes in meters?

7. **a.** What is the frequency of a square wave of amplitude $A = 3$ V and period $T = 2$ ms?
   b. If the speed of sound in a metal is 300,000 cm/sec, what is the wavelength of this wave in the metal?

8. **a.** Calculate the wavelength λ for 50 kHz, 100 kHz, 500 kHz, 1 MHz, 10 MHz, and 100 MHz ultrasound.
   b. What size object might each of these be useful in viewing? Give possible examples.

REFERENCES

GIANCOLI, D. *Physics: Principles with Applications*. Englewood Cliffs, N.J.: Prentice-Hall, Inc., Fundamental principles of physics, including waves and sound, for the undergraduate.

HALLIDAY, DAVID, and ROBERT RESNICK. *Physics*, New York: John Wiley & Sons, Inc., Popular physics text for students with mathematical and technical backgrounds.

HOBBIE, RUSSELL K. *Medical Physics: Selected Reprints*. College Park, Md.: American Association of Physics Teachers, 1986. This includes several articles discussing medical applications of ultrasound, along with a reprint of the AAPT *Resource letter MP-1: Medical Physics*.

———. "Resource letter MP-1: Medical Physics." *American Journal of Physics* 53 (1985):822. This excellent AJP resource letter contains a large section of publications on medical applications of ultrasound.
LINDSAY, R. BRUCE. *Acoustics: Historical and Philosophical Development*. Stroudsburg, Pa.: Dowden, Hutchinson & Ross, Inc., 1976. A large collection of papers on acoustics dating from the time of Aristotle to the turn of the twentieth century, including many of the important authors from the nineteenth century.

FAY, RICHARD. *Hearing in Vertebrates: A Psychophysics Data Book*. Winnetka, Ill.: Hill-Fay Associates, 1982. An excellent resource book containing everything you ever wanted to know about hearing in animals and birds.

HELMHOLTZ, HERMANN VON. *On the Sensations of Tone*. New York: Dover Publications, Inc., 1954. An English translation of the original German classic by one of the founding fathers of acoustics. This book gives a unique historical perspective to many aspects of acoustics and is to a great extent understandable by the general reader.

SETO, WILLIAM W. *Theory and Problems of Acoustics* (Schaum's Outline Series). New York: McGraw-Hill Book Company, 1971. Contains theory, plus 245 solved problems dealing with many aspects of waves and acoustics; mathematical level is generally well above that of this text.

SUSLICK, KENNETH S., ed. *Ultrasound: Its Chemical, Physical and Biological Effects*. New York: VCH Publishers, 1988. A summary of research in and applications of ultrasound.

Chapter 3

Standing Waves and the Overtone Series

We shall now study the properties of waves that are confined to a one-dimensional medium, such as transverse waves on a stretched string and sound waves in a narrow tube filled with air. After a discussion of how these waves behave at the end of such a one-dimensional medium and how waves add together to form standing waves, we shall see the special relationships among the waves that can exist in such a medium and how this naturally leads to the overtone series. This chapter will then conclude with a study of more complex and two-dimensional standing waves.

3.1 Transverse Standing Waves

A very special result occurs when two identical waves moving in opposite directions are added. Figure 3-1 shows segments of two such waves, along with the resultant sum wave, viewed at time intervals of one eighth-period. The figures represent a sequence of photographs at time intervals of one eighth-period showing a region two wavelengths wide. At time $t = 0$, when the two waves are in phase (that is, lined up peak to peak and trough to trough), the resultant is a wave of twice the original amplitude, as in (1) of Figure 3-1, but with the same wavelength. One quarter-period later, at time $t = T/4$, each of the waves has moved one quarter-wavelength, and they are exactly out of phase, producing zero displacement at all points, as shown in (3). One quarter-period after this, at time $t = T/2$, the two original waves are again in phase with each other, but the resultant wave, as shown in (5), is opposite in phase compared with the resultant wave at time $t = 0$. An additional quarter-period later, at time $t = 3T/4$, the waves are again out of phase, producing a zero displacement at all points, as shown in (7). Finally, after one full period, at time $t = T$, each wave has propagated one full wavelength in its respective direction, the resultant is the same as in (1), and the motion repeats.

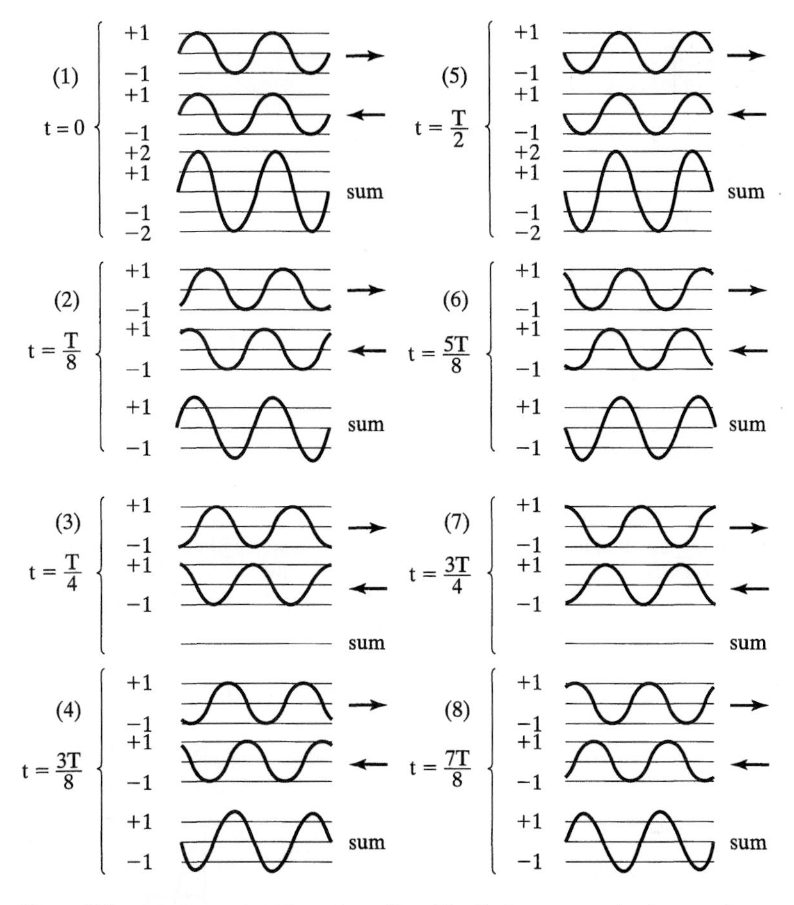

Figure 3-1 A two-wavelength section of two identical waves moving in opposite directions along a rope, at intervals of one eighth-period, showing the individual waves and the sum, the overall shape of the rope.

What is surprising is the shape of the sum wave between these times, as shown in (2), (4), (6), and (8) of Figure 3-1. For these cases, we must add the displacements of both waves at several points along the wave to obtain the resultant displacement at those points and draw through the points a smooth curve that represents the shape of the resultant wave. This is illustrated in detail in Figure 3-2, which is an expansion of (2) of Figure 3-1. Adding the two original waves at the positions labeled 0 to 8 along the rope gives the results shown in Table 3-1. The resultant is a wave of the same wavelength as the two component waves, but whose amplitude is 1.4 times the amplitude of either individual wave, not double, as when the two original waves are in phase with each other. Figure 3-3 shows, superimposed on one graph, drawings of the resultant wave at times (1) through (8) of Figure 3-1. The sum wave is standing—that is, progressing neither to the right nor to the left, but continually oscillating back and forth between configurations labeled (1) and (5). Its displacement varies between zero (no visible wave) and twice the amplitude of either of the two component waves. Figure 3-4

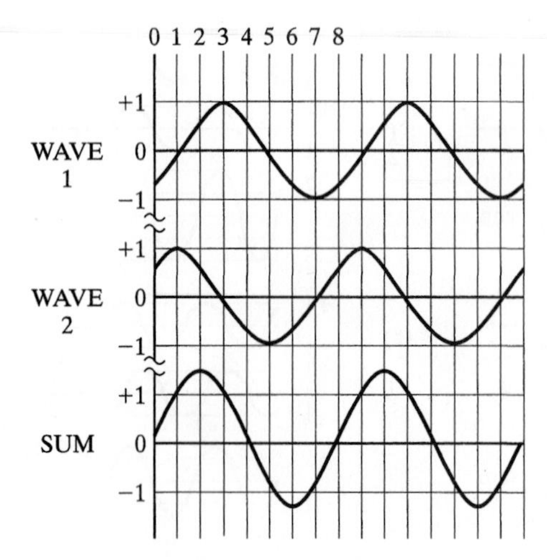

Figure 3-2 Expansion of drawing (2) of Figure 3-1 showing details of how the two component waves are added.

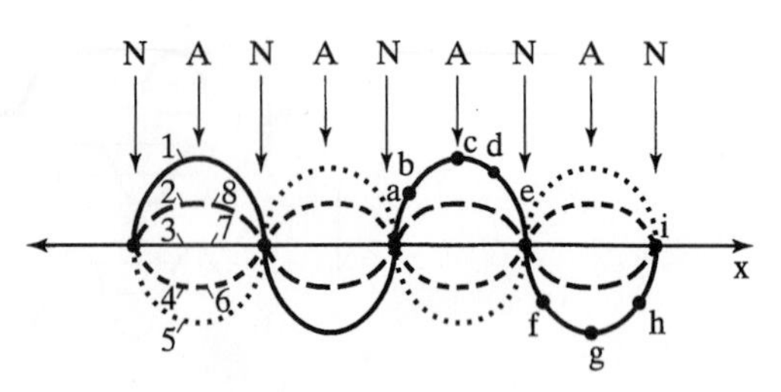

Figure 3-3 Consecutive "pictures" of two full wavelengths of the standing wave of Figure 3-1, shown at a sequence of equal time intervals of one eighth-period, labeled (1) to (8).

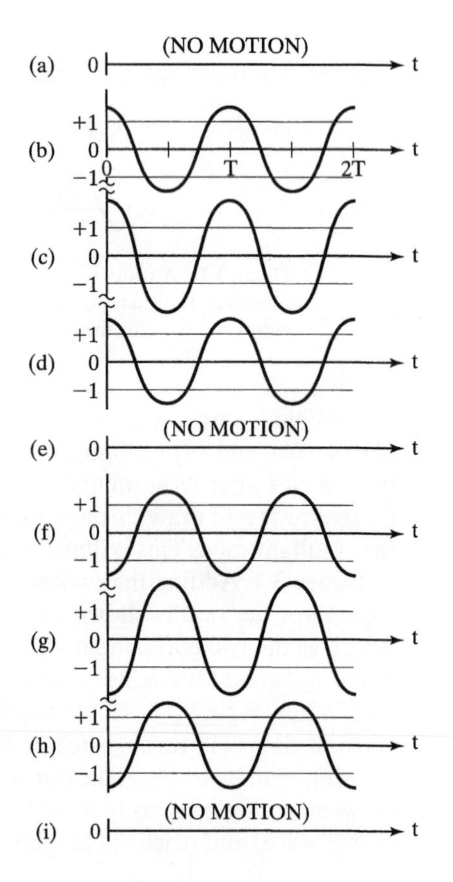

Figure 3-4 Motion of points (a) to (i) of the standing wave of Figure 3-3, shown over a two-period interval beginning at $t = 0$.

Table 3-1 Addition of Waves of Figure 3-2

	Position along rope								
Displacement	0	1	2	3	4	5	6	7	8
Wave 1	−0.7	0.0	+0.7	+1.0	+0.7	0.0	−0.7	−1.0	−0.7
Wave 2	+0.7	+1.0	+0.7	0.0	−0.7	−1.0	−0.7	0.0	+0.7
Sum	0.0	+1.0	+1.4	+1.0	0.0	−1.0	−1.4	−1.0	0.0

shows a graph of the motion of points (a) to (i) along the rope as a function of time. Points (a), (e), and (i), where there is no motion of the rope, are called nodes and are labeled N (remember no displacement) on Figure 3-3. Points (c) and (g), where the transverse motion and displacement attain their maximum value, are called *antinodes* and are labeled A. One section of the standing wave between two nodes is called a *loop.* The length of each loop of the standing wave pattern is one half-wavelength of the original wave.

This discussion has involved only transverse waves in a rope or similar medium. Longitudinal standing waves can also be produced in which the motion of the wave is along the one-dimensional medium. In later sections and chapters, we shall see that such standing waves are basic to the production of sound in many musical instruments.

3.2 Resonance and the Overtone Series

We have seen how two identical sinusoidal waves moving in opposite directions on a long rope combine to form a standing wave, and we have discussed the basic properties of such a standing wave. Now we shall look at what happens to waves at the end of a rope and then study the standing waves produced in a short section of such a medium.

Consider a single transverse pulse, consisting of one-half of an oscillation of period T, reflecting off the end of a rope. Figure 3-5 shows such a pulse for the case of a fixed end (we can think of it as attached to a wall and held motionless), as represented by the dot at the end of the rope in part (a), and the case of a free end (it can be moved by the wave) as in part (b). The "force" of the pulse hitting the fixed end of the rope causes the pulse to change from positive to negative, a 180° phase reversal. The wall pulls down on the end of the rope to ensure that there is no displacement at the end. This downward force can be thought of as generating a negative pulse traveling away from the end, the reflected pulse. No such phase reversal occurs when the pulse reflects off a free end, because there is no downward force exerted on the end of the rope.

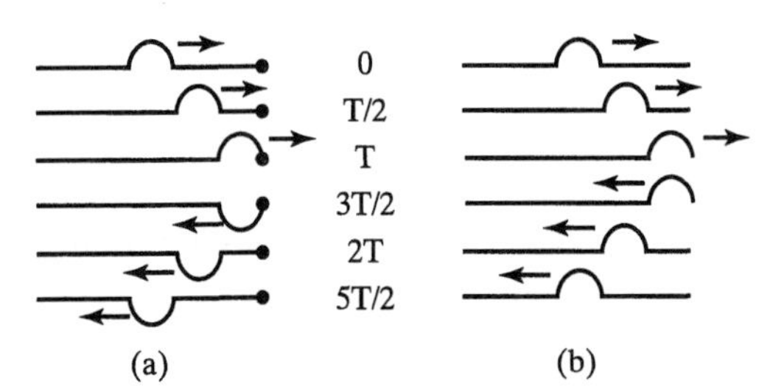

Figure 3-5 Reflection of a transverse pulse off a fixed end (a) and a free end (b) of a rope or coiled spring.

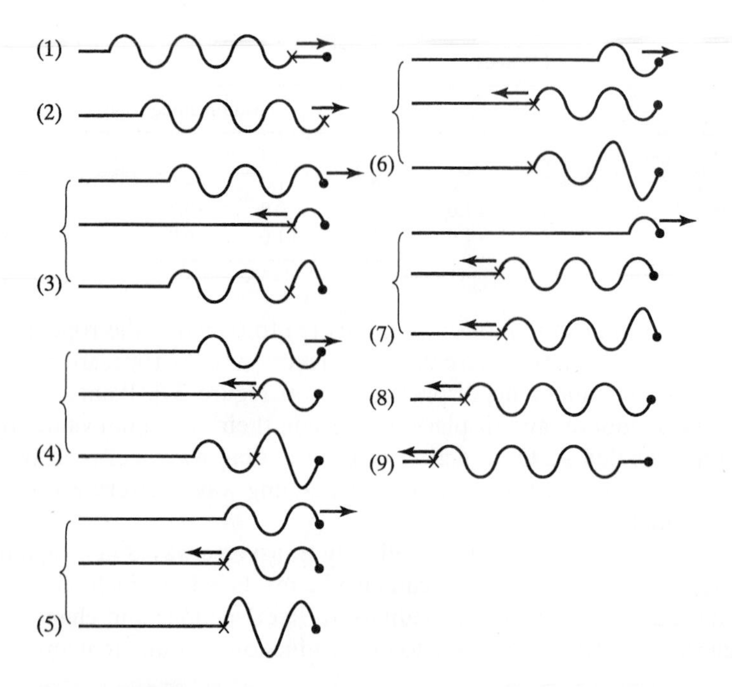

Figure 3-6 Reflection of a transverse pulse train, consisting of three periods of a sine wave, off a fixed end. The wave has a phase reversal upon reflection. The addition of incident and reflected waves produces a standing wave of twice the original amplitude. The x marks the leading edge of the incident wave as it reflects off the end.

Now let us expand to the case of three wavelengths of a sinusoidal wave reflecting off a fixed end, as shown sequentially in Figure 3-6. The x marks the leading edge of the wave. Consecutive drawings are spaced at time intervals of one half-period, so the wave moves one half-wavelength between drawings. Both incident and reflected waves are shown individually, followed by their sum, which is the overall shape of the rope. Notice the phase reversal (180° phase change) of the wave upon reflection; the first pulse of the incident wave is negative (drawing 1), whereas the first pulse of the reflected wave is positive (drawing 9). Also notice that the sum wave, obtained by adding the incident and reflected waves where they overlap, has twice the amplitude of the original wave shown in drawing (1) of Figure 3-6.

Figure 3-7 shows the development of a standing wave by reflection of an infinite sinusoidal wave off a fixed end at time intervals of one quarter-period. Figure 3-8 shows the incident, reflected, and sum waves in more detail for drawings (5), (6), and (7) of Figure 3-7. It is now apparent that to form a standing wave it is not necessary to have two initial waves; a single wave reflecting off an end will suffice. There must be a node at the fixed end, because that end is constrained to zero displacement at all times.

In fact, it is not even necessary to have a long wave reflecting off an end as in Figure 3-7. If a long pulse is started in a stretched rope fixed at both ends, it may reflect

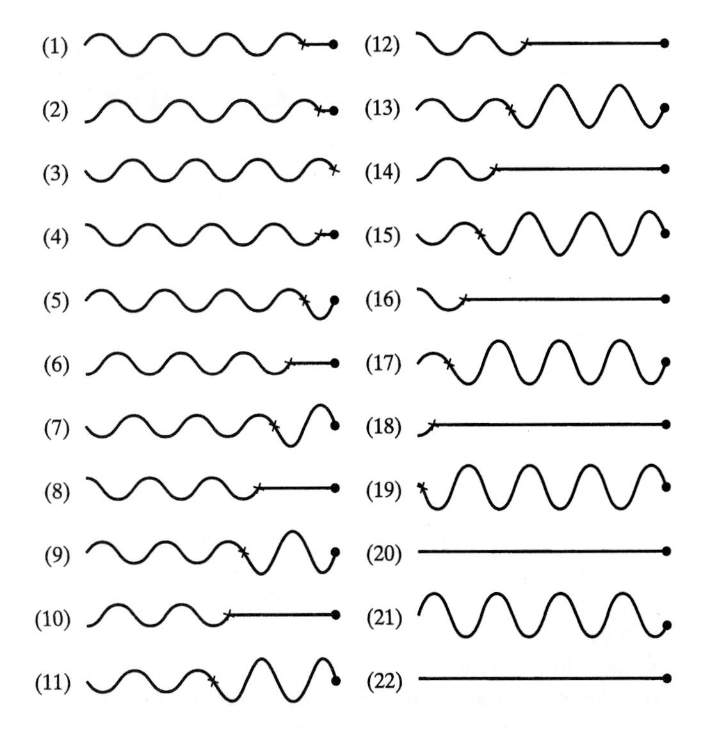

Figure 3-7 Reflection of a continuous wave train off a fixed end, creating a standing wave, shown at intervals of one quarter-period. The x marks the progression of the first pulse of the train.

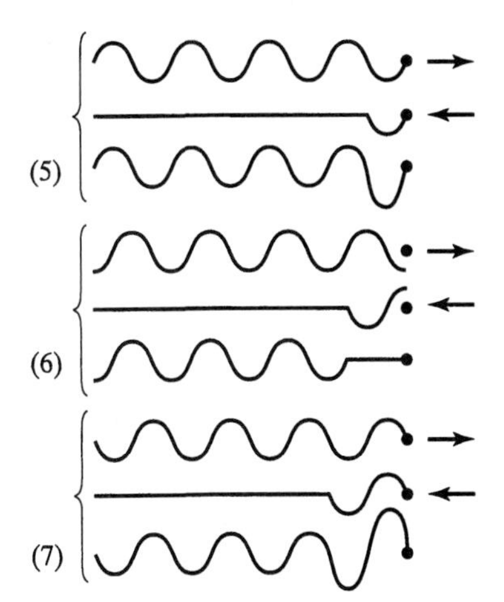

Figure 3-8 Incident, reflected, and sum waves for (5), (6), and (7) of Figure 3-7.

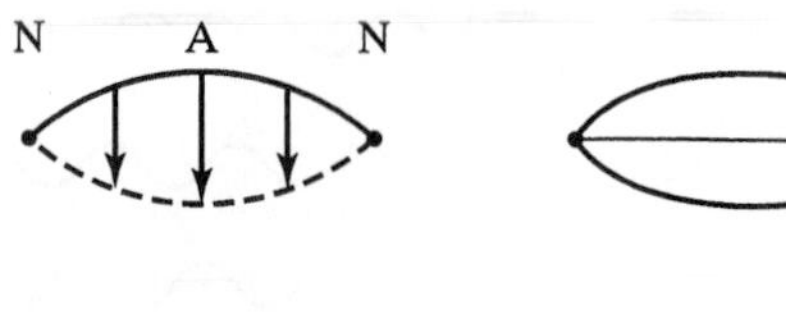

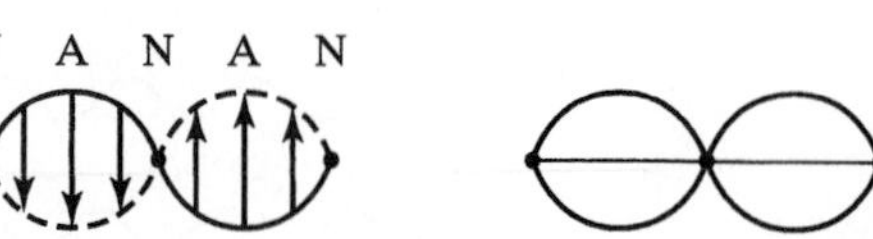

Figure 3-9 The two "simplest" standing waves on a stretched rope or wire and their representations.

off both ends to form a standing wave. Such a situation exists whenever a stringed instrument is plucked or bowed; plucking provides an initial pulse that dies rapidly, whereas bowing provides a continuous source of pulses.

What transverse standing waves can exist in a stretched rope fixed at both ends? The answer to this question can be found by observing that the constraints on any possible standing waves are (1) there must be a node at each end (because the ends are fixed) and (2) all the loops in the standing wave must be equal in amplitude and with length equal to one half-wavelength of some allowable standing wave. The two simplest possible standing waves in a stretched rope are illustrated in Figure 3-9. The drawings at the left show one extremum position of the rope for each wave, and the arrows show the direction of the ensuing motion toward the opposite extremum, which is indicated by the dashed lines.

The standing waves then continue to oscillate back and forth between these extremum positions. To the right of each standing wave is a drawing showing the extremum positions of each standing wave. This drawing is the graphical representation of the standing wave to its left; that is, by drawing such a graphical representation, we mean to convey the existence of the standing wave that it represents. Figure 3-10 shows some "standing waves" that violate the preceding constraints and therefore cannot exist.

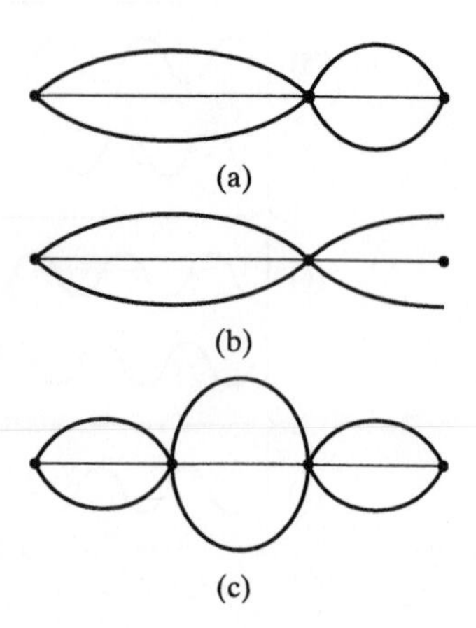

Figure 3-10 Three impossible "standing waves" in a stretched rope or wire with fixed ends. Wave (a) has unequal loop lengths, wave (b) violates the end conditions by having an antinode at one end, and wave (c) has unequal amplitude loops.

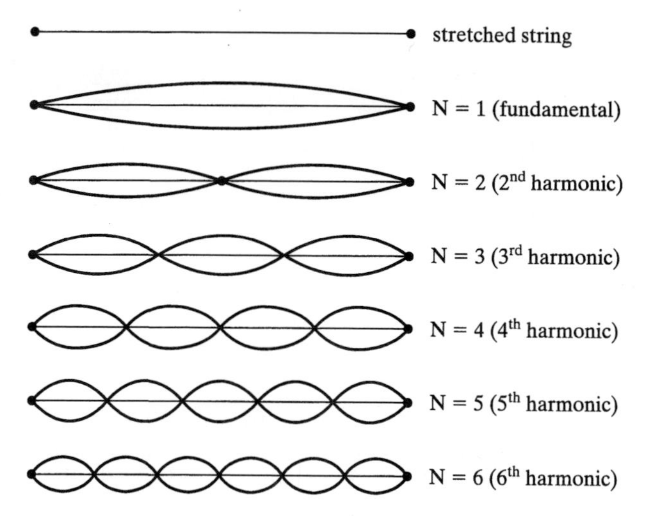

Figure 3-11 The representations of the first six possible standing waves in a stretched wire or rope.

Figure 3-11 contains the graphical representations for the first six possible standing waves in a stretched rope, labeled by their *harmonic number N*, the meaning of which will soon become clear. The nodes must be equally spaced along the rope, with an antinode midway between any two successive nodes. Recalling that one loop of a standing wave is one half-wavelength, we observe that the first standing wave is one half-wavelength long, the second one full wavelength (of a different value) long, the third one and one-half wavelengths long, and so forth, as given in Table 3-2.

In Sec. 2.1 we saw that the frequency f and wavelength λ of a wave are related by the equation $v = f\lambda$, where v is the speed of propagation of the wave in the medium. In this case, v is the speed of the wave on the stretched rope, which is assumed to be

Table 3-2 Wavelengths and Frequencies of the First Six Possible Standing Waves on a String of Length L

Harmonic number N	Wavelength λ	Frequency $\left(f = \dfrac{v}{\lambda}\right)$
1	$\lambda_1 = 2L$	$f_1 = \dfrac{v}{2L}$
2	$\lambda_2 = L = \frac{1}{2}\lambda_1$	$f_2 = \dfrac{v}{L} = 2f_1$
3	$\lambda_3 = \frac{2}{3}L = \frac{1}{3}\lambda_1$	$f_3 = \dfrac{v}{\frac{2}{3}L} = 3f_1$
4	$\lambda_4 = \frac{1}{2}L = \frac{1}{4}\lambda_1$	$f_4 = \dfrac{v}{\frac{1}{2}L} = 4f_1$
5	$\lambda_5 = \frac{2}{5}L = \frac{1}{5}\lambda_1$	$f_5 = \dfrac{v}{\frac{2}{5}L} = 5f_1$
6	$\lambda_6 = \frac{1}{3}L = \frac{1}{6}\lambda_1$	$f_6 = \dfrac{v}{\frac{1}{3}L} = 6f_1$

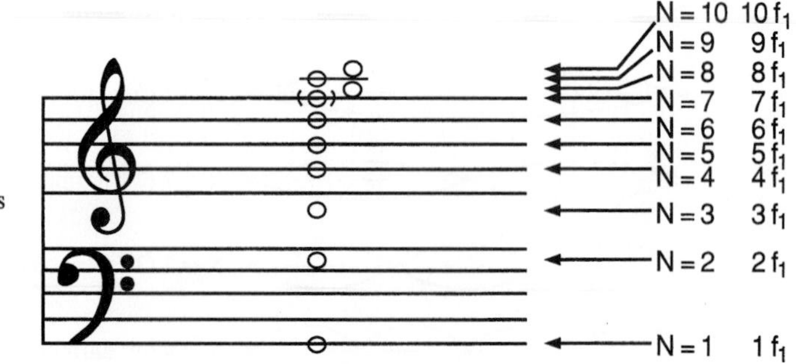

Figure 3-12 Notes on the musical staff having frequencies closest to the notes in the overtone series of G_2; each note is labeled by its harmonic number and the frequency of the corresponding harmonic.

constant for waves of all frequencies. We can now determine the frequencies of the various standing waves for a fixed length of rope, which are also given in Table 3-2. The harmonic number N, in addition to being equal to the number of loops in the standing wave, tells us the frequency of the standing wave. The frequency of each successive harmonic is equal to the frequency of the simplest mode (or fundamental) multiplied by its harmonic number N. If, for example, the simplest standing wave in a rope has a frequency of 100 Hz when vibrating in one loop, the rope will vibrate in two loops at a frequency of 200 Hz, in three loops at a frequency of 300 Hz, and so forth.

The simplest standing wave, where the rope vibrates in one loop at its lowest frequency, is called the *fundamental* or *first harmonic.* The other standing waves, called *harmonics* or *overtones,* are identified by their harmonic number N and have a frequency of N times the fundamental frequency. We shall use the term *harmonic* only when the frequencies of the higher standing waves are integrally related to the frequency of the fundamental. The different types of oscillation are called *modes*; each mode has its own frequency. A string may vibrate in many modes simultaneously, where the displacement of the string is the sum of the displacements resulting from the individual modes. (Those readers with little or no musical training should read Appendix A before proceeding.)

This complete set of sinusoidal standing waves with frequencies related by successive integers is known as the *overtone series.* The first ten musical notes of the overtone series whose fundamental frequency corresponds to the note G on the lowest line of the bass clef are shown in Figure 3-12. The sequence of musical intervals of the overtone series is the same no matter what note is chosen as the fundamental. The intervals associated with the first eight harmonics of the overtone series are given in Table 3-3.

The overtone series is important because complex waves (those with shapes other than pure waves) are composed of combinations of waves whose frequencies are in the overtone series of the note whose fundamental frequency is that of the complex wave. We shall study this in depth in Chap. 4.

A given musical interval, a type of "distance" between two pitches, is always obtained from two notes with a specific frequency ratio between them. For example, one octave (two notes with the same letter name) corresponds to a frequency ratio of 1 to 2. Thus, harmonics 1 and 2 form an octave, as do harmonics 2 and 4, 3 and 6, and 4 and 8. A true, or beatless, perfect fifth contains notes with a frequency ratio of 2 to 3, such as the ratio between harmonics 2 and 3 or harmonics 4 and 6. A true perfect fourth contains notes with a frequency ratio of 3 to 4, such as the ratio between harmonics 3 and 4 or harmonics 6 and 8. A true major third has a frequency ratio of 4 to 5, and a

Table 3-3 Musical Intervals Between the Fundamental and Other Notes of the Overtone Series

N	f	Interval
1	f_1	Unison
2	$2f_1$	One octave
3	$3f_1$	One octave + one perfect fifth
4	$4f_1$	Two octaves
5	$5f_1$	Two octaves + one major third
6	$6f_1$	Two octaves + one perfect fifth
7	$7f_1$	Two octaves + one minor seventh
8	$8f_1$	Three octaves

true minor third one of 5 to 6. These intervals, with the exception of the minor seventh, are approximately the same on the piano, which is tuned in equal temperament. The ratio of 4 to 7, although beatless, is somewhat less than a minor seventh on a piano.

Compared with the notes of the piano, the notes separated by octave intervals ($N = 1, 2, 4$, and 8) are the only notes of the overtone series that are perfectly in tune; harmonics 3, 5, and 6 are slightly different from the notes on the piano, and harmonic 7 is substantially lower than the note on the piano and is therefore enclosed by parentheses in Figure 3-12. This does not imply that the notes of the overtone series are out of tune; they are not. Rather, the frequencies chosen for the notes on the piano represent a compromise in which some of the frequencies differ from those of the overtone series. The choice of keyboard note frequencies, called the *temperament,* is treated in Chap. 9.

In the Renaissance, choral works in minor keys would often end in a major key, the third of the final chord being raised one half-step to make a major rather than a minor chord. The note two octaves plus a major third above the bass note in the final chord is present in the overtone series of the bass note. If the minor third, one half-step below the harmonic of the bass note, is being sung, these two notes will produce beating. This becomes particularly evident when the sound reverberates in a large room or cathedral. By replacing the minor third with a major third in the final chord, this beating is eliminated. This major third at the end of a piece in a minor key is called the *tierce de Picardie,* or Picardy third, after the region in France where it is believed to have developed.

3.3 Mersenne's Laws

The three physical laws governing the fundamental frequency of various stretched wires are called Mersenne's laws, after the French physicist and musical theorist Marin Mersenne (1588–1648). They relate the fundamental frequency f (or period T) to the linear mass density W (mass per unit length) of the wire, the tension F in the wire, and the length L of the wire. A summary of the three laws is contained in Table 3-4. For example, $T \propto L$ means the period is proportional to the length of rope. If we make the rope twice as long, the period of the fundamental mode becomes twice its original value (the frequency becomes half its original value), whereas if we make the rope

Table 3-4 Mersenne's Laws for the Fundamental Frequency f or Period T of a Stretched String

	Period		Frequency	Constants
Law 1	$T \propto L$	or	$f \propto 1/L$	W, F
Law 2	$T \propto 1/\sqrt{F}$	or	$f \propto \sqrt{F}$	W, L
Law 3	$T \propto \sqrt{W}$	or	$f \propto 1/\sqrt{W}$	L, F

W = mass per unit length of string
L = length of string
F = tension applied to string

one-third of its original length, the period becomes one-third of its original value (the frequency becomes three times its original value). $T \propto 1/\sqrt{F}$ or $f \propto \sqrt{F}$ means that it requires four times as much tension to raise the frequency to twice its original value or, equivalently, decrease the period to half its original value; this corresponds to a change of one octave. $T \alpha \sqrt{W}$ or $f \alpha 1/\sqrt{W}$ means, for instance, if we make the mass per unit length four times as great, the period will double or, equivalently, the frequency will become one-half its original value. Note that a wire of diameter d has a mass per unit length proportional to d^2, so $T \alpha d$ or $f \alpha 1/d$.

The three laws can be summarized in a single formula:

$$f = \frac{1}{2L}\sqrt{\frac{F}{W}}. \tag{3.1}$$

These laws tell us that low frequencies are achieved using long, heavy wires with little tension, and, conversely, high frequencies are achieved by using short, thin wires with high tension. The difference between sizes and strings in the orchestral string family arises from application of a combination of these three laws; the strings on the cello, for example, are longer, heavier, and held with less tension than those of the violin, and therefore yield a lower frequency.

An extreme illustration of Mersenne's laws involves the piano. A "simplified" piano might have all strings of the same type of wire under the same tension and obtain its frequency or pitch difference by changing only the length of the strings, thereby retaining similar tone quality over its entire range. Because a piano contains over seven full octaves of notes, and each octave interval increases the frequency by a factor of 2, the frequency of the highest note (almost 4200 Hz) is over 128 times the frequency of the lowest note (about 27.5 Hz). Thus according to Mersenne's first law, if the highest frequency string were 6 inches long, the lowest would be about 76 feet! Obviously some alternative method must be used; actual pianos have heavier strings (larger W) for the lower notes. Wrapping the center wire with an outer coil for the lowest strings increases the mass per unit length without greatly reducing flexibility. By obtaining the lower frequencies with a combination of increased length and increased mass per unit length, the tension in the strings can remain approximately constant over the entire piano to avoid unequal stress and warping of the frame.

3.4 Longitudinal Standing Waves

In Section 3.2 we discussed the nature of the transverse standing waves in a stretched rope or string. We shall now discuss the longitudinal analog to transverse standing waves in a string: standing sound waves in an air column or tube. We shall simultaneously develop a convenient transverse-wave representation for the longitudinal standing waves in a tube and observe the direct comparison between transverse standing waves in a string and standing sound waves in a tube. It will be helpful to make use of longitudinal waves in a long coiled spring (slinky spring) to visualize what is happening in a sound wave.

As in the case of transverse waves, longitudinal standing waves are formed when identical waves traveling in opposite directions are superimposed in the same medium. Figure 3-13 illustrates the motion resulting from such a situation. The columns of dots can be taken to represent either coils of a slinky spring or, for a very simple model of a sound wave in air, successive layers of air molecules. Shown in the figure are the positions of the air layers at $t = 0$, $T/4$, $T/2$, and $3T/4$, where T is the period of the wave; these configurations are then repeated. The arrows show the extent of the motion of each layer from its equilibrium position, the average position each molecular layer would assume in the absence of any wave. Certain layers are totally motionless; these points are the nodes. Midway between any two nodes is a position where the motion of the air layer (or spring coil) is at a maximum, the antinode. Because the nodes and antinodes are defined in terms of the velocity or displacement of the individual layers, they are sometimes called *velocity* or *displacement* nodes and antinodes.

Another quantity that can be used to describe locations along the wave is air pressure (which is analogous to the density of spacing of the spring coils). The air pressure is greater where the layers of air are closer together and less where they are farther apart. At a velocity antinode, the layers are moving rapidly back and forth together while maintaining about the same spacing, so the pressure remains approximately constant (no pressure fluctuation). At a velocity node, the spacing between the air layers changes,

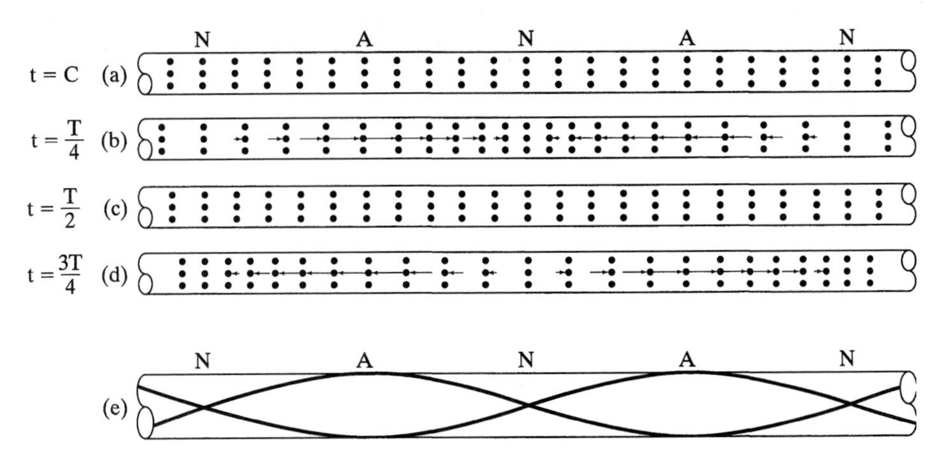

Figure 3-13 Motion of the layers of air molecules in a standing wave in a section of an infinite air tube, shown at four equal time intervals, (a) through (d), during one period. Arrows indicate displacement of molecular layers from their equilibrium positions. Drawing (e) is the transverse representation of this longitudinal standing wave.

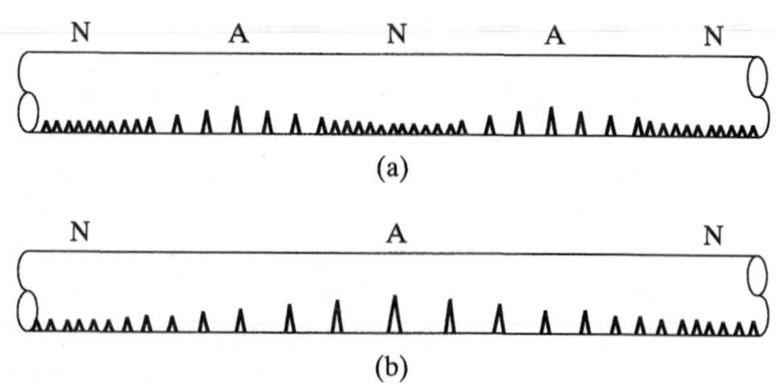

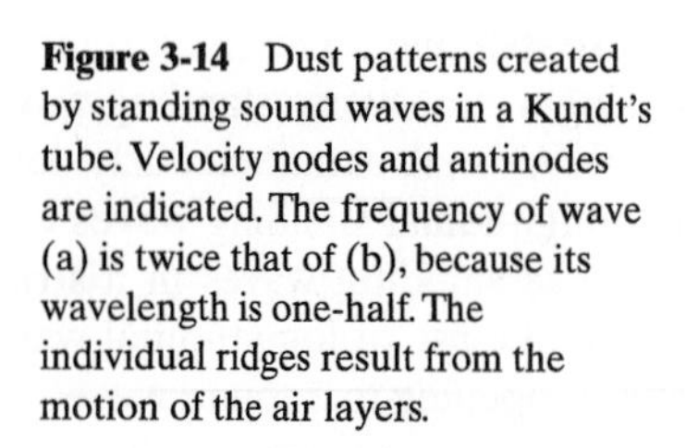

Figure 3-14 Dust patterns created by standing sound waves in a Kundt's tube. Velocity nodes and antinodes are indicated. The frequency of wave (a) is twice that of (b), because its wavelength is one-half. The individual ridges result from the motion of the air layers.

creating alternately high and low pressure. Thus a velocity node is an antinode for pressure fluctuation. Likewise, a velocity antinode is a node for pressure fluctuation (often called simply a *pressure* node). If the two original waves are sine waves, there will be sinusoidal variation of all the variables discussed: for example, positions of layers, velocities, pressure at any point.

It is clear that to draw a longitudinal standing wave in the manner of Figure 3-13(a) to (d) is very tedious. It is therefore helpful to make use of the transverse graphical representation shown in part (e). Both extrema in the corresponding transverse standing wave are drawn inside two parallel lines representing the tube. The nodes N are located at the crossover points and the antinodes A at the extreme positions. The maximum longitudinal velocity (or displacement) of the layer of molecules (or slinky spring layers) can be taken to be proportional to the distance between the two curves at any point along the standing wave.

We can verify that this molecular motion exists in the standing sound wave in an air column using a Kundt's tube, named after August Adolph Kundt (1839–1894), which consists of a simple horizontal tube with a small amount of fine dust lying along the bottom. Creation of a standing wave in the tube will result in violent motion of the dust at a velocity antinode, where the molecular layers are in rapid motion, alternating with quiescent regions at the nodes, where there is very little molecular motion due to the wave. The dust is swept into regions where the motion of the air layers is minimum. Typical dust patterns created by a standing sound wave in a Kundt's tube are shown in side view in Figure 3-14. It is the overall structure of nodes and antinodes that marks the loops of the standing wave.

Small, closely spaced striations are clearly visible, particularly at all points away from the nodes. These striations are kept in continual agitation by the motion of the air layers. Figure 3-14(a) is associated with a wave of higher frequency or shorter wavelength than Figure 3-14(b), resulting in a shorter distance between nodes in (a). The period of the oscillation for (a) is shorter than that of (b), so the total travel distance for molecular layers is less, resulting in a smaller spacing between the striations in (a). Figure 3-14(a) shows two loops, or one full wavelength; Figure 3-14(b) shows one loop, or one half-wavelength.

The most interesting standing waves are those in finite tubes. To describe such standing waves fully, we must first investigate how waves reflect at an end of the medium, as we did in the case of transverse standing waves.

Consider first a compressional pulse traveling down a slinky spring toward either a fixed or free end, as shown sequentially in Figure 3-15(a) and (b), respectively. When

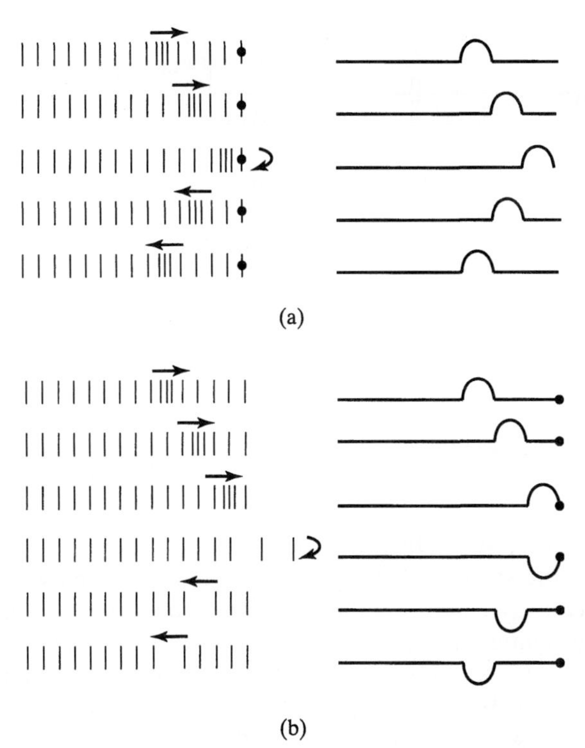

Figure 3-15 Reflection of a compressional pulse in a slinky spring off a fixed end (a) and a free end (b). There is a phase change in pressure or density upon reflection from a free end. The transverse representation of each is shown at the right.

the compression reaches the fixed end, it will reflect back as a compression. This is because when the coils adjacent to the wall are compressed and try to expand outward, they can do so only back in the direction of the spring; the wall does not allow them to continue to move in its direction. However, when compression reaches a free end, its motion causes the spring to extend, forming a rarefaction; it is this rarefaction that then travels back down the spring as the reflected pulse. Thus, unlike the case of transverse waves, for longitudinal waves in a slinky spring (and longitudinal waves in general), there is a phase reversal (in pressure) at a free end and no phase reversal at a fixed end. A transverse representation of the pulses is given to the right of each drawing in Figure 3-15, illustrating the described behavior in the reflection of a pulse. Notice that there is a phase reversal in pressure (compression to rarefaction, or vice versa) when the sound pulse reflects at an open end, but there is no phase reversal in pressure when it reflects at a closed end. This behavior is summarized in Table 3-5.

Table 3-5 Occurrence of a Phase Change When a Pulse Reflects off the End of the Wave Medium

	TRANSVERSE Displacement	LONGITUDINAL Pressure	
Fixed end	YES	NO	Closed end
Free end	NO	YES	Open end

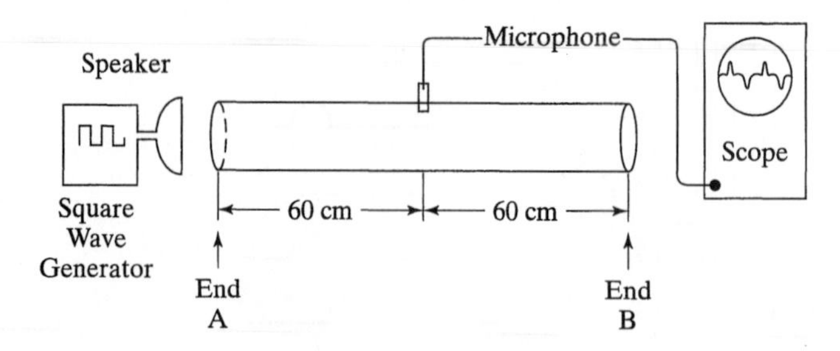

Figure 3-16 Apparatus for investigation of reflection of sound pulses off open and closed ends of air columns.

The following experiment shows the analogy between the open end of an air tube and the free end of a slinky spring or the closed end of an air tube and the fixed end of a slinky spring. A loudspeaker is placed at the end of a tube 120 cm long with a microphone inserted partway into the center of the tube, as illustrated in Figure 3-16. A 10-Hz square wave is applied to the speaker, causing the speaker cone to move rapidly toward the tube, then rapidly away from the tube, with 50 ms between each motion (100-ms period for the 10-Hz square wave). Whenever the speaker moves toward the tube, a compression is formed; rapid motion of the speaker away from the tube results in a rarefaction, as shown in Figure 3-17.

It takes about 2 ms for the pulse to travel 60 cm to the microphone, 2 ms more for it to reach end B, 2 ms more for the reflected pulse to travel back to the microphone, and so on. Thus any compressional pulse produced by the speaker can reflect back and forth in the tube many times before the speaker produces the next pulse 50 ms later.

Figure 3-18 shows the response of the microphone as a single pulse reflects back and forth in the tube. The pulse is produced at time $t = 0$, and 2 ms later a compression passes the microphone. This compression reaches end B 2 ms after passing the microphone, is reflected with a phase reversal in pressure, and returns to the microphone 2 ms later as a rarefaction. Four ms later, after reflection from end A with another phase reversal, the pulse again passes the microphone as a compression.

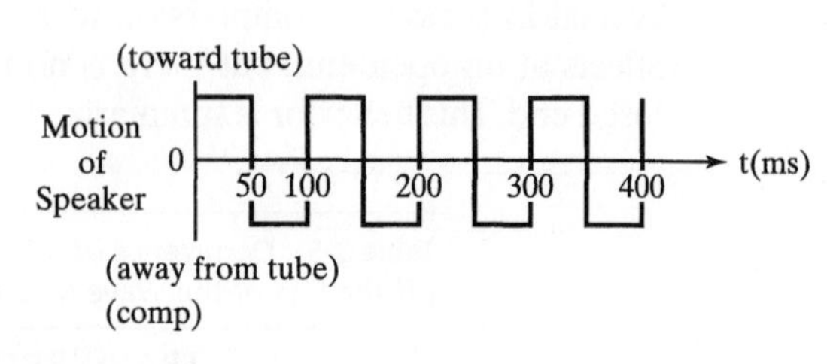

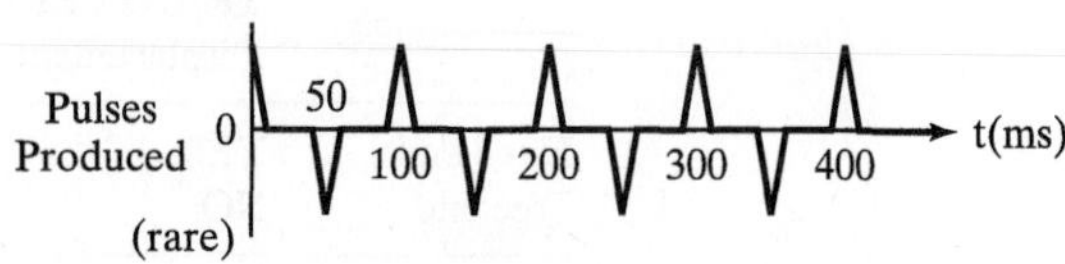

Figure 3-17 Application of 10-Hz square-wave motion of the speaker of Figure 3-16 produces a sequence of alternating compression and rarefaction pulses.

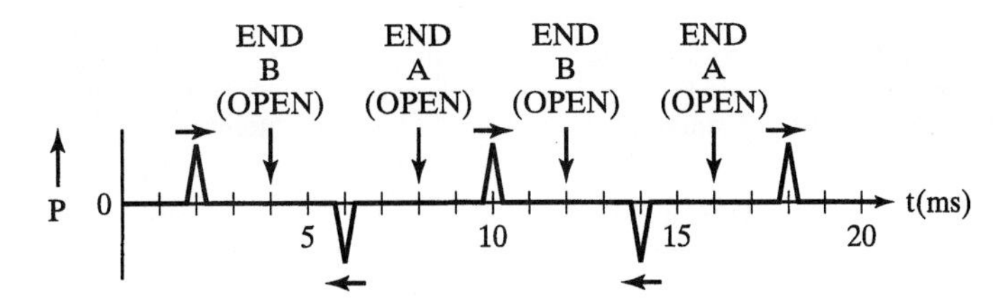

Figure 3-18 Pulses arriving at the microphone after reflection off ends of tube for experiment of Figure 3-16, with both ends open. At each end the pulse reflects with a phase reversal in pressure. The pulse began at end A at time $t = 0$. Arrows indicate the direction the pulse is moving past the microphone.

If end B of the tube is closed, there will be no phase reversal upon reflection from that end. Thus the first reflection from end B will return as a compression, as shown in Figure 3-19. In this case, when the pulse reflects from the open end (end A), there will be a phase reversal, and when it reflects from the closed end (end B), there will be no phase reversal.

Consider reflections of longitudinal waves under both types of end conditions. At an open end, there will be motion of the air layers, because the air is free to move there. Similarly, coils move at the free end of a slinky spring. Imagine a compressional pulse traveling along a spring and reaching a free end. The compression forces the spring to extend, a rarefaction forms at the end, and this rarefaction propagates back along the spring. Thus a phase reversal (in coil density) occurs at the free end of a spring. There is, therefore, a pressure node and a velocity (or displacement) antinode at the free end; the initial compression is always canceled there by the reflected rarefaction. Likewise, there is a pressure node and a velocity (or displacement) antinode at the open end of an air tube. In short, there is a pressure node and a velocity antinode at a free end or an open end for a longitudinal wave, and such waves undergo a phase reversal (in pressure) upon reflection.

Conversely, at a fixed end of a slinky spring or a closed end of an air column, there can be no motion of the layers. In fact, a thin wall could be put across the tube in Figure 3-13 at any nodal point, and the standing wave could still exist the same as

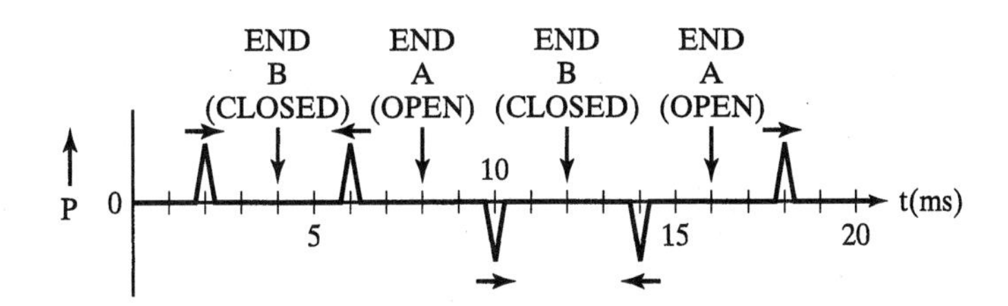

Figure 3-19 Pulses arriving at the microphone after reflection off ends of tube for experiment of Figure 3-16, with end A open and end B closed. When a pulse is reflected at an open end, there is a phase reversal in pressure; when a pulse is reflected at a closed end, there is no phase reversal in pressure. The pulse began at end A at time $t = 0$. Arrows indicate the direction the pulse is moving past the microphone.

Table 3-6 End Configuration for Standing Waves in a One-Dimensional Medium

	Transverse	Longitudinal		
	Displacement	Pressure	Velocity or displacement	
Fixed end	NODE	ANTINODE	NODE	Closed end
Free end	ANTINODE	NODE	ANTINODE	Open end

before. In that case, the standing wave would be formed from the original wave and its reflection. Thus, at a closed end of an air column or a fixed end of a slinky spring, there must always be a velocity (or displacement) node or a pressure antinode. Table 3-6 summarizes the standing-wave nodal and antinodal end configurations for the cases of transverse waves, slinky-spring longitudinal waves, and sound waves in tubes.

One obtains reflections from either end, so, as in the case of a standing wave in a stretched wire discussed previously, one can set up standing waves in air tubes subject to the proper end constraints. The first six such standing waves for an open tube (open at both ends) are shown in Figure 3-20, using the velocity (or displacement) diagrams of Figure 3-13. The open tube resonates at the fundamental frequency when the tube is one half-wavelength long, with an antinode at each end and a node in the middle. Successive harmonics each add an additional loop to the standing-wave pattern. As in the case of the rope, the nodes and antinodes must be equally spaced along the tube.

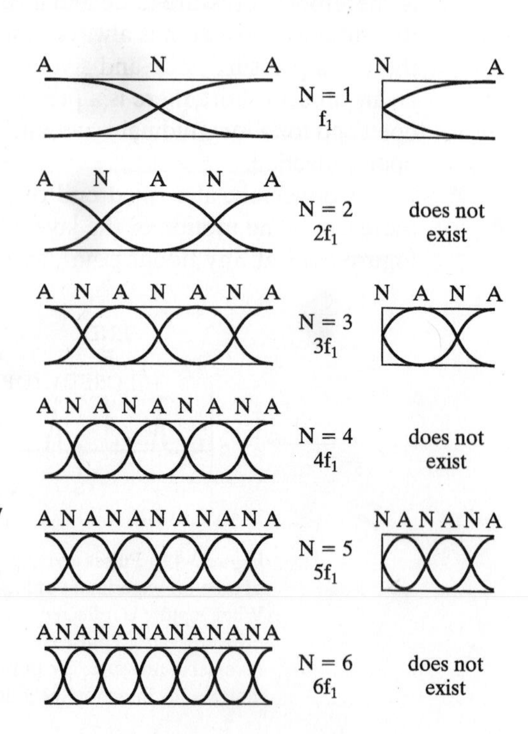

Figure 3-20 Transverse representations of velocity patterns for standing waves in open and closed tubes. Air displacement (or velocity) is graphed here. A closed end is a node just as a fixed end of a string is a node. (Compare with Figure 3-11.) A closed tube of a given length and an open tube of double this length have the same fundamental frequency.

The end constraints placed on the possible standing waves in a closed tube (open at one end and closed at one end) allow only the formation of odd harmonics in a closed tube, as shown in Figure 3-20. For the same fundamental frequency the closed tube is half the length of the open tube; the open tube is one half-wavelength long and the closed tube one quarter-wavelength long for their respective fundamental frequencies.

Table 3-7 summarizes the relationships between harmonic number, frequency, wavelength, and length of tube for the first six harmonics of open and closed tubes, respectively. If we choose the length of the open tube L_0 twice that of the closed tube L_c, their fundamental frequencies f_1 will be the same.

These ideas are directly applicable in determining the fundamental frequencies of wind instruments and the overtone (or harmonic) structure of their sounds. For any wind instrument, the bell end acts as an open end. An embouchure hole, such as that on the orchestral flute, or a fipple, such as that on the recorder family, is also an open end. Thus, these instruments can support standing waves of all harmonics. The reed end of a reed instrument (such as the clarinet) acts as a closed end, because the reed is slapped closed most of the time; it opens at regular time intervals to allow short bursts of air to enter the instrument. Because the clarinet has a cylindrical bore (the width of the air tube is constant along most of its length) and one closed end, low-numbered odd harmonics dominate, and it behaves acoustically like a closed tube, as shown in Figure 4-19. Although the oboe, bassoon, and saxophone families are reed instruments, they have conical-shaped bores and, for reasons discussed in Chap. 10, the conical bore allows all harmonics (both even and odd) to exist. Among the reed instruments, the clarinet family

Table 3-7 Wavelength and Frequency of Standing Waves in Open and Closed Tubes of Length L_o and L_c, Respectively. The Fundamental Frequencies f Are the Same if $L_o = 2L_C$.

Harmonic number	Open tube		Closed tube	
	Wavelength	Frequency	Wavelength	Frequency
1	$\lambda_1 = 2L_o$	$f_1 = \frac{S}{2L_o}$	$\lambda_1 = 4L_c$	$f_1 = \frac{S}{4L_c}$
2	$\lambda_2 = L_o = \frac{1}{2}\lambda_1$	$f_2 = \frac{S}{L_o} = 2f_1$	does not occur	
3	$\lambda_3 = \frac{2}{3}L_o = \frac{1}{3}\lambda_1$	$f_3 = \frac{S}{\frac{2}{3}L_o} = 3f_1$	$\lambda_3 = \frac{4}{3}L_c$	$f_3 = \frac{S}{\frac{4}{3}L_c} = 3f_1$
4	$\lambda_4 = \frac{1}{2}L_o = \frac{1}{4}\lambda_1$	$f_4 = \frac{S}{\frac{1}{2}L_o} = 4f_1$	does not occur	
5	$\lambda_5 = \frac{2}{5}L_o = \frac{1}{5}\lambda_1$	$f_5 = \frac{S}{\frac{2}{5}L_o} = 5f_1$	$\lambda_5 = \frac{4}{5}L_c$	$f_1 = \frac{S}{\frac{4}{5}L_c} = 5f_1$
6	$\lambda_6 = \frac{1}{3}L_o = \frac{1}{6}\lambda_1$	$f_6 = \frac{S}{\frac{1}{3}L_o} = 6f_1$	does not occur	

is unique in that it is the only one with a cylindrical bore, which not only gives mainly odd harmonic structure but also causes clarinets to overblow at the musical interval of a twelfth. By blowing harder into the instrument, or using a register key, one can excite the next possible higher harmonic. For the clarinet, the next allowable harmonic has a frequency three times the fundamental frequency, and is thus an octave and a fifth above the note played normally (see Appendix A). The other woodwinds overblow at one octave intervals. Whether the reed on a woodwind instrument is single (clarinet and saxophone families) or double (oboe and bassoon families) is irrelevant in determining which harmonics of standing waves are possible in the instruments. The only relevant factors are the shape of the bore (cylindrical or conical) and the existence of a node at a closed end for a cylindrical tube; however, the style of reed helps determine the relative *amplitudes* of possible harmonics.

Brass instruments are blown by buzzing the player's lips directly into a mouthpiece; such a mouthpiece is also a closed end. Here, again, the bore of the instrument as well as complicated effects resulting from mouthpiece and bell shape allow all harmonics to exist in these instruments.

Brass and woodwind instrument players are familiar with the following effect: As they play, their instruments warm up and rise in pitch. This rise in pitch can be explained using several of the principles described previously. The velocity of sound is greater in warm air than in cool air. As the air column in the instrument gets warmer, the frequencies of the fundamental and all harmonics increase. This is because the wavelength of each standing wave, related to the length of the instrument, remains the same while the velocity of sound in the instrument increases. By the formula $f = S/\lambda$ we can see that, if S increases and λ stays the same, the frequency f must increase.

3.5 Other Standing Waves and Applications

Thus far we have studied one-dimensional standing waves in strings and air columns. Very important standing waves exist in one-dimensional solids such as tuning bars or xylophone bars, and multidimensional standing waves become important in many applications.

Tuned percussion instruments, such as chimes, bells, and timpani, are interesting in that their resonant frequencies are not related by integers; that is, the resonances do not have frequencies of exactly N times the fundamental frequency. Several modes of oscillation are given in Figure 3-21 for a stretched circular membrane such as a drumhead or timpani. The frequency ratios differ greatly from integral values, as will be discussed in Chap. 14. In addition, the particular frequency that we hear as the "pitch" of the timpani is not the lowest one produced but rather the second lowest. The lowest is heard as a sort of dull low-frequency thud that dies away quickly.

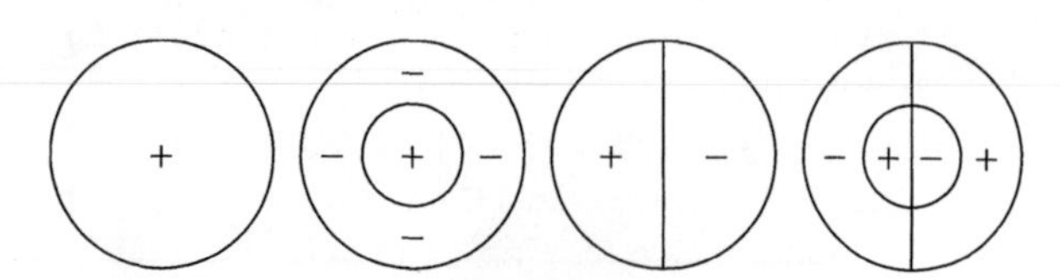

Figure 3-21 Some vibrational modes in a circular drumhead. Nodal lines are indicated, along with the relative phase of various parts of the drumhead, indicated by + (up) and − (down).

Figure 3-22 Chladni figures in $\frac{1}{8}$ inch thick blackened aluminum plates driven in the center. The frequencies of their oscillations are between 15 and 25 kHz. The scallops in the circular plate are caused by touching the plate at a point along its circumference; all other patterns allow the plates to vibrate freely.

A Chladni plate, named after the German acoustician Ernst Florens Friedrich Chladni (1756–1827), is a thin, regularly shaped metal plate that can be excited by drawing a violin bow across one edge or by some other, usually electrical, means. If salt or sand is sprinkled onto the vibrating plate, it will be jostled at the antinodal regions and collect along the nodal lines, revealing a two-dimensional standing wave pattern. The patterns thus produced are often very beautiful and can change radically if the excitation frequency is changed slightly. Some patterns are shown in Figure 3-22 for circular and square plates; the lines are the nodal lines. The tap tone, the tone produced by gently rapping with a knuckle the front or back plate of a stringed instrument like the violin, has a standing wave similar to these Chladni figures. The wood forming the plates of a violin must be carefully shaped so that the tap tone will be at exactly the right frequency to strengthen the notes in a specific frequency range of the instrument. The "singing" of a high-quality crystal wineglass when a moistened finger is rubbed around its edge is due to the existence of an audio-frequency standing wave in the glass, which is excited by the rubbing. The fact that a good wineglass can be broken by a loud sound is evidence that a mode of oscillation exists in the audible range. That is, an intense sound of that frequency can drive the standing wave to an amplitude greater than the elastic limit of the crystal. This effect is similar to the creation of standing sound waves around the inside walls of a circular room, such as a large cathedral dome. Walking around the edge of the dome, one hears alternating loud (antinodal) and soft (nodal) regions when an instrument near the edge of the dome plays a constant musical tone.

The Tacoma Narrows Bridge, discussed in Chap. 1, collapsed because of a resonance. We now see, using the vocabulary developed here, that a certain torsional or twisting mode of oscillation was driven in a complex manner by eddy currents created

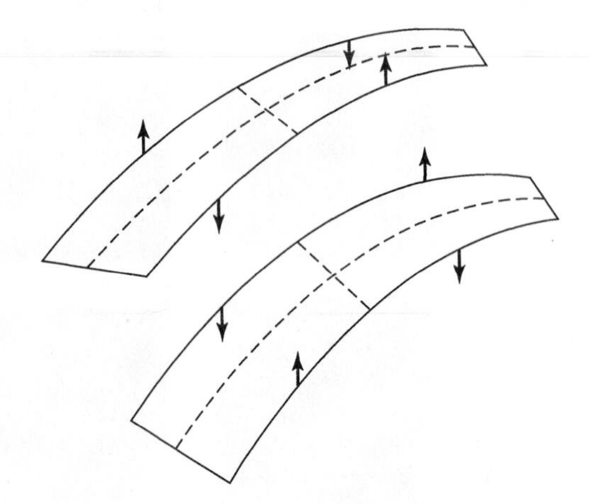

Figure 3-23 Oscillation mode of the center span of the Tacoma Narrows Bridge. Ends remain fixed; nodal lines are dashed.

when the wind passed over the bridge, producing a standing wave as shown in Figure 3-23. The amplitude of the oscillation eventually exceeded the elastic limit of the materials in the bridge. Figure 3-24 shows actual photographs of the Tacoma Narrows Bridge, during its large oscillations and after its collapse.

Soldiers marching across small bridges break step to avoid creating large periodic oscillations that might drive a resonance in the bridge and cause it to collapse.

Resonances in air tubes can be used to check for holes and discontinuities in inaccessible pipes in nuclear reactors, oil refineries, chemical plants, and other industrial situations. By observing the resonances in a long tube with sound waves and repeating the process over long time intervals, one can observe changes in the resonances with time, which indicate deterioration of the pipe. Long-wavelength (low-frequency) waves are used because they diffract easily and therefore bend with the pipe. If a hole appears in the pipe, this point will then act as an open end, creating a velocity antinode at that point and thereby changing the resonant frequencies of the pipe. Using this technique, holes as small as 2 mm diameter in pipes up to 10 m long have been detected.

Figure 3-24 The Tacoma Narrows Bridge during its large oscillations on November 7, 1940. A strong but steady wind created vortices that drove a standing wave in the bridge, creating a resonance. (The Camera Shop, Tacoma. Ed Elliott.)

SUMMARY

(***Section 3-1***) **Standing waves** are formed when identical waves move past each other going in opposite directions or when waves reflect back and forth from the ends of a finite medium. (***Section 3-2***) Standing waves in stretched strings held fixed at both ends consist of alternating **nodes** and **antinodes,** forming a uniform series of **loops**; each loop is one-half wavelength long. The vibrational **modes** for standing waves in stretched strings are its **harmonics**; the complete group is called the **overtone series.** Adjacent notes of the overtone series form basic musical **intervals.** (***Section 3-3***) **Mersenne's laws**, relating the fundamental frequency of a stretched string to the length, tension, and mass per unit length of the string, affect the design of string instruments and the piano. (***Section 3-4***) Standing waves in **open tubes** form the complete overtone series, whereas those in **closed tubes** consist of only odd harmonics. (***Section 3-5***) Two-dimensional and other complex standing waves, of interest because their **overtones** are not harmonics, are important in the design of wooden stringed-instrument bodies as well as tuned percussion instruments.

QUESTIONS

1. **a.** Draw two sine waves of equal amplitude and frequency, in phase, and add them graphically. What are the amplitude and frequency of the sum wave?

b. Repeat this procedure for two component waves with phase differences of 45°, 90°, and 180°. What are the amplitude and frequency of each sum wave?

2. For the first four resonant modes of a stretched string 100 cm long:

a. Draw the standing waves for each, and label the nodes N and the antinodes A.

b. Label each mode by its harmonic number and its frequency relative to the fundamental frequency f_1 of the string.

3. For the first four resonant modes of an open tube 345 cm long in air:

a. Draw the velocity (or displacement) standing waves for each, and label the nodes N and the antinodes A.

b. Label each mode by its harmonic number and its frequency relative to the fundamental frequency f_1 of the open tube.

4. For the first four resonant modes of a closed tube 345 cm long in air:

a. Draw the velocity (or displacement) standing waves for each, and label the nodes N and antinodes A.

b. Label each mode by its harmonic number and its frequency relative to the fundamental frequency f_1 of the closed tube.

5. What is a mode?

6. Which type of tube, open or closed, can never have a node in the exact center? Why?

7. Why does the pitch of a wind instrument rise when the instrument warms up?

8. Longitudinal standing waves can be produced in a 2-m aluminum rod by holding it at some point with one hand and stroking it with the other hand or tapping it gently with a hammer, causing it to oscillate with antinodes at each end.

a. Why are the ends antinodes and not nodes?

b. When the rod is pinched at the center while exciting the rod, the frequency emitted is 2500 Hz. What is the speed of sound in the metal if this is the fundamental mode?

c. Where should the rod be pinched to excite the next possible standing wave?

d. What is its frequency? (Hint: Longitudinal standing waves in a rod are similar to standing waves in an air column.)

9. A closed tube filled with air is 50 cm long.

a. What is the length of an open tube that has the same fundamental frequency?

b. What is this frequency?

c. These two tubes are placed in a gas with a speed of sound greater than the speed of sound in air. Do the frequencies of their fundamentals remain equal to each other?

d. Do they increase or decrease?

10. The frequency of the fundamental of a certain tube in air is 200 Hz. The frequency of the fundamental is 250 Hz when the tube is placed in an unknown gas.

a. What is the speed of sound in this gas?

b. What type of tube was used (open or closed)?

11. **a.** Draw a musical staff and write out the notes corresponding to the overtone series of G_2 (low G on the bass clef). Label each note by its harmonic number and frequency, if the G_2 has a frequency of about 100 Hz.

b. Which of these notes are resonances in an open tube whose fundamental note is G_2?

c. Which of these notes are in the overtone series of G_3, the note one octave above G_2?

12. **a.** Draw a musical staff and write out the notes corresponding to the overtone series of G_2 (low G on the bass clef). Label each note by its harmonic number and frequency, if the G_2 has a frequency of about 100 Hz.

b. Which of these notes are resonances in a closed tube whose fundamental note is G_2?

13. The fundamental frequency of a stretched string is inversely proportional to the square root of the mass per unit length. What is the relation between the diameter of the string and its mass per unit length? How is the frequency related to the diameter of the string?

14. Certain rope bridges can oscillate in the wind, or as a result of the motion of people walking on them with a constant march step. In general, the period of the oscillation is much longer (for example, even up to several seconds) than that of a stretched string. Explain this using Mersenne's laws.

15. Perform the graphical additions for Figure 3-1, parts (2), (4), (6), and (8).

16. Perform the simple wave addition for Figure 3-7, parts (8), (9), and (10).

PROBLEMS

1. **a.** Write the formula that relates the frequency and speed of a wave to its wavelength.
   b. What is the wavelength of a 500-Hz tone in air?
   c. What is the velocity of sound in a gas that has a 2000-Hz tone with wavelength of 1 m?

2. **a.** How long should a closed tube in air be such that its fundamental frequency is 100 Hz?
   b. What is the frequency of its next possible mode?

3. The second possible standing wave in a closed tube in air has a frequency of 1500 Hz. What is the length of the tube?

4. A certain stretched string has a fundamental frequency of 100 Hz. By what factor must its (a) length, (b) tension, (c) mass per unit length, or (d) diameter, individually, be changed to raise the fundamental frequency to 200 Hz?

5. A certain stretched string has a fundamental frequency of 175 Hz. What is the new frequency if one

   a. decreases the tension by a factor of 4
   b. doubles the length
   c. decreases the mass per unit length by a factor of 9
   d. triples the diameter

6. The speed of sound is proportional to the square root of the absolute temperature T_{abs}, where $T_{abs} = 273° + T_c$ and T_c is the temperature in degrees Celsius. If a flute, initially tuned to A = 440 Hz at a temperature of 20° C (68° F), warms up to 30° C (86° F), what will be the frequency of that A?

REFERENCES

BACKUS, JOHN. *The Acoustical Foundations of Music.* 2nd ed. New York: W. W. Norton & Company, Inc., 1977.

BENADE, ARTHUR H. *Fundamentals of Musical Acoustics.* New York: Oxford University Press, Inc., 1976.

These two books are the classics in the field of musical acoustics. Backus is somewhat more general and at a less sophisticated mathematical and physical level; Benade is much more detailed and at times requires considerable sophistication on the part of the reader. Both are excellent books.

———. *Horns, Strings, and Harmony.* Garden City, N.Y.: Doubleday & Company, Inc. 1960. Covers many aspects of acoustics at the level of advanced high school or nontechnical college students.

BERG, RICHARD E. "Sound," *Encyclopedia Britannica,* 2003 edition. This article is somewhat more advanced than the present text, covers several topics not discussed in this book, and gives an additional collection of references covering these topics.

Chapter 4

Analysis and Synthesis of Complex Waves

We have seen that the various notes of the overtone series can be produced as resonances in stretched strings, open tubes, and closed tubes. Now we shall consider how these harmonics are combined to form a single complex wave. The process of combining harmonics to form a complex wave is called *Fourier synthesis*, after the French mathematician Jean Baptiste Joseph Fourier (1768–1830). The inverse process, that of determining the harmonic content of complex waves, is called *Fourier analysis*, or spectral analysis. At the end of this chapter, we shall study various factors affecting the timbre, or quality, of a musical tone, which is related to the harmonic content of the complex waves.

4.1 Synthesis of Complex Waves

Consider the examples of wave addition given in Figure 4-1. Shown are the fundamental ($N = 1$), the second harmonic ($N = 2$), and the algebraic sum of these two (first and second) harmonics. These graphs are generally taken to represent the instantaneous shape of the standing wave in a freely vibrating string or the "shape" of a sound wave traveling through space. The difference between the sum wave in cases (a) and (b) results from the change in relative phase between the two component harmonics. Figure 4-2 shows the synthesis of a complex wave from equal amplitudes of first and third harmonics; again, the difference in the two resultant wave shapes is a result of the difference of the relative phase between the two components. Figure 4-3 shows complex waves formed from first and second harmonics with different amplitude relationships; in both cases the relative phase of the second harmonic with respect to the fundamental is the same.

We can now see that the shape of a complex wave, a wave formed by the addition of two or more harmonics, is determined by (1) the number and relative amplitudes of

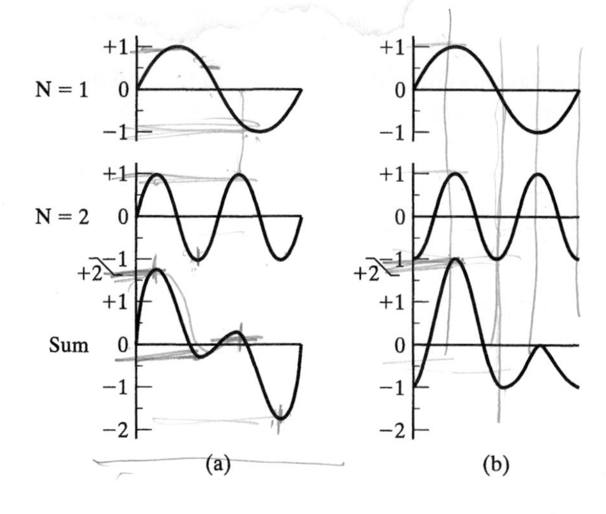

Figure 4-1 Fourier synthesis of complex waves from equal amplitudes of fundamental and second harmonic. Parts (a) and (b) differ in the phase of the second harmonic.

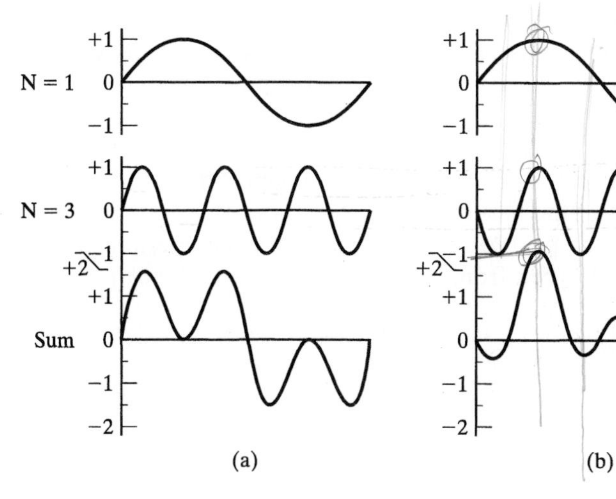

Figure 4-2 Fourier synthesis of complex waves from equal amplitudes of fundamental and third harmonic. Parts (a) and (b) differ in the phase of the third harmonic.

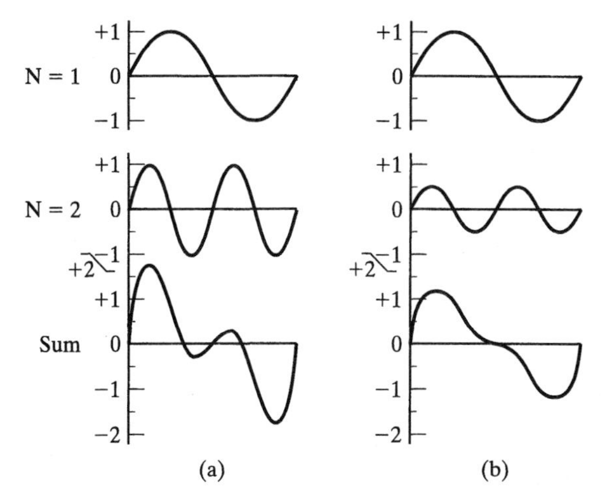

Figure 4-3 Fourier synthesis of complex waves from the fundamental and second harmonic. In part (a) amplitudes of the fundamental and second harmonic are equal; in part (b) the amplitude of the second harmonic is half that of the fundamental.

the component harmonics and (2) the phases of the higher harmonics, relative to the fundamental. It is interesting to listen to various complex tones formed from two harmonics while changing the amplitude and/or phase of the higher harmonic, relative to that of the fundamental. The tone quality or *timbre* is clearly affected by moderate changes in the amplitude of the higher harmonic but is hardly affected at all by rather large changes in the relative phases of the two harmonics. This property of the ear, summarized as *Ohm's law of hearing*, is one of the basic features of the place theory of hearing and will be discussed in Chap. 6.

The *standard* complex waves discussed in Chap. 1 (triangular, square, sawtooth, and pulse train) can be synthesized from the notes of the overtone series, as shown in Figures 4-4 through 4-7. Each figure contains, at the left, the sequence of harmonics present in the complex wave, with the appropriate amplitude and phase, and, at the right, the complex wave formed by the addition of each successive component. Notice that with the addition of each higher harmonic the complex wave looks more like the standard shape shown at the top. For reference, the relative amplitudes of the harmonics of each of the standard waves are given in Table 4-1. This table shows no relative phases, which must be chosen appropriately to produce the desired complex wave shape. However, the amplitudes are far more important than the phases in determining the timbre, or sound quality, of the wave.

The standard wave shapes (triangle, sawtooth, square, and pulse train) can be synthesized by combinations of sine waves of (1) the fundamental frequency of the complex wave and (2) some or all of the harmonics, each with the appropriate amplitude and phase. In general, it can be proved mathematically that *any* periodic wave with some frequency f can be synthesized from sine waves of the frequency f and its

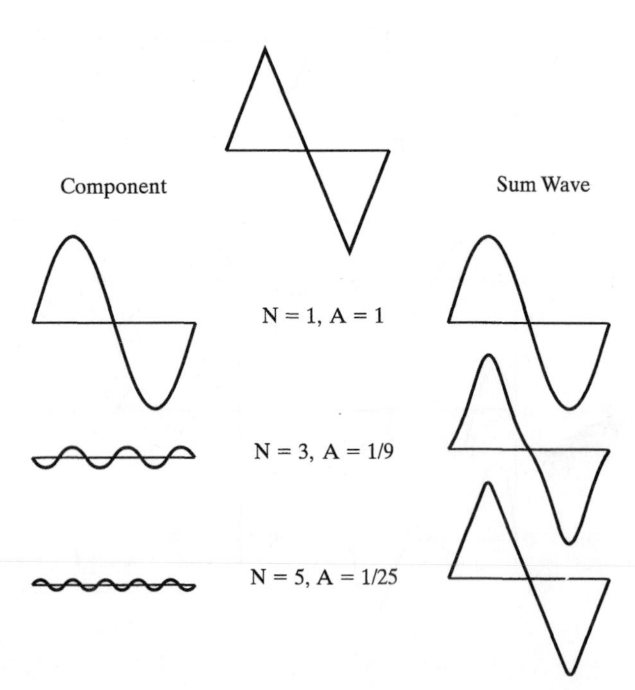

Figure 4-4 Fourier synthesis of a triangular wave. At the left are the successive harmonics; at the right are the sum waves, including each successive harmonic. The graph at the top is the wave being synthesized.

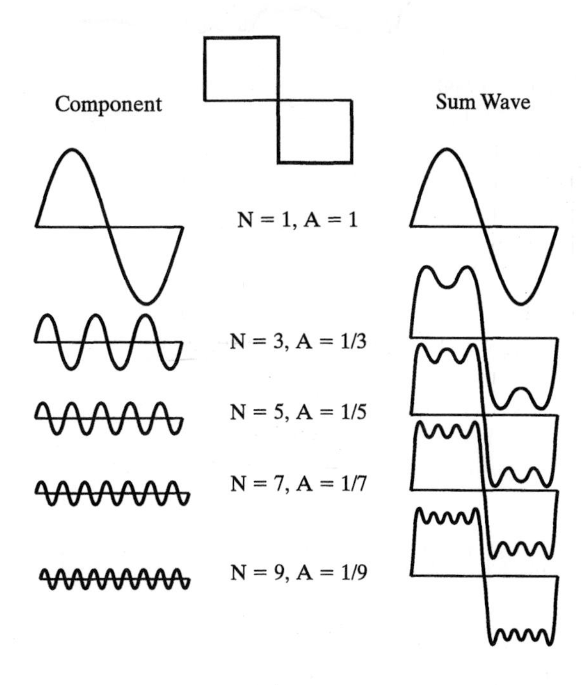

Figure 4-5 Fourier synthesis of a square wave. At the left are the successive harmonics; at the right are the sum waves, including each successive harmonic. The graph at the top is the wave being synthesized.

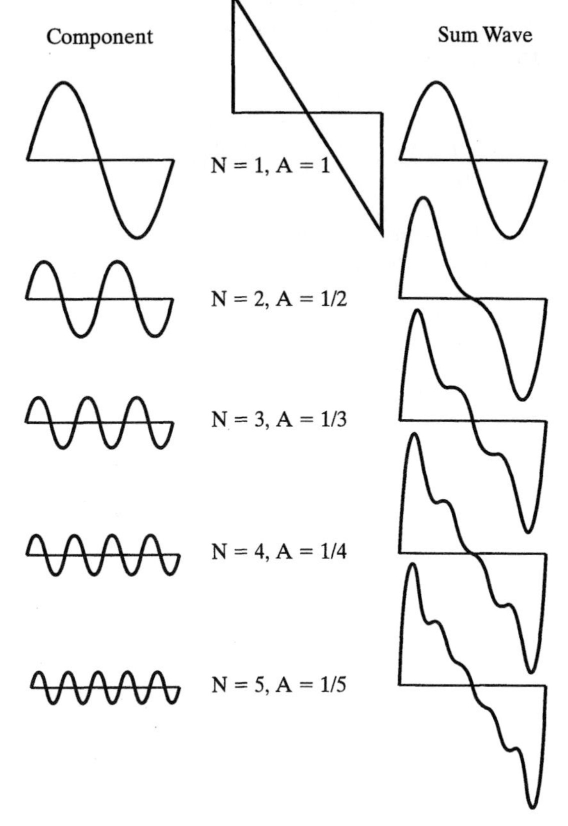

Figure 4-6 Fourier synthesis of a sawtooth wave. At the left are the successive harmonics; at the right are the sum waves, including each successive harmonic. The graph at the top is the wave being synthesized.

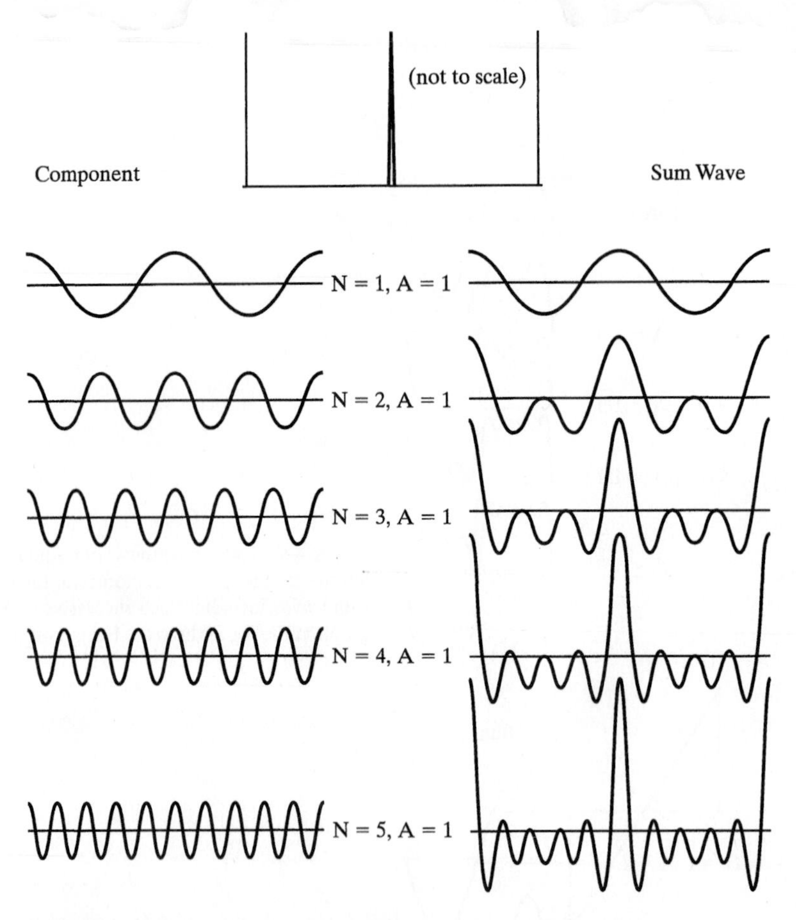

Figure 4-7 Fourier synthesis of a pulse train. At the left are the successive harmonics; at the right are the sum waves, including each successive harmonic. The graph at the top is the wave being synthesized.

Table 4-1 Relative Amplitudes of Harmonics Present in the Standard Waves

Wave	Harmonic amplitudes									
$N =$	1,	2,	3,	4,	5,	6,	7,	8,	9,	10,....
Sine	1,	0,	0,	0,	0,	0,	0,	0,	0,	0,..., $N = 1$ only
Triangle	1,	0,	$\frac{1}{9}$,	0,	$\frac{1}{25}$,	0,	$\frac{1}{49}$,	0,	$\frac{1}{81}$,	0,..., for odd N
Square	1,	0,	$\frac{1}{3}$,	0,	$\frac{1}{5}$,	0,	$\frac{1}{7}$,	0,	$\frac{1}{9}$,	0,..., for odd N
Sawtooth (ramp)	1,	$\frac{1}{2}$,	$\frac{1}{3}$,	$\frac{1}{4}$,	$\frac{1}{5}$,	$\frac{1}{6}$,	$\frac{1}{7}$,	$\frac{1}{8}$,	$\frac{1}{9}$,	$\frac{1}{10}$,..., for all N
Pulse train	1,	1,	1,	1,	1,	1,	1,	1,	1,	1,..., equal for all N

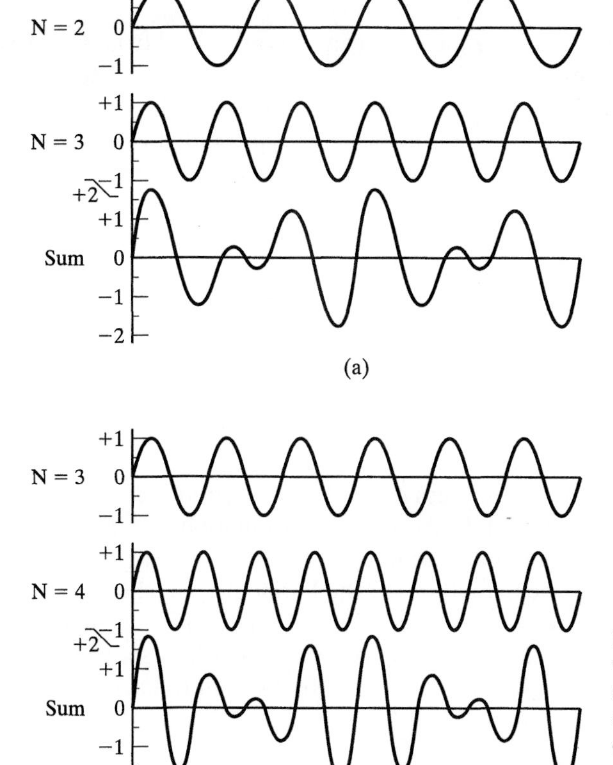

Figure 4-8 Fourier synthesis of complex waves from equal amplitudes of (a) second and third harmonics, and (b) third and fourth harmonics. Two periods of each complex wave are shown. The periodicity of both complex waves is that of the fundamental, although the fundamental is missing from the synthesis.

harmonics, with amplitudes and phases determined by the shape of the complex wave. This mathematical statement is known as *Fourier's theorem*. In general, for complicated wave shapes like the wave forms of real musical instrument tones, the amplitudes and phases are not related by any simple formula, as they are in the cases of the standard wave shapes just discussed.

Throughout the sequence of Fourier syntheses of complex waves, it was necessary to draw only one period of the complex wave and a number of periods of each harmonic equal to its harmonic number. After one period, the complex wave repeats itself, because each of the harmonics do too. As a rule, therefore, any wave containing harmonics of some fundamental frequency f will repeat itself with periodicity $T = 1/f$. This is true even if the wave has no fundamental in its harmonic structure, as shown in Figure 4-8. In these cases, although the fundamental is missing, the complex wave is periodic with the fundamental frequency, because both frequencies present in these complex waves are harmonics of the fundamental. Furthermore, owing to inherent properties of the ear to be discussed in Chap. 6, the pitch we actually hear is often that of the fundamental, although the tone quality of these notes is different when the fundamental is actually present with some significant amplitude.

4.2 Fourier Analysis and Fourier Spectra

We saw in the last section that any periodic wave can be synthesized from sine waves of the frequencies of the harmonics of the fundamental frequency, which is the frequency of the complex periodic wave itself. Because the tone quality of the complex wave is determined primarily by the amplitudes of the harmonics present, it is useful to display the harmonic content graphically. A graph called the *Fourier spectrum* is a useful means for displaying the harmonic content of complex periodic waves.

Figure 4-9 shows the Fourier spectra of two pure tones or sinusoidal waves, where the first (a) has a frequency of f_1 and an amplitude of 1, and the second (b) has a frequency of $3f_1$ and an amplitude of 0.5. These graphs illustrate the obvious fact that a sine wave is made up of just one sine wave—namely, one with its own frequency. The amplitude is in arbitrary units; what is important is the *relative* amplitudes of the harmonics. In each case shown in Figure 4-9, the frequency of the sinusoidal wave is its fundamental (f_1 and $3f_1$, respectively). Shown in Figure 4-10 are Fourier spectra of the complex waves of Figure 4-1. The graph simply presents the information that in the complex waves of Figure 4-1 there are two harmonics present, $N = 1$ (frequency f_1) and $N = 2$ (frequency $2f_1$), both equal in amplitude. Additional information—that is, the relative phases between the fundamental and the second harmonic—is missing, as in all similar Fourier spectrum graphs. Only the amplitudes of the harmonics are shown. Figure 4-11 shows the Fourier spectra of the two complex waves of Figure 4-2. Again, because no phase data are presented in this format, the spectra are identical, showing that both complex waves contain equal amplitudes ($A = 1$) of first and third harmonics.

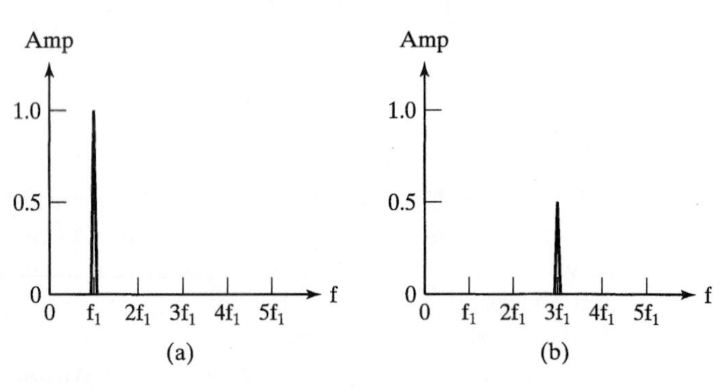

Figure 4-9 Fourier spectra of pure tones of (a) amplitude $A = 1$ and frequency $f = f_1$, and (b) amplitude $A = \frac{1}{2}$ and frequency $f = 3f_1$.

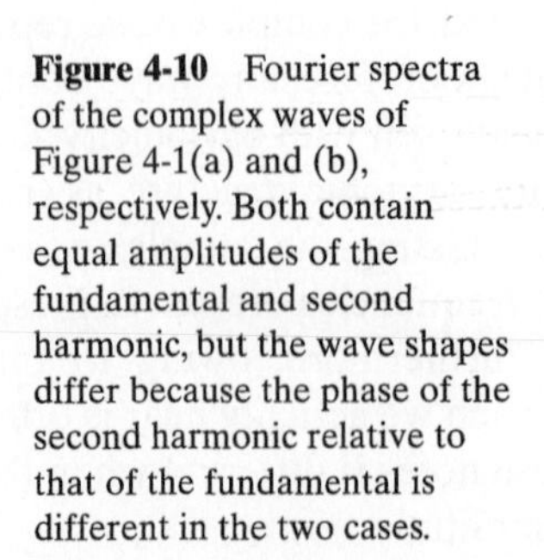

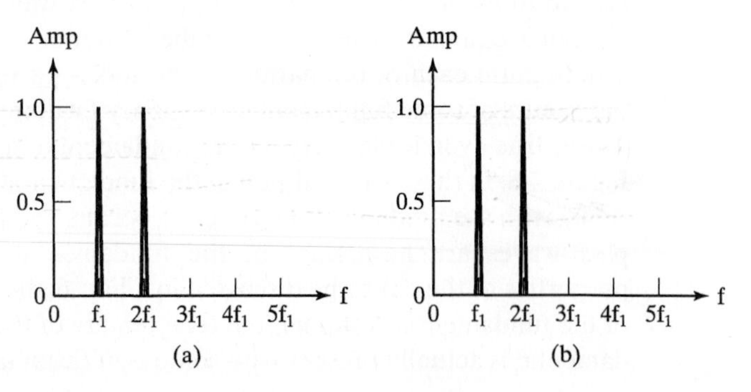

Figure 4-10 Fourier spectra of the complex waves of Figure 4-1(a) and (b), respectively. Both contain equal amplitudes of the fundamental and second harmonic, but the wave shapes differ because the phase of the second harmonic relative to that of the fundamental is different in the two cases.

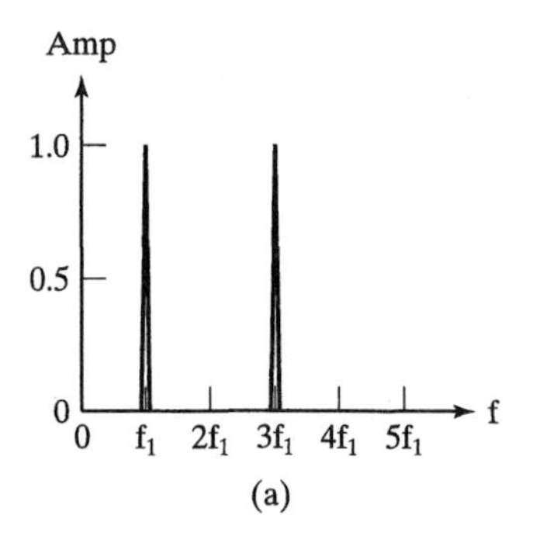

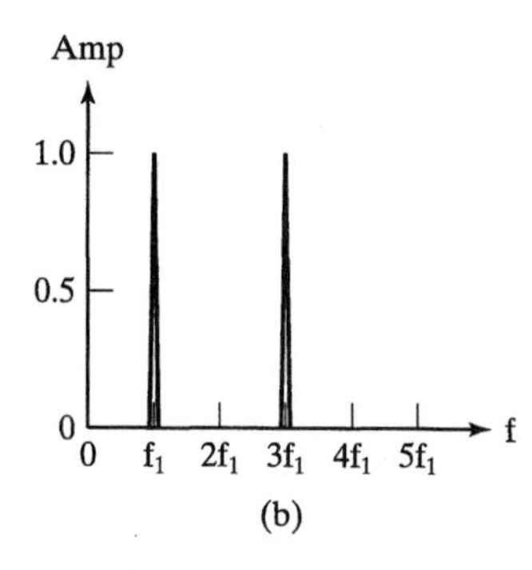

Figure 4-11 Fourier spectra of the complex waves of Figure 4-2(a) and (b), respectively. Both contain equal amplitudes of the fundamental and third harmonic, but their shapes differ because the phase of the third harmonic relative to that of the fundamental is different in the two cases.

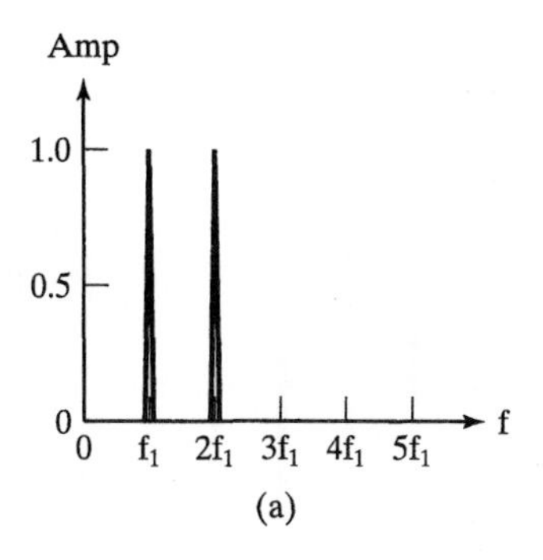

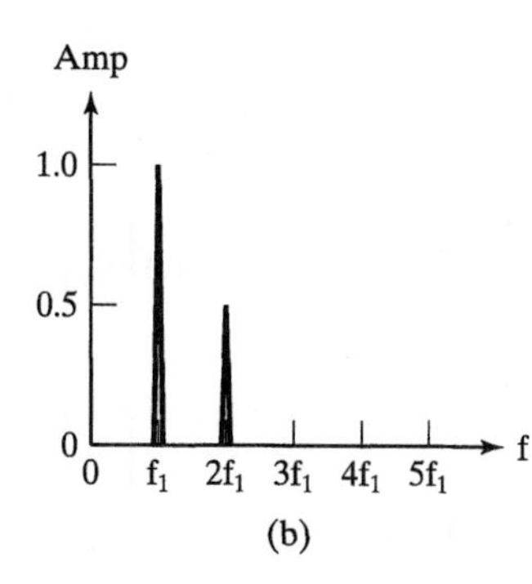

Figure 4-12 Fourier spectra of the complex waves of Figure 4-3(a) and (b), respectively. Both contain the fundamental and second harmonic, but the amplitude of the second harmonic is halved in graph (b).

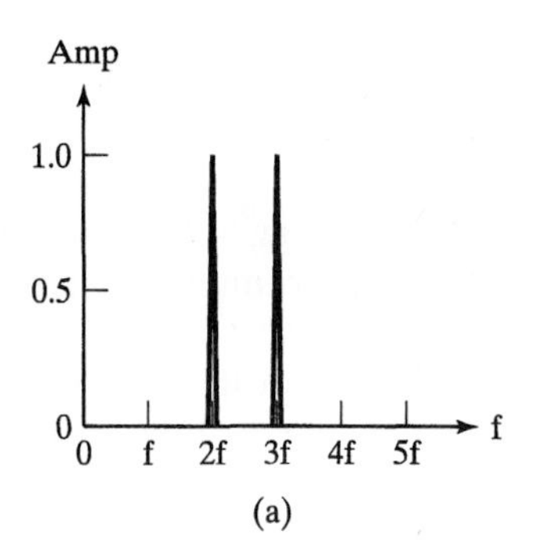

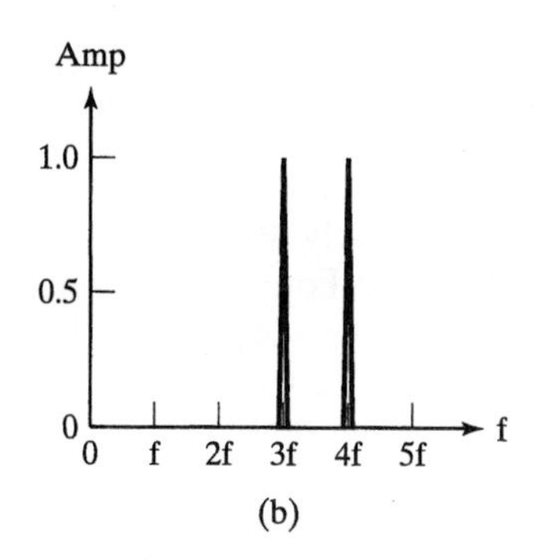

Figure 4-13 Fourier spectra of the complex waves of Figure 4-8(a) and (b), respectively.

Figure 4-12 shows the Fourier spectra for the waves in Figure 4-3. In this case, the graphs of the Fourier spectra reflect the differences in the amplitudes of the second harmonic between the two waves. Figure 4-13(a) and (b) show the Fourier spectra of the waves of Figure 4-8(a) and (b), respectively. In these cases, the fundamental frequency is absent in the complex wave and therefore does not appear as a peak in the Fourier spectrum graph. This does not violate Fourier's theorem; in this case, the amplitude of the fundamental is 0.

The Fourier spectra for the standard waves, whose harmonic amplitudes are tabulated in Table 4-1, are shown in Figures 4-14 through 4-17. These Fourier spectra present graphically the amplitudes of the harmonics present in the standard complex

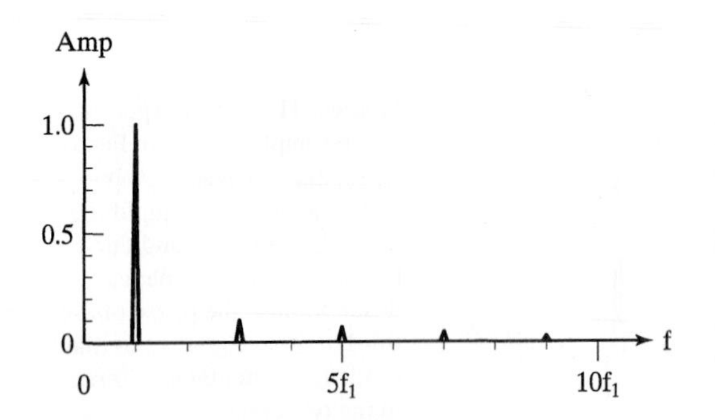

Figure 4-14 Fourier spectrum of a triangular wave of frequency f_1, up to the tenth harmonic.

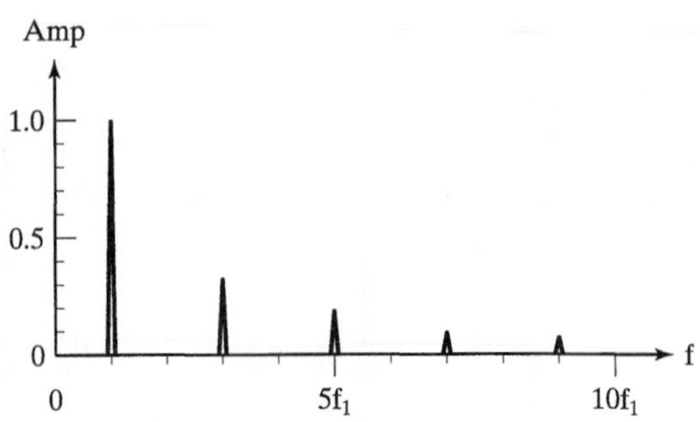

Figure 4-15 Fourier spectrum of a square wave of frequency f_1, up to the tenth harmonic.

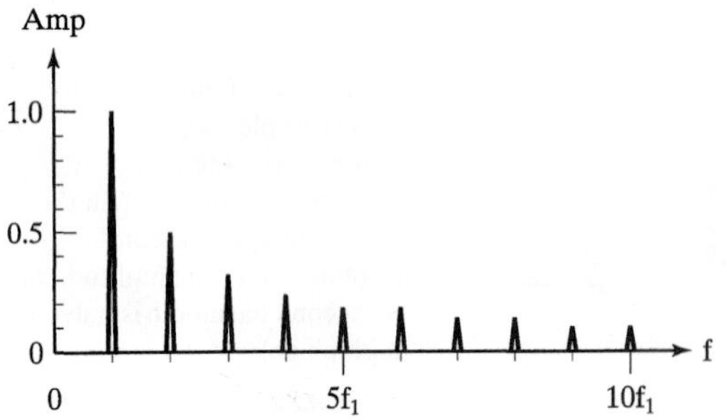

Figure 4-16 Fourier spectrum of a sawtooth wave of frequency f_1, up to the tenth harmonic.

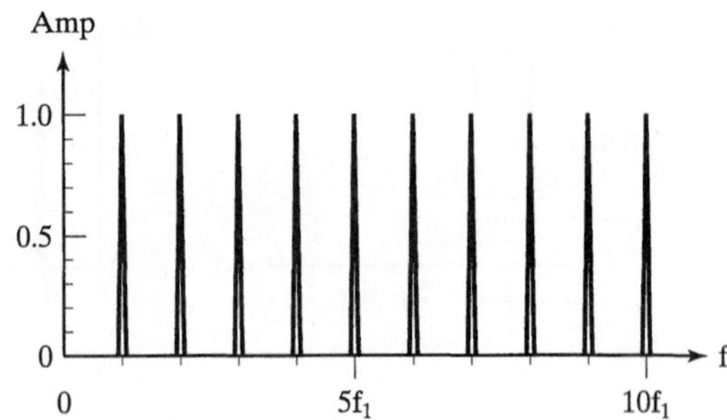

Figure 4-17 Fourier spectrum of a pulse train of frequency f_1, up to the tenth harmonic.

waves. The wave shape uniquely determines the Fourier spectrum, but because there is no phase information in the Fourier spectrum, the spectrum cannot uniquely determine the wave shape. In other words, as we have seen, many wave shapes may correspond to the same Fourier spectrum. Because the amplitudes of the harmonics have much more influence than the phases in determining the sound quality, or timbre, the Fourier spectrum tells us the most important information pertinent to the timbre of the sound wave. We can use an electronic device called a *Fourier analyzer*, or *spectrum analyzer*, to determine the harmonic content of any arbitrary musical tone or other waveform. For example, if we put a square wave of frequency f_1 into the Fourier analyzer, it will produce the Fourier spectrum of the square wave, plotted in Figure 4-15.

A very general correlation can be made between harmonic structure, as observed in the Fourier spectrum, and sound quality, or timbre. A simple sinusoidal wave (one harmonic, the fundamental) sounds pure or plain. On the other hand, as large-amplitude harmonics are added, the tone becomes richer, as in the sawtooth, or even raspy, as in the extreme case of the pulse train. Sounds between these extremes of harmonic content sound relatively more simple or rich. Square waves and other waves that have large-amplitude odd harmonics and little or no even harmonics produce a hollow or woody sound, like that of the clarinet. The triangular wave, containing only odd harmonics but with very small amplitudes, lies somewhere between the sine wave and square wave.

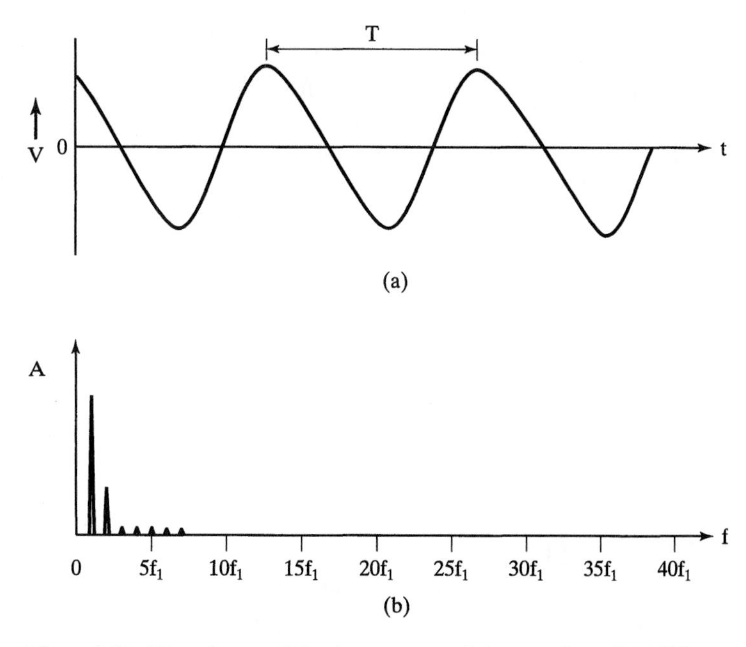

Figure 4-18 Wave form and Fourier spectrum of the note $C_5 = 523.25$ Hz played on an alto recorder.

Figures 4-18 through 4-21 show the wave forms and Fourier spectra for different notes played on a recorder, a clarinet, a violin, and a krummhorn, which are shown in Figure 4-22. The recorder tone is very simple, almost like a pure sinusoidal tone, as seen in its wave form, Figure 4-18(a). Its Fourier spectrum, Figure 4-18(b), is therefore very simple, consisting of the fundamental and only a few higher harmonics with relatively small amplitudes. On the other hand, the krummhorn has an extremely rich, reedy tone, and possesses a very complex wave form, Figure 4-21(a). Its Fourier spectrum, Figure 4-21(b), therefore contains a large number of harmonics (over 40 can be easily counted), some of which have amplitudes larger than that of the fundamental. Even if the amplitudes of the harmonics are much larger than that of the fundamental, the note sounds at the pitch of its fundamental, because all the frequencies are harmonics of that fundamental. This would be true even if the fundamental were missing.

The clarinet and the violin spectra and waveforms lie between these two extremes in complexity of wave shape (richness in timbre) and in abundance of large-amplitude overtones. The Fourier spectrum of the clarinet note (Figure 4-19(b)) has large-amplitude low-numbered odd harmonics, giving it a tone similar to that of the square wave. The sound curve of the clarinet note does not look like a square wave, because the amplitudes and phases of the harmonics are different from those of a square wave. The violin has a medium-complex sound curve, as shown in Figure 4-20(a), and therefore an intermediate number of harmonics in its Fourier spectrum, as seen in Figure 4-20(b).

From this brief study of instrument sounds and their Fourier spectra, we see that there is obviously a correlation between tones that sound pure (or simple) and their

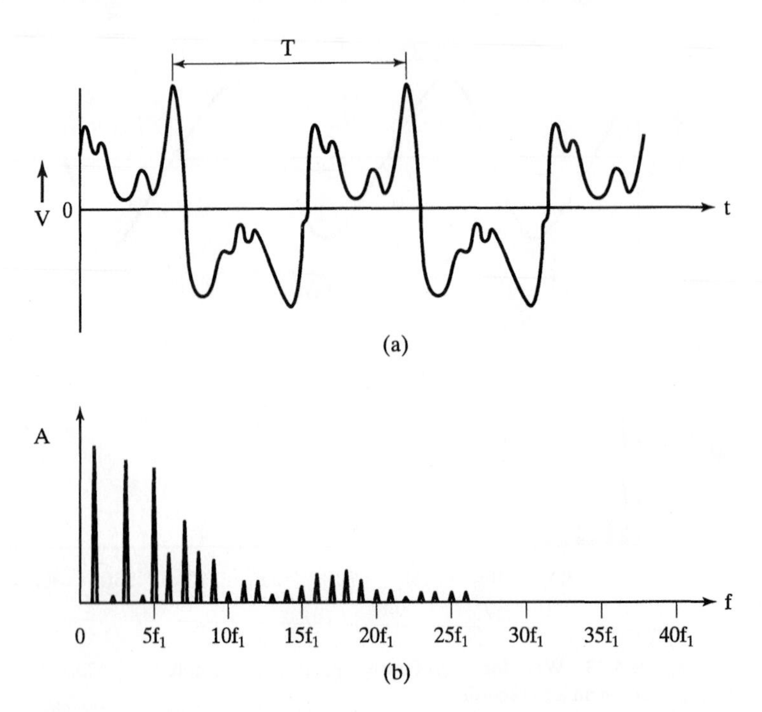

Figure 4-19 Wave form and Fourier spectrum of the note $B^{\flat}_3$ = 233.08 Hz played on a clarinet.

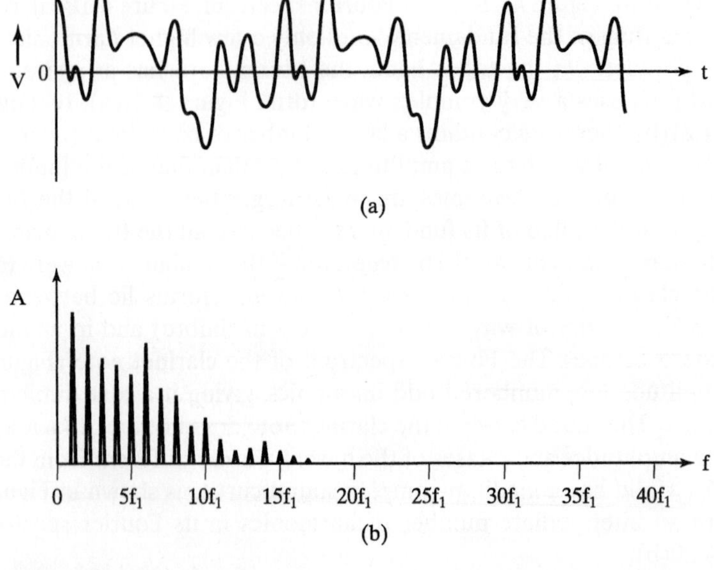

Figure 4-20 Wave form and Fourier spectrum of the note B_4 = 493.88 Hz played on a violin.

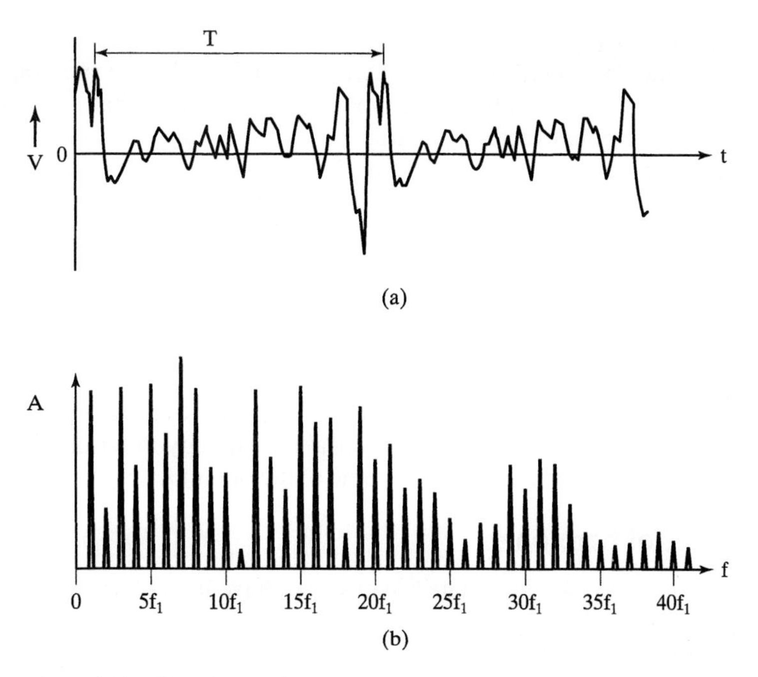

Figure 4-21 Wave form and Fourier spectrum on the note G_3 = 196.00 Hz played on a tenor krummhorn.

Figure 4-22 The instruments used to obtain the wave shapes and Fourier spectra of Figures 4-18 through 4-21. From left to right: a B-flat soprano clarinet, a tenor krummhorn, a Baroque alto recorder, and a violin.

simple Fourier spectra, which contain few or no large-amplitude harmonics. There is also a correlation between complex, rich tones and their Fourier spectra, which contain a large number of large-amplitude harmonics. It also appears that tones containing certain types of harmonic structure—characterized, for example, by odd harmonics only—can be identified by a particular timbre. However, one can deal in broad generalities only. As we shall see in the next section, any attempt to correlate Fourier spectra conclusively with instrument timbre breaks down when applied to details of subtle differences between instruments, even when the tonal variation is "obvious" to our ears.

4.3 Analysis of Tone Quality

We have studied, in a general way, the effect of harmonic content on the timbre of a musical tone. The explanation of varying musical tone quality in terms of harmonic structure is limited, and we will discuss next some of the other factors that affect tone quality.

Amplitudes of harmonics

As discussed previously, these data are basic but have limited use in explaining details.

Attack and decay transients

When a piano note is struck, a note on a wind instrument attacked by tonguing, a violin bow attack started, or any other musical sound begun, unique sounds occur that last just a moment. These sounds, called *attack transients*, possess a spectrum very different from the steady note that follows after the attack. Attack transients are very important in determining how one perceives the tone quality of an instrument; the auditory system responds to rapidly changing frequencies in the attack, and this response affects the tone quality associated with that note.

Several experiments illustrate this effect, as follows. Digital recordings are made with different instruments attacking and holding the same musical note. The digital file is then split and merged so that after the attack of one instrument the sustained note of another is heard. For example, the sustained tone of a violin can be chosen to follow the attack of an oboe. When the complete sound is played, most people will identify the sound as that of an oboe. In fact, the steady tones of most band and orchestral instruments are similar enough in their harmonic structure that the substitution of the attack of a second instrument will result in identification of the steady note as that of the instrument whose attack was used.

Another example of the effect of attack transients can be illustrated by playing a recording of piano music backward. Absence of the usual attack transients in the playback plus the unnatural increase in loudness results in a sound very different from that of a piano.

The *decay transients* of a musical tone also may have important effects on tone quality. This is especially true for some percussion instruments and plucked strings. These instruments have a common property that the vibration amplitude continually decreases after the initial attack.

Finally, it should be added that the subject of attack transients is of considerable contemporary research interest. It takes extremely costly and sophisticated electronic equipment, including high-speed computers, to analyze attack transients and reproduce common instrumental or vocal transients.

Inharmonicities

Thus far we have primarily studied only sounds that are composed of integrally related frequency components—that is, sounds whose overtone frequencies are the harmonic number times the fundamental frequency. Transients are, in general, not composed of simple harmonics only. Several exceptions to this basic rule are of importance in a discussion of tone quality.

In Chap. 1 we learned that we need a linear restoring force for SHM. For a vibrating piano string, this force results from the tension applied to the string; the tension tends to force the string back to its equilibrium position—that is, straight. There is another force also tending to make the string straight: the inherent stiffness of the metal strings. Even if there is no tension in the string, it will, when bent, tend to become straight. When the stretched string is vibrating in its fundamental mode, the effect of the stiffness is negligible, because the string does not have any sharp bends. For higher resonances, however, the bending becomes more pronounced because of the shorter wavelengths. When the piano string is played, it vibrates in many possible nodes simultaneously. Because of the extra restoring force resulting from the stiffness of the string, each successive resonance becomes higher in frequency with respect to the ideal value of Nf_1. The string stiffness causes the resonances of the string to deviate from an exact, integral relationship of f, $2f$, $3f$, and so on. This *inharmonicity* is more extreme in treble strings than in bass strings because a given displacement of a short string produces a greater bending than in a long string. In fact, in most cases the sixteenth harmonic of a piano note (up four octaves from the fundamental) is about one half-step higher in pitch than the exact harmonic, a significant amount. The high notes on a piano are usually tuned slightly high and the low notes tuned slightly low to eliminate beats and provide a better relationship between the adjacent octaves. Piano tone quality is believed to be created by a combination of its unique attack and decay transients, the inharmonicities of the overtones of the strings, and the detailed tuning relationship between the strings tuned to the same note.

In the case of the piano, the inharmonicities are generally not obvious to the listener, but can be revealed using electronic equipment. Even an untrained listener can hear the inharmonicity in some tuned percussion instruments, such as chimes. The frequencies of the overtones of such systems are not even close to bearing an integral relationship. The "note" we hear being played on a tuned drum like the timpani is usually the *second*-lowest-frequency standing wave, or resonance, in the system, rather than the fundamental. The fundamental of the chime can often be heard as a sort of dull thud at an intensity and frequency well below that of the note the instrument is sounding.

Two features are important in understanding and electrically reproducing (e.g., on a synthesizer) tuned percussion sounds, such as those of a chime. First, and possibly most obvious, is the sudden attack when the chime is struck, followed by the slow decay of the sound. Second is a unique set of nonharmonic overtones, including the discernible fundamental *below* the "pitch" of the bell.

The tabla, an Asian Indian drum often played as accompaniment to the sitar, is unique among tuned drums. Its drumhead is specially weighted to force several of the resonances to have an integral relationship and to remove the loud low fundamental sound. As a result, the tabla has an especially clear, ringing tone with a well-defined pitch.

Further discussion of inharmonicities of tuned percussion sounds will be given in Chap. 14 and when we study the electronic synthesizer—in particular, the ring modulator.

Formants

Even after numerous studies have been done on harmonic content, transients, and inharmonicity, significant unsolved problems remain in our understanding of tone differences between various instruments. Physicists like to stand back, view the phenomenon from some distance, and draw general conclusions from the data. For example, suppose a series of Fourier spectra are plotted for every note on some instrument, and harmonics that lie within a certain frequency region are found to be emphasized relative to the other harmonics. Such a frequency region is called a *formant region* of the instrument. A formant region can exist even if there is no other correlation or apparent similarity between the Fourier spectra of any groups of notes. In that case, the formant alone may be responsible for the tone quality.

Such a formant is believed by some to be responsible for the tone of the bassoon, where there is no apparent similarity between the Fourier spectra of different bassoon notes, other than an increase in amplitude of the harmonics in a high-frequency region. The resonator on an English horn may produce a formant at a high frequency that might contribute to the unique plaintive sound of the instrument.

Each human voice attains its unique tone quality (either speaking or singing) from the particular resonant frequencies of the *resonant cavities*: the larynx, the pharynx, the mouth, and the nasal cavity. The range, or group of frequencies emphasized in each individual's voice, and therefore the vocal quality, is largely dependent on details of the size and shape of these cavities, as well as the vibrations of the vocal folds. One can therefore view the particular groups of emphasized frequencies as formants leading to the unique quality of each individual's voice. Furthermore, each vowel sound has a particular character, which is determined by the resonant cavities that dominate the sound production or share in production of the sound, such as the nasal cavity for long "e," the mouth for "oo," and the throat for "uh." We can think of certain vocal formants being responsible for these vowel sounds, independent of the frequency or pitch at which the sound is uttered. We shall return to the human voice in Chap. 6.

Vibrato

Vibrato (periodic changes in the pitch of a musical tone) also adds a distinctive flavor to the tone. The vibrato of a singer's voice, for example, aids significantly in distinguishing the voice from other musical sounds. The term *vibrato* in general use refers not only to periodic changes in pitch, but also to periodic changes in amplitude, which should more correctly be called *tremolo*. The "diaphragm vibrato" of a flute player is close to pure tremolo; the vibrato obtained when a trombone player wiggles the slide in and out is almost a pure pitch vibrato. Singing vibrato is actually a mixture of true vibrato and tremolo. Vibrato on a violin or other string instrument is close to pure pitch vibrato.

Chorus effect

When many identical instruments play the same part in unison, the sound attains a quality considerably different from that of any of the individual instruments. This phenomenon is called the *chorus effect*, and results from the superposition of many similar tones with random relative phases and slightly different frequencies and tone qualities. It is particularly evident in the string sections of a symphony orchestra. The sound of the solo violin playing in a concerto is different from the sound of the whole violin section that accompanies it, and the two are therefore clearly distinguishable.

4.4 Resonance Curves and Musical Sound Production

There are two fundamental steps in the production of a musical sound: (1) production of the physical vibrations and (2) use of resonators to increase the amplitude of certain overtones. In this section, we shall see how the spectrum of the initial vibrations and the resonator characteristics determine the sound.

An elementary optics experiment is to pass a beam of white light (say from the sun or a very hot lamp) through a prism and observe that the white light is separated into its component colors, as shown in Figure 4-23. The reverse process can also be demonstrated: superposition of all the colors in the visible region results in white light. The audio analog to white light is white noise, which consists of all the frequencies of the audible spectrum superimposed with equal intensities; the graph of white-noise intensity versus frequency is shown in Figure 4-24. This curve (line) shows clearly that all frequencies are present and that all have the same intensity. White noise sounds like the static between stations of an FM radio or perhaps like rushing water; it is produced electronically in a synthesizer white-noise generator. A similar sound results from blowing into a flute mouthpiece or occurs when the air rushes out of the windway of a recorder and hits the sharp, wooden wedge near the mouthpiece.

A variation on white noise is called *colored noise* or *filtered noise*. This is a general label for any noise that does not have all audio frequencies represented in nearly equal intensities but instead has a very loosely defined pitch center caused by emphasis of some frequency range or ranges. Colored noise, like white noise, has a continuum

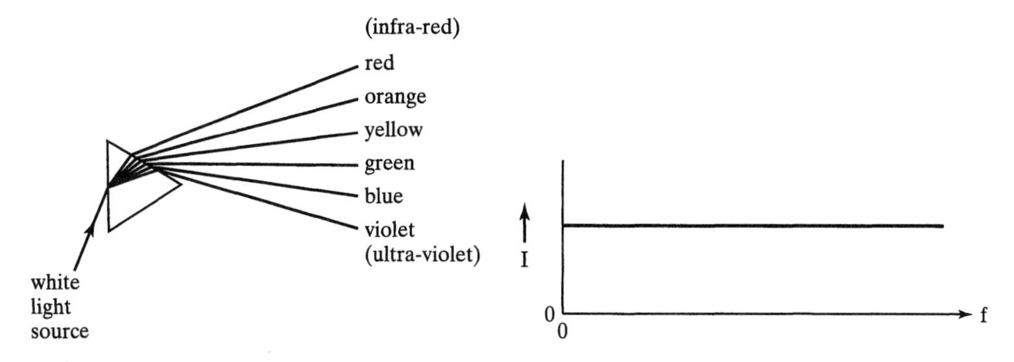

Figure 4-23 Separation of white light into its component colors using a prism.

Figure 4-24 Spectrum of white noise.

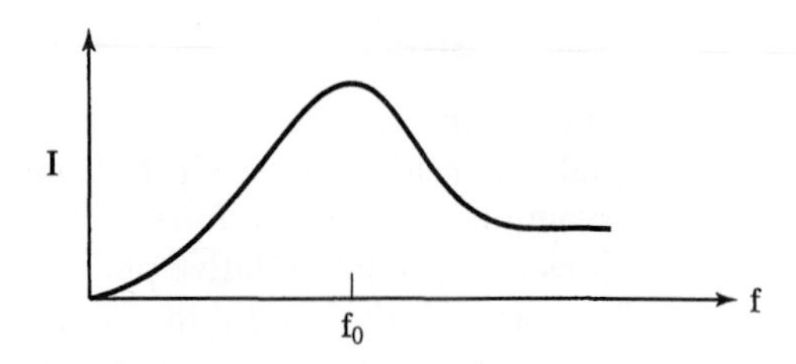

Figure 4-25 Spectrum of a type of colored noise.

of frequencies present, rather than only the frequencies in some overtone series or some other set of particular frequencies. The Fourier spectrum of one type of colored noise is shown in Figure 4-25. This noise would sound like a very fuzzy whistling tone of frequency f_0. An example of colored noise would be the howling wind, which has in its spectrum virtually the entire audible range, with an emphasis on the frequency region around the frequency of the howling tone. This region is, in general, rather poorly defined. It should be emphasized that the graph of Figure 4-25 illustrates only one type of colored noise and that the term *colored noise* applies in general to steady-state nonmusical sounds that contain a continuum of frequencies in unequal intensities.

Figure 4-26 shows the spectrum (a) and wave shape (b) of the broadband noise obtained by blowing gently across a microphone. No periodicity is discernible in the wave shape. This illustrates the importance of using a windscreen on the microphone both outdoors, to protect against wind noise, and when plosive word sounds directly impinge on the microphone.

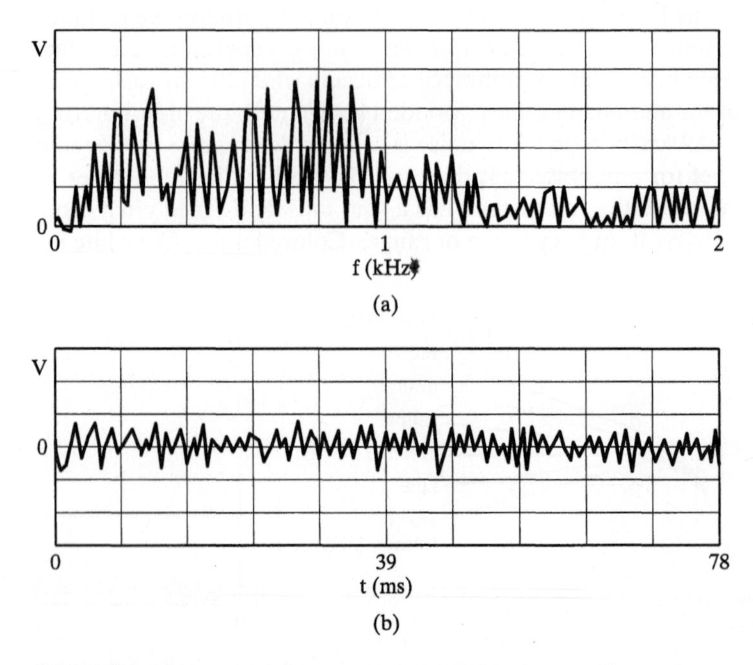

Figure 4-26 The spectrum (a) and wave shape (b) of the broadband noise obtained by blowing gently across a microphone.

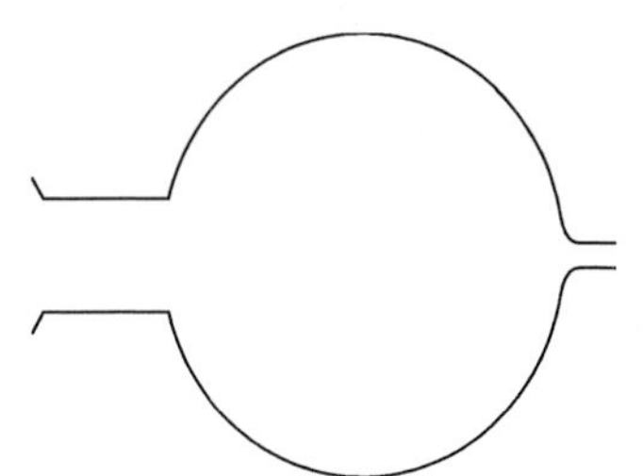

Figure 4-27 Cross section of a (spherical) Helmholtz resonator.

In electrical and audio engineering usage, the term *pink noise* means specifically noise with the same energy content in each octave of musical interval. This is equivalent to noise whose power per unit frequency drops off with frequency at a rate of 3 decibels per octave, a close approximation to the normal distribution of energy in orchestral and much other acoustic music. (We shall discuss the decibel intensity scale in Chap. 6.) This type of noise is useful in some tests of audio components, because it approximates the spectrum of much music.

White noise and colored noise are not generally classified as musical sounds. To produce a musical tone from white noise, a device is needed that will suppress most frequencies and strongly resonate those few that are to make up the desired musical tone.

The simplest such device, the Helmholtz resonator, named after the famous German physicist Hermann von Helmholtz (1821–1894), is illustrated in Figure 4-27. It is a simple spherical cavity with a wide mouth, like the resonator of Figure 4-27, with or without a small nipple on the side opposite the mouth. If a burst of air is blown into the cavity through the large neck, it will compress the air inside the cavity, which will then expand and rush outward, creating a low pressure inside the cavity. Outside pressure will then force the air back into the cavity at greater than atmospheric pressure. In this way the pressure in the cavity rapidly rises above, then drops below, the normal equilibrium atmospheric pressure. In fact, this resonance condition occurs for any such resonator at a frequency determined by the size and shape of the resonator. The small nipple on the right side of the Helmholtz resonator does not change the resonant frequency but is inserted into the ear to allow the listener to hear the resonance. Helmholtz used a large set of these resonators in the nineteenth century—before the era of electronic Fourier spectrum analyzers—to verify the existence of harmonics in complex tones. By holding successively smaller (higher-frequency) resonators to his ear with a musical note playing into the large opening, Helmholtz heard an increase in the amplitude of any frequency that was present in the harmonic structure of the instrument. Thus he could roughly determine the Fourier spectrum of the note.

The Helmholtz resonator is very important, because it possesses a single, low, isolated resonant frequency, as opposed to stretched wires, open and closed tubes, and other complex vibrating systems, which possess multiple closely spaced resonant frequencies. When a Helmholtz resonator is excited by white noise, the resonator causes an increase in amplitude of the very narrow band of frequencies near to its resonant frequency.

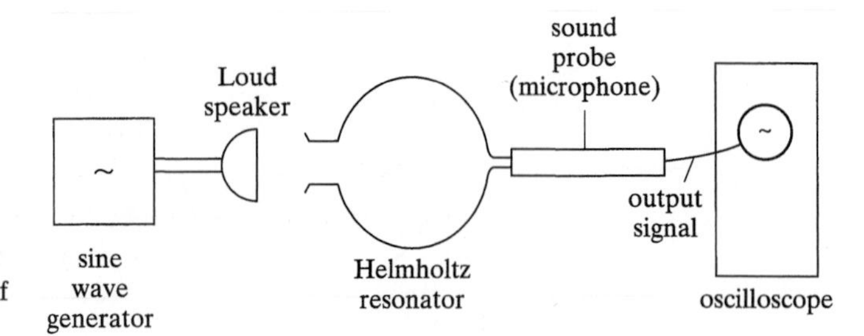

Figure 4-28 Equipment used to obtain the resonance curve of a Helmholtz resonator.

We can perform an experiment to demonstrate this resonant property using the equipment shown in Figure 4-28, where, instead of the ear, a sound probe and oscilloscope measure the resonant response. Rather than using a white-noise generator, we use a sine-wave generator, which produces all frequencies with the same amplitude, one at a time. As the frequency of the sine-wave generator is swept slowly across the audible range, the amplitude of the signal picked up by the sound probe will reach a maximum at the resonant frequency of the Helmholtz resonator and then decrease. The graph of this amplitude versus frequency is called the *resonance curve* for this Helmholtz resonator; it is shown in Figure 4-29. The graph clearly shows a single, rather narrow, resonance that occurs at the frequency that is produced by blowing across the opening of the resonator, just as one would blow across the mouth of a soda bottle.

In addition to its historical importance, the Helmholtz resonator has contemporary application to string instruments. The air cavity within the body of any string instrument acts as a type of Helmholtz resonator, creating emphasis of notes near its resonant frequency. In the case of the contemporary violin family, placement of the frequency of this "air resonance" is extremely important in producing uniform tone quality and loudness for all the notes in the low-frequency range of the instrument. The bottles of a bottle or jug band, particularly the bass bottles, also behave like Helmholtz resonators. In Chap. 7 we shall see that certain loudspeaker enclosures behave like Helmholtz resonators.

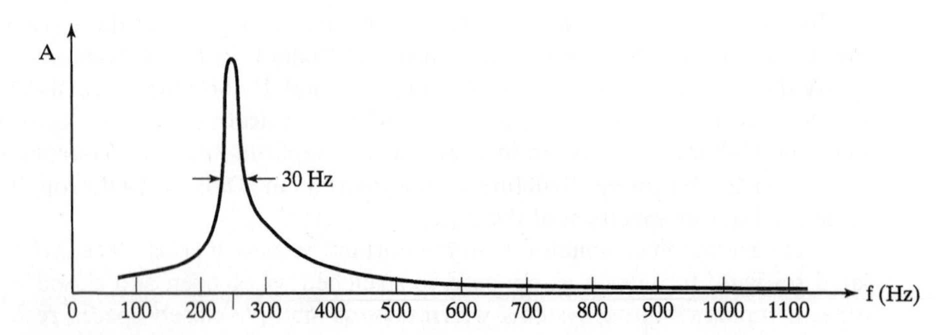

Figure 4-29 Resonance curve of a Helmholtz resonator whose resonant frequency is 250 Hz. The Helmholtz resonator has a single, narrow low-frequency peak with no other resonance structure below about nine times its fundamental frequency.

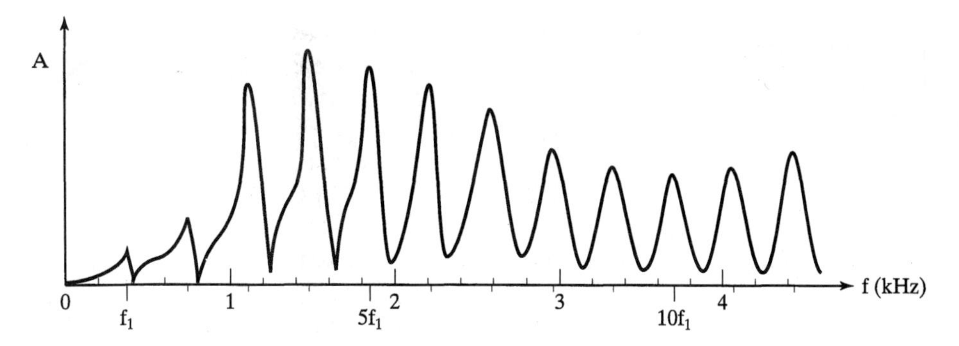

Figure 4-30 Resonance curve for an open tube whose resonant frequency f_1 is about 370 Hz. Harmonics of the fundamental frequency f_1 are indicated by the marks below the horizontal axis. All harmonics are present.

Drawing the resonance curves for other resonant systems is a convenient way to summarize the basic acoustical property of the system—that is, how the system responds to a stimulus at any frequency in the audible range. Figure 4-30 shows the resonance curve for an open tube. This graph reminds us that an open tube resonates at all harmonics of the fundamental frequency; it also tells us how well it resonates at each of the harmonics, which is important in determining the amplitudes of the various harmonics and therefore the tone quality in this "instrument." The closed tube can support only odd harmonics, as shown in the resonance curve of Figure 4-31. In both of these cases, the amplitudes, frequencies, and widths of the resonances are determined by the length and diameter of the tube. Resonance curves for real musical instruments tell us what overtones are possible in the sound of the instrument. Note that the resonance curve and the Fourier-spectrum curves are not the same, although they may be plotted on the same graph. The resonance curve tells which harmonics could be present in the tone of an instrument, if they were excited, whereas the Fourier spectrum gives the actual harmonic structure of a musical tone. The resonance curve describes the physical system; the Fourier spectrum describes the tone emitted.

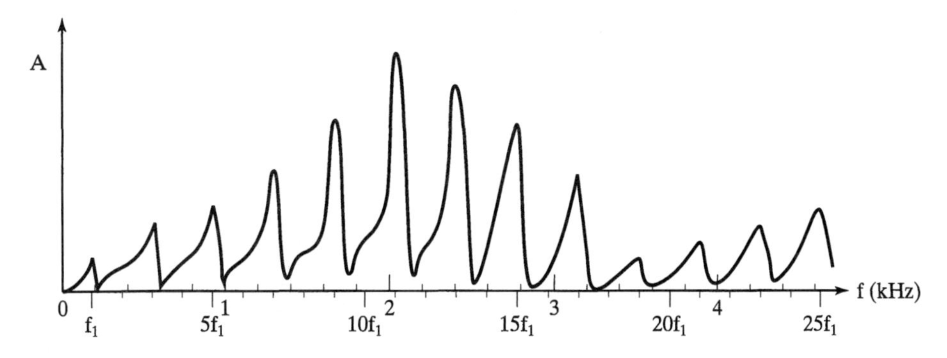

Figure 4-31 Resonance curve for a closed tube whose length is the same as that of the open tube of Figure 4-30, giving a fundamental frequency f_1 of about 185 Hz. Harmonics of the fundamental frequency f_1 are indicated by marks below the horizontal axis. Only odd harmonics are present.

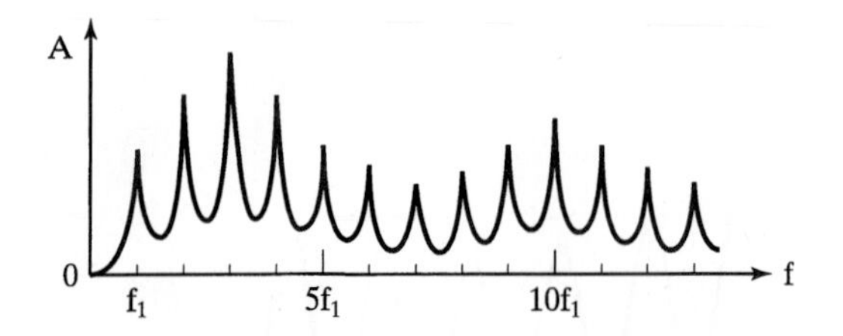

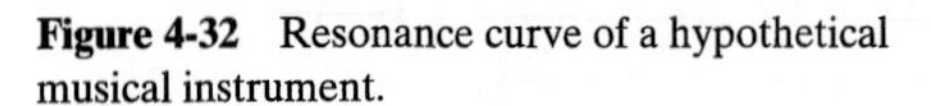
Figure 4-32 Resonance curve of a hypothetical musical instrument.

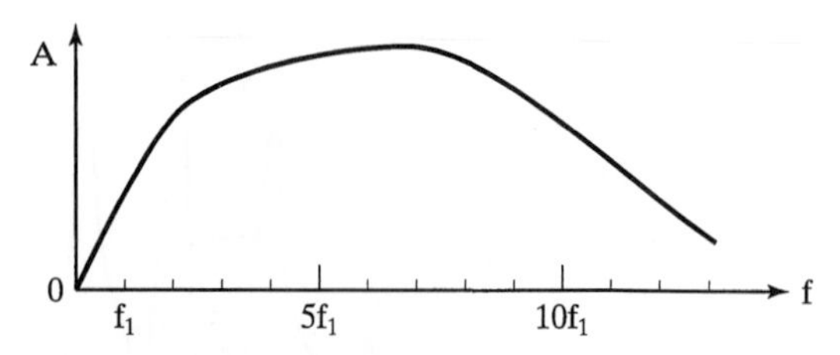

Figure 4-33 Noise spectrum exciting the hypothetical instrument whose resonance curve is shown in Figure 4-32.

Consider the hypothetical musical instrument whose resonance curve is shown in Figure 4-32. Because all harmonics of the fundamental frequency f_1 are present in the resonance curve, this instrument is a member of the open-tube type. Now suppose that this instrument is excited by the noise spectrum given in Figure 4-33. The strength of each harmonic present in the complex tone produced by the instrument will be a sort of product of the amplitude of the resonance curve at that frequency and the amplitude of the noise spectrum driving that frequency component, as shown in Figure 4-34. Because of the complicated interaction between the air column and the noise source, this product will only approximate the actual resulting spectrum. The harmonics in the resulting spectrum tend to be narrower and of higher amplitude than the simple product would predict, because there is always some interaction between the oscillating air column and the noise source. For example, both the noise spectrum and resonance curve amplitudes are greater at $3f_1$ than at $2f_1$ or f_1. Consequently, the amplitude of the Fourier spectrum is quite a bit larger at $3f_1$ than at $2f_1$ or f_1. Even though the noise spectrum between $7f_1$ and $10f_1$ is falling in amplitude, the increase in the amplitude of those frequencies in the resonance curve results in the amplitudes of components of the Fourier spectrum being approximately equal. In general, if either the noise spectrum or the resonance curve is zero at some harmonic frequency, the Fourier component at that frequency will be absent, or have zero amplitude. Although this procedure is neither quantitative nor rigorous, it does illustrate the qualitative relationship between the resonance curve, the noise spectrum, and the Fourier spectrum.

In reality, the production of sound by exciting resonances in musical instruments is more complicated. Even in the "simple" case just discussed, there is a

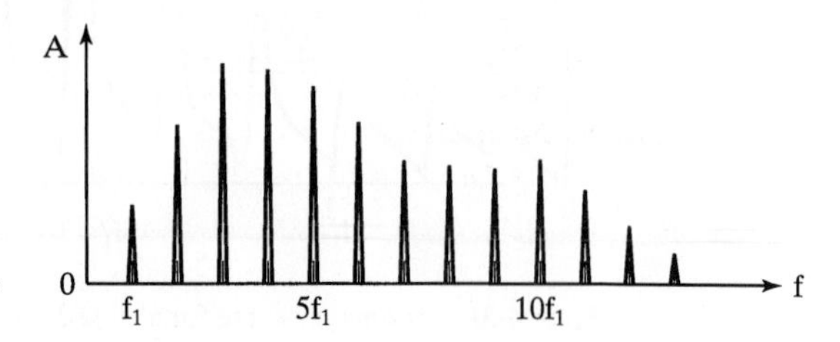

Figure 4-34 Fourier spectrum of the hypothetical instrument when the noise spectrum of Figure 4-33 excites the note whose resonance curve is shown in Figure 4-32.

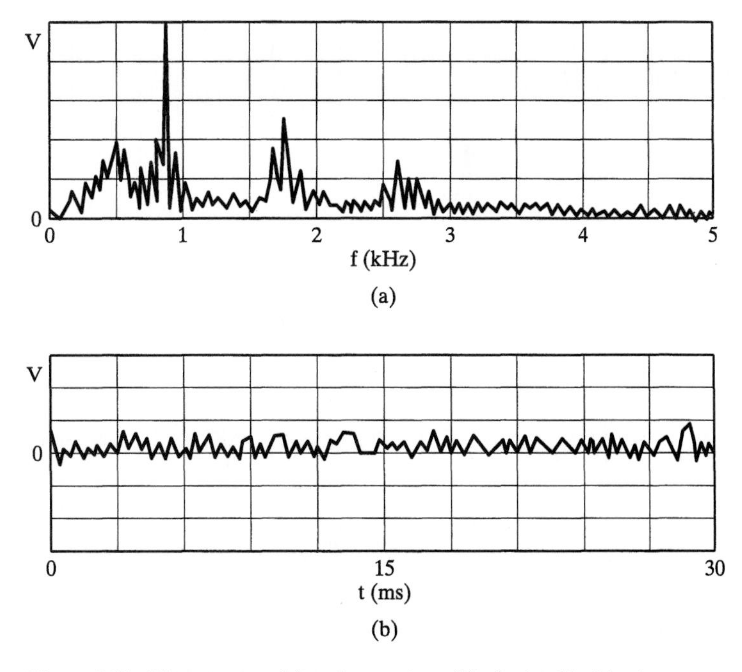

Figure 4-35 The spectrum (a) and wave shape(b) obtained by blowing across the end of an open tube with fundamental frequency 880 Hz. The source of the note produced is seen in the spectrum, although no discernible periodicity is observed in the wave shape.

feedback mechanism by which the harmonics present are strengthened in intensity and narrowed in frequency. This process happens very quickly and becomes part of the attack transient. The spectrum of the sound source can be affected by the standing wave. A more complete treatment of this topic will be given in Chap. 10, when we discuss the woodwind instruments.

An example illustrating the production of a "musical tone" by a resonant system with feedback driven by broadband noise involves blowing across the end of an open tube. Figure 4-35 shows the spectrum (a) and the wave shape (b) of the sound obtained by blowing across an open tube about 1.25 cm in diameter with a resonant frequency of 880 Hz (about 17.5 cm long). Because this is a very primitive musical instrument, it is not very efficient in feedback, so much noise remains in the sound. In fact, the periodicity cannot even be seen in the wave shape. However, our ears easily hear the 880 Hz note, the existence of which is clearly verified by the spectrum, which shows three harmonics.

Another illustration of these concepts involves the sound produced when water is poured into a tall cylinder. The sound of the gurgling water is nearly white noise, containing almost all frequencies of the audible spectrum, similar to the noise of Figure 4-26. The cylinder is a closed tube, with the rising water forming the closed end, and at any given

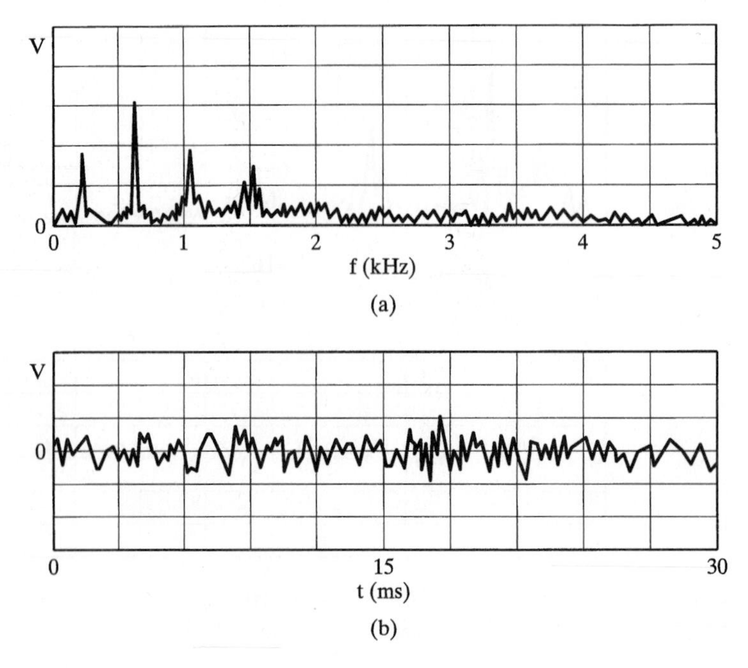

Figure 4-36 The spectrum (a) and wave shape (b) obtained at one point in time while pouring water into a 1000 ml graduated cylinder. The harmonics of the resonant tube can be seen in the spectrum, although no periodicity is observed in the wave shape because of the gurgling noise.

instant of time has a resonance curve consisting of odd harmonics of its fundamental f_1, as in Figure 4-31. The sound produced by the gurgling noise, with resonances from the closed tube superimposed on the noise, is shown at one particular time in Figure 4-36. As water fills the cylinder, the noise remains roughly the same, but f_1 and all of its odd harmonics increase in frequency because the length of the resonant air column decreases. The overall sound consists of a reasonably well-defined pitch center, composed of a fundamental and some odd harmonics superimposed on a continuous whooshing, gurgling noise. As the water fills the cylinder, all the wavelengths of the resonances decrease, and the pitch rises. Although the wave shape of Figure 4-36(b) shows no apparent periodicity due to the gurgling noise, the spectrum of Figure 4-36(a) at the same instant of time clearly indicates the source of the readily observed rising tone created in the resonant cylinder.

Figure 4-37 shows how the frequencies of the harmonics rise as a function of time. In this *sound spectrogram*, the frequencies present in the sound are plotted along the vertical axis, and time is plotted along the horizontal axis, with the darkness of the plot representing greater intensity for any particular frequency component. The individual harmonics of the resonant air column are clearly visible, superimposed on the broadband noise created by the gurgling water. The numbers at the left edge of the spectrogram identify the odd harmonics present in the sound. We will use sound spectrograms in more detail in our study of the voice in Chap. 6.

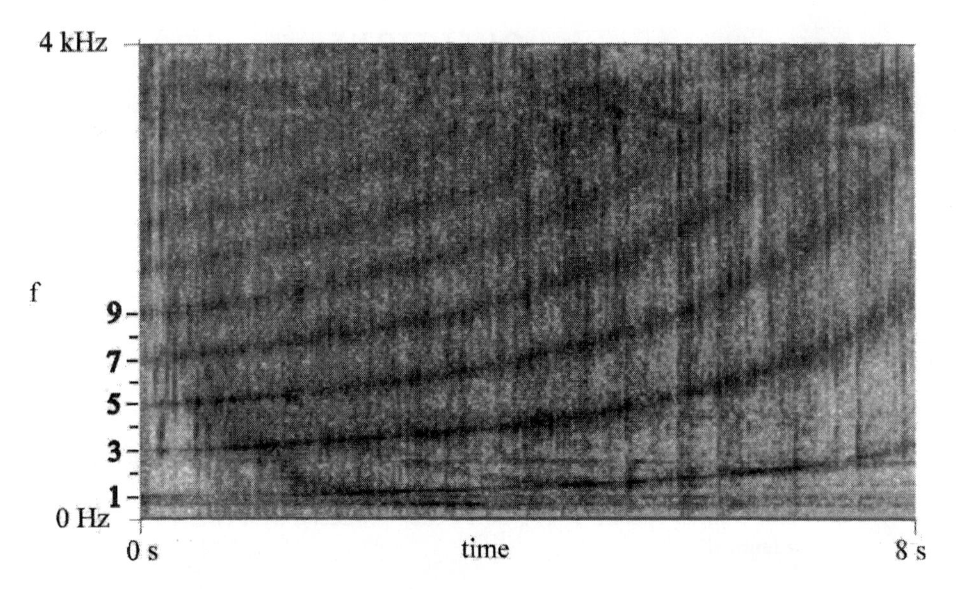

Figure 4-37 Spectrogram of water being poured into a one-liter graduated cylinder. Note the missing even harmonics of the closed tube and the increasing frequencies of the harmonics as the length of the air column decreases. (Luke Keller)

SUMMARY

(***Section 4.1***) **Fourier synthesis** is the process of producing **complex waves** by adding together all of their harmonics with the appropriate amplitudes and phases. The periodicity of the complex wave is the same as the periodicity of the fundamental, even if the fundamental does not exist in the complex wave. (***Section 4.2***) **Fourier analysis**, or **spectrum analysis**, is the process of extracting the amplitudes of the harmonics mathematically from the complex wave. The **Fourier spectrum**, a graph of the amplitudes of the harmonics as a function of frequency, does not contain information about the relative phases of the harmonics; knowledge of relative phases is necessary to reconstruct the complex wave from its spectrum. Simple waves, like the sine wave or the recorder, contain few harmonics, whereas very rich sounding waves, like the pulse train or the krummhorn, have many harmonics. Waves like the square wave or that produced by the clarinet emphasize odd harmonics, leading to a "woody" or "hollow" tone. The relative phases of the various harmonics have virtually no effect on the timbre of complex waves. (***Section 4.3***) Several factors help to determine the tone quality or timbre of a musical sound: amplitudes of harmonics, attack and decay **transients**, **inharmonicities**, **formants**, **vibrato**, and the **chorus effect**. (***Section 4.4***) **Resonance curves** of systems, such as tubes, plates, or musical instruments, tell us the frequencies of the modes at which the system can vibrate. **White noise** contains all audio frequencies with equal intensity. **Colored noise** has mostly unequal intensity for all frequencies. **Pink noise** drops off at a rate of 3 dB per octave. In a real musical instrument, noise or another source of energy applied to the system excites the modes of oscillation. This source, along with a feedback mechanism, will typically drive the various modes of oscillation to determine the harmonic content of the instrument.

QUESTIONS

1. Draw five identical pairs of graph axes numbered 1 to 5; one of each pair should have its horizontal axis labeled in milliseconds; and the second should have its horizontal axis divided into 100-Hz segments of frequency.

   a. On the first of each pair draw the following waves:

      i. a sine wave with a period of 10 ms
      ii. a triangle wave with a frequency of 200 Hz
      iii. a square wave with a period of 20 ms
      iv. a sawtooth wave with a period of 5 ms
      v. a pulse train with a frequency of 300 Hz

   b. On the second of each pair, draw the Fourier spectrum of each of these waves. Assume the amplitude of the fundamental frequency is 1.0 in each case.

2. The Fourier spectra of two notes, each produced by a different musical instrument, are identical.

   a. Will the notes sound identical, similar, or possibly very different?
   b. Will the waveforms look identical, similar, or possibly very different?
   c. What is Ohm's law of hearing and how can it be applied in this case?
   d. It is sometimes difficult to tell whether a certain note is being played on a violin, a viola, or a cello. Why is this?
   e. As the music continues, you can finally distinguish which instrument is being played. What might aid in this identification?

3. List six characteristics of musical sounds that have a bearing on their tone quality. State what each means, and give some examples in which each might be of use.

4. Why does one loud violin sound different from ten violins playing in unison such that the total sound-intensity level is the same as for the single violin?

5. **a.** What are white noise and colored noise? Give examples and describe how they sound.
   b. Draw graphs of intensity versus frequency for white noise and for pink noise.

6. **a.** What is a Helmholtz resonator?
   b. What is unique about the resonance curve for a Helmholtz resonator?
   c. What was its original use?
   d. Give examples of resonators that are similar to a Helmholtz resonator.

7. **a.** The clarinet is a cylindrical reed instrument whose reed end acts acoustically like a closed end. What harmonics might one expect to be emphasized in the resonance curve of the clarinet?
   b. The flute has the mouthpiece hole at one end and finger holes (or open tube) at the other end. What harmonics might one expect to be emphasized in the resonance curve of the flute?

8. a. A flute player plays a note. Then, while keeping her fingers in the same positions, she blows harder into the mouthpiece. The note sounded is one octave higher than the first one. Why is it an octave higher?
   b. When a clarinetist blows extra hard, while keeping his fingers in the same position, he "overblows" at an octave and a fifth. Why an octave and a fifth?

9. a. What is a formant?
   b. The human vocal system acts like a closed tube about 17.25 cm long. What are the frequencies of the first two formant regions?

10. a. Discuss the basic relationship between the resonance curve and the Fourier spectrum for a note produced on a wind instrument.
    b. How are the applied noise spectrum and the resonance curve used explicitly to determine the Fourier spectrum of the note produced?
    c. Draw two different noise spectra and two different resonance curves.
    d. Draw the Fourier spectra of the four possible notes produced by using each of the two noise spectra with each of the resonance curves.

11. Suppose that you blow on an open tube 20 cm long and about 2 cm in diameter to produce a musical tone. Draw a resonance curve that might belong to that tube, and a Fourier spectrum of the sound that might be obtained by blowing into one end across the tube edge. Draw similar curves for a tube 20 cm long and about 5 cm in diameter. Based on your knowledge of the physics of standing waves in tubes, speculate on possible differences between the curves. Obtain metal tubes the same length but differing in diameter and experimentally determine the Fourier spectrum of each tube. How does reality compare with your best guess? Try to explain the differences.

12. Obtain a computer with a Fourier synthesis program. Carry out the synthesis of the standard waves. Draw other periodic wave shapes and then try to synthesize them with your program.

13. Obtain a computer with a spectrum-analysis program. Experiment with a variety of musical instruments. Verify, and if possible extend, any conclusions regarding the relationship between the sound quality, the wave shape, and the spectrum.

14. Obtain a computer with a spectrum-analysis program. Locate your vocal formants for a number of different vocal sounds.

15. Obtain a computer with a sound-spectrogram program. Compare the spectrograms of different voices saying the same vowel sounds or words. Note the similarities and the differences.

PROBLEMS

1. What are the first four resonant frequencies of a closed tube 50 cm long?

2. What are the first four resonant frequencies of an open tube 50 cm long?

3. A flute is an open tube, so all harmonics are produced. With all holes covered, the flute sounds the note C_4 = 261.63 Hz. If all the finger holes are kept closed, what will be the next frequency sounded if the flute is overblown?

4. A clarinet is a closed tube, so only odd harmonics are produced in the lower register. With all holes covered, the clarinet sounds the note D_3 = 146.83 Hz. If all the finger holes are kept closed, what will be the next frequency sounded if the clarinet is overblown?

5. If the amplitude of the fundamental for a certain sawtooth wave is 2 volts, what are the amplitudes, in volts, of the next four harmonics? What are the harmonic numbers?

6. If the amplitude of the fundamental for a certain square wave is 3 volts, what are the amplitudes, in volts, of the next four existing harmonics? What are the harmonic numbers?

7. Examine the spectrum of the gurgling water shown in Figure 4-36, and determine the length of the air column at that time.

REFERENCES

BLOM, ERIC, ed. *Grove's Dictionary of Music and Musicians*. 5th ed. New York: St. Martin's Press, Inc., 1973. A ten-volume set containing a wealth of information on instruments, acoustics, and other areas related to music at the level of the nontechnical reader.

FLETCHER, N. H., and T. D. ROSSING. *The Physics of Musical Instruments*. New York: Springer-Verlag, 1990 (hardcover), 1993 (softcover). An excellent, up-to-date book containing a wealth of detailed information about virtually all types of musical instruments.

OLSON, HARRY F. *Musical Engineering: An Engineering Treatment of the Interrelated Subjects of Speech, Music, Musical Instruments, Acoustics, Sound Reproduction, and Hearing*. New York: McGraw-Hill Book Company, 1952. This is a classic book in the field of engineering applications of acoustics, written for the mathematically more advanced student.

RANDEL, DON MICHAEL, ed. *The Harvard Dictionary of Music*. 4th ed. Cambridge, Mass.: Harvard University Press, 1986, 2003. A classic dictionary of music, with emphasis on historical aspects.

ROEDERER, JUAN G. *Introduction to the Physics and Psychophysics of Music*. 3rd ed. New York: Springer-Verlag, New York, Inc., 1994. An excellent, basic introduction to psychoacoustics, written for a slightly more mathematical audience than the present text.

ROSSING, THOMAS D. "Resource Letter MA-2: Musical Acoustics." *American Journal of Physics* 55 (1987):589. This is one of many AJP resource letters dealing with various topics in physics, containing a good list of primary reference material.

———, ed. *Musical Acoustics: Selected Reprints*. College Park, Md.: American Association of Physics Teachers, 1988. This includes several articles dealing with a variety of topics in acoustics along with a reprint of the AAPT *Resource Letter MA-2: Musical acoustics*.

SACHS, CURT. *The History of Musical Instruments*. New York: W. W. Norton & Company, Inc., 1940. This is possibly the most complete book on the history of instruments, with many good drawings and diagrams.

SAUNDERS, FREDERICK A. "Physics and Music." *Scientific American*, July 1948. This is an early modern survey article on acoustics, presaging the contemporary interest in musical acoustics.

SAVAGE, WILLIAM R. *Problems for Musical Acoustics*. New York: Oxford University Press, Inc., 1977. Many problems, with some discussion, at or near the level of this book, with emphasis on music applications.

Acoustical References

Physics Demonstrations in Sound and Waves, Parts I, II, and III, Physics Curriculum and Instruction, three videotapes, each with eight segments about 3 minutes in length, for a total of 30 minutes apiece. The tapes are of excellent quality and cover several areas of vibrations and sound.

BERG, RICHARD E., and DAVID G. STORK. *Demonstrations in Acoustics*. College Park, Md.: University of Maryland, Department of Physics, 1980. This is a set of four 1-hour color videotapes, and gives brief explanations of many of the experiments described in the present text.

The Science of Sound, Folkways Record Album No. FX6136, Descriptive Literature by Bell Telephone Laboratories. Available from Frey Scientific Co., Mansfield, Ohio. This is an excellent recording, covering many areas of acoustics, which can be readily used in class lectures. The records contain short segments illustrating such topics as the overtone series, tone quality, filtering, distortion, reverberation, the Doppler effect, and others.

Periodicals

Several periodicals deal in part with topics in acoustics:
Journal of Acoustical Society of America (JASA)
Acustica
Physics Today
American Journal of Physics (AJP)
The Physics Teacher
The Instrumentalist

The first two journals contain primarily research papers of researchers in areas of acoustics written to and for each other and are therefore likely to be too difficult for the general reader.

Physics Today and the *American Journal of Physics* contain a mixture of types of articles, some of which are appropriate for the non-mathematically sophisticated reader. *The Physics Teacher* is used heavily by high-school teachers, and as such contains a high proportion of articles readable by elementary physics students. *The Instrumentalist* is written for musicians, but often discusses topics using physics.

Chapter 5

Electronic Music and Synthesizers

Electronic musical synthesizers are used in many settings: serious contemporary music, light background music, jazz, rock and roll, and commercials. Even some classic works of Bach, Beethoven, and others have been transcribed for synthesizer. Some ballet troupes, and even small opera companies, resort to synthesizers in response to economic pressures. Each advance in electronic technology is followed by a concomitant advance in synthesizers and new developments in the expressive range of electronic music. With changes in musical taste and further advances in electronics, the synthesizer will surely continue to play an increasing role in music.

In this chapter we shall discuss several aspects of musical synthesizers, beginning with the way in which signals can be combined to create musical sounds. Although analog synthesizers are now obsolete, the techniques they use are instructive in understanding how musical sounds are formed. Finally, we will review the modern digital synthesizer and some recent innovations and applications.

5.1 Combination of Waves and Modulation

It is often desirable to combine two or more waves to obtain a special sound or effect; modulation (unlike simple addition) includes several very specific techniques for combining two waves so that one of the waves changes some physical characteristic of the other. The particular methods of combining waves to be discussed here are (1) simple addition, (2) gating, (3) amplitude modulation, (4) balanced modulation, (5) frequency modulation, and (6) pulse-width modulation.

Simple addition

We covered simple addition of waves extensively in our treatment of Fourier synthesis and beats. Simple addition refers to the addition of two or more waves point by point to

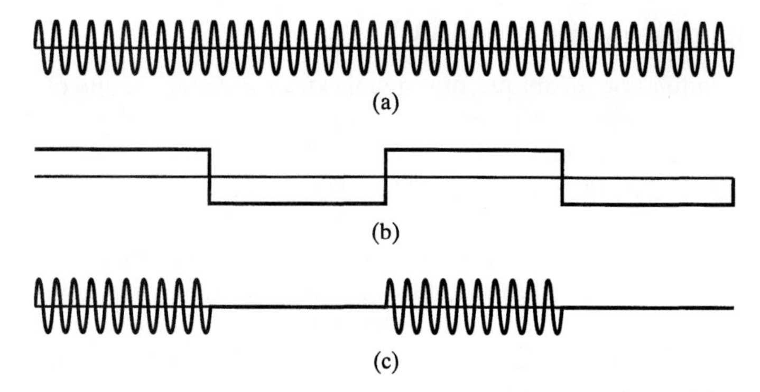

Figure 5-1 Gating of a high-frequency sine wave (a) by a low-frequency square wave (b) producing the gated signal (c). Here the gating level is chosen to be 0.

produce a complex wave. In general, the component waves need not be pure waves, as they have been in many previous examples. Recall that, if the two waves are in the harmonic series of a certain fundamental, their sum will have a period equal to that of the fundamental. If the two waves are not harmonically related, the sum wave will continuously change shape because of the changing of the relative phase between the component waves.

Gating

The technique of gating involves using one wave to turn another wave on and off. The gating signal switches on and off a higher-frequency signal. The higher-frequency signal is switched on whenever the value of the gating signal is above a certain preset value, called the *gating level.* Figure 5-1 shows a high-frequency sine wave gated by a lower-frequency square wave. The gating level in this example is 0. If the gating wave is a pulse that is on 10 percent of the time, the wave of Figure 5-2 results. (This is the way the signal was obtained for the experiment to determine the velocity of sound in Section 2.1.) One can consider a simple on-off switch or a synthesizer key pressed by a performer as producing a gating signal. The gating signal need not be periodic.

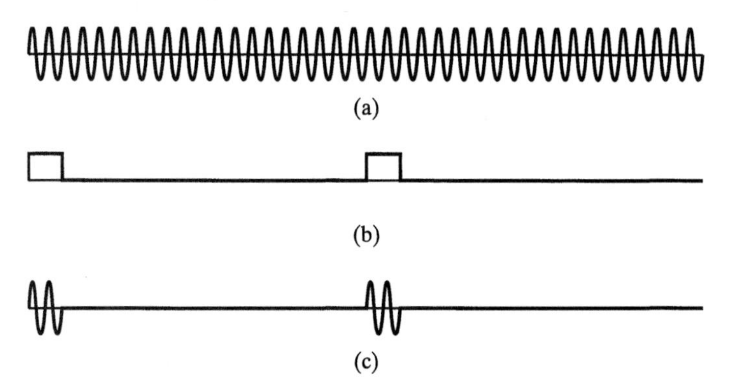

Figure 5-2 Gating of a high-frequency sine wave (a) by a pulse (b) that is on for 10 percent of its period to produce gated wave (c).

Amplitude modulation (AM)

In a modulation technique, one signal causes a change in one of the physical characteristics of the other. The higher-frequency signal is usually called the *carrier*, and the lower-frequency signal is the *modulator*. In amplitude modulation the modulator signal changes the *amplitude* of the carrier. The value of the amplitude of the modulated signal at any instant is dependent on the value of the modulator at that instant. Figure 5-3 shows a fast sine-wave carrier being amplitude modulated by a slow sine-wave modulator. The envelope of the modulated wave, the curve that defines its overall shape, has the shape of the modulator wave, and its axis is displaced upward from the carrier axis by a certain specified amount, the *offset level*. Because the amplitude of a wave at any instant defines the wave's limit above and below the axis, we can draw the mirror image of the displaced modulator curve to define the envelope below the axis. When the amplitude of the modulator is sufficient to reduce the amplitude of the carrier to zero at some time during each period, 100 percent modulation results, as shown in Figure 5-4. That is, the amplitude of the modulator is exactly equal to the offset level. Overmodulation, shown in Figure 5-5, occurs when the amplitude of the modulator is greater than the offset level.

If the carrier is an audio-frequency tone and the modulator has a frequency of about 1 to 10 Hz, the effect of amplitude modulation can be heard as a pulsing or throbbing tone, continually increasing and decreasing in loudness. The musical term for this is *tremolo*.

AM radio uses an audio-frequency modulator (speech, music, and so on) to modulate the amplitude of a radio-frequency (RF) carrier signal whose frequency lies in the AM radio-frequency band, about 540 to 1600 kHz. The modulated wave is the signal transmitted by the radio station. Our radios then receive the modulated RF carrier and perform the inverse process, demodulation, to reproduce the audio-frequency modulator signal, which we hear on the loudspeaker as music or speech.

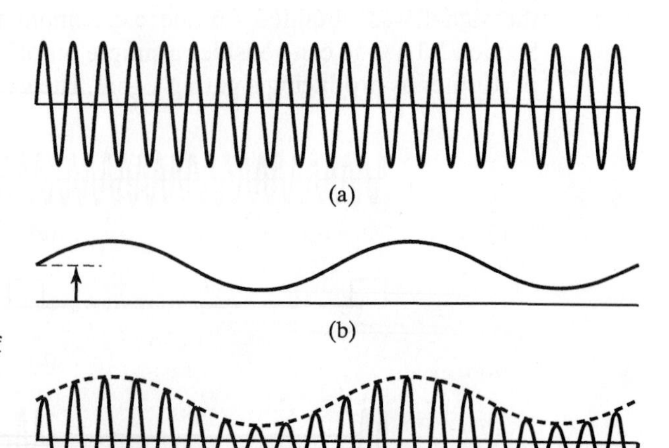

Figure 5-3 Amplitude modulation of a sine-wave carrier (a) by a sine-wave modulator (b). The envelope (dashed lines) of the modulated wave (c) is obtained from the modulator and its mirror image opposite the zero level of the carrier wave. An arrow marks the offset level.

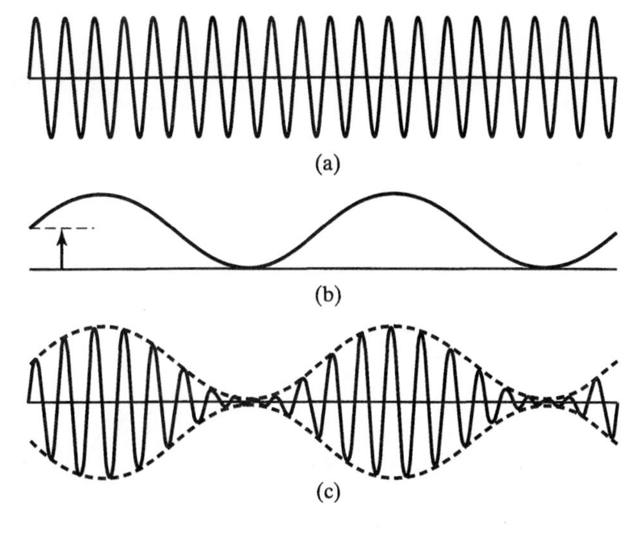

Figure 5-4 A sine-wave carrier 100 percent modulated by a sine-wave modulator. The carrier (a), the modulator (b), and the modulated wave (c) are shown. An arrow marks the offset level.

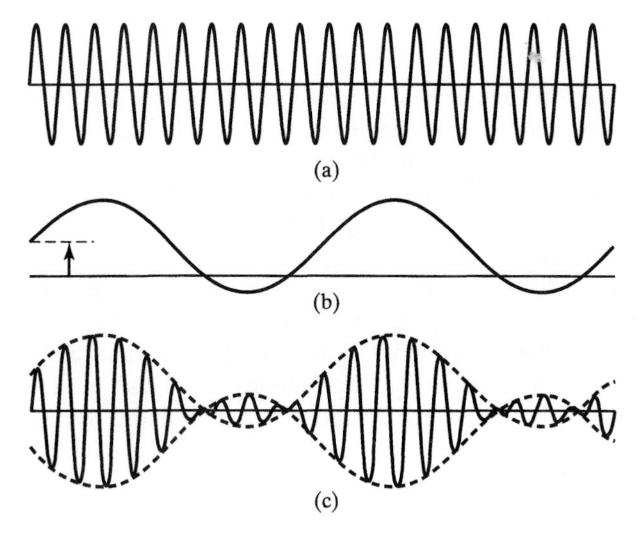

Figure 5-5 AM overmodulation of a sine-wave carrier (a) by a sine-wave modulator (b) to produce the modulated wave (c). An arrow marks the offset level.

Balanced modulation

In the case of amplitude modulation just discussed, you can see in Figures 5-3 through 5-5 that the zero, or offset, level of the modulator wave is displaced from the zero level of the carrier wave. For this reason, amplitude modulation is sometimes called *unbalanced modulation*. If the offset level of the modulator is 0, *balanced modulation* (BM) or *double-sideband modulation* (DSBM) results. In the world of musical synthesizers, the effect is sometimes also called *ring modulation* (RM), because it is used to produce the ringing of bells. Figure 5-6 shows a sine wave balance-modulated by a lower-frequency sine wave; the envelope is obtained again by reflecting the modulator wave about the axis of the carrier. Careful inspection of this graph and comparison to the graph of 100 percent

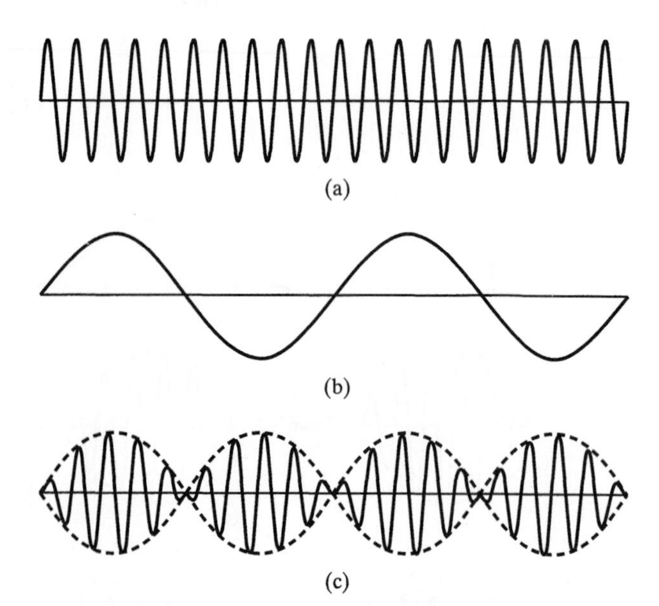

Figure 5-6 A sine-wave carrier (a) double-sideband modulated by a sine-wave modulator (b) to produce the modulated wave (c). The offset level of DSBM is 0.

amplitude modulation shown in Figure 5-4 will reveal a slight difference between the shapes of the envelopes, especially for values near zero. In addition, the final DSBM wave undergoes a phase inversion each time the modulator passes through zero. The graph of balanced modulation is identical to the graph of beats shown in Figure 2-37.

Let us consider more carefully the similarity between beats and balanced modulation. We saw in Sec. 2.4 that, when two equal-amplitude sine waves of frequencies f_1 and f_2 (f_2 larger than f_1) are added, we obtain a wave whose frequency F is the average value of f_1 and f_2: $F = (f_1 + f_2)/2$. The beat rate f_b is the difference between the two frequencies: $f_b = f_2 - f_1$. The balanced-modulation curve demonstrates that this same curve can also be generated by an audio frequency F double-sideband–modulated by a sine wave of low frequency $f_m = f_b/2$. Each period of the modulating wave produces two beats. Hence balanced modulation can be used to produce a complex wave containing only frequencies f_1 and f_2 from the frequencies F and f_m. The two frequencies thus produced are $f_2 = F + f_m$ and $f_1 = F - f_m$. This is of particular interest when beats are no longer heard as variation in loudness because f_m is in the audible range. For example, if $F = 500$ Hz and $f_m = 83$ Hz, the balanced modulator produces the frequencies 417 Hz = 500 Hz − 83 Hz and 583 Hz = 500 Hz + 83 Hz. These two frequencies are not members of a single overtone series; the ratio 583:417 is not reducible to a ratio of small integers. The balanced modulator produces nonintegral overtones like those in tuned percussion instruments. If the carrier wave is complex, containing harmonics, each of the harmonics will combine with the modulator to produce two new frequencies; all of these will, in general, be inharmonic.

Frequency modulation (FM)

In frequency modulation, the modulator causes the frequency of the carrier to vary while leaving the amplitude of the carrier wave unchanged. For positive values of the

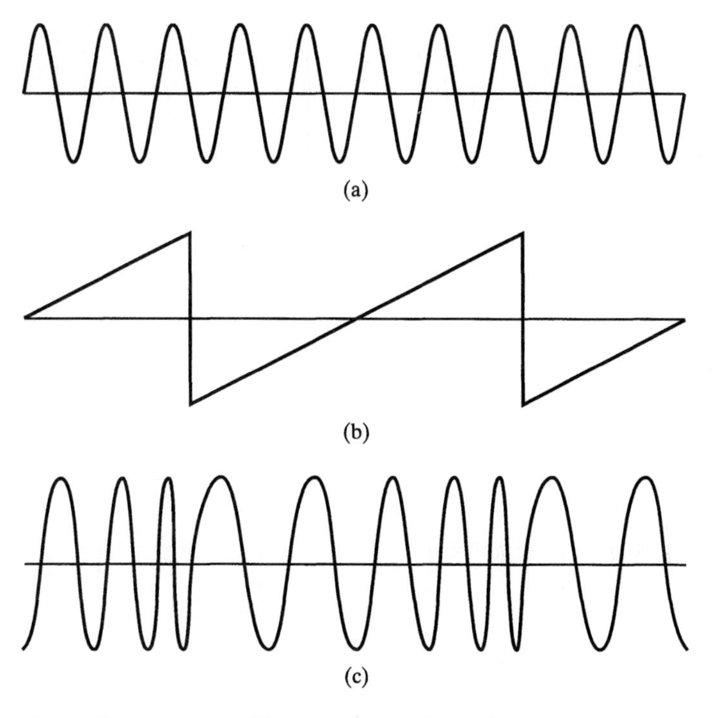

Figure 5-7 A sine wave (a) frequency-modulated by a sawtooth wave (b) to produce the modulated wave (c).

modulator signal, the (modulated) frequency becomes greater, and for negative values of the modulator signal, the frequency becomes less than its normal value; the greater the amplitude of the modulator, the greater is the deviation in the frequency. Figure 5-7 shows a sine wave frequency-modulated by a sawtooth wave. Figure 5-8 shows a square wave frequency-modulated by a triangular wave. Figure 5-9 shows a triangular wave frequency-modulated by square waves of various amplitudes and frequencies. A square-wave modulator has two values; therefore, the modulated wave has two frequencies, one below and one above the carrier frequency. With some thought one can understand that (1) the amplitude of the modulator can be increased, with the result that the frequency variation of the modulated carrier wave becomes greater, or (2) the frequency of the modulator can be increased, resulting in faster changes in the frequency of the carrier. The greater the amplitude of the modulator, the greater the frequency deviation, whereas an increase in the modulator frequency results only in more rapid alternation of the resulting frequencies.

When the carrier wave is in the audible range, and the modulator has a frequency of a few Hz and an amplitude sufficient to cause a few tenths of 1 percent change in the carrier frequency, we hear a sound of constant amplitude or loudness whose pitch oscillates slightly (somewhat less than one half-tone). Musicians call this effect *vibrato*. In fact, much of the "vibrato" produced in music is a combination of pure vibrato (frequency modulation) and tremolo (amplitude modulation).

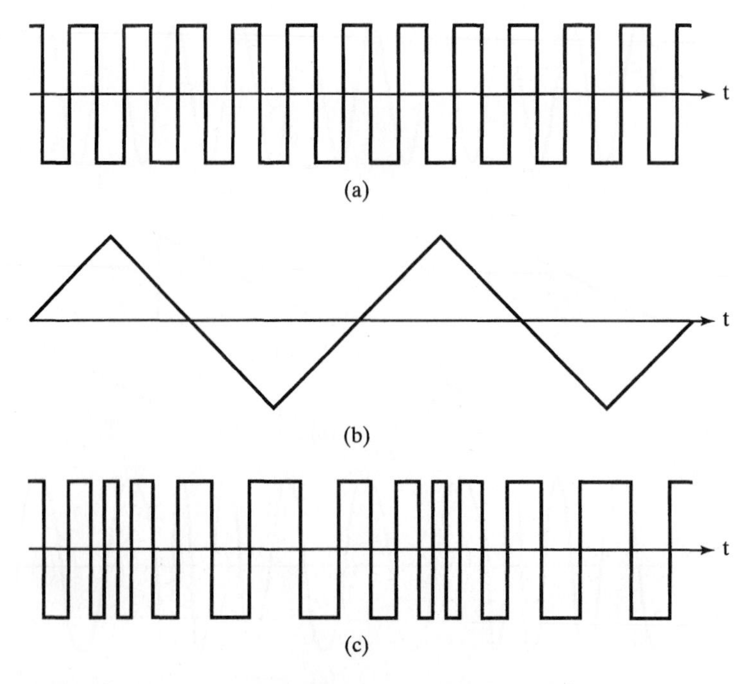

Figure 5-8 A square wave (a) frequency-modulated by a triangular wave (b) to produce the modulated wave (c).

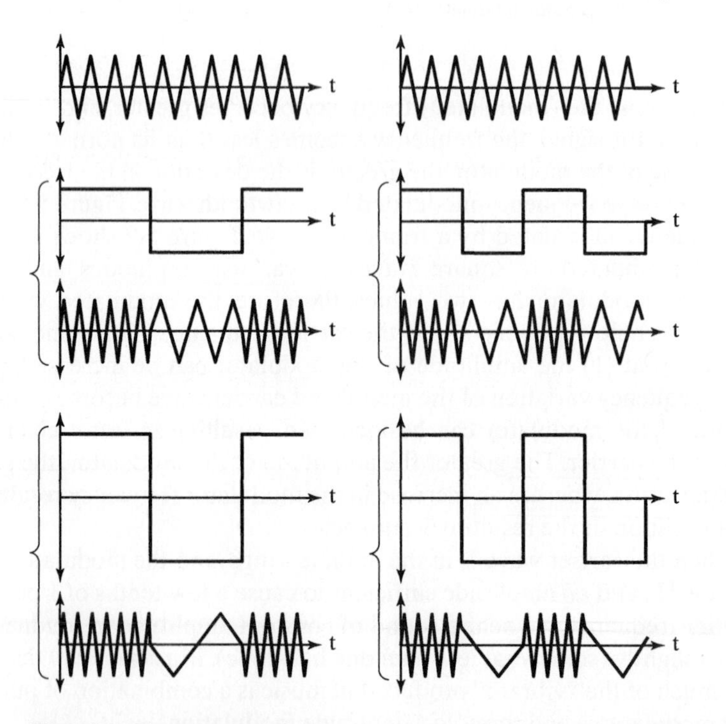

Figure 5-9 Frequency modulation of a triangular wave (top) by various square waves, producing modulated waves shown below each square-wave modulator.

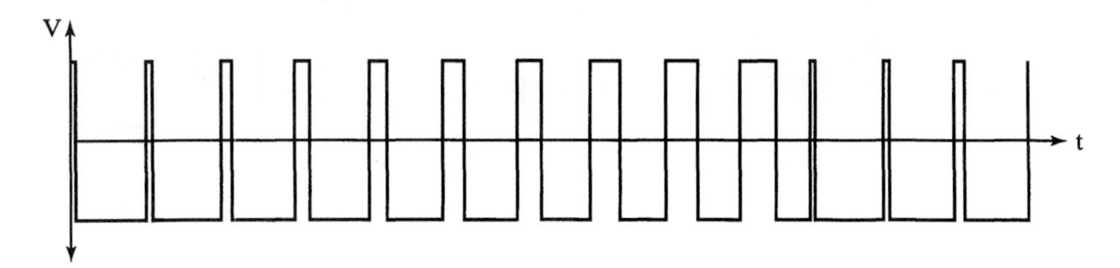

Figure 5-10 Pulse-width modulation (PWM). The period of the note remains constant, but the "on" time for the pulse ranges between 5 and 50 percent. Thus there is a change in harmonic structure and tone quality.

FM radio uses a carrier of about 100-MHz frequency modulated by speech, music, or other sound up to about 75 kHz above and below the carrier frequency. The improvement in the quality of FM radio over AM radio results from the use of the large frequency range over which the modulation occurs and the nature of the noise sources that affect radio waves. Most sources of radio noise affect the amplitude of radio signals, and thus AM signals will be greatly affected. Even though the amplitude of an FM signal is changed if noise is present, the noise has almost no effect, because the demodulation process deals only with the frequency of the signal. However, for subtle reasons, an amplitude threshold exists below which AM is less susceptible to noise than is FM.

Pulse-width modulation (PWM)

Pulse-width modulation, which produces a sound unique to electronic instruments, occurs when the duration of the upper level of a flat-top wave is determined by the value of a modulator wave. For instance, the modulated wave may change form between a square wave (50 percent on time) and a pulse train (very short on time) of the same fundamental frequency, as shown in Figure 5-10. In addition to changing the wave slope, pulse-width modulation changes the harmonic content and thus the tone quality of the wave. This unusual effect is not obtained on standard acoustic musical instruments played in the normal way.

5.2 Analog Synthesizers and Synthesis of Musical Sounds

The classic analog synthesizer consists of a collection of electronic modules, each of which performs some particular function, which can be connected together to create some desired musical sound, as shown in Figure 5-11. We shall now describe some of these components, and then study the formation of musical sounds by seeing how synthesizer modules can be used to create these sounds.

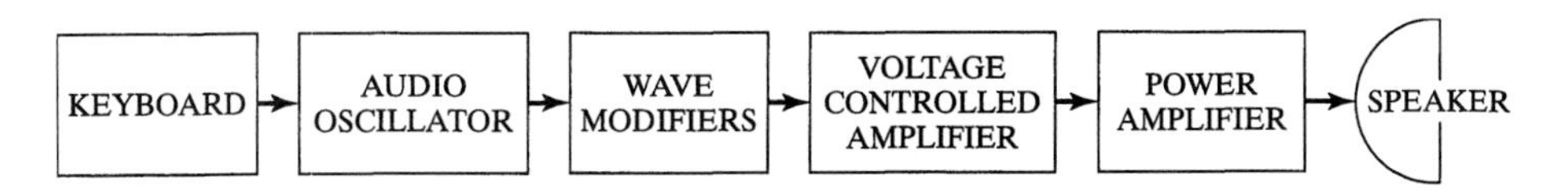

Figure 5-11 Block diagram for the flow of audio signals in a typical analog synthesizer.

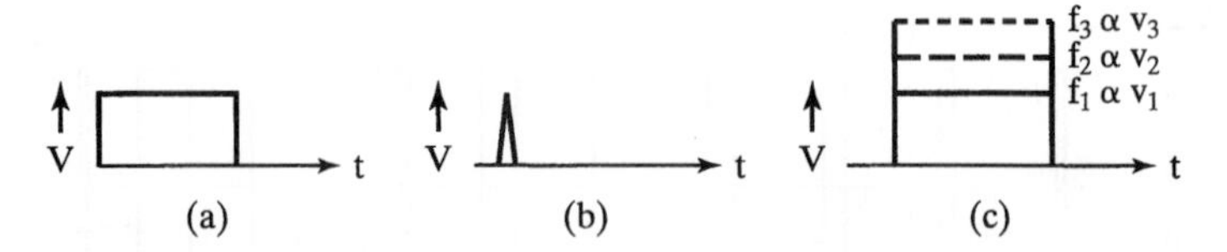

Figure 5-12 Three types of control voltages produced by depressing a synthesizer key. Signals (a) or (b) can by used to trigger the envelope generator or other control modules; signal (c) is used to control the pitch of the note produced by the voltage-controlled oscillator (VCO), and is different in level for each key—the higher the control voltage, the higher the frequency of the note.

Signals

Two types of signals are used in a synthesizer: audio signals and control signals. The audio signals used are electronic signals with frequencies in the audible range and amplitudes of about 1.5 V. They can be input to a speaker at any stage (after some amplification). Control signals can be either oscillatory or constant in voltage, depending on their function, and have amplitudes up to about 5 V, in keeping with the requirements of solid-state electronic modules. Control signals are used to control the operation of various components in the synthesizer. For instance, a control signal could control the amplifier and thus the loudness of the sound emitted by the speakers.

Keyboard

The electronic synthesizer keyboard looks similar to that of a piano, except that it typically contains fewer keys. The synthesizer keyboard provides two functions. First, it gives the control signal that determines the pitch of the note; second, it provides a trigger, or starting signal, that triggers the envelope generator, which can be used to create the attack transient of the tone. Figure 5-12 shows some typical control signals produced by pressing a key.

Sample and hold

In general, a keyboard pitch-control signal is input to the oscillator, and the audio output from the oscillator is sent (after further processing) to a speaker. Pressing a key will then cause the production of a tone with the pitch associated with that key. The oscillator remembers the pitch and continues to sound at this pitch even when the key is released. This is known as *sample and hold.* The tone will then continue at the same pitch until another key is pressed, at which time the oscillator will remember the new pitch.

Voltage-controlled oscillator (VCO)

The main oscillator for a synthesizer, often called the VCO, takes a control voltage from the keyboard to produce an audio signal whose frequency is determined by the amplitude of the control voltage. If the synthesizer is tuned like a piano (and they usually are), the keyboard pitch-control voltages will be set so that playing one half-step higher on the keyboard produces a voltage $1.05946 = \sqrt[12]{2}$ times that of the preceding

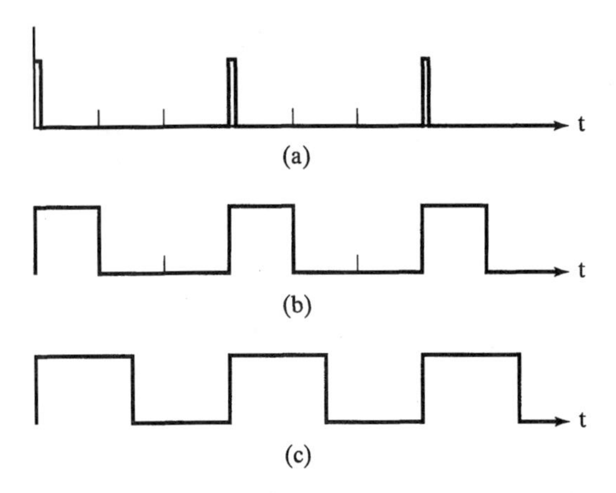

Figure 5-13 Flat pulse whose "on" time can be fixed by an external control to produce (a) a pulse train (very short on time), (b) one-third on time, and (c) a square wave (50 percent on time).

note. This is called *equal temperament* (as we shall see in Chapter 9) and guarantees that after one octave of 12 half steps the voltage will double. The VCO produces a frequency directly proportional to the voltage; for example, doubling the input voltage doubles the frequency and produces a one-octave change in pitch.

Most VCO units produce several standard waveforms: sine, triangular, sawtooth, square, and pulse train. Often the square wave and pulse train are produced by utilizing a flat pulse whose "on" time is variable between near zero (pulse train) and 50 percent (square wave). Some wave shapes that can be produced by such a system are shown in Figure 5-13. By means of an appropriate waveform containing the desired set of harmonics and the appropriate filters (discussed later), a large number of combinations of overtones can be produced.

Voltage-controlled amplifier (VCA)

A VCA simply takes the audio signal input to it and amplifies this signal (or adjusts its amplitude) by an amount proportional to some input control voltage. The keyboard, VCO, and VCA are the minimum components necessary to produce a musical line, including rests (moments of silence); a *block diagram* of the components used is given in Figure 5-14. The keyboard determines the pitch of the VCO using any desirable wave form. The keyboard trigger control causes the VCA to pass the signal when a key is depressed and stop when the key is released. Of course, this simple on-off gated tone is not very interesting, so we need to introduce additional elements: envelope generators, filters, and modulators, which are included in the box labeled "wave modifiers" in Figure 5-11.

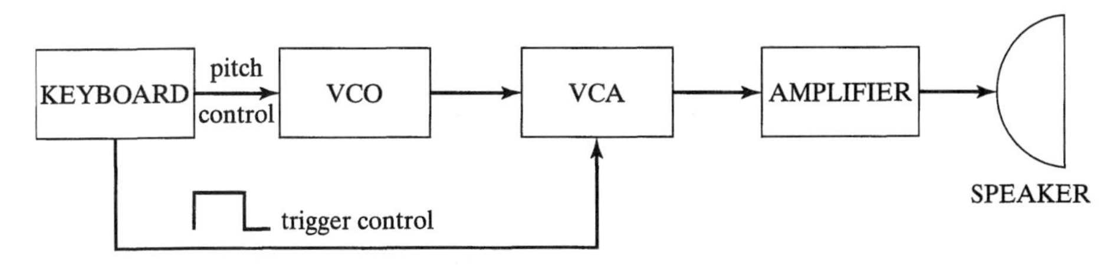

Figure 5-14 Block diagram indicating a simple type of keyboard control of pitch and duration of notes.

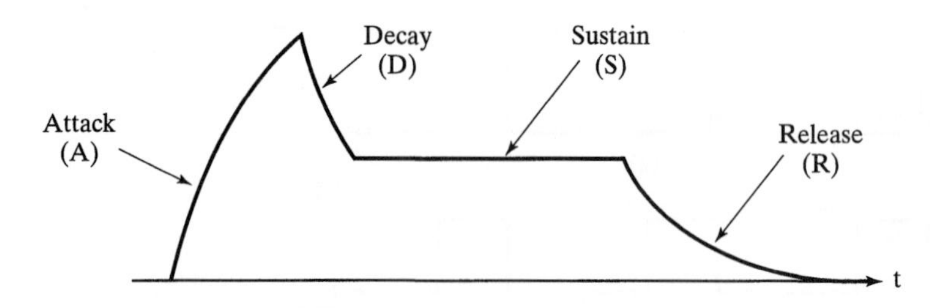

Figure 5-15 Definition of parts of the envelope produced by the envelope generator.

Envelope generator

Most musical sounds do not merely come on, remain at a constant amplitude, and stop abruptly, as does the tone discussed previously. Rather, for example, for a wind instrument, there is an attack period, during which the sound may be a bit louder than during the longer period when it is being held. For a plucked stringed instrument, the attack period is immediately followed by a continual decrease in the amplitude of the sound. An important part of the sound of these instruments is this "envelope," which determines the amplitude of the note at each time throughout its duration.

The envelope generator determines this envelope and has four sequential functions, as illustrated in Figure 5-15. The *attack* (A) is the time interval after depressing a key during which the initial rise of sound occurs. The *decay* (D) period is the time required for the amplitude to decrease from its attack peak to its steady state or *sustain* (S) level, at which it remains until the key is released. Upon release of the key by the performer, the *release* (R) period ensues. The attack, decay, and release periods and the amplitude level for the sustain are all independently controllable by knobs on the envelope generator; the duration of the sustain period is determined by the duration the keyboard key is depressed. Figure 5-16 shows the relationship between the keyboard trigger signal and the envelope generator for two typical cases. A reasonably long

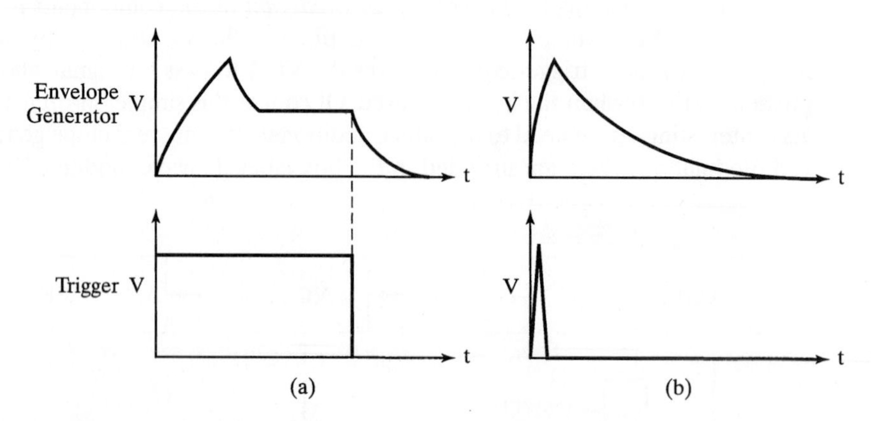

Figure 5-16 Trigger (control) signal from keyboard and envelope generator signal for a normal wind instrument sound (a) and a plucked stringed instrument (b).

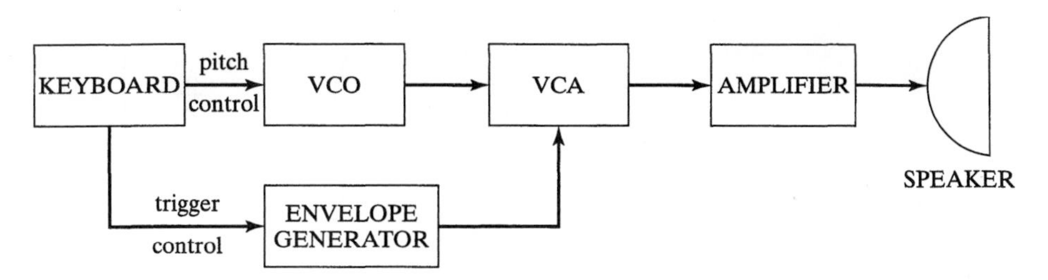

Figure 5-17 Block diagram of normal signal flow for control of the shape of notes by means of an envelope generator.

attack time followed by a high sustain level is typical for a wind-instrument envelope. A very rapid attack time followed by a long release with no sustain period is necessary to achieve a plucked-string sound. Release time may be used to simulate a long reverberation time, as is found in a cathedral. The envelope generator can also be used to control harmonic content in the wave by putting its signal into a filter in addition to, or in place of, its normal input into the VCA, thus allowing the production of transients. The normal flow diagram using an envelope generator is shown in Figure 5-17. For simple cases and in some elementary envelope generators, the decay control is missing, and the envelope is generated by attack, sustain, and release (sometimes called *decay*) only.

Low-frequency control oscillator

This unit, sometimes referred to as a control oscillator or a low-frequency oscillator, is used solely for control purposes. It will normally produce a sinusoidal control signal with a frequency between 0 and about 25 Hz, below the audible range. If the signal from this oscillator is sent to the VCO, it will cause periodic changes in the frequency (frequency modulation), or vibrato. If the signal is added to that from the envelope generator (sustain level) and input into the VCA, it will result in periodic changes in the amplitude (amplitude modulation), or tremolo.

It is also possible to input a control signal into the part of the VCO that controls the width of the pulse, as illustrated in Figure 5-13, producing pulse-width modulation.

Filters

A filter is an electronic device used to remove some frequencies or frequency bands from the frequency spectrum of the signal produced by the VCO. Several types of filters are normally available on a synthesizer.

The low-pass filter allows passage of low-frequency components while removing high frequencies, whereas a high-pass filter removes the low frequencies while allowing high frequencies to pass. Graphs showing the ratio of output-to-input signal amplitudes for typical low-pass and high-pass filters are shown in Figure 5-18. The transition frequency f_F of the filter can usually be set by the operator using an external control voltage or a knob on the unit, whereas the frequency range over which the transition occurs is usually an inherent property of the filter.

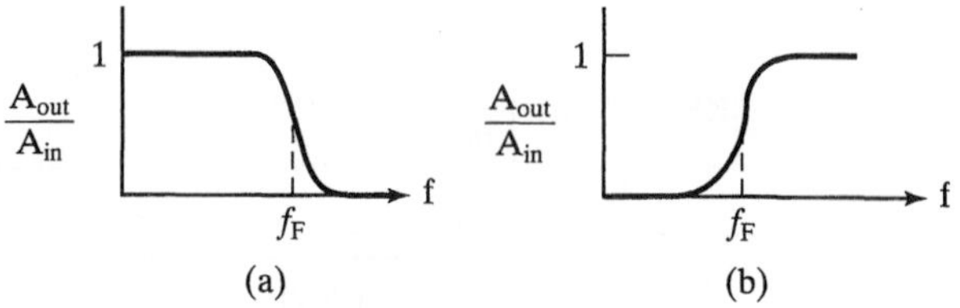

Figure 5-18 Ratio of output-to-input signal versus frequency for a low-pass filter (a) and a high-pass filter (b).

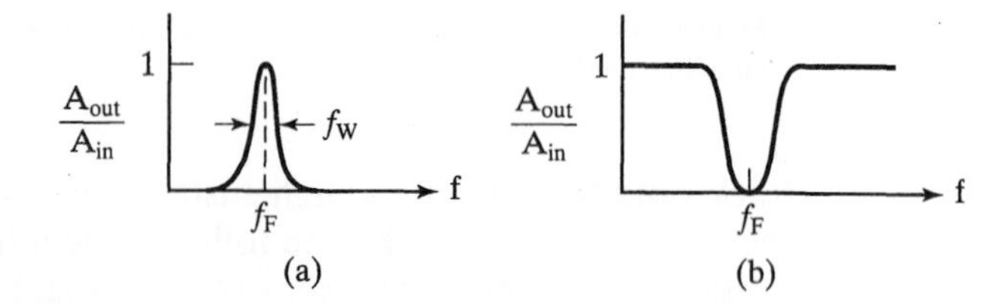

Figure 5-19 Ratio of output-to-input signal versus frequency for a bandpass filter (a) and a notch filter (b).

A bandpass filter passes only a small frequency band around its frequency f_F. The ratio of the frequency of the filter to the frequency range passed by the filter is called its *quality* Q; $Q = f_F/f_W$, so a filter with a small frequency width has a high Q. Both f_F and Q can be controlled by separate knobs on the unit or by a control voltage. A notch filter, or band-reject filter, is the inverse of the bandpass filter, and rejects all frequency components in a narrow band. The characteristics of bandpass and notch filters are shown in Figure 5-19.

For some filters, the frequency f_F is fixed. In more sophisticated filters, the frequency of the filter varies with the audio-frequency input, so that it always affects the same harmonic components of the complex wave, creating a more uniform Fourier spectrum for all notes. Such a filter, in which the frequency of the filter f_F varies with the frequency of the audio tone f_1 (that is, f_F/f_1 is a constant), is called a *tracking filter*. For instance, a tracking filter can be set to reject all harmonics above the third for a complex tone of any frequency.

Noise generator

A noise generator, or white-noise generator, produces white noise. It is most generally used for special nonmusical sounds but is required to produce the sound of a cymbal or a snare drum, for example.

A white-noise generator and a bandpass filter can be used to produce a very striking wind sound. The white noise is input into the bandpass filter and then sent to the VCA. Hand controls set the frequency and Q of the filter. By adjusting the Q of the filter, either a narrow or a relatively wide frequency band (pitch center) can be obtained; by adjusting the center frequency of the filter, this pitch center can be raised or lowered. Simultaneous adjustment of both controls causes the wind to howl and sound as if it were changing its speed.

Colored noise, consisting of white noise with some frequency ranges missing or reduced in amplitude, is obtained by filtering white noise.

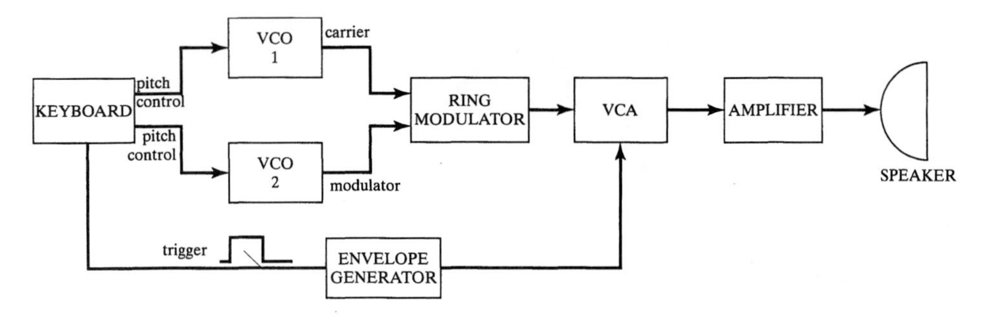

Figure 5-20 Block diagram for production of tuned percussion sounds using a ring modulator.

Ring modulator

A ring modulator performs balanced modulation with the two signals that are input into it. These two signals can be an audio signal and a control signal, or they can be two audio signals. Because, in general, ring modulation produces signals with non–integrally-related frequency components, the ring modulator is useful in synthesizing tuned percussion sounds, such as those from a chime or bell. Two features in the tuned percussion sound, such as a bell, are (a) nonintegral frequency components and (b) a sharp percussive attack with long release time and no sustain. These two features can be achieved using the synthesizer setup shown in Figure 5-20. Two audio signals from the keyboard are used in ring modulation while short-attack and long-decay envelope generator functions are used. It is necessary for the two frequencies to track; that is, the two frequencies must go up and down together as different notes are played. The (nonintegral) ratio of the frequency components in the signal produced by the ring modulator will then remain constant, producing similar tone quality for all bell notes.

5.3 Digital Synthesizers and Keyboards

Now that we have studied some ways in which various signals can be combined to create musical sounds, we can turn our attention to the way in which modern digital electronics are used in contemporary musical synthesizers. We shall then briefly review some of the important features of these incredibly diverse and flexible instruments, and then turn to some of the more recent advances in music and recording technology made possible by using this new generation of instruments and modern computers.

The fundamental feature common to all digital synthesizers is that they create the wave *digitally*—that is, the wave shape is stored as a series of discrete points, as illustrated in Figure 5-21. An electronic module called a *digital-to-analog* converter then converts this series of discrete digital electronic signals into smooth, continuously varying analog signals, as shown, forming the wave shape of the sound to be produced, as will be discussed in Section 7.9. The use of new, less expensive high-speed digital electronics, the accelerating development of *Musical Instrument Digital Interface* (MIDI) software, and the decreasing cost of computer memory have facilitated this new generation of digital synthesizers.

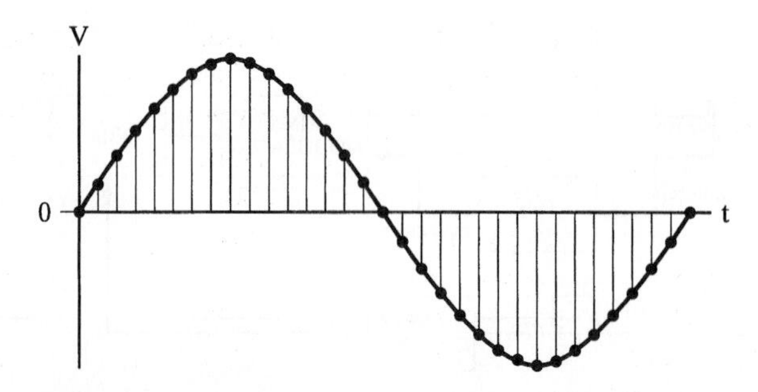

Figure 5-21 Idealized digital-to-analog process used by a digital synthesizer in forming a complex wave shape. The value of the wave at a series of equally spaced points in time, as shown, is read from memory, and the smooth wave shape is obtained by interpolation in the electronics of the digital-to-analog converter. Complications in the actual digital-to-analog process will be discussed in Sec. 7.9.

When any key is pressed on the keyboard, the key pressed determines the frequency, while the stored digital wave shape that has been previously selected determines the sound quality, or timbre, to be produced. Thus the desired wave is created in a single operation without the need for the series of oscillators, filters, and other signal modifiers used in the older generation of voltage-controlled synthesizers. Most digital synthesizers have a choice of dozens or even hundreds of standard sounds or voices, including sounds of acoustical musical instruments, classical electronic sounds, and weird or unusual sounds described by exotic titles, with the choice of musical sound available literally at the touch of a button. In the *FM synthesizer*, frequency modulation of one signal by another, the frequency of which may be close to that of the signal being modulated, is used to obtain a sound that is rich in harmonics. In addition, most modern synthesizers can access wave samples that are stored on disks, tapes, or computer memory.

Some digital synthesizers have retained the ability to mix various classical electronic wave shapes or to modulate these signals with each other to achieve a wide range of distinctive sounds.

Unlike the classic analog synthesizer, which could play only one note at a time, the modern digital synthesizer can play many notes simultaneously, as can a piano. The number of simultaneously playable notes is limited by the number of channels available in the synthesizer but is usually at least eight to ten. Many synthesizers allow two or more different wave shapes, or timbres, to be played simultaneously, but the instrument is generally not able to track the two different voices if they cross over each other—that is, when the "lower" musical line becomes higher than the "upper" line. Because choice of sound can be made by simply pressing a switch, rapid changes can be made in the sound of the instrument, an enormous convenience in live performance and an enormous improvement over the older analog synthesizers, which often required several switches or patch cables to be modified in order to change the sound. This increased flexibility, the variety of musical sounds, and the ease of operation typical of modern digital synthesizers helped to change the synthesizer from a purely studio instrument into one easily used for live performances.

Some synthesizers allow a choice of musical temperament, or type of scale to which the instrument is tuned. For example, some musicians have performed using a synthesizer tuned to one of the Baroque temperaments, as discussed in Chap. 9, using the harpsichord or a Baroque organ sound, and some modern performers use unusual temperaments in performing ethnic music. Some contemporary musicians might use a quarter-tone temperament or set the entire instrument to cover as little as one-half musical step to obtain special effects.

In addition to the variety of sounds available, synthesizers generally come with a large complement of effects such as vibrato, tremolo, and other modulators, glissando, various filters, and reverberation, in addition to a variety of keyboard sizes. Most synthesizers have percussion or rhythm accompaniment that can be programmed to play continuously throughout the performance or to act as a metronome. This feature can be of importance in coordinating and synchronizing the various voices when producing multitrack recordings.

The output of the synthesizer can be run into the auxiliary (or line-level) input of most consumer audio systems, or even used with a computer audio system, so the performer can use virtually any audio system with the synthesizer.

Depending on the particular purpose of the synthesizer or the particular group to whom it might appeal, these instruments go by a variety of names: synthesizer, keyboard, wavestation, music workstation, performance synthesizer, or composition synthesizer. In fact, *keyboard* is probably a more accurate descriptive term for these instruments, because in deriving their sound from digital waveform tables they no longer "synthesize" the sounds, as did their analog predecessors. Often the literature describing these synthesizers reads more like the description of a computer than that of a musical instrument.

As a result of several important advances, the synthesizer keyboard often replaces the piano in live popular music performance. Synthesizers are relatively inexpensive, more reliable than a piano under variable weather conditions (so long as electric power is available), and substantially more portable. Because they are basically electronic rather than mechanical, they require less maintenance, such as tuning, than a piano. In many schools, groups of synthesizers have replaced the traditional piano studio for group teaching. One relatively recent advance in synthesizers that has encouraged this development is a "touch-sensitive" key. When pressed, the key responds with resistance similar to that of a piano key and allows the performer to control the dynamics by striking the key harder or softer.

Perhaps the major weakness of the digital synthesizer in substituting for a piano in the past has been that a synthesizer could not easily duplicate the tonal characteristic of the piano because of the inharmonicities of the overtones. To overcome this weakness, a relatively sophisticated synthesizer known as the Synclavier was developed. This instrument uses digital samples of real piano tones as the source of its sound. To achieve the full dynamic sound of a piano, each note must be sampled at a series of dynamic levels; the touch-sensitive keyboard then activates the appropriate sampled sound. However, because of its sophistication, the Synclavier was very expensive. As electronics and computer memory have become less expensive, more modern synthesizers have incorporated both a touch-sensitive keyboard and digitized sound tables to obtain a more realistic piano sound. The digital sampler can also be used to

Figure 5-22 The Roland RD-700 Expandable Keyboard. (Courtesy of Roland Corporation.)

record virtually any acoustic instrumental or vocal sound for use with the synthesizer keyboard. Many keyboards now have digital samplers as an integral feature.

A moderately priced instrument, typical of high-quality keyboards now on the market, which includes many of the features discussed here is the Roland RD-700, shown in Figure 5-22. It possesses a full 88-note piano keyboard and large, expandable memory storage containing stereo-sampled piano sounds, as well as a large collection of other instrument sounds. The keyboard is touch sensitive, and includes an "arpeggiator" to produce the sound of strumming a guitar. Sounds may be selected from 128 voices, 45 arpeggiator styles, and 50 rhythm patterns. The range of sounds is easily expandable using a standard wave-expansion board. As with all modern keyboards, it is fully MIDI compatible.

5.4 MIDI, Computers, and Contemporary Electronic Music

One of the most important developments in synthesized music and music composition occurred when a group of digital-sound-equipment and computer manufacturers agreed to standardize their signals, creating the MIDI (Musical Instrument Digital Interface) standard. Synthesizers that use MIDI have three 5-pin DIN (an originally European standard specifying construction details) plugs on the back, labeled respectively, "MIDI IN," "MIDI THRU," and "MIDI OUT." These plugs are used to input or output signals to or from the synthesizer or, for example, when two or more synthesizers are used together, to pass the signal from one through another to a third synthesizer or to a computer. Using an interface box, these signals can be sent to computers that can perform a variety of operations on the signals using the appropriate software. Ultimately, any musical option that can be programmed with computers or that uses computers to control or to process the signals can be accomplished using some combination of synthesizers and computers.

A very economical synthesizer that includes many of these features is the VirSyn CUBE Software Synthesizer, shown in Figure 5-23. It is used in combination with a high-powered computer processor, such as the Pentium III 600 MHz or PowerPC G4 400 MHz, which carries out the primary synthesizer functions using software controlled by the synthesizer, thus reducing the cost of the synthesizer unit. Among the features of this synthesizer are four independent morphable sound sources controlled using the graphical interface on the synthesizer. It also allows synthesis of your own

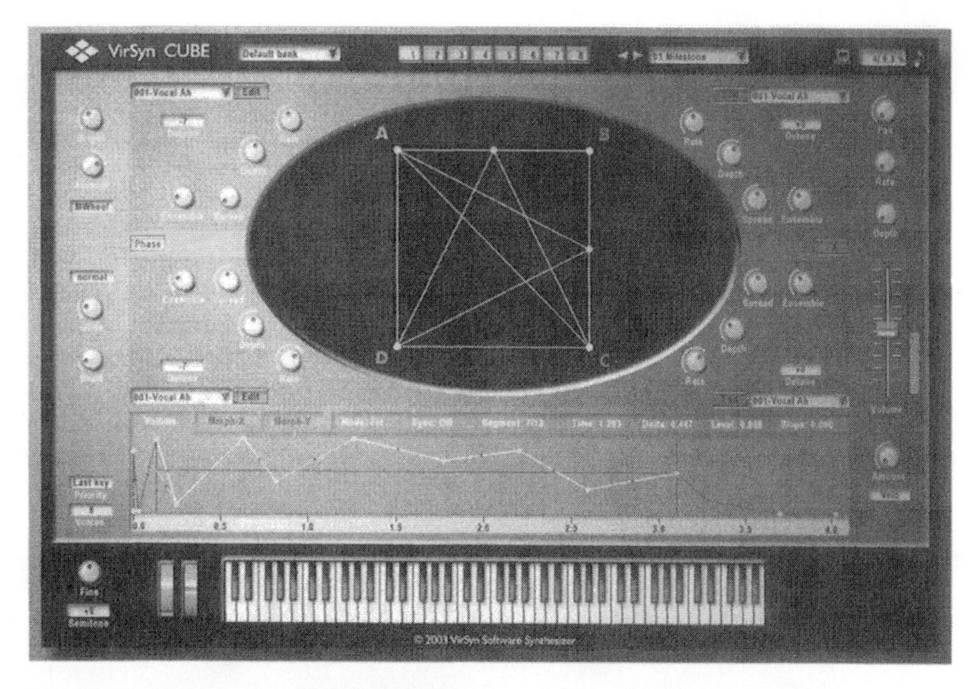

Figure 5-23 The VirSyn CUBE Software Synthesizer. (Courtesy of VirSyn.)

waves up to 512 partials or use of other wave shapes stored in memory, a large variety of other sounds and effects, and includes a MIDI learn function.

Some of the greatest labor-saving jobs for a music composer or arranger can readily be carried out using the computer, a keyboard, and MIDI system as follows: A person composes or arranges a piece consisting of several lines for different instruments. Upon playing the lines individually on the synthesizer, with the signals being sent into the computer, the computer writes the music in standard musical notation onto the screen in the form of a musical score. The composer can then edit the score to remove any errors, including those that may have been inadvertently played on the synthesizer. Alternatively, the composer can input the music into the computer manually using the computer keyboard by typing in each note and its time value. The computer will automatically assemble the score in easily readable form, allow the composer to make any appropriate modifications, transpose lines for the transposing instruments, and print out the score and a set of parts. Major computer software for carrying out these activities includes Finale and Sibelius.

Software such as Finale and Sibelius can also be used in learning music for performance. For example, much music has been recorded using different tracks for each part, and can be played back using a computer with one or more of the parts left out to be played live by a musician learning the part. The music can be played back at any speed while keeping the same pitch, so the learner may practice at a slower tempo. If desired, the music also may be replayed at a different pitch, to accommodate singers with limited range or to play the music with a different transposing instrument.

A MIDI lab is of immense significance both in the teaching of music and composition and as a tool for the professional composer. One MIDI workstation at the University of Maryland Department of Music is shown in Figure 5-24. Each workstation includes a Korg Triton Pro Digital Performer, a Macintosh G4 computer with color monitor, and Sibelius music software.

A professional electronic music laboratory will generally possess a collection of synthesizers, computers, mixers, and recording devices. Figure 5-25 shows the Electroacoustical Music Laboratory at the University of Maryland, College Park. Features include a Kurzweil K2600 workstation, a Kurzweil K2500RS rack sampler, MOTU (Mark of the Unicorn) 2408 and 1296 digital-to-analog converters, a Mackie 16-to-8-bus mixing console, a MOTU Digital Performer 4 with MIDI sequencing and digital audio handling, and a Macintosh G4 dual 500 MHz processor computer with Sibelius and other data-handling software. The mixer can accommodate up to 16 inputs at either microphone or line level, so any combination of acoustic and electronic sounds can be recorded separately and mixed in any desired manner. The resulting material can be replayed using four Dynaco BM6A studio reference monitors.

One of the most interesting recent developments in electronic music is the invention of new and unusual musical instruments. One such instrument is the Yamaha WX5

Figure 5-24 Student Sarah Pohl reviews one of her compositions using a MIDI workstation at the University of Maryland Department of Music.

Figure 5-25 The Electroacoustical Computer Music Studio at the University of Maryland Department of Music.

Wind MIDI Controller, shown in Figure 5-26. It is blown and fingered like an acoustic woodwind instrument, but through a MIDI output can be used to control any MIDI tone generator and therefore create virtually any electronic or acoustic sound. The instrument uses wind pressure to control and adjust the tone, can be blown like either a saxophone or a recorder by using interchangeable mouthpieces, and can be set to finger like a flute or a variety of saxophones. The fingering chart for the instrument is shown in Figure 5-27. Several of the operating functions can be adjusted in real time by the thumb (opposite the finger holes), including gain controls, setups, pitch-blend wheel, and programming of the MIDI tone generator. It comes with its own controller that allows the player to customize the instrument's response for his or her

Figure 5-26 The Yamaha WX5 Wind MIDI Controller. (Courtesy of Yamaha Corporation.)

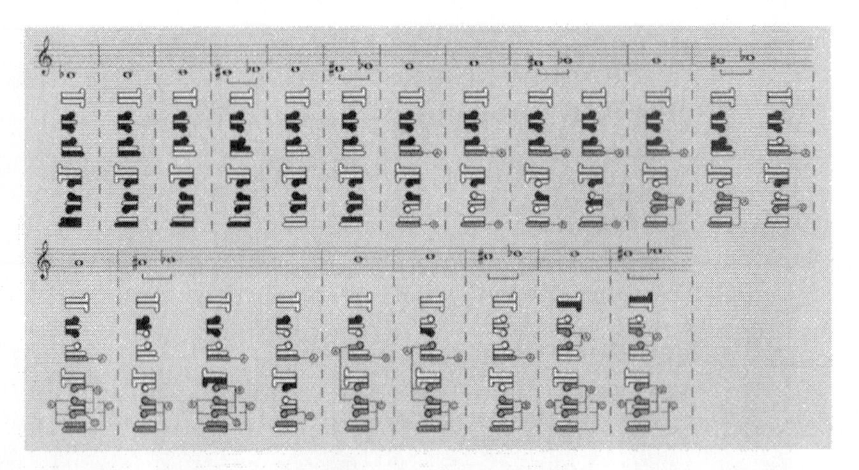

Figure 5-27 Fingering chart for the Yamaha WX5 Wind MIDI Controller. (Courtesy of Yamaha Corporation.)

embouchure and breath pressure, and its pitch can be changed so it will function as a transposing instrument.

Electronic music performance, ranging from the avant-garde to the truly bizarre, has become exceedingly diverse in recent years. With the development of high-speed computers and versatile synthesizers, a wide range of creative ideas has evolved using mixes of electronic and acoustic sounds, including interplay with a variety of Web-based materials. Electronic sounds can be presented alone, in combination with, or as accompaniment to singing or acoustic instruments. In some cases, recorded electronic sound is provided as an accompaniment to live performers; one challenge is to devise some way for the performer to interact with the recording as he or she would with another live performer.

Physics and music have been combined for musical performance with interesting results. One use of physics in music by composer David First involves development of a new "overtone series" based on *Schumann resonances*. Schumann resonances include the sequence of frequencies of circular electromagnetic standing waves encircling the earth in a spherical shell between the earth's surface and the ionosphere, believed to be caused by lightning strikes or other electromagnetic disturbances. Their frequencies, the first seven of which are 7.8 Hz, 14 Hz, 20 Hz, 26 Hz, 33 Hz, 39 Hz, and 45 Hz, are multiplied by a constant factor to raise them into the audio-frequency region. They form an ethereal combination that can be used as the tonality for electronic music. Acoustic instruments can join in the music as the inspiration comes. In music to be heard through stereo headphones, Mr. First has employed electronic tones to create 7-Hz binaural beats (see Section 6.9), and suggests that this leads the listener's brain to relax into an "alpha" state.

The avant-garde Kronos String Quartet has used signals from outer space obtained from solar physics research at the University of Iowa as a source of musical inspiration for the multimedia production "Sun Rings," composed by Terry Riley. Radio-frequency signals obtained during space research and converted to the audio range provide a variety of interesting sounds that can be emulated by the quartet or used as background for composed music. The Kronos quartet has also performed in a number of other nontraditional settings, including a CD recorded jointly with Huun-Huur-Tu, an ensemble of Tuvan throat singers.

Other composers have used the ethereal nature of the quantum-mechanical Schrödinger equation as an inspiration for their music, and even attempted to develop a brain–computer interface to help them create their music.

An important long-term goal for music-library organization involves indexing literally all music using computer representation or coding of their notes, rhythms, and sounds. Using such an index, a performer might identify music from the coded melodies, harmonies, or rhythms, or a researcher would be able to identify all uses of certain melodies in musical compositions over a period of one thousand years. The computerized classification of sound necessary for such a project is being actively investigated. Research such as this, in which tools of physics such as modern computers and electronics play dominant roles, is becoming more important in the musical world.

SUMMARY

(*Section 5.1*) Waves can be combined using several processes, including **simple addition, gating, amplitude modulation, balanced modulation** (sometimes called **double-sideband modulation** or **ring modulation**), **frequency modulation**, or **pulse-width modulation**. (*Section 5.2*) An **analog synthesizer** makes use of a series of electronic modules, each of which has a special function, to create and process musical sounds, but its awkward operation limits its use to studios. Signals in analog synthesizers can be either control signals or audio signals. Control modules include **voltage-controlled oscillator**; **voltage-controlled amplifier**; **envelope generator**; low-frequency **control oscillator**; low-pass, high-pass, and other **filters**; **noise generator**; and ring modulator. (*Section 5.3*) Modern **digital synthesizers**, extremely versatile instruments that are readily used in live performance as well as in studios, owe their popularity largely to the development of high-speed digital processing and inexpensive computer memory. Often called **keyboards**, they differ from earlier synthesizers in that the audio is processed as **digital** signals. A new set of electronic components required for this type of processing includes **digitizers, digital-to-analog converters, analog-to-digital converters**, and digital mixers. (*Section 5.4*) **MIDI** systems are now a standard for composing, writing, reading, and transferring music between mediums. MIDI labs with computer software such as Sibelius or **Finale** are used for teaching purposes as well as for composition of modern **electroacoustic music**.

QUESTIONS

1. **a.** Describe the wave shape and the effect on the sound of an audio tone undergoing

- **i.** gating
- **ii.** amplitude modulation
- **iii.** frequency modulation
- **iv.** balanced modulation

b. Define in context the carrier and modulator waves.

c. What are the musical terms for amplitude modulation and for frequency modulation?

2. Draw the following waves, indicating for each how you proceeded in setting up the axes for each carrier wave and its modulator wave:

   a. sine wave gated by square wave
   b. triangular wave amplitude modulated by a sine wave
   c. square wave frequency modulated by a square wave
   d. sine wave balance modulated by a sine wave

3. Differentiate between audio and control signals as used in musical synthesizers. For each, give two examples of how they might be used. What are the frequency ranges for the two types of signals?

4. Describe what each of the following synthesizer functions does:

   a. sample and hold
   b. voltage-controlled oscillator
   c. voltage-controlled amplifier
   d. envelope generator
   e. control oscillator
   f. low-pass filter
   g. high-pass filter
   h. bandpass filter
   i. notch filter
   j. noise generator

   Give an example of where each of the functions might be used.

5. Draw a synthesizer block diagram that would be used to produce the sound of a

   a. guitar
   b. bell
   c. flute
   d. drum
   e. toy slide whistle rising in pitch

6. Select a weird sound from the options, using a modern digital synthesizer. Draw the block diagram for an older analog synthesizer that you might employ to obtain the same sound.

7. Obtain a modern digital synthesizer and learn to operate it. If you also have an older analog synthesizer, compare the capabilities of each in the following areas:

   a. simultaneously playing two or more notes
   b. changing sounds easily and quickly
   c. developing your own new sound
   d. playing in unusual temperaments

8. Using a modern digital synthesizer, perform music using the following sounds:

   a. harpsichord
   b. trumpet
   c. clarinet
   d. bells

   How do the sounds of the instruments compare with the synthesized sounds? What might you suggest to make the sound more realistic?

9. Use a modern digital synthesizer set to its piano sound to perform standard piano music. How does the sound of the synthesizer compare with that of the piano?

PROBLEMS

1. Two sinusoidal tones of equal amplitude and frequencies of 500 Hz and 504 Hz respectively are added together. Determine the frequency of the tone produced, the beat frequency, and the equivalent carrier and modulator frequencies for a ring modulator to produce the same effect.

2. A ring modulator has a carrier frequency of 400 Hz and a modulator frequency of 6 Hz. Determine the two sinusoidal frequencies that can be added together to produce the same sound.

3. A 500-Hz sine wave is used as the carrier in a ring modulator, with a 50-Hz sine wave as the modulator wave. List the frequencies in the modulated wave.

4. A 440-Hz sine wave is used as the carrier in a ring modulator, with a 75-Hz sawtooth as the modulator wave. List the four most prominent frequencies in the modulated wave; show your work.

5. Suppose that you are using a fixed-frequency low-pass filter in producing your music with an analog synthesizer. If the filter passes up to the 16th harmonic for the note C_3, what harmonic will it pass for the note C_5?

6. Suppose that you are using a tracking low-pass filter in producing your music with an analog synthesizer. If the filter passes up to the fifth harmonic for any note, what is the maximum frequency passed for the note C_5? for the note F_6?

REFERENCES

ANDERTON, CRAIG. *MIDI for Musicians.* New York: Amsco Publications, 1986. One of several books written in a clear, readable style, providing an excellent survey of MIDI applications.

APPLETON, J. H., and R. H. PERERA, eds. *The Development and Practice of Electronic Music.* Englewood Cliffs, N.J.: Prentice-Hall, Inc., 1975. A collection of articles dealing with many aspects of electronic music, synthesizer, and recording techniques.

FRIEDMAN, DEAN. *Synthesizer Basics.* New York: Amsco Publications, 1986. A practical introduction to modern digital synthesizers at a modest level in a very readable style.

FRIEND, DAVID, ALAN R. PEARLMAN, and THOMAS D. PIGGOTT. *Learning Music with Synthesizers.* Milwaukee, Wisc.: Hal Leonard Publishing Corporation, 1974. A review of fundamental physics (Chaps. 1 to 4 in this book), an introduction to synthesizers similar to that of this chapter, and further details, including use of block-diagram patch sheets for the ARP Odyssey.

Synthesizer Basics, revised edition, revised by the editors of *Keyboard Magazine.* Milwaukee, Wisc.: Hal Leonard Publishing Company, 1987. This is a revised edition of an earlier Hal Leonard publication including a good summary of basic physics and an excellent survey of modern synthesizers.

YAVELOW, CHRISTOPHER. *Macworld Music and Sound Bible.* Harrisburg, Va.: IDG Books (division of International Data Group), 1991. This contains a wealth of information about synthesizers and MIDI applications.

Periodicals

Journal of New Music Research
Computer Music Journal

Recordings

BEAVER, PAUL, and BERNARD L. KRAUSE. *The Nonesuch Guide to Electronic Music.* Nonesuch Records, HC-73018 Stereo. A two-record set containing a great number of synthesizer sounds and special effects; it is accompanied by a booklet with explanations of the sounds and how to achieve them.

CARLOS, W. *Switched-on-Bach.* Columbia Records, Stereo MS7194. This is a classic in the field of electronic music, containing selections from the works of Bach performed on the musical synthesizer.

———. *A Clockwork Orange.* Warner Brothers, Stereo Record WB2573. From the sound track of the movie by the same name, illustrating the use of the synthesizer in combination with other musical instruments. Noteworthy is the voice synthesis in the Carlos version of Beethoven's Ninth Symphony.

Chapter 6

The Human Ear and Voice

In this chapter we shall first study the human ear and the process by which sound waves are received and transmitted to the nerve endings that convert mechanical vibrations into electrical impulses.

Our study will relate the basic anatomy of the peripheral auditory system to the important features of the place theory of hearing, the most widely accepted basic explanation of how the ear functions. We shall cover features of the frequency and amplitude responses of the ear and the mathematical concept of logarithms, which is related to the nonlinearity of both the amplitude and frequency responses of the ear. A brief discussion of some examples of hearing phenomena will conclude this study. Detailed discussion of the central auditory system, such as the auditory cortex, will be generally avoided.

We shall then study the parts of the vocal system and their relationship to the production of speech and musical sounds. We shall analyze speech patterns as we previously did musical-instrument sounds.

6.1 Peripheral Auditory System

The human ear is one of the most amazing organs of the body; it possesses an incredible range of sensitivity in both its frequency and amplitude responses. For example, the ear responds to vibrations over the range of frequencies from about 20 Hz to 20 kHz, a factor of about 1,000 in frequency, or almost ten octaves, whereas the eye is sensitive only to the range of electromagnetic waves having wavelengths between about 400 and 700 nanometers (billionths of a meter), a factor of less than 2 to 1 in frequency. By analogy, we might say that the eye sees less than an octave. The ear adequately responds to a range of pressure variations of about 1,000,000 to 1.

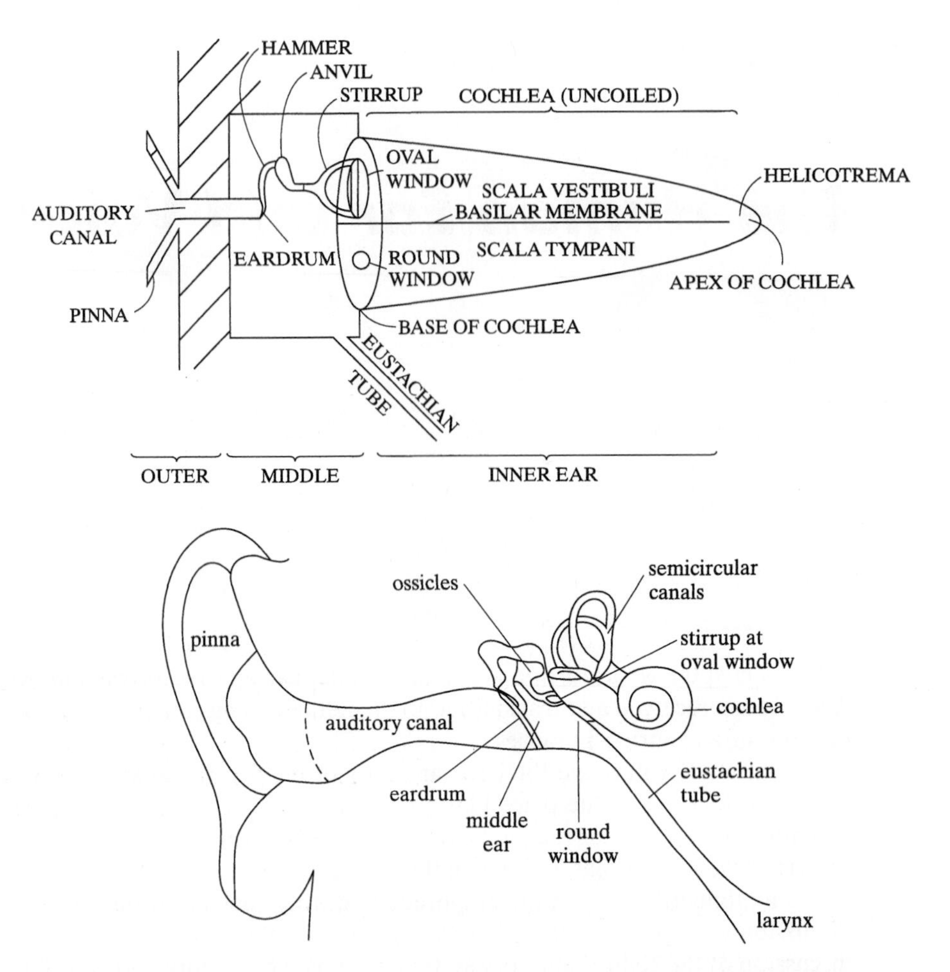

Figure 6-1 Schematic cross section (a) and approximately scaled drawing (b) of the peripheral auditory system.

The peripheral auditory system is divided into three parts: the inner ear, the middle ear, and the outer ear, as shown schematically in Figure 6-1(a) and in the approximately scaled drawing of Figure 6-l(b). The outer ear performs the function of focusing the sound waves onto the eardrum. The pinna acts as a collector, gathering sound waves and concentrating them into the auditory canal to a limited extent. It also aids in sound localization, by imposing subtle structure to the sound entering the ear canal as a function of position. In some animals, such as cats, a movable pinna further aids this end. The auditory canal transmits the sound waves to the eardrum. In addition, the length of the auditory canal helps to protect the very sensitive eardrum against some shocks and intrusion by external objects.

The eardrum separates the outer and middle ears. Perhaps it is inappropriately called a "drum," because the waves in the eardrum resemble traveling waves in a flap or a loose drumhead more than standing waves with resonances in a tight drumhead. A

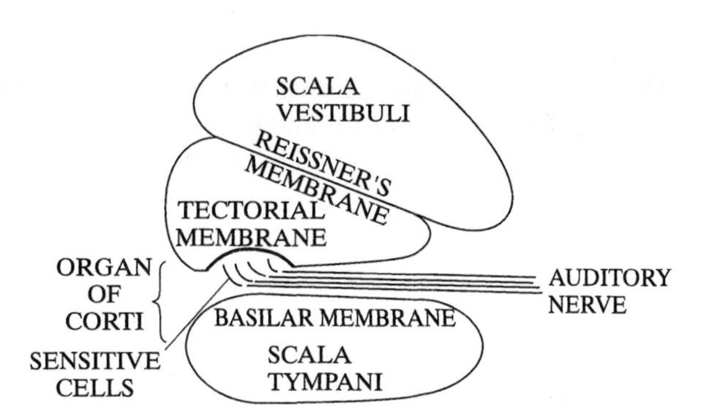

Figure 6-2 Schematic cross section of the cochlea, as if uncoiled.

very important function of the eardrum is to separate the outer and middle ears physically, so that the air pressure will not rapidly equalize between the two. This would reduce the amplitude of the vibration of the drum caused by the pressure changes of the impinging sound wave. Even a very small hole in the eardrum will reduce the pressure difference, particularly at low frequencies. The slower changes in air pressure caused by changes in the weather are equalized by air flow from the throat to the middle ear through the eustachian tube, which can be seen in Figure 6-1.

The middle ear consists primarily of the bone chain of the three *ossicles*: the hammer, the anvil, and the stirrup. These bones convert the small-amplitude vibrations of the eardrum into the larger-amplitude pressure oscillations required to set up waves in the fluid of the inner ear. Attached to the ossicles are muscles that help to limit the vibration of these bones for very-large-amplitude continuous sounds, and thus prevent damage to the middle ear. Unfortunately, sharp noises such as gunshots or loud music may occur too quickly for the protective mechanism to prevent damage to the middle ear.

The principal hearing organ of the inner ear is the cochlea, which is actually coiled on itself, like a snail, but is shown schematically in Figure 6-1 uncoiled for simplicity. The cochlea contains the nerves that convert the physical vibrations into electrical signals. The width of the coiled cochlea decreases along its length from base to apex; the typical cochlear cross section is shown schematically in Figure 6-2.

Vibrations from the stirrup enter the inner ear through the oval window, creating traveling waves in the fluid inside the scala vestibuli. The wave passes through an opening at the end of the cochlea called the helicotrema, and returns through the scala tympani to the round window, a flexible region of the base of the cochlea. The round window provides a point of pressure relief for the impinging traveling wave and damps the wave so that it will not reflect.

The scala vestibuli and the scala tympani are separated by the basilar membrane, which runs the entire length of the cochlea. The region along the basilar membrane containing the nerve endings that convert the waves to electrical impulses and transfer the vibrations to the auditory nerve is called the organ of Corti. The basilar membrane is approximately 3.5 cm long and contains about 30,000 nerve endings, often called *hair cells* because of their physical appearance, distributed fairly uniformly along its length. In actuality, the force that activates the hair cells results from a rather complicated shearing motion between the tectorial membrane and the basilar membrane, which occurs whenever a wave travels down the scala vestibuli.

Electrical impulses from these hair-cell nerve endings are transmitted by the auditory nerve to the brain, which relates the sound heard to those previously experienced and interprets the signals as words, music, noise, and so on.

In addition to its hearing functions, the inner ear contains the semicircular canals, which are responsible for detecting gravitational fields and acceleration and play an important role in balance.

6.2 Place Theory of Hearing

The position along the basilar membrane at which the maximum hair-cell and nerve response occurs is correlated with frequency. The *place theory of hearing* is based primarily on this observed correlation of frequency with position of response along the basilar membrane.

An important experimental observation is that, over the entire range of audible frequencies, two tones separated by an interval of one octave (a factor of 2 in frequency) excite regions equally spaced along the basilar membrane, slightly over 3.5 mm. This can be illustrated by the graph of frequency versus position of response shown in Figure 6-3. The

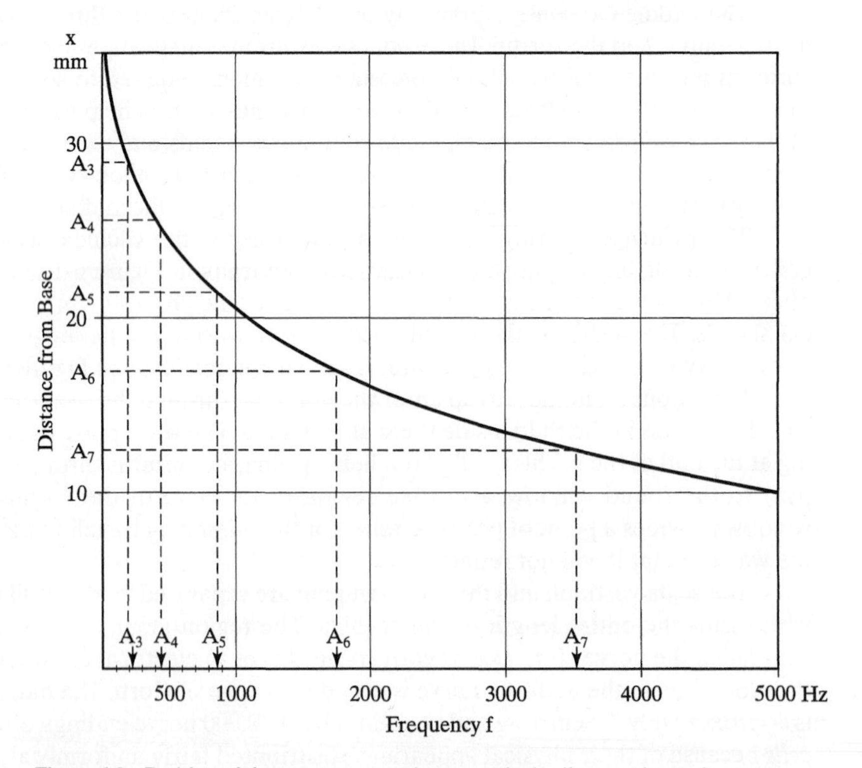

Figure 6-3 Position of the resonance maximum on the basilar membrane for a pure tone of frequency f (after von *Békésey*, 1960). (From *The Physics and Psychophysics of Sound* by Juan G. Roederer, Figure 2.8, page 25, © 1995 by Springer-Verlag, New York. Used with permission.)

curve in Figure 6-3 is not linear, because equal distances along the vertical axis do not correspond to equal frequency intervals but rather to octaves, that is, equal frequency *ratios*. The full range of human hearing, 20 Hz to 20 kHz, encompasses just under ten octaves; the full length of the basilar membrane of about 3.5 cm (35 mm) is divided into ten approximately equal intervals of one octave, each about 3.5 mm in length. It is interesting to note that equal musical intervals "sound" the same irrespective of the pitch level at which they are sounded. For identical musical intervals (at any pitch level) the physical spacings along the basilar membrane are approximately equal. It thus appears that the most important factor in determining how a musical interval between two pure tones sounds is the spacing along the basilar membrane of the points that respond to the two tones.

The response of the hair cells along the basilar membrane to a sinusoidal tone is not limited to a single receptor or even to a narrow band of receptors. Rather, the response is spread out over a relatively wide region along the basilar membrane. The region along the basilar membrane where the nerve endings produce a large response to a sinusoidal audible signal is called the *critical band*. Over most of the audible range the critical band corresponds to about 1.2 mm along the basilar membrane and includes about 1300 hair cells of the total of about 30,000. This corresponds to about 15 percent in frequency but is larger at lower frequencies. Musically, this is slightly less than a minor third (three half-steps) over most of the frequency range and almost one octave at frequencies below 200 Hz. The neural mechanism is capable of receiving this broad range of nerve impulses and narrowing this range down so that a single-pitch tone is perceived. This process, called *sharpening*, will be discussed shortly.

Listening to two pure tones sounding simultaneously, we can further investigate the effect of the critical band and experimentally determine its width. Suppose we set one of the frequencies at some value, say 1000 Hz, and vary the other continuously between 20 Hz and 20 kHz, keeping the two amplitudes equal. If the two tones differ greatly in frequency, we will simply hear them as separate. On the other hand, if their frequencies are very close, we cannot hear either tone individually. Instead, we hear a single tone, whose frequency is the average value between the two original tones, and beats that have a frequency equal to the difference between the two original frequencies. Starting with the two frequencies the same, we hear the single tone, with no beats, and, as one frequency is raised or lowered, we hear beats that become faster as the frequency difference increases. About the time that the beats become too rapid to distinguish, both tones will begin to become distinguishable, with a rather low-frequency roughness or coarseness heard in the background. As the frequencies become farther separated, the coarseness effect decreases and finally disappears. The frequency range over which the coarseness is heard is the region in which the critical bands of the two pure tones overlap. If we make the simplifying (though not exactly true) assumption that the response of the hairs in the critical band is symmetric in frequency about the center frequency (the frequency of the pure tone), the width of the critical band for a pure tone is just the difference between the two frequencies where the roughness begins to appear as the tones are brought together in frequency.

This effect can also be demonstrated by playing two high notes on recorders or flutes. A dull, low-frequency roughness can be heard when the two tones are a musical interval of a minor third (three half-steps) or less apart, indicating that the critical band is at least the width of a minor third in that frequency range. At low frequencies the critical band is large; a minor third played around G_2 sounds rough. This is a partial explanation of why composers rarely use small musical intervals in the bass range.

What is the smallest frequency change that is perceptible, and is it limited in some way by the very large width of the critical band? We can perform another experiment to determine how sensitive our hearing mechanism is to small frequency changes. A sine wave of audible frequency is frequency modulated by another sine wave with a frequency of about 1 Hz; this produces a tone that slowly varies up and down in frequency, centered on the frequency of the original sine wave, an effect called vibrato. The greater the amplitude of the modulating wave, the greater the change in the frequency. If the amplitude of the pitch variation is decreased, there will be a point at which the changes in frequency become imperceptible. Likewise, if we start with a very small or zero frequency change and increase the frequency limits very slowly, there will be some point at which the pitch change becomes barely perceptible. The frequency difference between the two extreme values at that point is called the frequency *just noticeable difference* (frequency JND). The JND can be measured over the complete range of audible frequencies by changing the original frequency and repeating the experiment. The frequency JND is between about 0.5 percent and 0.6 percent over most of the audible range but is greater at extreme high or low frequencies. The frequency JND is about one-tenth of one half-step of musical interval over most of the audible range, increasing to about one quarter-step (half of a half-step) for low bass notes.

The frequency JND is much smaller than the critical band, which is more than a factor of 10 larger at most frequencies. That is, despite the rather large width of the region excited by a single frequency tone, we can distinguish changes in the position of the peak of that response of less than one-tenth of its width. This neural process of assigning a single frequency to the wide band of excitation along the basilar membrane is called *sharpening*. The sharpening effect is enhanced for complex waves (such as square waves) by the additional frequency information in the harmonics. As a result of sharpening, two square-wave tones can become much closer in frequency before the occurrence of coarseness resulting from the overlap of their critical bands; complex tones at musical intervals of less than one half-step are normally easily resolved by the ear.

We have seen that the human ear can generally discriminate two very closely spaced frequencies sounded sequentially, owing to the sharpening of the neural network; however, it is much more difficult to resolve two pure tones whose frequencies are very close when they are sounded simultaneously. The minimum frequency separation between two sinusoidal tones sounded simultaneously, which can be perceived individually, is known as the *limit of frequency discrimination*. For two pure tones, this is between one and two half-steps of the scale (about 7 percent) for low frequencies and increases to about a minor third (three half-steps, or about 15 percent) for high frequencies. This limit of discrimination can be demonstrated by using two sine-wave oscillators, which produce beating when the spacing between the two tones becomes too small for them to be distinguished individually. However, as most musicians are aware, two notes played only one half-step apart *can* usually be perceived as separate notes. This results primarily from the additional sharpening effect of the neural system because of the existence of higher harmonics in complex musical tones.

The relationship of the JND and the limit of discrimination can be demonstrated by analogy to the sense of touch on the skin of the inner forearm (between the wrist and elbow), using two pencil erasers touching the forearm. Have a friend touch your forearm at either one point or at two points simultaneously to illustrate the following effects. If you are touched once, then again a short distance along the arm, it is easy to feel relatively small changes in the position at which you were touched. If your arm is touched simultaneously in two closely spaced points, you will find it very difficult to resolve the two touches; that is, it is difficult to feel that there are two separate, simultaneous touches. In fact, the spacing between the two touches must be somewhat larger than the minimum discernible change of position of a single touch before the two contact points can be resolved. That is, the sense of touch position along the arm (analog to the JND) is much finer than the sense of resolution of the arm to two separate simultaneous touches (analog to the limit of frequency discrimination). This can be viewed as resulting from the very large region of response along the arm to a touch at a single point, analogous to the width of the critical band. This effect is similar on skin anywhere on the body. The basilar membrane can thus be viewed as a very highly sensitive piece of skin with a particular sensitivity to vibrations in the audible frequency range.

It should be noted that all the experiments discussed here have involved judgments and are basically subjective in nature. In those cases in which a more exacting experiment is not available to measure a physical property of the ear, the result is taken to be the average value determined from a number of people with "normal" hearing. Lack of ability to hear some effects as well as other people does not necessarily imply a hearing defect, but can be due to a wide variation even among individuals with normal hearing. Nevertheless, despite differences between subjects in musical talent, training, and experience, these experiments should not be considered "unscientific." The important psychoacoustic results are to be related to physical properties of the ear mechanism and as such remain largely independent of the observer; in that sense, they can be considered "objective" results.

One additional point that should be emphasized is that the concept of the critical band is basically a psychological one, not a neurological one. The experiments described do not *demand* an explanation in terms of the place theory as described. However, recent research, including some experiments using the Mössbauer effect, has verified the correlation between frequency and position of response along the basilar membrane. In these experiments, a minute amount of radioactive Co^{57} is implanted on the basilar membrane, and motion of the membrane Doppler shifts the frequency of the emitted radiation, allowing identification of the regions of response to various frequencies.

6.3 Amplitude Response of the Ear

The *threshold of hearing* is the intensity of the smallest oscillation that can be perceived by the ear, and the *threshold of pain* is that intensity at which a wave becomes painful. The ear can sense pressure variations as small as 1 part in 10,000,000,000 of atmospheric pressure; an oscillation does not become painful to the ear until its amplitude reaches about 1 part in 10,000 times atmospheric pressure (1,000,000 times greater). These thresholds are plotted in Figure 6-4 as a function of frequency over the audible range.

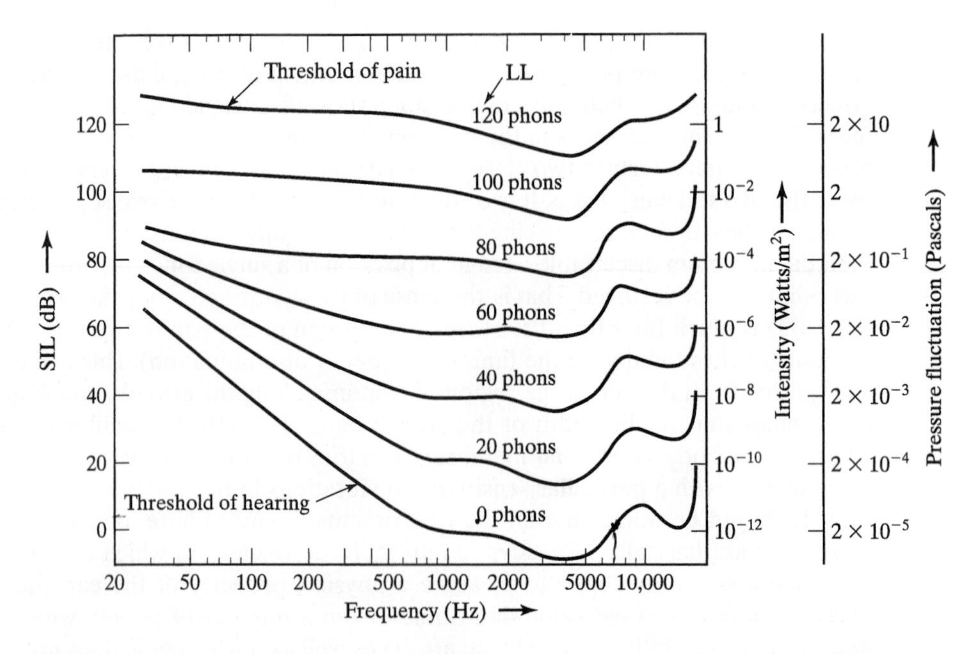

Figure 6-4 Equal loudness curves. The curve with the lowest level marks the threshold of hearing; the curve with the highest level marks the threshold of pain.

The range of pressure fluctuations over which the human ear is sensitive is significantly greater than the variation in pressure associated with changes in the weather. Our range of pressure sensitivity of over 6 orders of magnitude (factors of 10) is truly amazing. The minimum pressure level is roughly equivalent to placing the wing of a fly on the eardrum and corresponds to an oscillation that causes the eardrum to move about one-tenth of the diameter of an atom. This is the atmospheric pressure difference at sea level between one point and another about $\frac{1}{3}$ inch above or below it.

The extreme sensitivity of the ear is illustrated by the "seashell" resonance effect. When you hold a seashell to your ear, what you hear are the ambient, soft sounds that have been resonated and thereby amplified by the seashell cavity acting as a Helmholtz resonator. If the ear were slightly more sensitive to pressure variations, we would be continually bothered by the soft noises created within the body itself. In fact, in an anechoic chamber these sounds are audible, because they are not masked by background noise. The threshold of hearing is high at frequencies below about 100 Hz; if it were not, we would hear the many low-frequency oscillations in our bodies, such as those associated with our heartbeat and blood flow. Another low-frequency oscillation can be observed by listening to muscles of the fingers twitching at frequencies in this range. The low-frequency noise heard when the little finger is inserted into the ear canal is just such a muscle vibration that has been amplified slightly.

The intensity of a sound wave is measured in watts per square meter (W/m^2) and is proportional to the square of the pressure fluctuation, which is measured in Pascals (Pa). For instance, tripling the pressure variation corresponds to increasing the intensity of the wave by a factor of 9. At 1000 Hz, the threshold of pain has a pressure fluctuation 10^6 times as large as that of the threshold of hearing. Thus the intensity at the threshold of pain is 10^{12} (or 10^6 squared) times greater than that of the threshold of hearing. This can be verified on the two right-hand axes in Figure 6-4. The minima in the loudness level curves near 4 kHz and 12 kHz are the effect of resonances in the ear canal acting as a 2 cm closed tube.

Other units shown on the figure, dB and phons, will be discussed shortly.

6.4 Logarithms and Sound-Intensity Scales

In our analysis of the frequency response of the ear, we saw that the ear did not respond linearly to frequency stimuli but rather that the same *ratio* between two frequencies is always perceived to be the same musical interval. A similar type of nonlinearity occurs in the ear's response to signals of varying amplitude. To express these ideas simply and quantitatively, we will need to use logarithms.

By definition, if $y = 10^x$, then the logarithm of y, written $\log y$, is equal to x. In other words, $x = \log y$. For example, if $y = 100 = 10^2$, then $\log y = 2$; if $y = 1 = 10^0$, then $\log y = 0$. The logarithm of 1,000,000 or $\log 10^6$ is 6, and the logarithm of 0.001 or 10^{-3} is -3. Table 6-1 gives the logarithms of the numbers 1 through 10. These logarithms are fractions between 0 and 1.

Logarithms have two properties that make them particularly useful when describing the amplitude response of the ear: (1) numbers with large ranges have logarithms that do not cover a large range, and (2) steps in *equal ratios* have logarithms that differ by *equal steps*.

Because the range of intensity to which the ear is sensitive is so very large (about a factor of 1,000,000,000,000, or 10^{12}), dealing directly with intensities is cumbersome. It is convenient to define a new scale in which this enormous range can be expressed in a more compact manner. The *sound intensity level* (SIL) scale, with the fundamental unit of the bel, named after Alexander Graham Bell (1847–1922), is such a scale. A decibel (dB) is one-tenth of a bel. The sound intensity level of a tone of intensity I relative to a tone of intensity I_0 is defined as follows:

$$\mathrm{SIL} = \mathrm{SIL}_0 + 10 \log \frac{I}{I_0}, \tag{6.1}$$

where I_0 is the intensity (in watts per square meter) of the reference tone whose sound intensity level is SIL_0 (in decibels). The choice of this reference tone intensity is equivalent

Table 6-1 Logarithms of Integers

$\log 1 = 0.00$
$\log 2 = 0.3$
$\log 3 = 0.5$
$\log 4 = 0.6$
$\log 5 = 0.7$
$\log 6 = 0.8$
$\log 7 = 0.85$
$\log 8 = 0.9$
$\log 9 = 0.95$
$\log 10 = 1.00$

to setting the zero point of the scale. When dealing with human hearing, the threshold of hearing at 1000 Hz is chosen as the reference level. Its intensity has been measured as about 10^{-12} W/m^2, and its sound intensity level defined as 0 dB. The logarithm (power of 10) definition of the sound intensity level implies that for every increase in the intensity by a factor of 10 the sound intensity level increases by 10 dB; likewise, for every increase in the intensity by a factor of 100, the sound intensity level increases by 20 dB, a factor of 1000 corresponds to 30 dB, and so on.

The sound intensity level of the threshold of pain at 1000 Hz (intensity of 1 W/m^2) can be directly calculated using the formula

$$\text{SIL} = \text{SIL}_0 + 10\log\frac{I}{I_0} = 0\text{ dB} + 10\log\frac{1\frac{\text{W}}{\text{m}^2}}{10^{-12}\frac{\text{W}}{\text{m}^2}} \tag{6.2}$$

$$= 10\log 10^{12} = 10 \times 12 = 120\text{ dB}. \tag{6.3}$$

This can be confirmed by checking the vertical axes at the right of Figure 6-4.

We can go from intensity to SIL, and vice versa. Suppose that we know that the SIL of a given tone is 20 dB. What is its intensity?

$$\text{SIL} = 20\text{ dB} = \text{SIL}_0 + 10\log\frac{I}{I_0} = 0\text{ dB} + 10\log\frac{I}{I_0}. \tag{6.4}$$

Therefore,

$$\log\frac{I}{I_0} = 2 \text{ or } \frac{I}{I_0} = 100 \tag{6.5}$$

and thus

$$I = 100 \times 10^{-12}\frac{\text{W}}{\text{m}^2} = 10^{-10}\frac{\text{W}}{\text{m}^2}. \tag{6.6}$$

What is the SIL if two such sources are played simultaneously? The total intensity of the combined sound is just the sum of the two individual intensities, or 2×10^{-10} W/m^2. The SIL is then

$$\text{SIL} = 10\log\frac{2 \times 10^{-10}\frac{\text{W}}{\text{m}^2}}{10^{-12}\frac{\text{W}}{\text{m}^2}} = 10\log 200 = 10(\log 100 + \log 2) \tag{6.7}$$

$$= 10(2 + 0.3) = 23\text{ dB}. \tag{6.8}$$

Here we have used the fact that $\log(ab) = \log(a) + \log(b)$, where $a = 100$ and $b = 2$. We can also check this result by applying the formula with $\text{SIL}_0 = 20$ dB and

$I = 2I_0$. We are comparing the SIL of two sources to that of the original 20-dB tone. Then

$$\text{SIL} = 20\text{ dB} + 10\log\frac{2I_0}{I_0} = 20\text{ dB} + 10 \times 0.3\text{ dB} = 23\text{ dB}. \tag{6.9}$$

In doubling the intensity, the SIL has only increased by 3 dB. This is true no matter what the original intensity of the sound.

Combination of three identical 20-dB sources gives

$$\text{SIL} = \text{SIL}_0 + 10\log\frac{3I_0}{I_0} = 20\text{ dB} + 10 \times \log 3 \tag{6.10}$$

$$= 20\text{ dB} + 10 \times 0.5\text{ dB} = 25\text{ dB} \tag{6.11}$$

and the combined SIL increases 5 dB to give a total of 25 dB. Combination of ten identical 20-dB sources gives

$$\text{SIL} = 20\text{ dB} + 10\log\frac{I}{I_0} = 20\text{ dB} + 10\log\frac{10I_0}{I_0} \tag{6.12}$$

$$= 20\text{ dB} + 10\log 10 = 20\text{ dB} + 10 \times 1\text{ dB} = 30\text{ dB}. \tag{6.13}$$

The combined SIL increases 10 dB to give a total of 30 dB. Again we see this convenient result: An increase in the intensity by a factor of 10 increases the SIL by 10 dB.

Combination of 100 such sources of equal intensity gives

$$\text{SIL} = \text{SIL}_0 + 10\log\frac{100I_0}{I_0} = 20\text{ dB} + 10 \times 2\text{ dB} = 40\text{ dB}, \tag{6.14}$$

raising the SIL by 20 dB to a total of 40 dB. Thus, if one source produces a SIL of 20 dB, it requires 10 such sources to produce a SIL of 30 dB (a 10-dB increase), and 100 sources to reach a SIL of 40 dB (another 10-dB increase). It would require 1000 such sources to provide a 50-dB SIL, 10,000 for 60 dB, and so on. Similarly, it would require 10^{12} identical incoherent or random sources, each with SIL of 0 dB, to raise the SIL from the threshold of hearing (0 dB) to the threshold of pain (120 dB).

The decibel scale of sound intensity level is important, because our ears tend to interpret changes in intensity of a certain number of decibels as being roughly equal changes in loudness, independent of the actual intensity of the signal. That is, increasing the SIL of a 1000-Hz tone in 10-dB steps will be interpreted by the listener as increasing the loudness of the tone by roughly equal increments. This is based on subjective responses by a large number of people with normal hearing.

We can perform an experiment to determine the minimum observable intensity variation of a pure tone as we determined the minimum observable frequency difference; similarly, this quantity is called the intensity JND (just noticeable difference). Such a measurement, made over the entire audible frequency and intensity range, shows that the intensity JND is approximately 1 dB over most of the audible range, varying between about 0.5 and 1.5 dB. The JND is not a certain amount in *intensity,* but is an almost constant *fraction* of the intensity (about 25 percent) of the tone that is being varied and therefore approximately a constant in the logarithm of the intensity, or a constant in SIL units (dB). This result reinforces our choices of the dB SIL scale as being fundamental in interpreting our response to stimuli of different intensity. Not only is it a more convenient scale with which to work, but decibel units are also more useful and consistent in viewing and interpreting the response of the ear (equal loudness increments and intensity JND) over a wide range of audible stimuli.

One limitation of the decibel sound intensity level scale is that it does not take account of the subjective changes in loudness of different-frequency tones whose intensity is the same. Such changes in apparent loudness with frequency arise from the variation of the threshold of hearing and human sensitivity with frequency. Curves of equal loudness are shown in Figure 6-4. The *loudness level* or *phon* scale *does* take this variation into account. The unit of loudness level is the phon. One phon is the loudness level of a pure tone of 1000 Hz whose sound intensity level is 1 dB; 1 phon is equal to 1 dB at 1000 Hz. The loudness level of a pure tone of arbitrary frequency and intensity is equal in phons to the sound intensity level in decibels of a 1000-Hz tone that is judged to be of the same loudness as the original tone. The curves on Figure 6-4, called *Fletcher-Munson* curves, represent tones whose loudness is the same, two of which happen to be the threshold of hearing (0 phons) and the threshold of pain (120 phons).

One additional scale of loudness should be mentioned: the scale of *subjective loudness*, measured in units of *sones*. The preceding scales can be considered as physical scales, derived from physical measurement of some quantity like intensity or by comparing an arbitrary tone with one whose intensity is known with respect to a physical scale. On the other hand, the subjective scale is set up to take account of judgments as to what intensity sounds twice or three times as loud as a reference intensity. The subjective scale employs a new unit, called the sone, which marks equal steps in subjective loudness. Because of the subjective nature of the sone scale and the objective nature of the decibel scale, the two are not simply related.

This difference can be illustrated in the following way. When two 50-dB tones of arbitrary frequency are added, the sum is 53 dB. It has been shown, however, that the resulting tone will be judged as twice as loud only when the two tones differ greatly in frequency. When the two tones are close in frequency and their critical bands overlap, subjective loudness additivity breaks down; the combined tone is judged to be considerably less than twice as loud as either component tone.

Each scale discussed has features that make it useful in certain specific fields of study. The sound intensity level in decibels, the scale that is most closely associated with physically measurable quantities, is very convenient and is therefore used most often in the world of physics and engineering. We shall use this scale regularly in our study of recording and reproduction of sound and in our study of room acoustics.

6.5 Periodicity Pitch and Fundamental Tracking

If two or more tones whose frequencies are successive harmonics in some overtone series are sounded simultaneously, an additional frequency will be "heard" by most listeners, particularly trained musicians. The neural mechanism of the ear is able to identify the two frequencies as being members of an overtone series and assigns to the total stimulus the frequency of the fundamental. This phenomenon is called the *periodicity pitch*, *missing fundamental*, or *subjective fundamental*. If one plays a sequence of pairs of tones that have a frequency ratio of 3 to 2 but have different fundamental frequencies, the ear will recognize the periodicity of each pair of notes and supply the fundamental tone for each; this is called *fundamental tracking*. For instance, if tones with frequencies of 200 Hz and 300 Hz are played simultaneously, followed by simultaneous 300 Hz and 450 Hz tones, you will track a 100 Hz tone changing to a 150 Hz tone. Fundamental tracking is used in the production of very-low-frequency organ tones by sounding the second and third harmonics together and allowing this auditory mechanism to insert the fundamental. By using two pipes in the 16-foot octave, the designer can eliminate the need for bulky and costly 32-foot fundamental pipes. This effect is distinct from the phenomenon of difference tones, which will be discussed next.

6.6 Aural Harmonics and Combination Tones

Let us consider in more detail some effects of the nonlinearity of the response of the ear to intensity. Figure 6-5 shows a large-amplitude sine wave. Beneath that is a graph related to the logarithm of this sine function. This change of shape might represent the distortion of an incoming sine wave by the nonlinearity of the ear mechanism. That is, the top graph might be the graph of the pressure reaching the ear; due to the nonlinear response of the outer and middle ear mechanism to intensity, the wave in the cochlear fluid might look more like the lower graph. Clearly, a Fourier analysis of this fluid wave would show other frequencies present in addition to the frequency of the original sine wave. These additional frequencies generated by the nonlinear amplitude response of the ear are harmonics of the original wave, because the period of the distorted wave is the same as that of the original wave. These frequencies, called *aural harmonics*, play an important role in hearing. Aural harmonics of a pure tone become significant when the tone is "loud"—that is, when the pressure varies over several orders of magnitude.

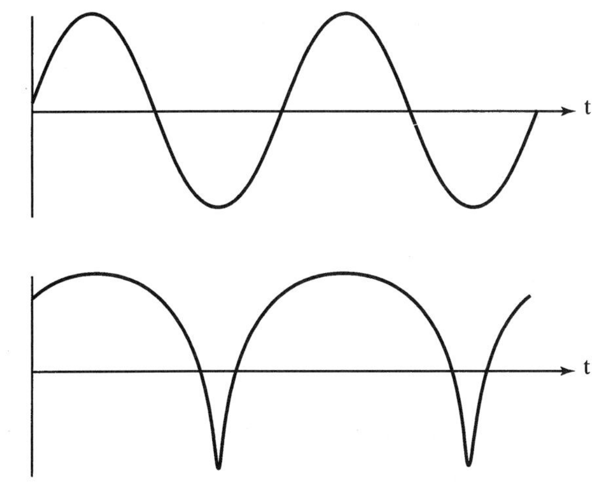

Figure 6-5 A sine curve and a curve logarithmically related to it.

Particularly important are the additional frequencies generated when two tones are played simultaneously at about the same intensity level. These frequencies, called *combination tones*, or *sum and difference tones*, occur when the intensities of the incoming waves again are great enough that the nonlinear response of the ear becomes important. In general, if the original frequencies are f_1 and f_2, combination tones will be generated at the frequencies $f_c = |nf_1 \pm mf_2|$, where n and m are any integers (1, 2, 3,...) and the vertical lines denote the absolute value of the quantity inside (for example, $|5.1| = 5.1$, $|-3.6| = 3.6$). If the plus sign is used, the tones $f_+ = nf_1 + mf_2$ are called *sum* tones; if the minus sign is used, the tones $f_- = |nf_1 - mf_2|$ are called *difference* tones. The larger the numbers n and m, the softer are the combination tones.

The difference tone $f_1 - f_2$ was first pointed out by the violinist and composer Giuseppe Tartini (1692–1770) in the early eighteenth century. He listened for difference tones as an aid in tuning double stops during his violin performances.

As an example, for two tones having frequencies of $f_1 = 500$ Hz and $f_2 = 700$ Hz, respectively, the combination tones are listed in Table 6-2 along with the numbers n and m. In general, these tones are significantly softer than the original tones; nonetheless, the first few difference tones can be heard clearly if they are well within the audible region and lower in frequency than either of the original tones. On the other hand, the sum tones are very difficult to hear, because in addition to being somewhat softer, they are always higher in frequency than the louder original tones and are effectively covered up or masked by the louder, lower-frequency tones (see Sec. 6.8 on masking).

Combination frequencies exist as components in the complex wave moving through the cochlear fluid, and as such are physically real—that is, actual physical vibrations. The nonlinearities in wave shape are introduced into the wave by the entire ear mechanism. We can therefore verify the existence of any sum or difference tone by producing beats with a pure tone slightly different in frequency from the combination tone. For example, playing a very-low-amplitude 1201-Hz sinusoidal tone simultaneously with the loud 500-Hz and 700-Hz tones will create beating at 1 Hz between the sine wave and the sum tone $f_+ = 500 + 700 = 1200$ Hz. Combination tones can be

Table 6-2 Combination Tones Produced from 500- and 700-Hz Pure Tones

n	m	f_+(Hz)	f_-(Hz)
1	1	1200	200
1	2	1900	900
2	1	1700	300
2	2	2400	400
3	1	2200	800
3	2	2900	100
1	3	2600	1600
2	3	3100	1100
3	3	3600	600

$f_1 = 500$ Hz $f_+ = nf_1 + mf_2$
$f_2 = 700$ Hz $f_- = |nf_1 - mf_2|$

contrasted with the periodicity pitch or missing fundamental, which exists only in the nervous system as an electrical impulse and not as a real physical wave.

We "hear" bass notes coming from very small loudspeakers, such as those on small transistor radios, particularly when the music is loud, even though the speakers cannot physically produce such frequencies. Two mechanisms contribute to this phenomenon. Difference tones created by adjacent harmonics in the overtone series of the music reinforce the fundamental, and the fundamental tracking mechanism aids the perception of the sequence of overtones as being associated with the fundamental frequency and therefore supplies the sound of the fundamental.

In the experiment in which we varied one sinusoidal oscillation in frequency about another fixed oscillation, we heard beats that rapidly became higher in frequency until they "felt" like a coarse buzzing effect. The frequency of this buzzing or coarseness, or of the beating, is just the difference frequency between the two oscillations; further separation of the two oscillator frequencies increases the frequency of the coarseness into the audible region, where it is heard as a difference tone if the sound is "loud."

6.7 Ohm's Law of Hearing

We saw in Chap. 4 that the principal factor in determining the quality of a steady complex wave is the intensity of the various overtones. If the peripheral auditory mechanism introduces nonlinearities into the incoming sound wave, and the central ear mechanism creates and inserts fundamental frequencies into the observed spectrum, what effect does this have on the quality of any particular sound? In the late nineteenth century, the two physicists Georg Simon Ohm (1787–1854) and Hermann von Helmholtz (1821–1894) first formulated the theory, known as Ohm's law of hearing, which states that the sound quality of a complex tone depends only on the amplitudes of its harmonics and *not* on their relative phases. This assumption is basically valid, as we have seen in our experiment involving the sounds of various shapes of standard waves. We have also seen that distortion nonlinearities occur when the amplitude of a wave is large, leading to slight changes in quality because of the additional harmonics.

Although Ohm's law well describes the relative effects of amplitude and phase of the higher harmonics on the quality of a complex audible tone, we can perform an experiment to demonstrate its limitations. If two pure tones at an interval of nearly, but not exactly, one octave (just about a factor of 2 in frequency ratio) are sounded simultaneously, a type of "beating" somewhat different from that discussed in Chap. 2 will occur, called *second-order beats* or *quality beats*. Quality beats are particularly strong when the intensity of the higher-frequency tone is reduced somewhat below that of the lower-frequency tone. In the case of regular (first-order) beats, the two tones alternatively interfere constructively and destructively, creating a change of amplitude that is perceived as variation of the intensity of the sum wave. In the case of second-order or quality, beats, as shown in Figure 6-6, the amplitude of the sum wave remains approximately the same, but the wave shape is continually changing at a rate that depends on the deviation of the two frequencies from the exact ratio of 2. Although the sound of the beating is somewhat different from that of first-order beats, it is clear that the mechanism of the ear is quite capable of detecting changes in the wave form.

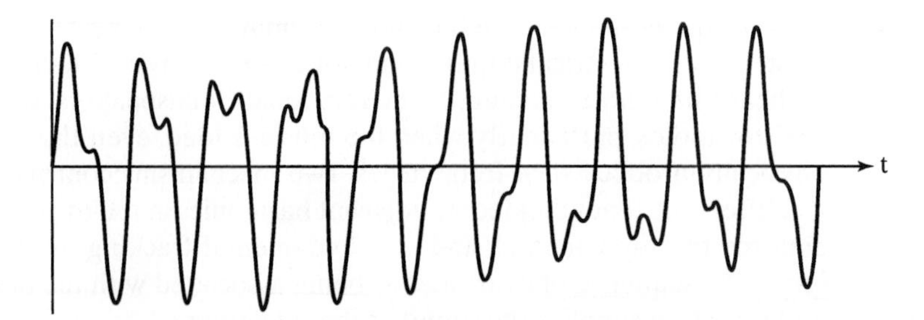

Figure 6-6 Complex wave formed from fundamental and slightly mistuned second harmonic whose amplitude is one-half that of the fundamental. Slight changes in tone quality occur as the pattern changes shape and are called second-order, or quality, beats.

Second-order beats are not due to nonlinearities in the ear, which might produce a second harmonic of the lower tone that beats, in the conventional sense, with the higher tone. We know this because second-order beats can be heard even when the two original tones are soft and nonlinearities are insignificant.

The relative phase of the two components changes continually and correlates with a perceived tone-quality change. Thus Ohm's law of hearing breaks down in this example. This phenomenon is only audible when the lower frequency is below about 1500 Hz.

Fortunately, Ohm's law of hearing is for the most part valid under conditions normally obtained when listening to recorded music. Because of the laws of electromagnetic induction, recording and reproduction processes that make use of magnetic fields typically introduce phase changes, which are a function of frequency, into the music. As a result, the phases of the various harmonics of a musical tone are changed during the recording and playback processes, leading to a distortion of the wave shape called *phase distortion*. Despite myriad phase-distortion problems, most high-quality audio systems render what even an audio expert would consider reasonably faithful sound reproduction. Recent advances in digital electronics allow more careful control of the phase at all stages of the recording and reproduction processes. As greater control of phase has become possible, more research is being carried out regarding the effect of phase distortion on audio sound, as will be discussed in Chap. 7.

6.8 Masking

The existence of one tone affects the perception of an *identical* tone with less intensity. In fact, because the JND of intensity is about 1 dB, the second tone must be raised in intensity to about one-quarter of that of the first before the listener can detect its presence:

$$\mathrm{SIL} = \mathrm{SIL}_0 + 10\log\frac{I}{I_0} = \mathrm{SIL}_0 + 10\log\frac{1.25 I_0}{I_0} \tag{6.15}$$

$$= \mathrm{SIL}_0 + 10\log 1.25 = \mathrm{SIL}_0 + 10 \times 0.1\ \mathrm{dB} \tag{6.16}$$

$$= \mathrm{SIL}_0 + 1\ \mathrm{dB}. \tag{6.17}$$

This effect, called *masking*, plays an important part in the perception of combinations of tones and in our ability to reject unwanted sounds in favor of those in which we have some interest.

To investigate the phenomenon of masking, we observe a soft pure tone of variable frequency, called the *masked tone*, in the presence of a louder tone that is fixed in frequency, called the *masking tone*. The *masking level* is the sound intensity level to which the masked tone must be raised for it to be heard in the presence of the masking tone. In the case of tones with identical frequencies, the masking tone is very effective: that is, the masking level is relatively high. In general, masking occurs readily for frequencies above the masking frequency; that is, low-frequency tones readily mask higher-frequency tones. On the other hand, the masking levels for frequencies less than that of the masking tone are relatively low, and even low-intensity tones of low frequency can be easily heard. In general, the masking level drops as the masked tone gets farther in frequency from the masking tone. The first difference tone between two loud sine waves less than an octave apart lies below the louder tones in frequency and can readily be heard. Conversely, the first sum tone has a frequency above those of the louder tones and generally is masked effectively.

It is important for the ear to be able to reject unwanted aural stimuli and noise in favor of the desired information. A standard hearing test includes a test of the ability of the subject to pick out a pure wave signal in the presence of others as well as background noise. Each ear can be checked individually by playing white noise into the other ear. This procedure eliminates the possibility of the second ear hearing the sound being presented to the first ear by bone conduction, or transmission of the vibrations through the skull. Masking helps explain our ear's ability to pick up a single conversation in the presence of others.

Masking is relevant to the perception of the timbre of complex tones. We perceive complex tones as a whole, and not as a series of individual pure tones (the harmonics present in its overtone series), for several reasons: the difference tones between adjacent harmonics reinforce the fundamental; the high harmonics are spaced closer together than the critical band and cannot be perceived as individuals; the high harmonics tend to be masked by the lower harmonics, including the fundamental; and the fundamental tracking mechanism reinforces the fundamental as the pitch of the note.

Masking is related to the critical band, in that tones mask each other when their critical bands overlap. This effect can be observed in some types of music, such as Bach's fugues. When the melodic lines are closely spaced, we hear a melody with rich harmonic texture. As the lines become more openly spaced, each can be heard more clearly as an individual melody line.

6.9 Binaural Effects

Thus far we have considered effects dealing with both ears together or that apply identically to both ears. We have seen that the paths of the sound from each ear to the brain are similar. In certain binaural effects it is also important to realize that the paths from the two ears are separate; the actual vibrations entering each ear do not mix physically. Rather, it is the *information* about these vibrations that "mixes" as neural signals in the brain.

If two very soft pure tones of about the same frequency are played simultaneously, one into each ear, a different type of beat, called *binaural beats*, may be heard. In the case of regular first-order beats, there is a physical mixing of the two beating tones. In binaural beats, however, the two tones must be soft enough that no bone conduction or direct physical transmission of the vibration through the skull occurs. This ensures that there will be no physical mixing of the two waves. In the case of binaural beats, the only mixing of the two waves occurs in the brain, where they are no longer actual physical waves, but rather sequences of neural impulses that will be interpreted by the brain as pure tones. The two tones cannot interfere in the physical sense, and thus no physical beating can occur. When the two tones are presented in this way, some people are able to hear a type of beating that sounds very different from either first-order (regular) or second-order (quality) beats. Binaural beats are often characterized as movement of the sound either within or around the head. This is true even when the two sounds are *very different* in intensity, in contrast with monaural beats.

Whereas it is possible for most people to match two pitches using first-order beats, it is extremely difficult to match pitches using binaural beats. If the two pitches are heard in separate ears, it is common for even a trained musician to be in error by over one half-step, and less experienced subjects often have larger errors.

The binaural experiment illustrates that the paths of the aural network to the brain are separate, and clearly demonstrates the necessity of actual physical mixing (that is, addition of displacements) of two waves to obtain normal (first-order) beats. A significant mismatch in pitch perception between the two aural networks, known as *diplacusis*, occurs to some extent in everyone and is rarely a serious handicap to a musician. A musician with severe diplacusis, however, may perceive the pitch for the part of the orchestra on his or her left as different from the part on the right, and this makes matching pitches with other members of the ensemble very difficult.

Our ability to localize sound sources in space is another binaural phenomenon. Two different processes are used, one for low frequencies (between 500 and 800 Hz) and another for higher frequencies.

Localization of low frequencies involves the phase, or time, difference between the arrival of a wave crest at one ear and its arrival at the other. Because of the spatial separation of the ears, a crest of a low-frequency wave will, in general, arrive at the ears at different times. The ear-brain system is sensitive to this small difference and uses it to localize the source. Simultaneous arrivals would indicate that the source is directly in front of (or behind) the observer, whereas an earlier arrival in the right ear would indicate that the source is somewhere to the right. The time difference corresponds to the angle of the source position relative to the observer.

The phase of high-frequency tones *cannot* be used for localization because there is ambiguity as to the relative phase. Imagine a sound source placed in front of and somewhat to the side of an observer. The path difference from the source to the different ears could equal one wavelength of the particular tone emitted. Crests would consequently arrive at the two ears simultaneously, even though one was emitted one period before the other. Based on the phase information alone, the ear-brain system might erroneously judge the source as directly in front of the observer.

Instead of phase information, the system uses intensity information for high-frequency localization. Because the wavelengths of the high-frequency tones are smaller than the diameter of the human head, they do not diffract as readily around the head "shadow" on the side away from the source. Thus one ear will receive a more intense signal than the other. The ear-brain system is sensitive to this intensity difference and can use it to localize the source. Because the longer-wavelength, low-frequency waves *readily* diffract around the head, the intensity difference between low-frequency waves at the two ears is small and unreliable for localization.

Binaural phase sensitivity is also important in sound localization when selectively listening to one of many conversations. Both binaural phase sensitivity and masking effects can be used to select which of many sound sources an observer will attend to. Binaural phase (as well as intensity sensitivity) is fundamental for stereophonic musical perception.

An experiment illustrating the extreme sensitivity of the aural system to differences between time of arrival of a signal at the two ears can be performed using a stethoscope. Tapping on the face of the stethoscope can be used to locate the center of the device. If the center of the stethoscope membrane is tapped, waves strike the two ears simultaneously. If, however, a point to the side is tapped, the waves strike one ear slightly before the other. In this manner, the center of the membrane can be located to an accuracy of about 1 mm.

As a sound wave reflects from the complex convolutions of skin in the pinna, subtle structure of the timing of reflections is imposed on the wave entering the ear canal. This structure is different for a wave coming from the front of the head than from the rear, for example, and the ear-brain system can distinguish these in many cases. This monaural effect works in conjunction with binaural effects to aid in sound localization, including elevation, the height above or below the head.

Another important phenomenon involving sound localization is called the *precedence effect*. In a shopping mall, there may be a large number of overhead loudspeakers, but the music always seems to emanate from the nearest one, regardless of its relative intensity. This demonstrates that sound localization is determined by the wave arriving earliest. In Chapter 8 we shall see that one must consider the precedence effect when designing auditoriums in which speaker systems are used.

6.10 Hearing Loss

Hearing loss can be divided into two basic types: temporary threshold shifts and permanent loss. Hearing loss resulting from damage or aging of the central hearing mechanism is truly permanent, but certain types of outer or middle ear problems can be corrected, at least partially, through surgery. Permanent hearing loss can be caused by physical damage to the ear mechanism, by disease, by drugs, or by the natural aging process.

Considerable study has been done recently on the effects of loud noises on the functioning of the ear. It is clear that prolonged exposure to loud sounds, particularly around or above 80 or 90 dB, can produce a temporary threshold shift. It is less clear what permanent damage, if any, is done by the multitude of loud noises to which most of us are regularly exposed. Continual exposure to amplified rock music and other loud sounds has been strongly implicated as a source of hearing damage.

Although there is a built-in muscular control in the middle ear to limit damage to the ear by sustained loud sounds, the ear is incapable of reacting rapidly enough to protect itself against very short bursts of sound, such as gunshots. Such sharp noises, if not excessively loud, may cause temporary threshold shifts in certain frequency ranges; if extremely loud, they can cause permanent loss of hearing by forcing some part of the ear mechanism beyond its elastic limit.

Diseases, such as infections of the middle ear, can also result in permanent impairment of hearing. A disease-related hearing impairment may occur to a fetus when a woman in the first (or possibly early second) trimester of pregnancy contracts the three-day measles, or rubella. When this occurs, immediate and irreversible hearing loss in the baby may result, apparently from malformation of parts of its inner ear. Rubella often carries only light symptoms and goes away very quickly, so the disease can escape detection but still affect the baby. For this reason, most doctors recommend vaccination against rubella for all women of childbearing age.

Certain drugs, such as streptomycin, have produced permanent hearing loss when consumed by small children. Ringing in the ears while under medication can be an early indication of such hearing loss. Fortunately, these side effects are well known, and the use of such drugs has been appropriately limited.

Normal aging of the skin and hair cells in the organ of Corti results in decreasing flexibility and springiness and therefore in a lessened response, particularly in the high-frequency range; such a normal aging process is called *presbycusis.* In its extreme, it can raise the threshold of hearing even at lower frequencies and necessitate the use of a hearing aid. One effect of presbycusis is a widening of the critical band (particularly at high frequencies) and thus a widening of the masking pattern. As a result, background noise more readily masks the incoming desired audible signal in some persons suffering from presbycusis. Because simple amplification by a hearing aid increases the intensity of both the desired signal and the noise, hearing aids do not improve the clarity of signals in such cases. Amplification of the desired source, such as talking louder or using a microphone at the speaker's lips, must therefore be used to obtain significant improvement.

Another common disorder is *tinnitus,* the presence of neurally generated sound sensations, sometimes described as a constant "ringing" in the ear. A common cause is believed to be exposure to intense narrow-band sound, which selectively damages hair cells in some small region on the basilar membrane.

If the hearing loss was caused by damage to the outer or middle ear, hearing can often be restored, at least in part, by surgery. In particular, there are techniques for replacing or bypassing the ossicles, with restoration of nearly full hearing. The eardrum can also be repaired or, in some cases, replaced by a skin transplant. On the other hand, damage to the inner ear mechanism is often irreversible, and thus far it has not been possible to restore hearing loss resulting from damage in the central auditory system. Loss of hearing because of inability of the mechanism in the ear to produce electrical impulses along the basilar membrane can often be overcome with a cochlear implant, as will be discussed shortly.

A technique for identifying hearing loss is for the person to obtain an *audiogram,* which shows the threshold of hearing for each ear over the entire audible spectrum. Other audiograms can be made later to determine whether further hearing loss has occurred. Such tests can only reveal certain types of hearing impairment. It is believed by some that hearing articulation (the ability to distinguish sounds with different transients) may suffer significantly even when the audiogram of a subject has shown only slight degradation.

6.11 Cochlear Implants

The development of the cochlear implant has had an enormous influence on the treatment of hearing loss. Over 100,000 of these devices have been implanted worldwide, approximately half of which are in children, down to the age of less than one year. Much hearing loss is due to cochlear damage, when the motion between the membranes is unable to generate electrical impulses. If the auditory nerve still functions, these implants are able to restore functional hearing to even profoundly deaf persons by providing that electrical stimulus along the cochlea. Experience has shown that children are capable of adjusting to the devices very quickly and effectively, and most develop normal language and speech. The success of the cochlear implant has even resulted in a significant reduction of enrollment in schools for the deaf. Cochlear implants are now routinely covered by health insurance for children and most adults, and are evaluated for cost to benefit for older adults.

The components of a typical cochlear implant are shown in Figure 6-7. A directional microphone (a) picks up the desired sound and sends it to a speech processor (b), worn behind the ear, that filters the sound into a series of frequency bands. A more versatile speech processor (c), including controls for adjusting levels, may be worn on the belt. The

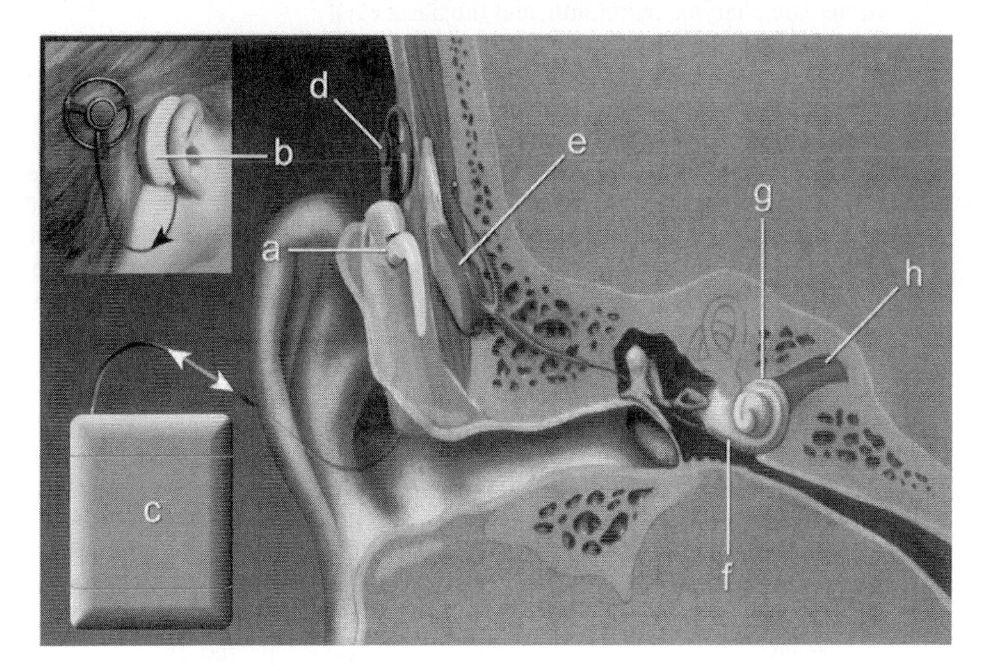

Figure 6-7 Components of a cochlear implant: directional microphone (a), speech processor (b) or optional more versatile speech processor (c), antenna (d), receiver-stimulator (e), and stimulating electrodes (f). Electrodes are implanted in the first turn of the cochlea (g); the auditory nerve (h) is also shown. (Courtesy of Professor Graeme Clark, The Bionic Ear Institute and The University of Melbourne.)

speech processor's output is sent to a small radio transmitter (d), held in place on the head by a magnet, that broadcasts the signals to a receiver-stimulator (e) implanted in the mastoid bone. These signals contain the amplitude of the electrical stimulus to be sent to each of the electrodes (f) that are implanted along the basilar membrane in the cochlea (g). This artificial stimulation thus replaces the stimulation of the nerves in the organ of Corti that would occur in the normal ear. In practice, the signal intensity, including the minimum and maximum sound intensity that will sent to each of the electrodes, is adjusted to be most effective and comfortable for the user. This implant uses 22 separate electrodes covering the frequency range below 100 Hz to above 6000 Hz. They are uniformly located along the basilar membrane so they are distributed at equal frequency ratios. The resulting pattern of auditory nerve activity in response to the sound provides a close representation of speech and other environmental sounds. The stimulation is then conducted to the brain by the auditory nerve fibers (h).

6.12 Anatomy of the Vocal Tract

The sound produced by a musical synthesizer is obtained by filtering or modulating a source of sound, such as a standard wave or some type of noise. As in the case of the synthesizer, vocal sounds are produced by a source of sound followed by a filtering and resonating system. The source of sound for the human voice is the vibration of the vocal folds, and the filter consists of the resonant cavities forming the vocal tract: the larynx, the pharynx, the mouth, and the nasal cavity.

A physical principle relevant to the operation of the vocal folds, as well as the lips of a brass instrument player and the reed of a reed instrument, is Bernoulli's principle, named for the Swiss mathematician and physicist Daniel Bernoulli (1700–1782). Bernoulli's principle states that in a fluid-flow situation (such as water or air) the pressure in the moving fluid is lower at places where the speed of flow is greater than at places where the flow is slower. An example is the "draw" of a chimney. Wind blowing across a chimney results in a low-pressure region just above the chimney, which pulls the smoke up the chimney. The pressure is lower just above the chimney than it is near the fireplace below, and the smoke is drawn out of the chimney. Some perfume atomizers employ the same principle.

In the case of air rushing through the vocal folds, there is a low-pressure area in the constricted region between the folds, leading to a Bernoulli force. The Bernoulli force and tension in the vocal folds cause the folds to close. Similarly, the Bernoulli force tends to close the lips of a player playing a brass instrument or the reed of a reed instrument.

The basic anatomy of the vocal tract is sketched in Figure 6-8. Air coming from the lungs is pushed through the vocal folds by muscles in the chest and stomach area, including the diaphragm. When the air begins to pass through the vocal folds, the Bernoulli force causes the folds to close. Immediately after the vocal folds close, air pressure builds up in the trachea, rapidly forcing the folds open once again. The burst of air through the vocal folds again creates the Bernoulli force, and the cycle is repeated. This process repeats at a rate determined by the tension in the folds, which in turn is controlled by muscles. The rate of opening and closing determines the frequency of the resulting vocal sounds. High muscle tension can be used by men to force the vocal folds to vibrate at a high frequency, producing a very pure, almost sinusoidal tone called *falsetto* voice.

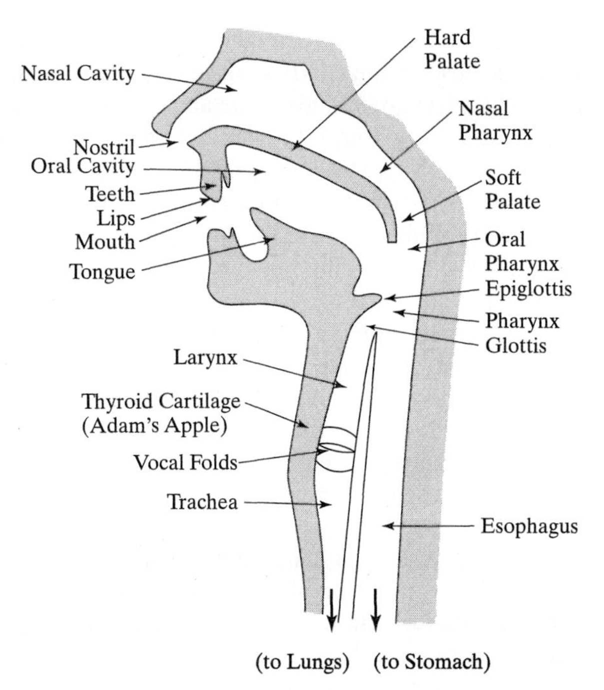

Figure 6-8 Schematic cross section of the human vocal tract.

The sound from the vibration of the vocal folds is modified according to the resonance characteristics of the other elements of the vocal tract. The most salient factor determining the Fourier spectrum of sustained vocal sounds is the resonant behavior of the air column formed by the larynx, the pharynx, and the oral cavity. This column acts as a closed tube, the closed end at the vocal folds and the open end at the lips. As we saw in Sec. 3.4, the resonant frequencies of such a tube are its odd harmonics, and in this case these frequencies can be manipulated by adjusting the shapes of the several cavities. The resonances associated with these frequency regions, called the *vocal formants,* are described in the next section.

The nasal cavity and the oral cavity, which can be adjusted in size and shape using the tongue, lips, and the soft palate, lead to additional formant regions, with typical sounds resulting from their use. The epiglottis protects the larynx against intrusion of food or other matter, and the thyroid cartilage, or Adam's apple, protects the vocal folds and the throat area in general from external damage.

6.13 Vocal Formants

As described in the previous section, the sound source for the vocal system is the vibration of the vocal folds; this vibration is brought about by a combination of the force of the air from the lungs causing the folds to open, followed by a Bernoulli force that causes them to close. The frequency of oscillation of the vocal folds is determined primarily by the controlled tension in the vocal folds, whereas the amplitude of the oscillation is affected by increasing or decreasing the rate of the airflow between the folds.

In general, the length of the "tube" forming the primary vocal resonator, between the vocal folds (its closed end) and the lips (its open end), is between 17 and 18 cm. Vocal formants are frequency regions in which harmonics have large amplitudes. The resonant frequencies of such a closed air column are at about 500, 1500, 2500, and 3500 Hz, and so on. The principal vocal formants are then at approximately these frequencies; however, they can be manipulated simply by either constricting or enlarging the diameter of the column at the position of one of the antinodes or nodes in the formant standing wave. Any component frequencies of the vocal-fold oscillations near one of these formants will be resonated by the vocal column.

For example, the fundamental, or first, formant has a velocity antinode at the lips. When the mouth is wide open, the velocity antinode is farther back in the mouth than when the lips are nearly closed. The effective length of the resonant tube, and therefore the lowest formant frequency, can be controlled by the size of the mouth opening. The amplitudes and frequencies of higher formants are determined by the complicated shape of the vocal tract. The lips are generally used to adjust the lowest formant frequency, whereas the higher formants are affected by the tongue and other parts of the vocal tract. The manipulation of the lips need not affect the frequency at which the folds vibrate. By manipulating the tongue and lips, an average adult male can vary the first formant frequency by almost 1000 Hz and the second formant frequency by almost 2000 Hz. Fortunately, one need not think about this as one speaks; it comes to us automatically. When the mouth is closed, the nasal cavity determines the formant structure. Because the shape of this cavity cannot be changed significantly, the vocal formant structure cannot be changed as much when the mouth is closed. Consequently, the range of sound qualities obtained with the mouth closed is limited.

Several of the vowel sounds, including "o͞o," "aw," "ah," "ĕ," "ĭ," and "ē," differ primarily by changing the relative frequencies of the first two formants. If you slowly and continuously say "o͞o-aw-ah-ĕ-ĭ-ē" (sort of a really long "why?") you are simultaneously (a) increasing the frequency of the second formant while (b) raising the first formant frequency (until ah), and then lowering it to its original frequency. Typical tongue and lip motion used to adjust the frequencies of the first and second formants can be observed by this sequence of sounds. A good example of a vocal formant occurs in the vowel sound ē. In this case there is a very strong emphasis of harmonics around the frequency of 2700 Hz. Table 6-3 lists the average frequencies of the first and second formants of a male voice for the above sequence of vowel sounds.

Table 6-3 Average Frequencies of First and Second Formant Regions of a Male Voice for a Sequence of Vocal Sounds

	1	2
o͞o	300 Hz	800 Hz
aw	600 Hz	800 Hz
ah	1000 Hz	1200 Hz
ĕ	600 Hz	2000 Hz
ĭ	450 Hz	2400 Hz
ē	300 Hz	2700 Hz

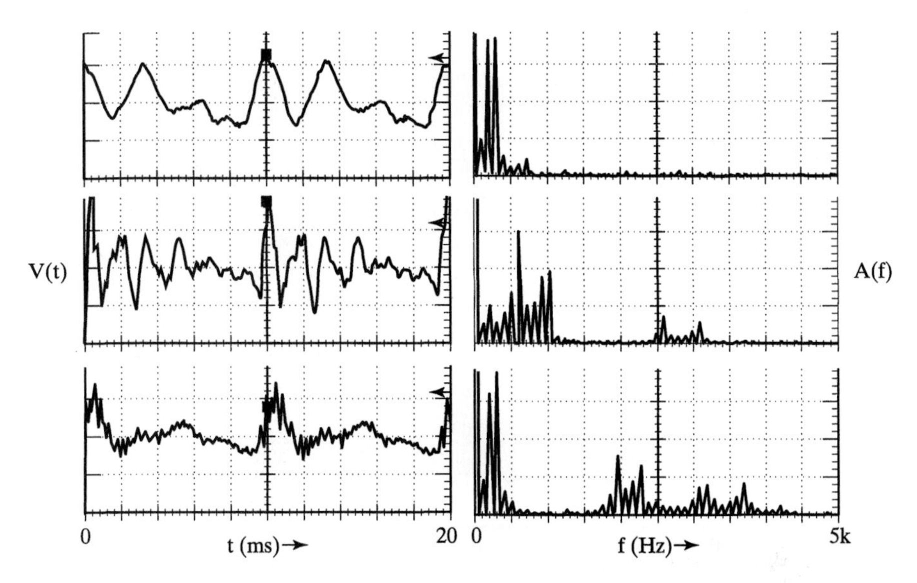

Figure 6-9 The wave shape (left) and spectrum (right) of three steady-state vowel sounds sung by a male voice at a frequency of about 100 Hz: o͞o (top), ah (center) and ē (bottom). Peaks in the spectra define the frequencies of the vocal formants.

Figure 6-9 shows wave shapes (at the left) and spectra (at the right) for a male voice singing o͞o, ah, and ē at a frequency of about 100 hertz. For the sound "o͞o" (top), the first formant is at about 250 Hz and the second at about 650 Hz. The first formant has risen to about 550 Hz and the second to about 950 Hz for the "ah" sound (center), and for the "ē" sound (bottom) the second formant has continued to rise to about 2000 Hz while the first formant has returned to about 250 Hz. In this case, the formants are below the average formant frequencies given in Table 6-3, because the vocal-tract components of the singer are above the average size. The region of emphasized frequencies between 2500 Hz and 4000 Hz arises from the nasal cavity acting like a complex Helmholtz resonator, and its envelope remains very similar over a large range of fundamental frequencies.

The data in Table 6-3 and Figure 6-9 can be presented using another type of graph called a *sound spectrogram* or *vocal spectrogram*. The voice spectrogram in Figure 6-10 shows the same voice singing the series of vowel sounds at a frequency of about 100 Hz. Frequency is plotted on the vertical axis and time on the horizontal axis; the darkening of the horizontal lines shows emphasized groups of harmonics. The entire sequence of vowel sounds takes a time of about ten seconds, seen on the horizontal axis; the vowel sounds are indicated at the top of the graph. The frequencies of the first two formant regions can be clearly seen; the first formant moves up and then back down as the second formant rises monotonically throughout the sequence of vowels. The higher formants are resonances of the nasal cavity. The little squiggles of the horizontal lines are vibrato. Voice spectrograms will be discussed further in the next section.

The effect of these formants is significantly different between men's and women's voices. In particular, when a soprano sings in the upper part of her range, the fundamental

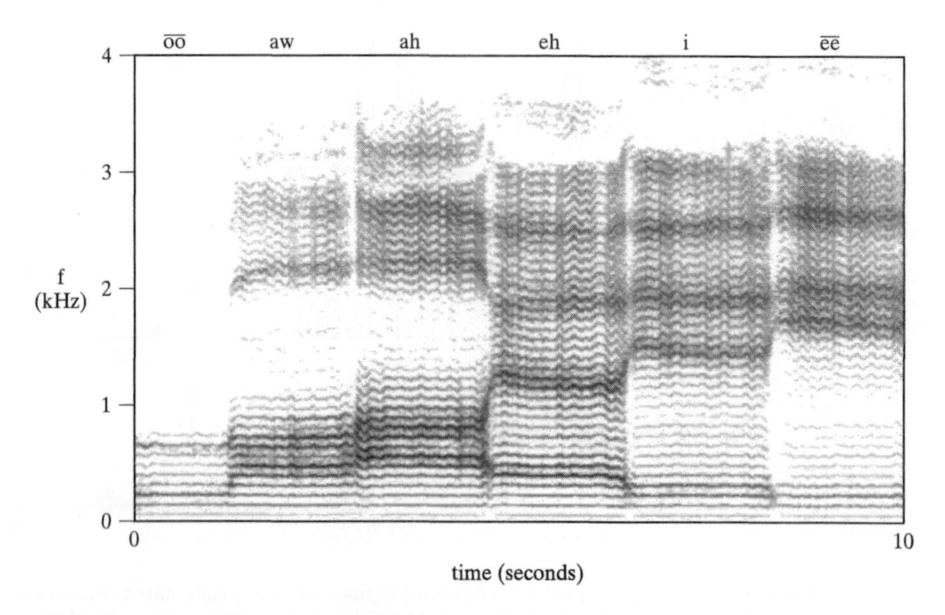

Figure 6-10 Voice spectrogram of vowels sung at a frequency of about 100 Hz. The darker lines define the frequencies of the vocal formants.

is generally significantly higher than the normal frequency of the first formant. The best "resonance" in a female voice, however, occurs when the frequency of the first vocal formant is adjusted to be equal to the fundamental frequency of the note sung. A soprano must therefore "drop her jaw" to obtain a very large lip and mouth opening in order to decrease effectively the length of the vocal tract and thereby raise the frequency of the formant. Unfortunately, this makes pronunciation of words while singing in the high range extremely difficult.

An additional feature of the male voice, known as the *singing formant,* is relevant in determining the quality of a male singer's voice. One difference between excellent and mediocre male voices is their ability to *project,* or to be heard with a rich tone above the accompaniment, such as an orchestra. An important part of this ability to project is the existence of the singing formant, which lies in the region between about 2500 and 3000 Hz. Emphasis of vocal harmonics in this range increases their amplitudes above that of the average of the orchestral accompaniment and allows the voice to be heard above the orchestra. Those male voices having a dominant singing formant have better projection than singers lacking such a formant. The normal speaking voice generally lacks emphasis of frequencies in the region of the singing formant.

An unusual example of manipulation of the vocal formants is the case of certain Tibetan monks who can produce the illusion of singing two notes simultaneously. Singing the note whose frequency is about 63 Hz, the first formant frequency is adjusted to about 315 Hz and the second to about 630 Hz. These correspond to the fifth and tenth harmonics of 63 Hz. Because every harmonic in the overtone series of the frequency of 63 Hz is present, that note is heard. The emphasis of 315 and 630 Hz creates the illusion that a note of 315-Hz frequency is also being sung simultaneously. The fifth harmonic of any note is up a musical interval of two octaves and one major third, so the combination of the two notes makes a pleasant-sounding, though unusual-sounding, chord.

During the late 1980s, a group called the Harmonic Choir became known for the ability of its members to perform a type of chant in which they are able to emphasize single vocal harmonics. As they sing a single sustained low note, the listener can hear individual harmonics in the overtone series of that note being emphasized in an upward or downward sequence. *Harmonic singing* has also achieved some scholarly attention (see References).

An amusing, if familiar, experiment involves inhaling a lungful of helium and singing or speaking. The vocal tract remains the same length. Because helium gas is considerably lighter than air, the velocity of sound is greater in helium. Using the relation $f = S/\lambda$, we see that, when the speed of sound increases and the wavelength remains the same, the frequency must increase. The frequency at which the vocal folds oscillate does not change significantly, but the vocal formants are increased in frequency, leading to the squeaky "Donald Duck" sound. The opposite effect of reducing the frequencies of the vocal formants can be achieved using a heavy gas such as sulfur hexafluoride, which has a density about five times that of air.

6.14 Analysis of Vocal Sounds

We have now seen the effects of the vocal formants on certain vocal sounds, particularly the vowel sounds. Vowel sounds are the simplest type of vocal sounds to analyze, because they are generally held for a longer time than the consonants, particularly the "plosives," which require the closure of the oral passage, such as the "p" sound in "pat."

"Continuant" sounds, such as "n" and "m," can also be easily analyzed, because they are long lasting. In both these cases, the mouth end of the vocal tract remains closed, and the sound is therefore dominated by the formants of the nasal cavity. The nasal cavity acts like a Helmholtz resonator, with a formant at the resonant frequency of the resonator. Other continuants can be similarly analyzed, as in the case of ordinary vowel sounds.

"Sibilant" sounds (for example, "s" and "z") include a component of almost white noise. In the case of voiced sibilants, like the letter "z", a discrete frequency spectrum is superimposed on the continuous spectrum forming the letter "s." For example, in the case of "z", the vocal folds are producing sound. Figures 6-11 and 6-12 show the Fourier spectra and wave shapes for typical "s" and "z" sounds; the voiced effect of the "z" sound is the origin of the discrete Fourier components superimposed on the noise spectrum of the "s" sound.

Thus far we have dealt only with steady state, continuous sounds, which can be characterized by a single Fourier spectrum graph giving their Fourier components at any time. If the sound changes, and we wish to describe a sequence of sounds forming a real word or diphthong, we must increase the information in the graph to account for the change of the spectrum with time. In this case, we are interested not so much in each harmonic as such but rather in groups of harmonics; in particular, we want to look at the formant emphasis versus time. A sound spectrogram plots the intensity of the various harmonics or formants on the vertical axis versus time along the horizontal axis.

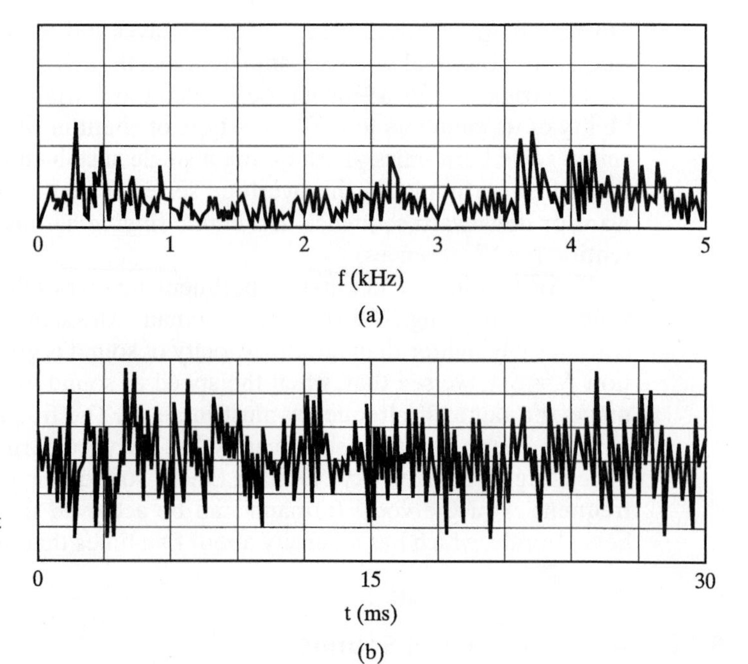

Figure 6-11 The Fourier spectrum (a) and wave shape (b) for the sibilant sound of "sss". Notice that the spectrum shows no indication of a spectral harmonic series but only a few isolated peaks, and the wave shows no apparent periodicity.

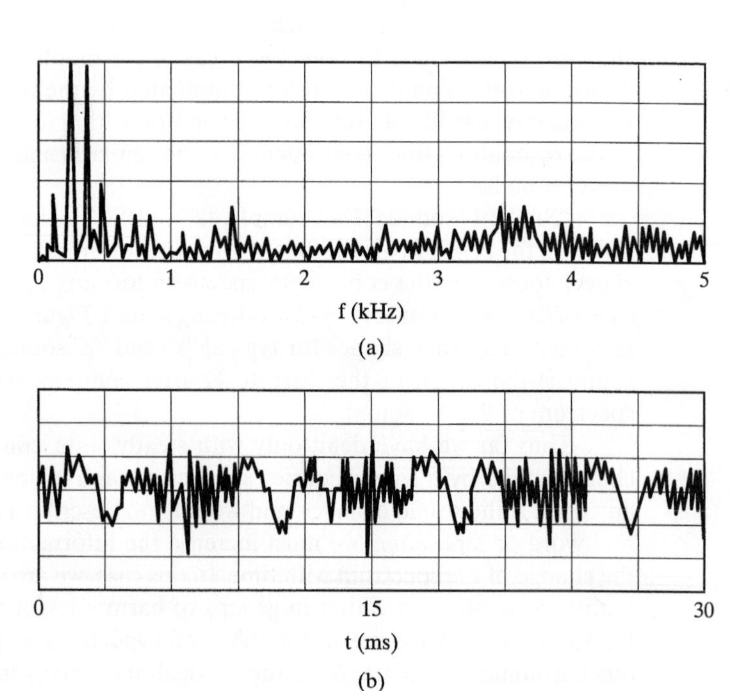

Figure 6-12 The Fourier spectrum (a) and wave shape (b) for the voiced sound of "zzz". Notice that the spectrum shows the harmonics of the voiced "z" sound superimposed over the broadband noise of the "s" sound of Figure 6-11. The wave shape shows some concomitant degree of periodicity.

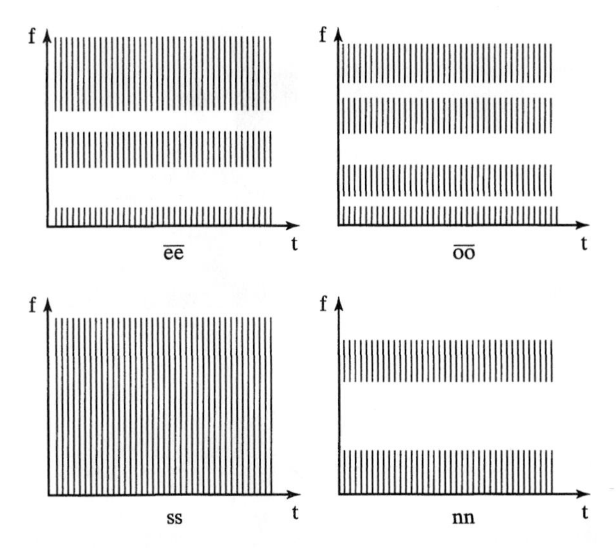

Figure 6-13 Schematic audio spectrograms of the sounds "$\overline{\text{ee}}$", "$\overline{\text{oo}}$", "ss", and "nn".

Simplified spectrograms of the vowel sounds "$\overline{\text{ee}}$" and "$\overline{\text{oo}}$" are shown in Figure 6-13. The vertical line at a given time is denser for Fourier components of greater intensity. There is no change with time for these sounds, so the graphs remain the same as time passes. Also shown are spectrograms for the consonants "s" and "n." The sibilant "s" is characterized by its noise spectrum; roughly equal amplitudes for all frequencies, whereas the "n" possesses a range of frequency components determined largely by the resonant characteristics of the nasal cavity.

Figure 6-14 shows simplified sound spectrograms for composite sounds, "n$\overline{\text{oo}}$", "s$\overline{\text{oo}}$", "n$\overline{\text{ee}}$", and "s$\overline{\text{ee}}$", all sung at a constant pitch. At the point where the sound

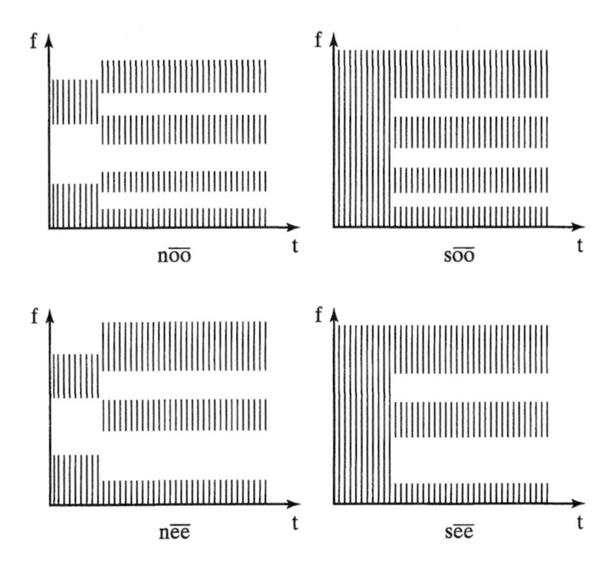

Figure 6-14 Schematic spectrograms of the sounds "n$\overline{\text{oo}}$", "s$\overline{\text{oo}}$", "n$\overline{\text{ee}}$", and "s$\overline{\text{ee}}$". Each of the horizontal bars represents a vocal formant.

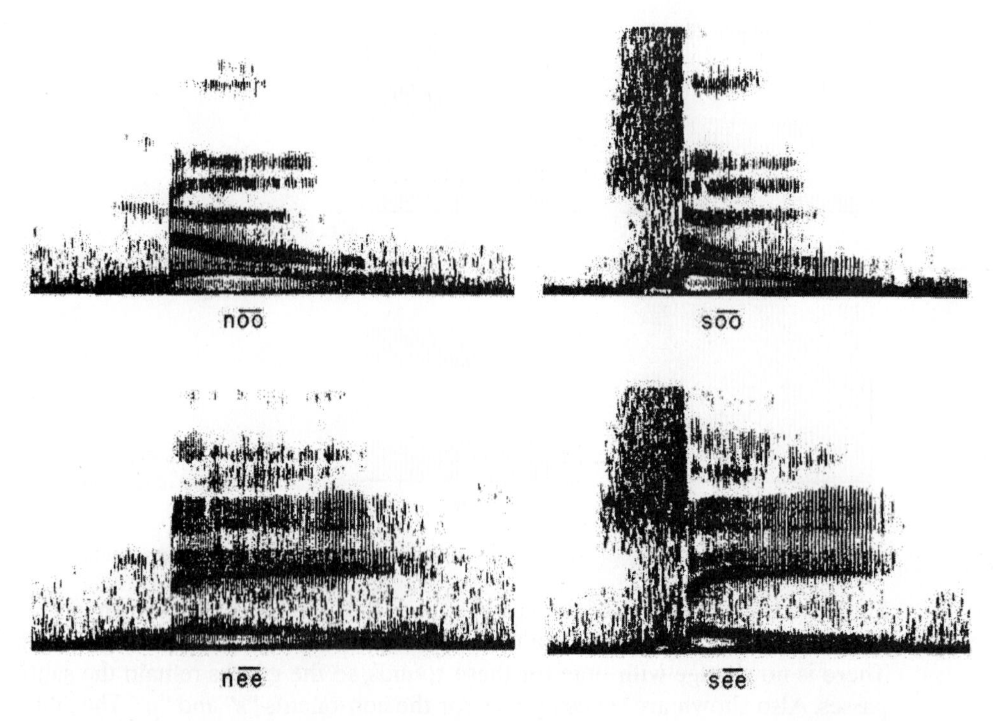

Figure 6-15 Actual audio spectrograms of the sounds "nōō", "sōō", "nēē", and "sēē", spoken by an adult male.

changes, the spectrogram changes form. Figure 6-15 shows real sound spectrograms of the same four composites shown in simplified form in Figure 6-14. As the intensity of any formant decreases, the line becomes lighter and finally disappears. In these cases the pitch and intensity of the voice naturally fall off toward the end of the vowel sound, as can be seen in the spectrograms. The maximum frequency in each spectrogram is 8000 Hz, resulting in a cutoff of the sibilant sound spectrum. Figure 6-16 is a comparison of a real voice and a musical synthesizer producing the word "wow". In this case, as in Figure 6-15, the continuum of changing vowel sounds, as well as the frequency and intensity contours, can be identified.

Plosives, consonants such as "b", "p", and "d", are more difficult to analyze, because they occur quickly. To describe a plosive, a graph of harmonic content as a function of time can be drawn similar to the preceding sound spectrograms, where the entire sound occurs within a period of a few milliseconds. Because this time interval is not significantly longer than the period of the Fourier components involved, it is very difficult to analyze the sound and obtain the amplitudes of the Fourier components.

With the recent development of compact, inexpensive, and durable computer memory with fast digital electronics, the field of voice recognition is experiencing an enormous expansion. Small portable language-translation devices are available for many modern languages. Voice-programmed cell-phone controllers are available, which, after being taught to recognize voice patterns of the

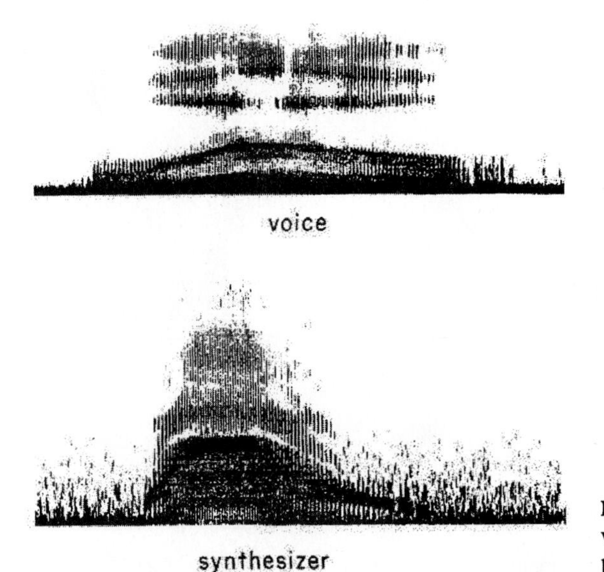

Figure 6-16 Audio spectrogram of the word "wow" spoken by an adult male and produced by a musical synthesizer.

operator, can be programmed by voice input. Most voice-recognition devices use an analog-to-digital sampling procedure, similar to that described in Section 5.3, to record the voice pattern, and a digital-to-analog procedure, such as that described in Section 7.10, to reconstruct the sound. Because of the important applications of this work to telecommunications, considerable work is now being done in the area of voice synthesis and pattern recognition at research facilities worldwide.

An important application of sound spectrograms lies in teaching the art of speech to those with serious speech impairments, or even to the profoundly deaf. One such device, the Computerized Speech Laboratory manufactured by Kay Elemetrics, is shown in Figure 6-17. The complete system includes a box containing electronic instrumentation and some computer software, with the computer screen acting as the display, as seen in the photograph. In one application, a person whose voice is to be emulated makes sample spectrograms of vocal sounds and, using the electronic spectral-analysis system and the computer display, the patient attempts to match this sample spectrograph, thereby learning correct pronunciation. The system can be used in a number of different ways, designed to meet the needs of the particular patient. Other applications allow the patient to match vocal frequency inflections, locate the average frequency of vocal formants, and plot graphs of the ratio of the second formant frequency to that of the first formant. For younger children, the vocal exercises are programmed as interactive games, and the spectral displays can be presented using attention-grabbing geometrical figures. The software allows a range of options in assessing and treating a number of vocal problems.

This system is used in a large number of audiology clinics throughout the United States, and has been strongly supported by Americans with Disabilities Act programs at elementary schools. Research into this procedure is continuing at a number of university clinics.

Sound spectrograms are also used in several other areas. The audio spectrograms of Figures 6-15 and 6-16 were taken using a spectrograph that was originally purchased

Figure 6-17 The Computerized Speech Laboratory system, by Kay Elemetrics. The system includes a chassis containing electronics (next to the computer CPU box) and software to manipulate the data and create the sound spectrogram display. (Courtesy of Kay Elemetrics Corp.)

by an entomologist to study the sounds of crickets. Spectrograms of bird calls are included in *Birds of North America: A Guide to Field Identification,* two of which are reproduced in Figure 6-18. Before reading on or reading the figure captions, see if you can guess the birds whose classic calls are shown in the figure.

Sound (a) is the screeching "jay-jay" cry of the blue jay; sound (b) is the gentle "ooah-ooo-oo-oo" cooing of the mourning dove.

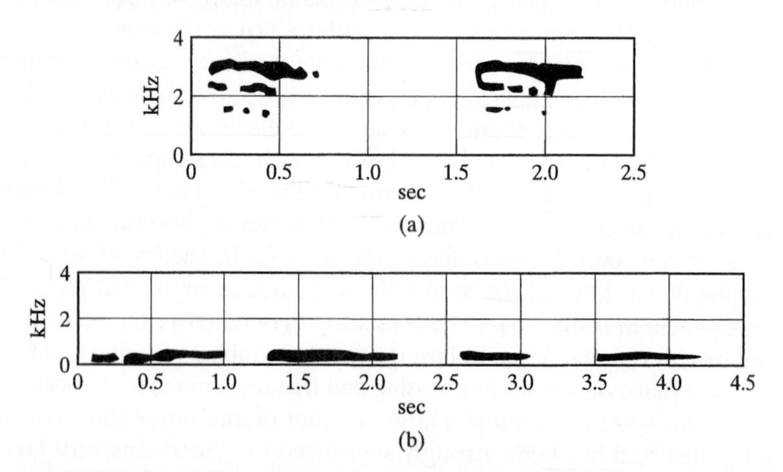

Figure 6-18 Audio spectrograms of (a) the cry of the blue jay, and (b) the cooing of the mourning dove. (From *Birds of North America: A Guide to Field Identification*, by Chandler S. Robbins, Bertel Bruun, and Herbert S. Zim. Copyright © 2001 by the authors. Reprinted by permission of St. Martin's Press, LLC.)

SUMMARY

(***Section 6.1***) The **peripheral auditory system** has three sections. The **outer ear** includes the **pinna** and the ear canal; the **middle ear** includes the three **ossicles**: the **hammer**, the **anvil**, and the **stirrup**; and the **inner ear** includes the **cochlea**. The **eardrum** separates the outer ear from the inner ear; the **base** of the cochlea separates the inner ear from the middle ear. Sound vibrations enter the cochlea through the **oval window** and move along the **basilar membrane** in the fluid of the **scala vestibuli**, pass through an opening called the **helicotrema** at the end of the cochlea, and return through the **scala tympani** to the **round window** at the base of the cochlea. Sensitive nerve endings, called **hair cells**, along the basilar membrane in the **organ of Corti** convert the vibrations into electrical impulses that are sent along the **auditory nerve** to the **central auditory system**, which is located in the brain. (***Section 6.2***) The **place theory of hearing** correlates frequency with location of the response along the basilar membrane. Although a very wide region along the membrane, called the **critical band**, is excited by a single frequency, through the process of **sharpening** the ear can distinguish a change in frequency, called the **just noticeable difference**, that is much smaller than the critical band, and even resolve two tones that are very closely spaced, called the **limit of frequency discrimination**. The ear has a **logarithmic frequency response**, so the quantities above are mostly constant fractions over the entire range of hearing. (***Section 6.3***) The ear also has a **logarithmic intensity response**, with a constant factor, the **decibel**, being the unit of **sound intensity level**. The ear possesses an enormous range of about 120 dB. **Equal loudness curves** are graphs of the decibel level of pure tones, as a function of frequency, that sound equally loud to your ears. (***Section 6.4***) Equal steps of loudness are described by equal SIL changes in decibels. Decibel units, which add, describe more accurately how sounds combine in the ear mechanism. (***Section 6.5***) **Periodicity pitch** involves a series of overtones heard by the ear as a complex tone at the fundamental frequency. **Fundamental tracking** occurs when a series of notes is produced by sounding only their overtones. (***Section 6.6***) **Combination tones**, including **sum tones** and **difference tones**, created by **nonlinearities** in the ear response, affect how we hear loud sounds by adding the fundamental to sounds lacking bass frequencies. (***Section 6.7***) **Ohm's law of hearing** suggests that the amplitudes of the harmonics are the most important feature in our perception of complex tones, although counterexamples, such as **second-order beats**, show that it is not entirely true. **Phase distortion** in audio systems changes the sound slightly but is not considered serious in quality reproduction systems. (***Section 6.8***) **Masking**, in which a louder tone tends to inhibit audibility of a softer tone, plays a part in the way we perceive complex waves, influences our ability to distinguish desired sounds from unwanted background noise, and affects our perception of music. (***Section 6.9***) **Binaural** hearing involves both ears, and differs from monaural perception in that the mixing of sounds from the two ears occurs in the brain and is not physical mixing. **Binaural beats** involve spatial perception rather than loudness, and can be heard even when the amplitudes of the two tones are very different. **Spatial localization** of sound is a binaural process involving phase difference between the two ears for low frequencies and intensity difference for high frequencies. (***Section 6.10***) Hearing loss, either **temporary threshold shifts** or **permanent hearing loss**, can be caused by disease, medication, or exposure to loud and/or sharp sounds. **Presbycusis**, the widening of the critical band with age, results in an inability to distinguish sounds or to reject background noise. An **audiogram** of the response of the ear over the entire audible frequency range can be used to evaluate hearing loss. **Tinnitus**, or ringing in the ear, can be caused by exposure to loud noises, disease of the ear, or reaction to some medicines. (***Section 6.11***) **Cochlear implants** have allowed profoundly deaf people to hear and to speak with great success. The device is implanted in the cochlea to stimulate the nerve fibers in a way similar to its natural stimulation in a functioning ear. (***Section 6.12***) The vocal folds are rapidly drawn together by the **Bernoulli effect** and pushed apart by the pressure of the air to create a sound very rich in harmonics; it is then filtered by the **larynx**, the **pharynx**, the **mouth**, and the **nasal cavity** to form vocal sounds. (*Section 6.13*) The primary **vocal formants**, created as odd-harmonic resonances of the vocal tract, can be adjusted in frequency by changing the position of the lips, tongue, and throat. Vowel sounds differ primarily in the location of the first and the second formants. Good male singers have a region of emphasized harmonics in the range of 2500–3000 Hz called the **singing formant**. Sopranos may have difficulty with singing diction in their high range, because the lower formants lie below the notes in their high range. (*Section 6.14*) The unique character of various speech or other sounds can be recorded on a graph of harmonic intensity versus time called an **audio spectrogram**. Audio spectrograms can be used as voiceprints and as examples for teaching speech to people who are deaf or have hearing impairments.

QUESTIONS

1. **a.** Trace a sound wave through the peripheral and central auditory systems to the stage in which sounds are identified in the brain.
b. Describe each component part and its individual function.

2. **a.** Discuss the fundamental hypotheses of the place theory of hearing.
b. Where are the nerves that respond to high frequencies? low frequencies?

3. Define the following and discuss the relationship among
a. critical band
b. just noticeable difference
c. limit of frequency discrimination
d. sharpening

4. **a.** Define periodicity pitch and fundamental tracking.
b. Give an example in which this effect is relevant to the perception of some musical stimulus.

5. From the graph of equal loudness contours (Figure 6-4) determine the following:
a. the loudness level of the threshold of hearing at 200 Hz
b. the sound intensity level of the threshold of pain at 10 kHz
c. the sound intensity level of a tone of 5 kHz whose intensity is 10^{-6} W/m^2
d. the loudness level, sound intensity level, and intensity of a 1500-Hz tone that is perceived to be the same loudness as a 7-kHz tone whose intensity is 10^{-12} W/m^2

6. **a.** How do second-order (quality) beats and binaural beats differ from normal (first-order) beats?
b. How are each of the three types of beats produced?

7. **a.** What is Ohm's law of hearing?
b. Give examples and applications and a counterexample.

c. Define masking.

d. In general, which tone, 800 or 1200 Hz, would more easily mask a 1000-Hz tone?

8. a. Discuss two processes involved in spatial localization of sound sources.

b. Which depends on high-frequency information? Which depends on low-frequency information?

9. a. What are audiograms and how are they used?

b. What should a normal audiogram look like?

c. Draw a "typical" audiogram for someone with a permanent threshold shift.

d. What might cause such a shift?

e. What is presbycusis?

10. a. What is Bernoulli's principle?

b. How is it used to explain the vibrations in human vocal folds?

11. a. What are vocal formants, and how are they important in determining vocal sounds?

b. What is the singing formant, and how does it help a singer?

12. a. What is a sound or audio spectrogram?

b. Draw possible audio spectrograms for the vocal sounds "o͞o", "e͞e", and "ah".

c. Draw simplified audio spectrograms of the sounds "zz" and "ss".

d. Describe how they differ.

13. Obtain a copy of the book *Birds of North America* (see References). Attempt to identify birds in your area by their sound spectrographs.

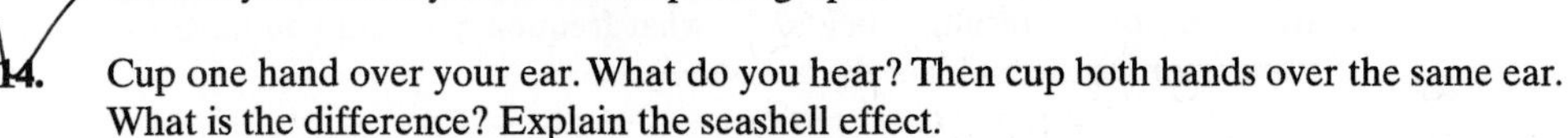

14. Cup one hand over your ear. What do you hear? Then cup both hands over the same ear. What is the difference? Explain the seashell effect.

15. Obtain computer-based sound-spectrogram software, and use it to observe your vocal formants. Compare the frequencies of your formants with those in Table 6-3 and Figures 6-9 and 6-10, and discuss any differences. Compare and contrast formants among students—for example, two female students, a male and a female, and two males.

16. Obtain a spectrum analyzer and use it to view your formant regions for a number of vowel sounds, as discussed in Section 6.13.

17. Obtain the spectrum of your voice for the vowel sound "e͞e" sung at a low pitch and at a high pitch in your singing range. Sketch the spectrum for each and identify the formant regions. Why should the envelope of the upper formant region be similar for "e͞e" sounds at any pitch for the same voice?

18. The ear canal can be viewed as a tube about 2 cm long with the eardrum acting as a closed end. What is the resonant frequency of that tube? Speculate on how this might be related to the frequency at which the ear is most sensitive.

PROBLEMS

1. Suppose the SIL of a single 5-kHz oscillator of intensity I_0 is 24 dB.

   a. What is the SIL of 100 such oscillators sounding together randomly or incoherently?

   b. 500 such oscillators?

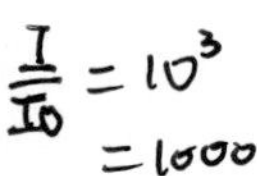

2. A clarinet plays a soft note with a SIL of 40 dB. By what factor must the intensity be increased in order to play the same note at a SIL of 70 dB?

3. If the whirring of a single fan produces a SIL of 50 dB, how many identical fans would need to be operated simultaneously to make a SIL of 63 dB?

4. A single violin plays a note at a SIL of 50 dB. If the violin section contains ten players, at what SIL would each have to play in order for the section to produce a level of 70 dB?

5. Tones of 650 and 800 Hz are sounded simultaneously at high intensity levels. Determine the frequencies of several low-order (m and n value up to 3) sum and difference tones, and state whether each would be likely to be audible. (Be sure to consider masking.)

6. Tones of 400 and 700 Hz are sounded simultaneously at high intensity levels. Determine the frequencies of several low-order (m and n value up to 3) sum and difference tones, and state whether each would be likely to be audible. (Be sure to consider masking.)

7. A 100-Hz sawtooth wave and a 202-Hz sine wave are sounded simultaneously. What is the frequency of the resulting beats?

8. A 100-Hz square wave and a 306-Hz sine wave are sounded simultaneously. What is the frequency of the resulting beats? To what frequency would you have to tune the square wave in order to eliminate the beats?

REFERENCES

BÉKÉSEY, GEORGE VON. *Experiments in Hearing*. New York: McGraw-Hill Book Co., Inc., 1960. The magnum opus of the modern theory of the ear, including the basic experiments validating the place theory.

KRELL, JOHN. *Kincaidiana, A Flute Player's Notebook*. Culver City, Calif.: Trio Associates, 1973. General book on flute playing and technique, containing the music for the "Trio for Two Flutes."

MOORE, B. C. J. *An Introduction to the Psychology of Hearing*. New York: Academic Press, 1982. An excellent modern introduction to psychoacoustics.

OLIVE, J. P., A. GREENWOOD, and J. COLEMAN. *Acoustics of American English Speech: A Dynamic Approach*. New York: Springer-Verlag, 1993. An excellent, modern technical review.

ROBBINS, CHANDLER S., BERTEL BRUUN, and HERBERT S. ZIM, illustrated by Arthur Singer. *A Guide to Field Identification, Birds of North America*. Racine, Wisc.: Western Publishing Company, Inc., 1983. Contains audio spectrograms of bird calls to help in their identification.

ROEDERER, JUAN G. *Introduction to the Physics and Psychophysics of Music*, 3rd ed. New York: Springer-Verlag, 1995. The best modern source of up-to-date information on the ear and psychoacoustics at the undergraduate level.

SCHAEFER, JOHN. *New Sounds: A Listener's Guide to New Music*. New York: Harper and Row, 1987. See especially Chapter 10: "The Oldest and Newest Instruments," which discusses the music and the technique of the Harmonic Choir.

SUNDBERG, JOHAN. "Formant Technique in a Professional Female Singer." *Acoustica* 32:8.

———. "The Acoustics of the Singing Voice." *Scientific American*, March 1977. These excellent articles describe the fundamental principles relating to voice production and vocal formants for female and male singers.

HUDSPETH, A. J., and VLADISLAV S. MARKIN, "The Ear's Gears: Mechanoelectrical Transduction by Hair Cells." *Physics Today*, February 1994. This excellent survey article provides an up-to-date summary of our understanding of the conversion of mechanical vibrations into electrical signals by the hair cells, and gives an excellent set of recent technical references.

CLARK, GRAEME. *Cochlear Implants: Fundamentals and Applications*. New York: Springer-Verlag, 2003.

Excellent up-to-date reference by researcher at the University of Melbourne and The Bionic Ear Institute.

VOROBA, BARRY. *Experimenting in the Hearing and Speech Sciences*. Eden Prairie, Minn.: Starkey Laboratories, Inc., 1978. Describes many experiments illustrating various features of the ear and the voice and serves as lab manual for the Starkey Labs Hearing Science Laboratory.

Recordings

HOUTSMA, A. J. M., T. D. ROSSING, and W. M. WAGENAARS. *Auditory Demonstrations*. (Phillips CD #1126-061) Woodbury, New York: Acoustical Society of America, 1987. A compact disc containing a large number of auditory effects and illusions in easily accessible format. Excellent material with excellent record quality.

DEUTSCH, DIANA. *Musical Illusions and Paradoxes*, Philomel Records, Inc., La Jolla, Calif., 1995. Several interesting illusions.

The Music of Tibet; The Tantric Rituals. An Anthology of the Worlds' Music, No. 6, Society for Ethnomusicology, Anthology Record and Tape Corporation. This exotic record presents an unusual mode of chanting by Tibetan monks who appear to sing two notes simultaneously; accompanied by reprints of articles explaining the religious significance of the chanting and the acoustical features of the sounds produced.

Voice of the Computer, Recording DL 710180, Decca Records. This is the original record using the computer to generate sound. It contains Shepard's tones (1) ascending and descending by half-steps in an example and (2) used in glissando form in a suite of electronic music.

The Harmonic Choir, DAVID HYKES, Director. *Hearing Solar Winds*, Ocora/Radio-France, 1983. Recordings illustrate harmonic chant, in which the performers are able to control the harmonics of their voices.

Chapter 7

Sound Recording and Reproduction

In this chapter we shall describe the principal features of high-fidelity audio systems. Following a brief discussion of basic electrical circuits, Ohm's law of electrical circuits (which is different from Ohm's law of hearing), and electrical power, we shall apply these concepts to audio reproduction systems. We shall then discuss properties and characteristics of the individual components in high-fidelity sound recording and reproduction systems. The aim of this presentation is to familiarize the reader with the physical principles on which the operation of this equipment is based, to acquaint the reader with the operational aspects of audio components, and to aid the reader in any future choice of reproduction systems or components.

7.1 Electrical Circuits and Ohm's Law

To understand electronic sound recording and reproduction we must first review the basics of electrical circuits. A simple electrical circuit is analogous to a hydraulic system consisting of a water pump, a tube carrying the water around in a complete circuit from the exit of the pump back to its entrance, and a constriction in the tube. The analogous hydraulic and electrical systems are drawn schematically in Figure 7-1. In the case of a hydraulic system, the pump pushes the water around the tube with a force that is dependent on the operation of the pump. The constriction provides a resistance to the water flow, limiting the rate at which water can circulate through the system. The water flow can be increased by enlarging the diameter of the constricted region of the pipe or by increasing the pressure with which the water is pumped. This can be expressed as an equation:

$$\text{water flow} = \frac{\text{pumping pressure}}{\text{resistance due to the constriction}}. \tag{7.1}$$

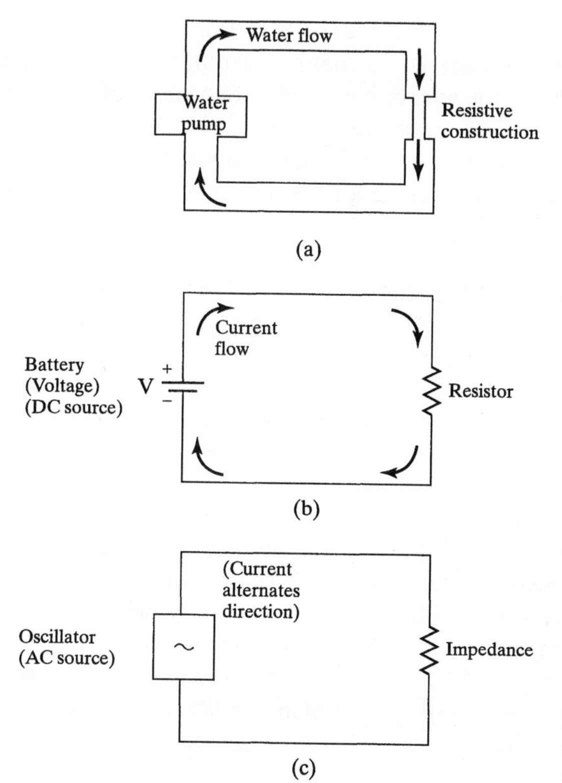

Figure 7-1 (a) Simple hydraulic circuit, and its electrical analog for the cases of (b) direct current, or dc, and (c) alternating current, or ac.

An electrical circuit can be thought of as a hydraulic system in which a battery or power supply replaces the water pump, an electrical resistor replaces the constriction in the tube, and the flow is the flow of electrons, or electrical current, rather than water. The equation relating these three quantities, known as *Ohm's law* (of electrical circuits), can be written as

$$I = \frac{V}{R} \quad \text{or} \quad V = IR, \tag{7.2}$$

where I is the electrical current in amperes (symbol A), V is the electromotive force or voltage in volts (symbol V), and R is the electrical resistance in ohms (symbol Ω, capital Greek letter omega). The battery provides the source of electrical current, which is effectively limited by the resistor. Increasing the applied voltage by adding more identical batteries in series or replacing the one shown with a battery of larger voltage will cause more current to flow through the circuit. Decreasing the resistance of the resistor will also allow more current to flow, consistent with both our hydraulic analog and the Ohm's-law equation.

This simple system is a direct current (dc) system, because the (positive) electrical current always flows in one direction, from the plus to the minus battery terminal. In reality, it is the negatively charged electrons that flow from the minus terminal on the battery to the plus terminal. The same laws apply with a slightly different mathematical formalism in the case of an alternating current (ac) system. The electrical power available at wall sockets in American homes is alternating current; 110-volt 60-cycle ac means that the electrical voltage at the socket has an effective value of 110 volts, and oscillates sinusoidally at 60 hertz. This voltage thus causes an electrical current that reverses itself at a rate of 60 hertz and is also sinusoidal in shape when a resistance is placed across the two sides of the circuit. Such a resistance might be a light bulb, an iron, or an electrical motor.

Another ac system is an audio system. When a microphone picks up a musical sound, the electrical signal it produces (which follows the wave form) varies about the zero in voltage, between positive and negative values. In this way current is forced first in one direction and then in the reverse direction.

Alternating-current circuits are generally more complicated than dc circuits, and the ac analog to Ohm's law is also more complicated. In particular, the phase of the signal can be affected by electrical components. We shall, however, ignore such complications and treat ac circuits as if Ohm's law held as described earlier. Because according to Ohm's law of hearing these phase alterations have little audible effect—at least for continuous sounds—our simplification is still informative.

An ac electrical circuit is also shown in Figure 7-1. In the simplest case, an oscillator replaces the battery, producing an oscillatory electrical voltage, which causes an oscillating electrical current of the same frequency to exist in the resistor. In the case of alternating current, the analog to resistance is called *impedance*. The current is limited by the impedance of the resistor and can be increased in amplitude by either increasing the amplitude of the voltage applied by the oscillator or decreasing the impedance of the resistor, or both.

When electrical current exists in an electrical appliance, the electrical power of the battery or oscillator is transformed into another type of energy, such as heat, light, sound, or other mechanical energy. For example, if a light bulb is connected to a battery as shown in Figure 7-1, the electrical power is converted into light and heat energy.

The mathematical formula for power in an electrical system is

$$P = VI, \tag{7.3}$$

where P is the power delivered to the resistor in watts (symbol W), V is the voltage applied, and I is the current in the resistor, as before. Using Ohm's law, $V = IR$, the formula for electrical power can also be written as

$$P = I^2R = \frac{V^2}{R}. \tag{7.4}$$

Thus we see that the greater the voltage and current applied to the resistor, the greater the power it consumes. Increasing the power available to an application can be achieved by increasing the applied voltage or decreasing the resistance of the device.

In an audio system consisting of many individual components, it is necessary to match the signal-level output by one component to the level that the next component expects. To transfer power most efficiently from the amplifier to a speaker it is also necessary to *match the impedance* of these two components. We shall discuss this in more detail later.

7.2 Reproduction and Amplification Systems

The goal of an audio system is to reproduce music as closely as possible to the original sound; a system that does this well is said to possess good *fidelity*. We shall now look at entire audio systems to study their general properties and functions, and investigate the ways in which the components are matched to achieve optimal sound production. Following this will be detailed discussions of each of the components.

The typical configuration of audio components in a complete sound system is shown in Figure 7-2. Signal sources might include microphone, tape recorder, $33\frac{1}{3}$ RPM disc record player, compact disc player, sound from a DVD video player, and an AM-FM tuner. The signals from these sources are input into a *preamplifier* that accepts each signal and produces an output of the appropriate amplitude for subsequent input to the amplifier. The *power amplifier*, or simply *amplifier*, accepts the low-voltage signal from the preamplifier and increases its amplitude and power level so that it can drive the speaker or speaker system.

We saw earlier that a sound wave will reflect whenever it passes from one medium to another of a different sound velocity. Likewise, alternating electrical signals may pass inefficiently from one audio component to the next one in the system. In the electrical case, it is more appropriate to consider this a result of the impedances of the two devices. The impedances of the two devices are what determine how much of the signal will be transmitted and how much will be reflected. Signals will pass most efficiently when the output impedance of the first component equals the input impedance of the second component. Such *impedance matching* results in greatest efficiency as the signal passes from component to component through the system.

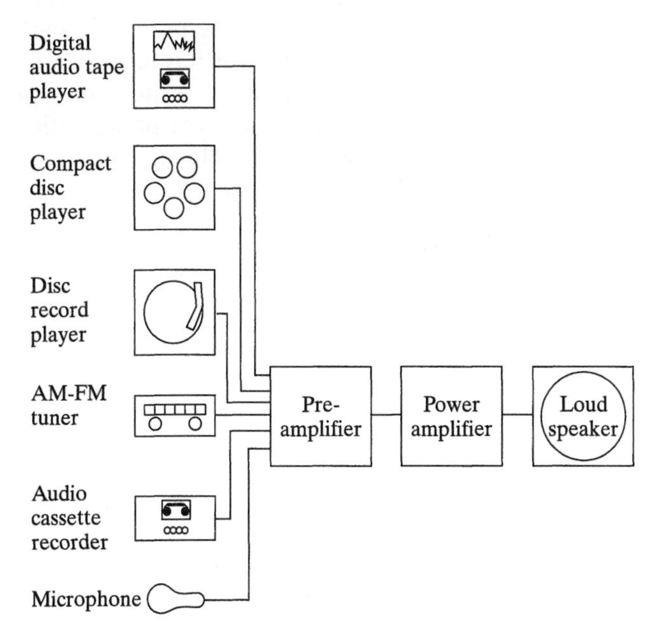

Figure 7-2 Layout of the components in a typical audio amplification and reproduction system.

The several components act as links in a chain, each accepting signals from the previous component, processing the signals, and passing them to the next component. The input impedance and the output impedance of a single component can be different, and they should be chosen to match the impedances of the neighboring components. For example, the output impedance of a microphone must match the input impedance of the preamplifier into which its signal is fed; the output impedance of the preamplifier must match the input impedance of the power amplifier; and so on.

If any component does not have the proper output impedance, or the source output impedance does not match the input impedance of the component into which the signal is fed, energy transfer will be inefficient. In extreme cases, the signal may be almost entirely lost. If the input impedance is too high, the output current will be low. If the input impedance is too low, the output voltage will be low. Because the power transferred is the product of the current and the voltage, an intermediate input impedance maximizes the power transferred to the device. Because the power requirements of the speakers are generally large, impedance matching is most crucial between the power amplifier and the speakers. Because the power output from a microphone is very low, impedance matching between the microphone and the preamplifier is also crucial.

An additional requirement is that the signal provided by any component be of the proper voltage level and capable of delivering sufficient power (or electrical current) to make the next component function. The range of voltages input to a component is important, as well. If too low a voltage is used, the ratio of signal to electrical noise from the component may be too small; if too high a voltage is used, the component may be overdriven and distort the signal. For example, a preamplifier output generally has neither a sufficient voltage level nor adequate output power to drive a loudspeaker.

Several types of microphones are commonly available; they can be subdivided into those with high output impedance and those with low output impedance. If the input to the preamplifier is not of the appropriate impedance, it is necessary to match impedances with an impedance-matching transformer or a low-noise audio amplifier, which is inserted into the line between the microphone and the preamplifier input. Microphone signal amplitudes are typically only a few millivolts, and a microphone is capable of delivering only minuscule levels of power, as can be seen from the equation for power discussed previously. The preamplifier into which the signal is fed must therefore have great sensitivity to small signals and the ability to amplify these low-voltage signals.

Most audio components produce output signals of around 1.5 V maximum amplitude, which is called *high level* or *line level*. This is done to reduce the accumulation of noise that can occur when very low-level signals are transported over wires between separate units. The signal is then fed into the preamplifier and on to the amplifier and speaker. These high-level signals are generally input into an impedance of about 50,000 ohms, with a resulting power transfer of a fraction of a milliwatt. Because power levels on the order of at least 10 watts are required to drive most loudspeakers, it is clearly necessary to introduce a power amplifier into the system to drive the speakers.

An AM-FM tuner receives the signals from either an AM or an FM radio station, demodulates the radio-frequency (RF) signal to obtain the audio-frequency signal (e.g., the music or announcer's voice), and amplifies this signal to the 1.5-volt amplitude level. Again, this signal is fed into a preamplifier whose impedance is about 50,000 ohms; thus the power transferred is only a fraction of a milliwatt.

The preamplifier is the most versatile component in an audio system and performs numerous functions. It is capable of accepting signals of various impedances and voltage levels and providing the appropriate output signal level and impedance for the power amplifier. It also contains the controls (such as volume, loudness, filters, and balance) for adjusting the output signal to various listening conditions and input signals. The preamplifier produces a 1.5-volt output, which is sent into the power amplifier, which has an input impedance of about 50,000 ohms.

The power amplifier is an extremely simple unit; its sole purpose is to accept a signal of about 1.5 volts at an impedance of about 50,000 ohms and increase its voltage and current such that the power is sufficient to drive a loudspeaker. It usually has no controls; all controls are on the preamplifier. Whereas the power level input into the amplifier is in the milliwatt range, typical power levels from power amplifier's input into speakers are between 10 and 100 watts. It typically requires less than 1 watt of acoustic power to obtain a comfortable audio level in a normal listening room. Speaker systems typically vary in their efficiency to convert electrical power into acoustic power between about 1 and 10 percent. Thus, for example, 1 watt of acoustic power could be obtained by using a 100-watt power amplifier driving a speaker with an efficiency of 1 percent or a 10-watt amplifier driving a speaker with an efficiency of 10 percent.

The choice of an amplifier and loudspeaker system must reflect the efficiency and impedance of the speaker and the maximum volume of sound desired. Most loudspeakers have an impedance of 8 ohms, but a number have a 16-ohm impedance, and a few have an impedance of 4 ohms. If the impedance match is not proper, efficiency will be lost, and in some cases the speaker will produce a distorted sound and possibly be damaged. Overdriving a speaker with too powerful a power amplifier can also cause damage.

Some complete units are available that contain tuner, preamplifier, and power amplifier in a single chassis, eliminating the need for concern about matching these components. In this case, the prime consideration is matching amplifier output with the correct speaker input.

In general, virtually any commercially available CD player can be used with any preamplifier; the same is true for tape recorders. However, there are substantial differences among microphones, and to achieve best efficiency with least noise level, it is necessary to choose properly among various types of microphones.

Another factor important in the operation of audio systems involves the concept of *linearity*. An element in the recording or reproduction chain is said to be linear if the signal that it produces is equal to the input signal multiplied by a constant factor; thus a linear element maintains the same wave shape. It is desirable that all of components of an audio system be linear. Most modern preamplifiers and power amplifiers easily meet linearity requirements. Loudspeakers are the least linear element in the audio system; microphones vary, but the better ones are generally more linear than most loudspeakers.

7.3 Microphones

The two most popular microphones are the dynamic microphone and the condenser or electrostatic microphone. We begin by discussing the physical principles fundamental to the operation of each.

The principle fundamental to the operation of a *dynamic microphone* is the law of magnetic induction, or *Faraday's law*, named after the English physicist Michael Faraday (1791–1867). According to this law, a changing magnetic field within a coil of wire induces an electrical voltage in the coil, as shown in Figure 7-3. The magnitude of the induced voltage is proportional to the rate of change of the magnetic field; reversing the direction of the field or the sense of field change (increasing to decreasing, or vice versa) changes the sign of the induced voltage. The correlation between induced voltage and the position of the magnet is shown in Figure 7-3. If the motion of the magnet corresponds to that of some complex wave, the induced-voltage-versus-time curve will correspond to the waveform of the complex wave. It is important to remember that it is the *change* in magnetic field through the coil, not the magnetic field itself, that induces the voltage, so there is a phase change between the motion of the magnet and the signal this motion creates. This leads to phase distortion for complex waves, as will be discussed in Section 7.5.

A dynamic microphone works on this principle, as shown in Figure 7-4. The sound wave, incident from the left, strikes a movable diaphragm to which a coil is attached, moves the coil about the stationary magnet and thereby induces a voltage in the coil. A variation of this microphone could have a movable magnet and fixed coil. The result is the same, because the relative motion between the magnet and the coil is identical.

The *condenser* (or *electrostatic*) *microphone* makes use of the electrical capacitance between two parallel plates. A capacitor is an electrical circuit component consisting of two parallel conducting plates on which electrical charge is stored; its purpose is simply to store charge. Capacitance refers to the capacity of the two plates to store charge. If too much charge is placed on the plates, a spark will discharge them. In general, the larger the plates, the greater the capacitance. If equal and opposite electrical charges are placed on the two plates of a parallel plate capacitor, as shown in Figure 7-5, the voltage between the plates decreases as the distance between the plates decreases, and the voltage increases as the distance between the plates increases. In a condenser, or electrostatic microphone, the

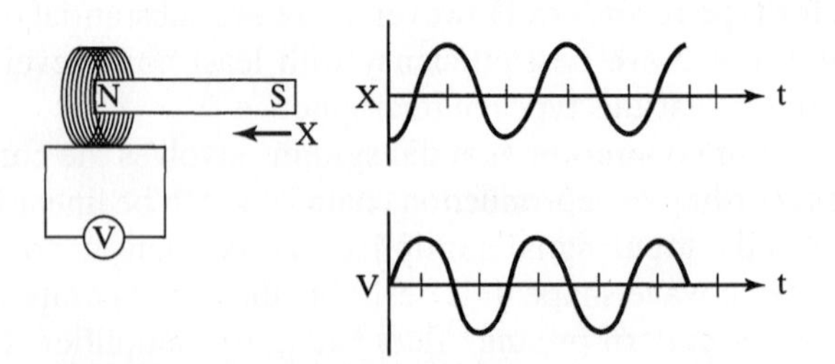

Figure 7-3 Electrical voltage versus time generated by the motion of a magnet through a fixed coil of wire. As the magnet is moved back and forth through the coil sinusoidally, the voltage across the coil varies sinusoidally.

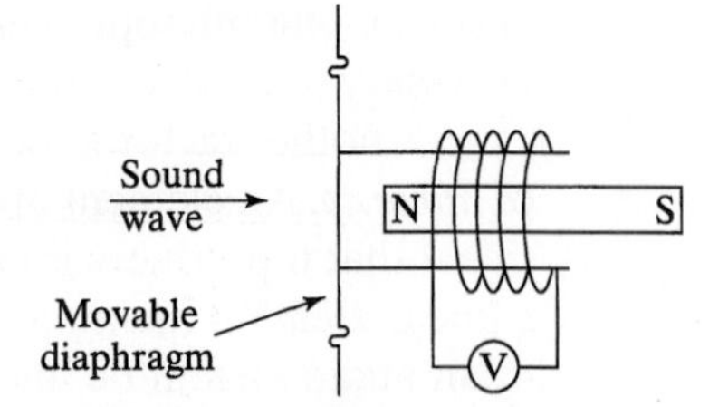

Figure 7-4 Mechanism of a dynamic microphone. The coil moves about a fixed magnet, producing voltage measured by meter V.

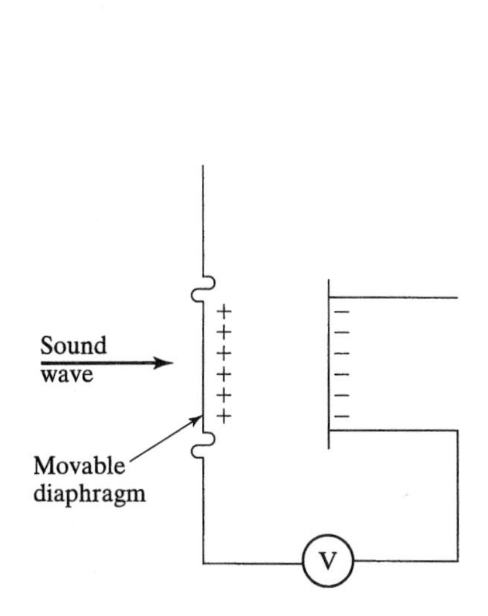

Figure 7-5 Mechanism of a condenser or electrostatic microphone. A fixed charge on plates produces a voltage that varies as the front plate is moved by sound waves, changing the distance between the two plates.

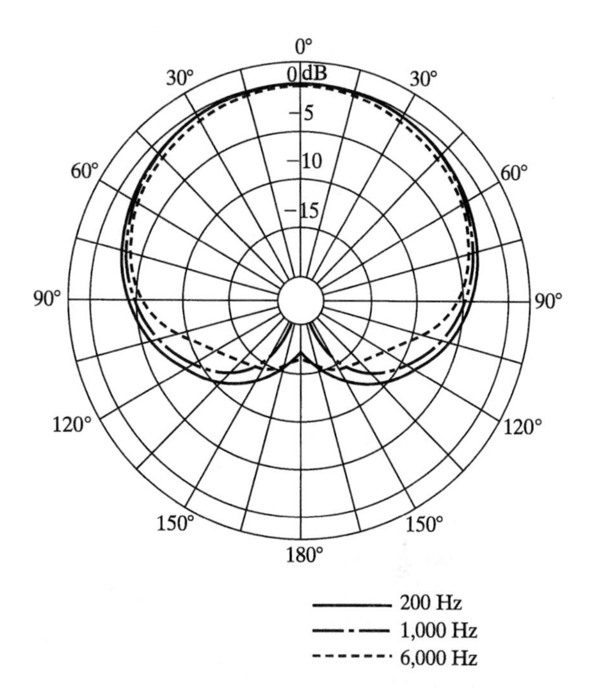

Figure 7-6 Directivity characteristics of a high quality electret condenser microphone.

sound wave is incident on one of the plates, which is a movable diaphragm. The changing voltage created by the changes in spacing between the two plates follows the wave form of the incident sound wave. The electret condenser microphone is a modification of the condenser microphone with excellent frequency and transient response characteristics (see the following). It employs permanently charged capacitor plates, eliminating the need for external voltage sources. This reduces noise and increases the useful range of signal amplitudes, or dynamic range. Electret condenser microphones generally use a battery to power a small amplifier in the microphone.

Other types of microphones use yet other physical principles to convert sound into electrical signals, such as the crystal microphone, the carbon microphone, and the ribbon microphone. Crystal microphones are sometimes used in inexpensive systems. Ribbon microphones can be of extremely high quality but are very expensive and sensitive to damage from shock. Because these microphones are not in general use, we shall not describe their properties. Most telephones use carbon microphones.

Microphones are divided into two principal types by the directional characteristics of their response. A microphone that responds uniformly to sounds coming from any direction is called *omnidirectional*. Some microphones respond preferentially to signals incident from one direction. The most common of these *directional microphones* is the *cardioid* microphone, so named because the plot of response versus angle looks like a heart shape, as shown in Figure 7-6. The relative magnitude of the microphone's response to a standard sound source at a given angle is plotted as the distance

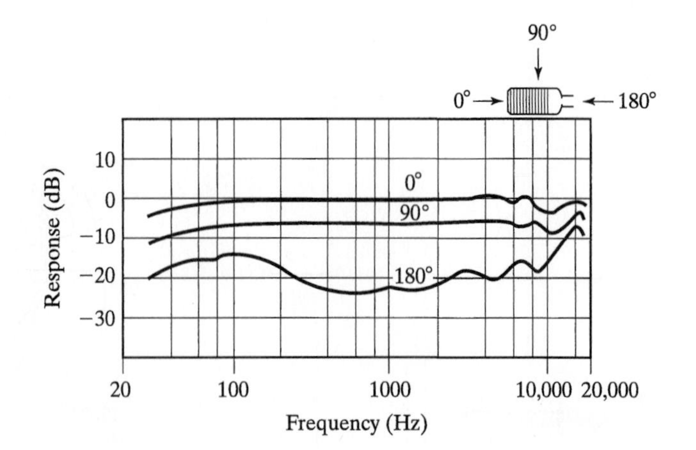

Figure 7-7 Frequency-response characteristics of a high quality electret condenser microphone.

from the center of the graph to the response curve. Response curves are shown at 200, 1000, and 6000 Hz, where the 0° position is directly in front of the microphone. An advantage of the cardioid microphone is its ability to pick up a musical performance while rejecting other sounds—for example, audience noise. For the microphone shown, sounds from the backward direction are reduced 15 dB below those incident from the forward direction.

A "shotgun" microphone has a response that falls off very rapidly within a very narrow cone about the forward direction. Such microphones are used to pick up a question from one member of an audience, for instance. Extreme directionality can be attained using a parabolic reflector with a microphone at its focus. Such a *parabolic microphone* is used for picking up the sound of a quarterback at a football game or for isolating the call of a bird in the woods.

Figure 7-7 shows the frequency-response characteristics of a high-quality electret condenser microphone over the frequency range from 20 Hz to 20 kHz. Its response in the forward direction (0°) is relatively uniform, or flat, over the entire audio-frequency range from below 30 to above 17,000 Hz. Remember that the ear can just barely hear a change in sound intensity level of about 1 dB. Microphones of lesser quality have response curves that drop off more rapidly at the low- and high-frequency limits of their range or have large wiggles or resonances in their frequency response; that is, their frequency response is not flat. For the microphone illustrated in Figures 7-6 and 7-7, the frequency response is relatively linear even in the direction (180°) opposite to the direction in which the microphone is pointed.

In addition to the frequency response to steady sound waves, the response to very fast pulses or fast changes in wave shape, called the *transient response*, is also very important. A microphone that is incapable of responding to rapid signals will change the overtone structure of the sound wave, so the electrical signal does not identically match the wave form. Condenser microphones are superior not only in their frequency linearity but also in their transient response, as illustrated in Figure 7-8. This is primarily because the low mass of the movable diaphragm allows it to respond quickly to the changes in air pressure of the sound wave. The low mass also permits rapid damping of the diaphragm motion, preventing *overshooting*.

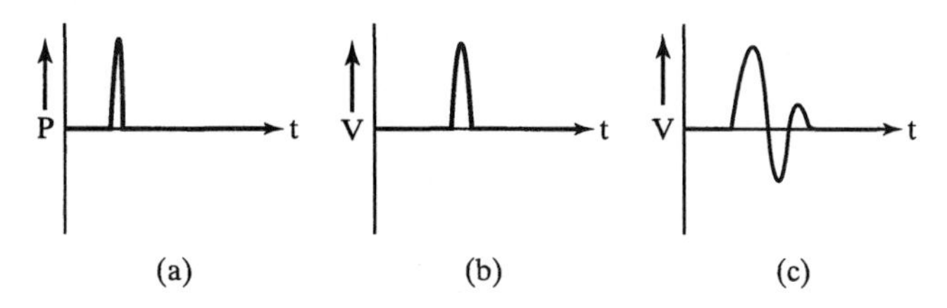

Figure 7-8 Typical transient response to a sharp sound pulse (a) by a condenser microphone (b) and a dynamic microphone (c).

The windscreen, which looks like a wire mesh, is often placed around the microphone to eliminate the noise of wind rushing past the microphone diaphragm. The wind screen is often used in recording speech, in which very low frequencies are unimportant, and in which close proximity of the speaker to the microphone could lead to puffs of air from certain vocal sounds, causing loud extraneous noises. The wind screen also attenuates other unwanted low-frequency noises, such as those from fans, air conditioners, and humming transformers or fluorescent lights. Some wind screens are constructed from foam rubber balls.

Microphones can also be classified as to their type of response. Most, like condenser microphones, respond to the instantaneous pressure. Others, such as ribbon microphones, respond to the pressure gradient, or difference in pressure at two nearby points. The latter type produce a "throaty" sounding recording and are not as well liked, generally.

7.4 Loudspeakers

Just as with microphones, loudspeakers can be divided into two classes depending upon the mechanism used to convert electrical signals into sound: electromagnetic and electrostatic. Recall that in a dynamic microphone the motion of the coil through the magnetic field induces the electrical signal in the coil. In an electromagnetic speaker, on the other hand, an ac electrical signal in a coil in a magnetic field is used to produce motion of the coil. Likewise, recall that in an electrostatic microphone the relative motion of the plates creates the electrical signal. In an electrostatic speaker, an ac electrical voltage between two plates is used to move one of the plates. Most commonly used speakers are of the electromagnetic class. The names for each type of electromagnetic speaker are generally descriptive of the type of enclosure in which the speaker is mounted and the types of speakers mounted in the box.

A loudspeaker converts the electrical energy from a power amplifier into sound waves. Performing this task efficiently and linearly over the entire range of audio frequencies is extremely difficult, and speakers are generally the weakest link in the entire chain in musical recording and reproduction. Whereas undesirable nonlinearities in CDs, amplifiers and preamplifiers are nearly absent, and in microphones are negligibly small, nonlinearities in even "good" loudspeakers are often quite large.

The physical principle governing the operation of a dynamic loudspeaker is illustrated in Figure 7-9. If electrical current is passed through a wire in the presence of a

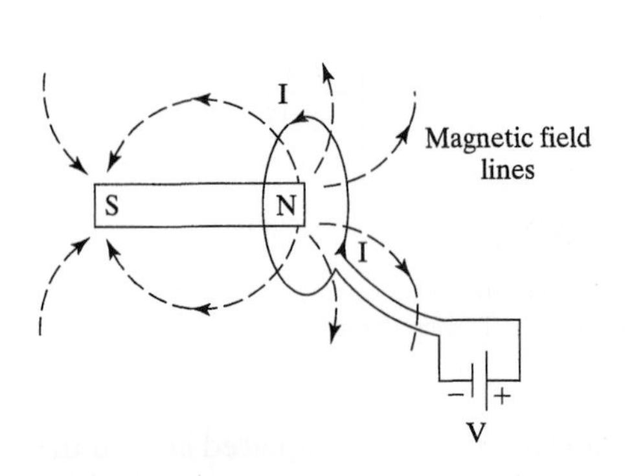

Figure 7-9 Passing electrical current through a wire in the presence of a magnetic field creates a force on the wire. In this case, the force on the current-carrying loop is to the left.

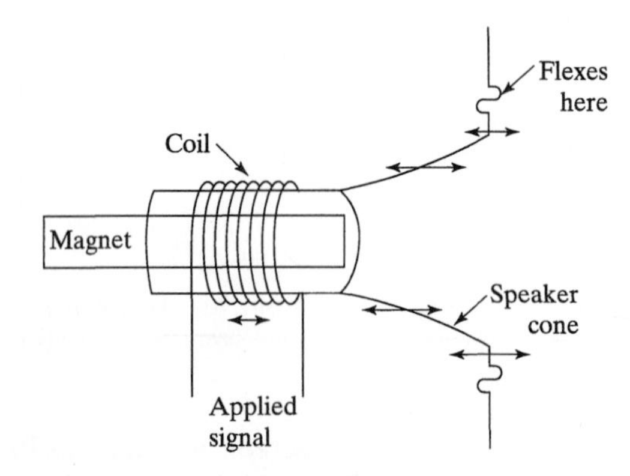

Figure 7-10 Mechanism of an electromagnetic loudspeaker.

magnetic field, a force will be exerted on the wire. This force will be greater for larger currents through the wire. In the case shown in Figure 7-9, the force on the loop is to the left; if the current in the wire were reversed, the force would be to the right. A simplified diagram of a loudspeaker is shown in Figure 7-10. When the current is in one direction, the force is to the right, causing the coil and cone to move in that direction, producing a compression. Reversing the current in the coil reverses the force on the coil and the cone will move to the left, producing a rarefaction. Application of a sinusoidal electrical signal causes sinusoidal motion of the speaker cone, resulting in a sinusoidal sound wave; applying complex electrical waves to the speaker coil produces complex waves. In the absence of any applied signal, the cone will remain at its equilibrium position.

In general, because of the large volume of air that must be moved to produce an audible wave of very low frequency, a large speaker must be used to adequately reproduce low frequencies. As a result of the large mass of material that must be moved in order to move the speaker cone and produce a sound wave, speakers are in general incapable of good response to very fast pulses, and their transient response is similar in character to the transient response of a dynamic microphone.

An unmounted loudspeaker simply set on a shelf and used as is will produce poor sound. As seen in Chap. 2, an enclosure is necessary to prevent low-frequency waves from the rear of the speaker, which are out of phase with those from the front, from diffracting around to the front and interfering destructively with the waves from the front. Three important types of speaker enclosures that accomplish this are illustrated in Figure 7-11.

The simplest such "enclosure" is simply a wall with a hole in it into which the speaker is placed. Such a speaker arrangement is called an *infinite baffle*, because the size of the baffle is very large, infinitely large in the ideal case, and there is therefore absolutely no mixing of the waves from the rear and the front of the speaker. Ceiling speakers are often infinite-baffle speakers. Sometimes very large speaker enclosures are called infinite baffles.

Because most people do not have a spare room or an extremely large wall to use as their speaker enclosure, it is necessary to reduce the size of the speaker box to more

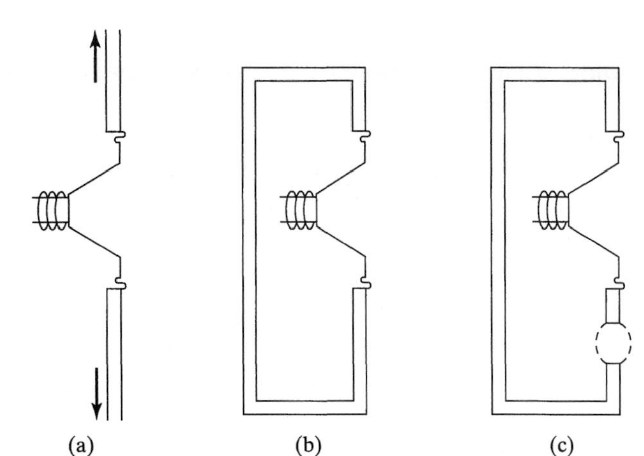

Figure 7-11 Cross-sectional views of (a) infinite-baffle, (b) acoustic-suspension, and (c) tuned-port loudspeaker systems.

reasonable limits. The speaker system with the simplest type of smaller enclosure is the *acoustic suspension speaker*, in which the speaker is placed in an airtight box. This prevents interference of the waves from the front and rear of the speaker, because the waves from the rear cannot get out. The air pressure in such a speaker enclosure creates an additional restoring force to return the speaker cone to its equilibrium position. However, acoustic-suspension speakers are relatively inefficient in converting electrical energy into sound energy.

Another limitation of this type of enclosure is the relatively rapid fall-off in output level at frequencies below about 60 to 100 Hz. The speaker cone itself can have a resonant frequency of 100 Hz or higher and must be damped. In designing a speaker box, one must be careful to avoid box geometries that yield strong resonances from the waves inside the box; this would lead to distortion of the sound produced. Figure 7-12(a) shows the frequency response of a typical loudspeaker in an acoustic-suspension baffle.

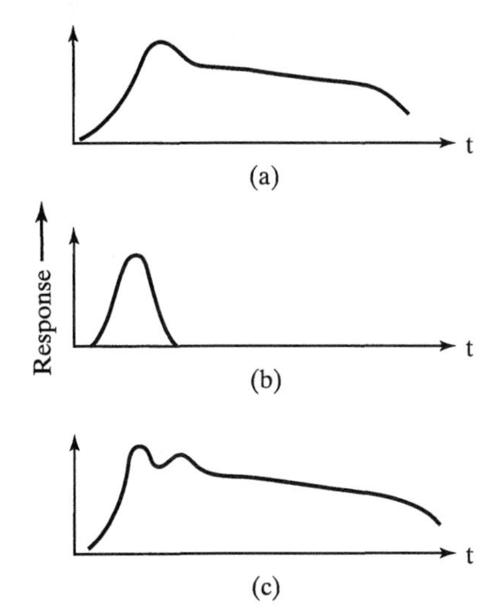

Figure 7-12 Frequency response for (a) an acoustic-suspension speaker, (b) a speaker box with tuned port alone, and (c) the response of speaker and tuned-port enclosure combined.

The graph shows the output of the speaker for a variable-frequency sinusoidal signal of constant amplitude. The region of large response at the low-frequency end shown in graph (a) is due to a resonance in the speaker.

The low-frequency range can be increased and the efficiency of the speaker over the entire audio range improved by a simple modification of the cabinet; the result is the *tuned-port* or *bass-reflex* speaker enclosure, shown in Figure 7-11(c). If a port, or hole, is cut into the speaker box, generally in the front below the main speaker, the box itself will act as a Helmholtz resonator, with a resonant frequency that can be chosen to be slightly below the resonant frequency of the loudspeaker itself. A typical resonance curve for a speaker enclosure with a port is shown in Figure 7-12(b). If the speaker response is taken with the speaker in the box with the hole, the sum response curve will look like that in Figure 7-12(c).

The hole is adjusted in size and position so that the Helmholz-type resonance is at an appropriate frequency and has an appropriate amplitude to make the total response curve look like that in part (c); hence the name "tuned port." In this way, the bass range is effectively extended to a frequency well below the resonant frequency of the speaker alone. Tuned-port speaker systems are generally made using speakers of high efficiency, and consequently tuned-port speaker systems are more efficient than acoustic-suspension speakers.

High-quality speaker systems use at least two speakers: a small speaker, called a *tweeter*, which responds better to high frequencies, and a larger speaker, called a *woofer*, which responds better to low frequencies. Such a *two-way speaker system* needs a built-in electronic *crossover network* that automatically routes the high-frequency electrical signals to the tweeter and the low-frequency signals to the woofer, and appropriately balances the signals to each speaker at the midrange, where both speakers respond. By using two speakers, the response over the entire frequency range can be improved significantly, and by directing low and high frequencies into their own speakers, the transient response can be greatly improved. In more sophisticated speaker systems, one or more midrange speakers are added to the usual woofer and tweeter, requiring a more complicated electronic crossover network. For a given speaker system, proper electronic equalization can improve (flatten) the frequency response.

Occasionally the bass speaker is housed in a separate enclosure from the other speakers, thereby enabling the smaller speakers to be placed on a shelf or table top. Because the listener's ability to localize low-frequency sounds is relatively poor, there is no perceived splitting of the location of the sound source.

Even in a good speaker system, some of the greatest sources of nonlinearity involve the crossover network. Here again, digital electronics provide a solution with the introduction of the "smart speaker." A normal crossover network produces changes in the crossover frequencies as the power level of the music varies. A computer-controlled feedback system can be used both to adjust the crossover frequencies and to limit the power level to any individual speaker in the system and thus avoid distortion. Some of these systems include integrated electronics with speaker arrays or provide additional phase control, equalization, and other effects. Figure 7-13 is a photograph of one such system, the Infinity Kappa 8.1 four-way speaker. The system includes woofer, midbass, midrange, and tweeter with built-in adjustable crossover networks; the opening at the bottom is the bass-reflex port.

Figure 7-13 The Infinity Kappa 8.1 4-Way Loudspeaker System, with cover off (left) and cover on (right). Visible features include (top to bottom) tweeter, midrange speaker, midbass speaker, woofer, and bass-reflex port. (Courtesy of Infinity Systems, Inc.)

Stereo headphones have several advantages over loudspeakers. First, they effectively eliminate most outside sound, and, second, stereo separation is superb. Because the mass of the moving diaphragm is small, the transient response is excellent in high-quality stereo headphones. The frequency-response characteristics of good stereo headphones are very good. Although they require less power than most speakers, they have the obvious drawback that only one person can listen at a time. They are also more likely to cause hearing damage when used by unwary persons, because it is easy to produce 130 dB inside a headphone cupped over the ear.

As speaker systems and audio electronics have improved, the effects of phase and polarity on the sound have received more attention. Modern digital electronics has allowed greater control of the wave shape of a musical tone by controlling the phases of the harmonics. Contrary to the hypothesis of Ohm's law of hearing, it appears that phase can, in certain cases, play a significant role; some audiophiles believe that careful control of phase has improved the fidelity of the sound of certain instruments. Recent studies of the effect of the polarity of speakers seems to indicate that a significant change in the sound may occur if both speakers are wired *in phase with each other* but *out of phase with respect to the originally recorded sound*. Reversing the wires on both speakers, keeping them in phase with each other but inverting the phase of both, so as to change a compression into a rarefaction and vice versa, has often not been considered a serious problem. Indeed, phase information has been lost both in the recording process and the reproduction process, because of their use of electromagnetic devices. By allowing greater control of the phase, digital electronics has opened up the field of investigation into the effects of phase and polarity on sound reproduction.

7.5 Preamplifiers

The preamplifier serves as the control center for the audio system, accepting the signals from the various sources and processing these signals for input to the power amplifier, which then leads to the speakers. A preamp must have inputs that match the output impedance and signal level of each of the signals required in the system, and must have output impedance and output voltage level that match that of the power-amplifier input. Typical controls on a high-quality preamp include input selection, volume, bass and treble boost or attenuation, loudness, filtering, and balance. Often it is possible to choose speaker arrangement—either stereo, monaural, or either channel—individually. On many units it is possible to input a signal from one tape recorder and record on another with the option to play it simultaneously through the speakers. The input selector simply chooses the unit that will serve as the source of the signal for the system; because of the selector (typically a knob), it is possible to leave all signal sources connected at all times.

The volume control varies the level of the signal output to the power amplifier; normally, there are no volume controls on the power amplifier itself. Volume is not the same as "loudness," which will be discussed shortly. Volume controls are often made to be logarithmic, because the ear responds logarithmically to increases in intensity. As a result, the volume control is perceived as approximately linear.

Both high-pass (i.e., low-frequency-attenuating) and low-pass (i.e., high-frequency-attenuating) filters are normally available on good preamplifiers. Figure 7-14 shows the transmission of low- and high-frequency filters for one particular solid-state stereo preamplifier. In this case, there are a single low-frequency filter and three separate high-frequency filters, labeled by the cutoff frequency of the filter in kilohertz. High-frequency filters are useful in removing high-frequency noise such as tape hiss, whereas low-frequency filters can eliminate electrical hum or other low-frequency noise. Here, as elsewhere, the filter is either in or out (on or off). Limited continuous adjustment of relative signal levels at various frequencies is obtained using the tone controls, so that the sound may be adjusted to accommodate the preferences of the user.

Continuously variable bass and treble tone controls are a fundamental feature of any preamplifier. The maximum effect of the tone controls for the same preamplifier is shown in Figure 7-15. Adjusting the bass control changes the amplification factor between the limits shown at the low-frequency end of the graph. Maximizing the bass raises the level at 20 Hz by nearly 20 dB. Similarly, adjustment of the treble control

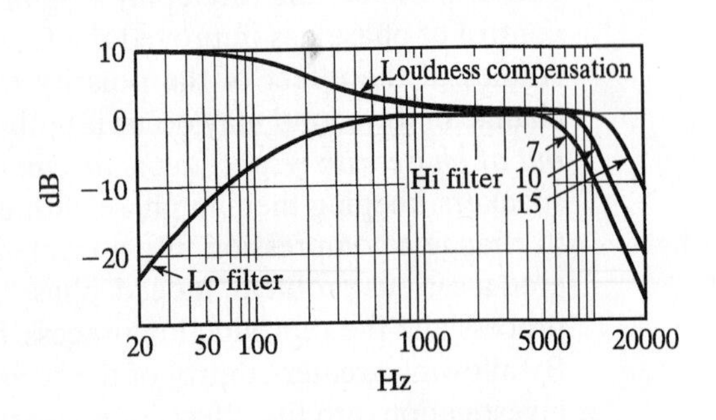

Figure 7-14 Attenuation factors for three high-frequency filters and a low-frequency filter, along with amplification factor for loudness control for a high quality stereo preamplifier. Each high-frequency filter is labeled by its cutoff frequency in kilohertz.

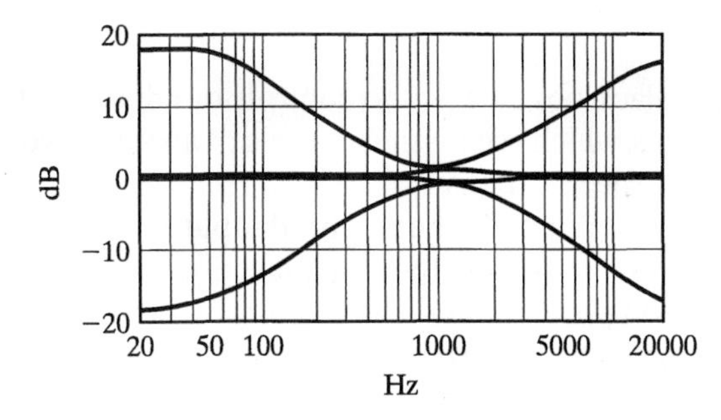

Figure 7-15 Limiting amplification and attenuation factors for bass and treble tone controls (with controls set at maximum and minimum values, respectively) for a high quality stereo preamplifier.

changes the amplification factor at the high frequencies. Setting both controls at their center leaves the levels unchanged for all frequencies.

The *loudness* control does not refer to volume but rather deals with the production of sounds of constant loudness level (in phons) to the ear at low sound intensity levels. In Chap. 6 we saw that there is a considerable difference between the sound intensity level in decibels for the threshold of hearing at various frequencies. For high sound intensity levels, say about 50 or 60 dB, all frequencies are heard when they are produced at the same level, as can be seen from Figure 6-4. However, at low overall sound intensity levels, the low-frequency sounds, and to a lesser extent the high-frequency sounds, may be reduced below the threshold of hearing. This can occur when music from ordinary recordings is played softly by turning down the volume control. In such a case, it is necessary to boost the extreme frequency ranges, particularly the bass, so that the loudness level for all frequencies is approximately the same, and the music will sound the same at the lower level as it did at the higher level. This is accomplished by switching on the loudness control. The amplification factor for the loudness control is shaped to compensate for the increase in the threshold of hearing at low frequencies, as shown in Figure 7-14. In some preamplifiers, the high-frequency end of the spectrum is also emphasized by the loudness control.

A final control generally available on preamplifiers is the balance, which adjusts the relative output level between the two speakers of a stereo system. In the cases of the controls discussed, either the control applies equally to both channels, or separate controls are provided for each channel.

For a high-quality preamplifier, the distortion—either total harmonic distortion (THD) or intermodulation distortion (IMD)—should be below about 0.05 percent. Intermodulation distortion is a combination tone (sum or difference) effect in which two tones played together in one channel combine and generate others. This is a nonlinear phenomenon similar to the combination tones produced in the ear. The linearity should be within 1 dB (the intensity JND) over the entire audible range for equalized input signals and slightly better for high-level input signals. Most name-brand audio equipment exceeds these specifications.

The choice of a preamplifier and amplifier system should be made on the basis of the features available, the specifications of the unit, and, of course, the price, but most important is the subjective listening response: How good does it sound to you?

7.6 Power Amplifiers

Most power amplifiers are simply "black boxes" with no controls or adjustments, with input impedance of about 50,000 ohms, which accept signal levels of about 1.5 volts and produce an output according to their specifications.

Specifications for a power amplifier include the required input level and impedance, the power output at full-output level, and the impedance of the compatible speaker system. The linearity of a good power amplifier should be within about 0.5 dB over the entire audio-frequency range. The distortion should be less than 0.5 percent (both THD and IMD) for any level between zero and the maximum output specification. Noise caused by the amplifier circuit itself should be nearly 100 dB below the normal signal level.

The linearity is often specified over a frequency range extending well above the audible limit. Although this is not particularly useful, the farther the linear region extends, the more likely it is to retain its linearity in the audible-frequency range even at high signal-output levels; thus many amplifiers have a specified linearity of well above 20 kHz.

If the impedance of the speaker is not appropriately matched to the output impedance of the power amplifier, the power transferred to the speaker will be reduced. For example, the Dynaco Stereo 120-A amplifier, an amplifier of high quality, will deliver up to 60 watts per channel into 8-ohm speakers, the specified impedance. Use of 4-ohm speakers limits the power transfer to 50 watts per channel, and use of 16-ohm speakers further reduces the maximum power output to 40 watts per channel.

Although it takes only a few watts of audio power to fill even a large listening room with sound, most power amplifiers provide considerably more. A greater amplifier power will be required if low-efficiency loudspeakers like acoustic suspension speakers are used, whereas high-efficiency speakers, such as tuned-port speakers, require less power. A 100-watt amplifier providing only 20 watts of power to a speaker will sound better than a 20-watt amplifier working at nearly full power, because it is working well within its linear operating range. Any amplifier operating near peak power output will produce more distortion, especially in transient peaks like percussion sounds, where the level may actually rise instantaneously above the steady-state peak-power-output level. It is therefore wise to use an amplifier with maximum power output somewhat greater than the desired operating input level of the speaker system.

7.7 AM-FM Tuners

A tuner receives radio-frequency signals from a radio-station broadcast, demodulates the signal to obtain the information stored in the modulator signal, and amplifies it to a level appropriate for input into a high-level, high-impedance preamplifier. If the tuner also contains the preamplifier, amplifier, and speaker, it is called a radio.

AM (amplitude modulation) radio stations are assigned carrier frequencies every 10 kHz over the AM radio band, which extends between 540 and 1600 kHz. The audio bandwidth, or (maximum) range of usable frequencies available to any station, in order to avoid interference with adjacent stations, is about 5 kHz. Only frequencies up to about 5 kHz can be broadcast, and true high fidelity is therefore impossible for most AM stations. Furthermore, noise caused by the electronic components and static will cause changes in the amplitude of the signal and lead to noise and distortion of the sound. Although AM radio is adequate for voice transmission, most AM stations are not capable of high-fidelity music reproduction, as are FM stations.

FM (frequency-modulation) radio stations are assigned frequencies every 0.2 MHz (or 200 kHz) over the FM radio band, which extends from 88.1 to 107.9 MHz. This frequency range lies between Channels 6 and 7 of the television frequency range. In FM the frequency of the signal changes; in the case of FM radio, the maximum allowable frequency excursion is ±75 kHz, or a total excursion of 150 kHz about the carrier frequency. It is therefore readily possible for an FM station to transmit audio frequencies up to 15 kHz. Because noise and static affect only the *amplitude* of the radio signal received, while the information content of FM signals is stored in *frequency* variations, FM radio is relatively noise free.

FM radio stations can transmit stereophonically. Two separate audio signals are used to modulate different regions of the frequency band, the second signal being transmitted on a subcarrier 38 kHz above the frequency of the main carrier. The subcarrier is simply a 38 kHz sine wave, amplitude-modulated by the second signal. An additional 19-kHz audio tone that is transmitted acts as a signal to the FM tuner to find the second signal and use the stereo decoder in the tuner to obtain the two channels of sound. To obtain compatibility between monophonic and stereophonic FM broadcasts, the sum signal (L + R) is broadcast in the usual frequency range and can therefore be picked up monaurally. The difference signal (L − R) is transmitted on the subcarrier. The sum and difference of these two signals are used to obtain the left and right speaker signals, respectively. The entire signal, consisting of (L + R), the 19-kHz pilot signal, and the 38-kHz AM subcarrier containing (L − R), is used to frequency-modulate the main FM carrier around the frequency of the station.

One important noise problem in FM transmission is high-frequency hiss. This can be reduced significantly by preemphasis of the high-frequency range and equalization by the tuner circuitry, just as is done in tape recording. A control labeled FM Muting is often found on good tuners; this attenuates the loud static obtained when tuning between FM stations but allows the station signal to come through unhindered.

Some specifications on which the choice of a tuner are based are sensitivity to low-level stations (which may be crucial for those living far from the radio stations), signal-to-noise ratio, distortion, frequency linearity of audio output from the tuner, and stereo separation. The existence of an easily readable indicator, to determine whether a station is properly tuned, is important. In metropolitan areas, where stations may be assigned to adjacent frequency bands, selectivity, or the ability of the tuner to tune in one station while suppressing signals from adjacent stations, becomes important.

7.8 Tape Recorders

A great variety of tape recorders is now available. We shall limit our discussion to some of the more popular types, summarize some of the basic features in the operation of tape recorders, and discuss some of their more important properties.

The most common type of audiotape system for consumer use is the $\frac{1}{8}''$, quarter-track, stereo audiocassette. Audiocassettes are prepackaged reels of tape about $\frac{1}{8}''$ wide; the normal cassette lasts from about 30 to 45 minutes per side at a speed of $1\frac{7}{8}''$ per second, for a total playing time of between 1 and $1\frac{1}{2}$ hours per cassette (both sides).

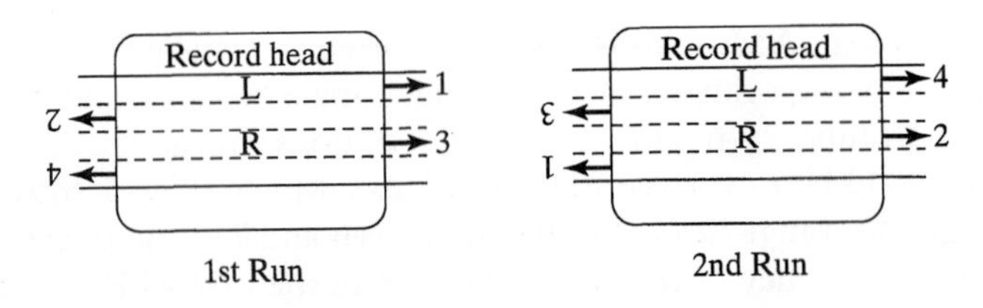

Figure 7-16 Channel configuration of recording and playback heads for quarter-track stereo tape recording.

The layout of the tracks on the tape is shown in Figure 7-16. Recent engineering improvements have raised the quality of cassette fidelity and reliability to the point at which they can be considered for use in high-quality audio systems. Both audiocassettes and CDs are commonly used in cars.

Magnetic tape is constructed from a thin (typically 0.0005 or 0.001″-thick) plastic base coated with a thin layer of magnetic material. This coating, usually iron oxide, chromium dioxide, or ferrichrome, is in the form of small particles; each of these, in turn, consists of even tinier volumes of the magnetic substance, called *magnetic domains*. For the sake of clarity, in the following discussion we shall assume that the material is iron oxide; the principles are the same for the other coatings. Each of the tiny domains acts like a magnet; that is, each has north and south poles that respond to the presence of other magnets. These domains are in random orientations on a blank unrecorded tape. Because of the lack of domain alignment, there is, therefore, no net magnetic field on the blank tape.

We can idealize the recording head of a tape recorder as a C-shaped piece of iron with a coil of wire around one section, as shown in Figure 7-17. The electronic signal to be recorded passes through the coil and creates a magnetic field in the gap of the magnet. The strength of the magnetic field is proportional to the current in the coil. There are only two possible directions for the resulting magnetic field; either side of the gap can be a north pole or a south pole, with the other side of the opposite polarity, depending on the direction of the current in the coil. As the current reverses direction, so does the orientation of the magnetic domains on the tape. Consequently, the periodicity of the audio signal is reflected by the periodicity of the alignment of the magnetic domains. The amplitude of the field in the gap, proportional to the amplitude of the audio signal, determines the degree of alignment of the magnetic domains and therefore the magnetic fields they produce.

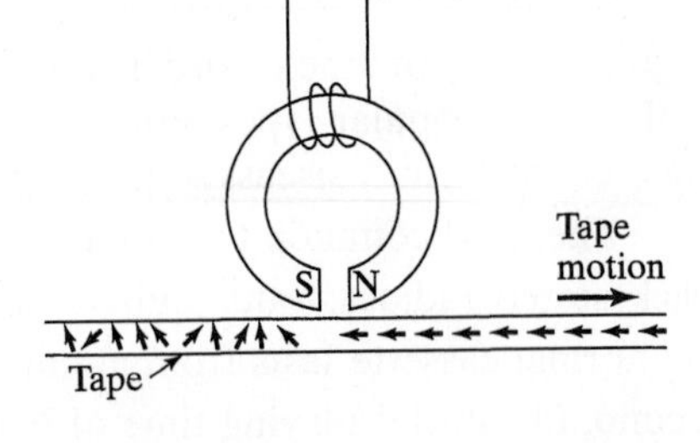

Figure 7-17 Schematic view of the recording head of a tape recorder.

Most of the better tape decks have three heads—one to erase, one to record, and one to play back the signal—all of which are located in the head unit through which the tape passes. During recording, the tape is drawn at constant speed past the record head. The magnetic field in the head gap, modulated by the audio-frequency signal in the coil, forces the tiny magnetic domains to reorient. The number of domains that are lined up and the extent to which they are lined up—that is, the angle that they make with respect to the direction of motion of the tape—is determined by the strength of the magnetic field applied. If the signal current varies sinusoidally with time, the magnetic strength, determined by the number of crystals lined up and the extent to which they are lined up, will also vary sinusoidally along the tape.

To play back the tape, the modulated tape is drawn past the playback head, which is similar to the recording head. As the tape passes by the pickup in the playback head, the head senses the magnetic field produced by the sum of the tiny magnetic domains on the tape. A voltage is induced in the coil around the head. The strength and periodicity of the magnetic field on the tape determine the strength and periodicity of the induced voltage.

In a magnetic system like the tape recorder, the magnetism of the tape is proportional to the electrical current in the coils in the tape head. The current is proportional to the amplitude of the sound being recorded. However, on playback the output of a tape playback head is proportional to the *rate of change* of the magnetic field. Because all the Fourier components in a complex wave are changing at different rates, the playback head will introduce a phase change that is different for each frequency. This leads to a change in the shape of the wave, known as *phase distortion*. Fortunately, according to Ohm's law of hearing, the quality of the sound, to a good approximation, is not dependent on the relative phase of the various harmonics, so to a large extent phase distortion leaves the sound quality unchanged. Nevertheless, on some of the better and more expensive tape decks, additional electronic circuits are used to eliminate phase distortion.

The magnetic domains in an oxide particle have an inherent resistance to change of their orientations, a sort of inertia, called *hysteresis*. Because of this, very-low-intensity signals in the record head coil do not produce a corresponding modulation in the tape. An extra signal, called the *bias signal*, or simply bias, is superimposed on the audio signal to overcome this magnetic inertia. The bias signal, a constant-amplitude sine wave with a frequency of about 100 kHz, well above the audible range, supplies an extra "push" or "shove" to the magnetic domains to overcome their magnetic inertia, which allows them to respond to the lower-amplitude audio-signal modulation. The bias signal is not heard on playback because of its extremely high frequency. Different types of tape respond differently to external magnetic fields and therefore require different bias-signal amplitudes. Most tape decks have switches that allow the user to choose the bias that is most suited to a particular type of tape.

Tapes are erased by an erase head, which is similar to the record and playback heads. In the erase head, however, an extremely high-frequency, high-amplitude sinusoidal signal is used. The magnetic crystals first feel a strong force tending to orient them in one direction, and then, soon thereafter, a strong force tending to orient them in the opposite direction. The net effect is that the magnetic domain orientations are random or scrambled, leaving the tape with no net magnetization and hence no recorded signal.

There are nonlinearities and noise problems in tape recording and playback that require the use of *preemphasis* and *equalization* to obtain the highest fidelity and the least noise. When the sound is impressed on the tape the level is attenuated for lower frequencies and amplified for higher frequencies. This is necessary so that the low-frequency signals will not become too large in amplitude for the tape to respond linearly, and so that the higher frequencies will be larger in amplitude than the inherent *tape hiss* resulting from residual random magnetization of the magnetic domains in on the tape. Most tape recorders use the National Association of Broadcasters (NAB) standard for equalization. Other techniques can be employed to remove most of the tape hiss, which is the most serious type of noise generally occurring in good tape systems.

A typical audiocassette tape deck contains stereo microphone and high-level inputs, sometimes with mixers in which the relative amplitudes of the signals can be adjusted before recording. Separately adjustable outputs from each channel are normally available. Sound level monitors, called *volume unit* (VU) *meters*, are normally included and can be switched to monitor either the signal being recorded or that being played back. On most of the better machines, there is provision for choice of bias for normal and low-noise tapes as well as the Dolby noise-reduction option. Less often, the option of output from the tape head or tape amplifier is available. Tape head output requires equalization when input into the preamplifier, whereas tape-amplifier output has been equalized by the preamplifier in the tape deck and can be fed directly into a linear amplifier or a preamp tape-amplifier input. Some tape units come with a power amplifier and speaker, but this is intended primarily as a convenience in monitoring the sound when portability is required.

Several systems have been devised to eliminate tape hiss from reproduced sound. The most popular system is the Dolby noise-reduction system. Inherent tape noise is most prevalent and audible at frequencies above 5 kHz. Most normal music passages are much louder than the tape noise and upon playback effectively mask the noise. When a soft musical passage is reproduced, however, the noise can be heard. A Dolby noise-reduction system boosts the input-signal components above 5 kHz as much as 10 dB whenever the musical passage is soft. The boosted signal is then recorded on the tape. During playback, the Dolby device automatically compensates and reduces the intensity of all high-frequency components. This reduces the audio signal to its original level and reduces the tape noise to 10 dB below its original level. A signal-to-noise-ratio improvement of 10 dB can thus be achieved. When a non-Dolby tape is played, the Dolby decoder should be switched off; otherwise, the high frequencies will be attenuated.

7.9 Digital Sound Reproduction and the Compact Disc

An almost incredible metamorphosis in consumer sound reproduction occurred during the five-year interval between 1980 and 1985. In 1980 research and development were being completed for the basis of digital sound reproduction, and no single standard for the production of digital discs had been established; virtually all contemporary recordings were on the 12″ plastic $33\frac{1}{3}$ RPM stereophonic disc. A new technique to reduce the noise inherent on these discs, the dbx compander system, was becoming established.

The quadraphonic disc had been developed, and $33\frac{1}{3}$ RPM discs were available using two standards for quadraphonic sound, the quadradisc discreet (CD4) format and the SQ matrix format. During 1980 and 1981 the standards for compact discs were established, and the digital compact disc was first released to the consumer market in 1983. By 1985 virtually all of the large record manufacturing companies were producing compact discs, and most had abandoned the older format. All new recordings were being made on digital compact discs, and the older LP discs were rapidly becoming unavailable in record stores. Within a period of only a few years the analog record disc had become virtually obsolete! By 1990 many of the older recordings had been reissued on compact discs, although the high-frequency noise typical of the older format, created during the original analog recording process, can sometimes be heard on the reissued discs.

Why were the new digital discs so well accepted, and what technical considerations led to this incredible change in the consumer market? The newer technology, based on lasers and digital electronics, is more reliable. Although the playing time for stereo discs was about the same for a single-side compact disc as for a two-sided analog record, the CD is much smaller and lighter. There is no physical contact between the CD surface and the reading mechanism, so it is free from wear, and the danger to the CD from warpage is virtually negligible. It is believed that a CD should last many decades. The CD can be used in a car stereo or with a portable machine similar to portable audiocassette decks, and there is even a CD player for joggers. The dynamic range of a CD player is over 90 dB, far exceeding any previous medium. The compact disc is virtually free from noise, frequency variation, and distortion. In contrast with any other reproduction medium, random access is readily available for the CD.

Before studying some features of compact disc technology, we should learn something of how the compact disc stores its data—we must learn to count in binary. Our common decimal numbering system is a "base 10" system. The digits in our numbers represent up to ten units (10^0), then up to ten sets of ten (10^1), then up to ten sets of 100 (10^2), then up to ten sets of 1000 (10^3), and so on. For example, the number 1694 represents one set of 1000 plus 6 sets of 100 plus nine sets of 10 plus 4 sets of 1. Counting in binary is very similar to this, except that it is simpler, because each digit in the number is either 0 or 1; a comparison of counting in binary (base 2) and base 10 through 32 (base 10) is given in Table 7-1. But why should one bother to count in this (to us) very different format? Because the digits in the binary system—namely either 0 or 1—are very simple, they can be very simply represented in a computer. For example, this binary system can be either a voltage (1) or no voltage (0), or on the surface of a compact disc as either a reflective area (0) or a nonreflective area (1). In binary mathematics, operations such as addition and multiplication are also very simple, so computers do their mathematical operations in base 2. We call such binary digits "bits."

Because of several technical considerations, the actual system on compact discs uses the *edges* of the pits to represent the "1s" and either smooth surface (out of the pits or in the pits) to represent the "0s" in the binary code. This choice also enhances the ability of the system to compress (i.e., represent in a more compact way) the data and to check itself for a variety of disc defects and errors that may occur in the reading

Table 7-1 Numbers up to 32 in Base 10 (Decimal) and Base 2 (Binary)

Base 10	Base 2
0	000000
1	000001
2	000010
3	000011
4	000100
5	000101
6	000110
7	000111
8	001000
9	001001
10	001010
11	001011
12	001100
13	001101
14	001110
15	001111
16	010000
17	010001
18	010010
19	010011
20	010100
21	010101
22	010110
23	010111
24	011000
25	011001
26	011010
27	011011
28	011100
29	011101
30	011110
31	011111
32	100000

process. Figure 7-18 is a microphotograph of part of the surface of a CD. In the photograph, the lengths of the pitted and unpitted areas along a row of its can be seen to vary, a result of the encoding of different combinations of ones and zeros. The pits are about 0.5 micron (1 micron = 10^{-6} meters) wide and 0.2 micron deep, with adjacent rows of pits 1.6 microns between centers. The spacing between adjacent rows is *diffraction limited*; that is, it is limited by the inherent quantum mechanical limitation of the light to be focused to a point. If successive rows of pits were closer together, the focus of the laser beam could not be made small enough that it would view a single row of pits. Present CD systems use infrared light, so the diffraction can be decreased by using light with a shorter wavelength such as visible or even ultraviolet light, allowing the rows of pits to be spaced more closely. This has been done with the digital video disc (DVD), which provides about 16 times more storage than the CD.

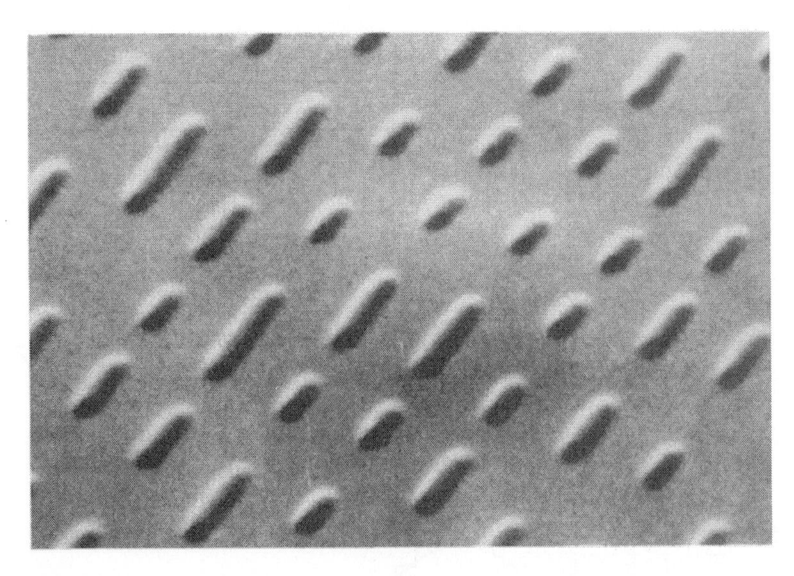

Figure 7-18 A microphotograph of a section of a compact disc, showing the geometry of the pits. (Courtesy of Philips Electronics)

The musical wave shape is encoded in a binary format on the surface of the CD. Figure 7-19 illustrates the concept of digitization of a wave: At a series of equal time intervals the value of the voltage is sampled, and those sampled values are stored on the disc. In Figure 7-19, a sine wave is sampled, in a slightly simplified manner, at rates of 16 and 32 times per period. Analog-to-digital conversion involves a type of sample-and-hold process similar to that discussed in Sec. 5.2. An analog wave, such as that picked up by a microphone, can be digitized and stored in digital form without ever being stored in analog form, thus eliminating the noise typically obtained using a magnetic medium. The electronic device used for the digitization process is called an analog-to-digital converter. In playback, the sampled data is read off the disc, and the shape of the wave is reconstructed using a digital-to-analog converter. The conversion process, as shown in Figure 7-19, however, distorts the reconstructed wave. Because a greater sampling rate produces a wave more nearly like the original analog signal, a high sampling rate is very desirable. That wave is filtered to remove the high frequencies present in the distorted wave and eventually fed into the amplifier system, just as a signal from other sources, such as audiotape players or microphones, would have been. Because CDs are recorded in stereo, alternating sample points are for the left and the right channels. A switcher called a *multiplexer* is required to send the signal to the appropriate channel, where it will be processed as described later. To define a wave, it must be sampled at least at its peak and at its trough. Therefore, to obtain a frequency response of up to 20 kHz with the CD system, the sampling frequency must be at least 40 kHz, twice that of the highest frequency to be reproduced on the disc. This is also a mathematical result of the spectrum-analysis process. In fact, the actual sampling rate is chosen to be 44.1 kHz, somewhat above twice the highest frequency for the system, the reason for which we shall discuss shortly.

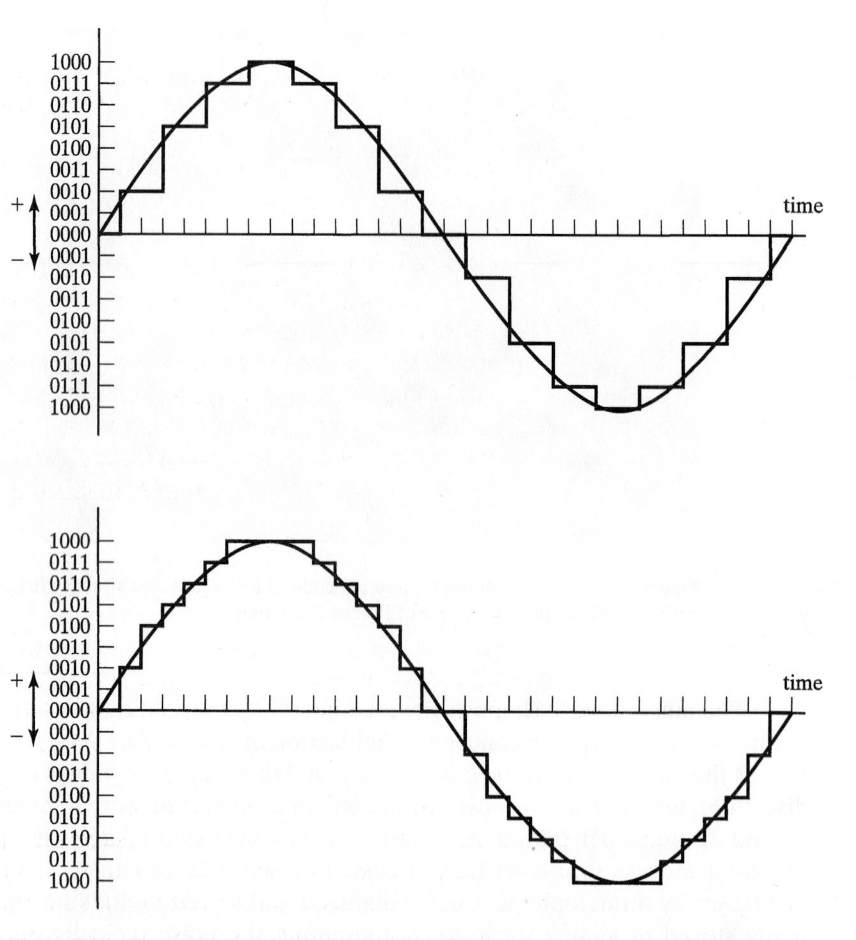

Figure 7-19 The concept of analog-to-digital and digital-to-analog conversion. Shown here are simplified analog-to-digital conversion of a sine wave to four-bit binary code using sampling frequencies of 16 (upper) and 32 (lower) times the frequency of the wave. The stairstep sample-and-hold signal, obtained in the analog-to-digital recording process, is reproduced by the CD laser reader and later filtered to obtain a smooth representation of the original wave form. The goal of the analog-to-digital process is to reproduce the sine wave shown, so a greater sampling rate will produce a more accurate representation.

The CD rotates at a *constant linear velocity*; the laser reader scans the same *length of track* per second whether it is reading an inner or an outer track. Therefore the disc rotates slower as the scanner reads from the inside (500 RPM) to the outside (200 RPM) and the music progresses. Upon being read by the laser system, the sampled data points are fed into a computer memory known as a *first-in-first-out buffer*. The data are then read out and sent to the digital-to-analog converter at a rate determined by a very accurate oscillator. Using this technique, small changes in the rotational speed of the disc have literally no effect on the music, and variations in the frequency of the disc are virtually undetectable.

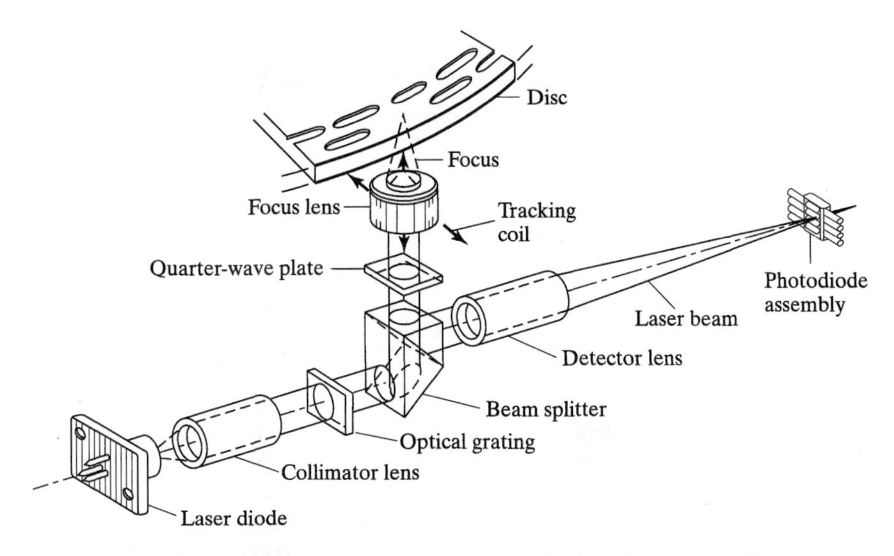

Figure 7-20 A common system for laser scanning of compact discs described in the text. The infrared laser beam is created by the laser diode at the lower left and focused through the optical system onto the CD. The reflected beam is focused through the optical system to the photodiode at the upper right, which converts the infrared signals into electrical signals representing the binary coded data. (Clifford, Martin, *Complete Compact Disc Player*, 1st Edition, © 1987. Reprinted by permission of Pearson Education, Inc., Upper Saddle River, N.J.)

The choice of 16 binary bits in each sampled point means that the entire range of amplitude is broken up into 2^{16} or 65,536 (actually + and − 32,768) different steps. The dynamic range of the disc would then correspond to the ratio between signals of amplitude 1 and amplitude 32,768, which is over 90 dB. Thus the dynamic range of the compact disc covers virtually the entire dynamic range of the ear, almost twice the dynamic range of the older analog disc. In practice, this analysis is a bit oversimplified. When the disc is digitally recorded, some space is required on the upper end to assure that the 32,768 bits are never exceeded, so the dynamic range is somewhat less than 90 dB in actual practice. Some CD players have been made less expensive by using only 14 bits, dropping the two least significant bits (the "one" and "two" bits.). The better CD players use all 16 bits.

Figure 7-20 shows one of the typical optical systems used to read data from a compact disc. The beam of monochromatic light from an aluminum gallium arsenide infrared laser diode (at left) is guided through the optical system shown by the series of optical elements. The expanding beam from the laser diode is focused into a nearly parallel beam by the collimating lens. The beam then encounters a diffraction grating, which separates the incoming beam into a strong central spot, which in turn is used for tracking the binary data, and a series of side spots, two of which (one on each side) are used to guide the main spot along the line of pits which contain the data. The beam splitter serves two functions. First, it bends the beam upward, where it is then focused by the focus lens onto the surface of the compact disc. This wave reflects off the surface of the disc and returns back through the focus lens in the opposite direction. Second,

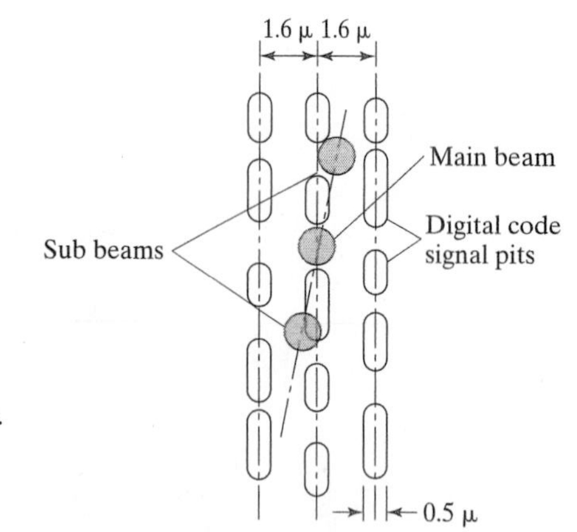

Figure 7-21 Three-beam laser tracking system. The three laser beams are produced when the initial laser beam passes through the optical grating (see Figure 7-20). The main beam is used to scan the signal pits, while the two sub beams are used to guide the main beam along the row of pits.

the beam splitter reflects the returning beam through the detector lens, which focuses it onto the photodiode assembly (at right), which converts the infrared signal into the binary code, as indicated earlier, for processing.

Tracking of the main laser beam along a row of pits is aided through use of the two side beams of the laser diffraction pattern. The spatial relationship between the main laser beam and two side beams, when focused onto the surface of the disc, is shown in Figure 7-21. As the disc rotates the reflections of the two side beams are monitored. If either of them begins to lose or gain intensity a feedback circuit tells the drive mechanism to move so as to correct the imbalance between the two side beams. This keeps the main beam on track by moving the laser beam a tiny amount perpendicular to the row of pits. If the disc is warped and moves away from the focal point of the laser beam, the system also corrects by moving the focus lens along the optical axis of the beam. These corrections occur so rapidly that the digital information being transmitted by the center beam remains virtually unaffected.

An intricate sequence of operations is employed to correct for errors in the data as read from the compact disc. In actual practice, the binary coding described earlier is not used directly on the disc. Rather, a code known as the *Cross Interleave Reed-Solomon Code* (CIRC) is employed. Each set of 8 bits (one-half of each 16-bit binary number) is encoded separately in the CIRC format, so that localized damage to the disc will affect only half of the coding in any data point. The most important result of the CIRC is the ability of the system to correct for *burst errors*—that is, a string of damaged or missing data points caused by problems such as a scratch on the surface of the CD. The CIRC system used by CD players can correct for scratches up to about 2.5 millimeters (about 3,500 binary bits) wide, and can even correct for small holes drilled in the CD. The CIRC system also automatically fixes single bits that might be incorrect because of reading errors, manufacturing defects, or slight damage to the disc. In fact, the CIRC is not directly encoded on the CD, but rather each string of 8 bits of CIRC code is expanded to 14 bits through a process known as *eight-to-fourteen modulation* (EFM). EFM compensates for the inability of the system to read accurately a series of alternating 1s and 0s. By adding an additional 0 bit between each of the original bits in every 8-bit sequence, creating a 14-bit sequence, the reliability of the laser reader is increased significantly. In addition to these error-correction techniques, interpolation checks are employed so that the wave shapes are realistically smooth.

Some details of the digital-to-analog process are of interest, because they are often involved in the specifications of CD players. As indicated earlier, to obtain an audible frequency range up to 20 kHz with the CD, a frequency of 44.1 kHz was chosen. Upon reconstructing the analog waves from the digital samples, we find that the analog wave includes an unexpected amount of frequency components above 22.05 kHz. This results from the inability of the 44.1 kHz system to adequately define the frequency components above 22.05 kHz, resulting in *aliasing*, which causes a roughness in the digital-to-analog signal. The actual digital-to-analog converter functions rather like a sample-and-hold device, creating small steps in the analog voltage representing the wave shape, as shown in Figure 7-19. It is therefore necessary to remove these high-frequency components by passing the analog wave through a low-pass filter before it is amplified and sent to the loudspeakers. Unfortunately, two bad things happen in this procedure. First, no electronic filter has an infinitely sharp cutoff, so in removing components at 22.05 kHz, some of the components below 20 kHz are attenuated. Second, filters have the effect not only of reducing the amplitude of the components above the cutoff frequency but also of changing the relative phase of the higher-frequency components. This leads to phase distortion in the final wave, which many audiophiles believe reduces the fidelity of the reproduction.

The way to avoid these problems is through a procedure known as *oversampling*. As the digital sample points are read from the disc and stored for later conversion into the waveform, mathematically interpolated values can be calculated and stored at points between the sample points. By following this procedure before introducing the digital-to-analog process, the electronics will believe that the sampling rate is much higher. For example, by doubling the sampling rate the spurious frequencies will lie above 44.1 kHz. The low-pass filter required to remove the spurious high harmonics will have a cutoff frequency much higher than 20 kHz, so it will affect the audible frequencies much less. Some CD players advertise 8-fold oversampling, in which the sampling rate is increased by a factor of eight, to 352.8 kHz, raising the required filter cutoff frequency proportionally. An interesting case of oversampling, sometimes called Sigma-Delta, occurs when the number of interpolated points is so large that the amplitude of the signal never changes by more than one step in any time interval. As much as 384-fold oversampling has been used in some CD players. One result of such a high sampling rate is that the requirements for the digital-to-analog converter and the low-pass filter are eased, leading to what some listeners believe is a more faithful sound. Exactly how much oversampling is required to produce acceptable results is a matter of personal choice.

The technical considerations presented here for the CD present a good case for why the compact disc medium has become so popular. Some compact disc players can hold at least five discs and can be programmed to play selections from the discs in any sequence, a feat not possible with any previous medium.

Compact-disc recording devices form the pits in blank discs when provided the appropriate digital data. In recent years they have become both reliable and inexpensive, and are now used in a large number of applications, including recording live performances directly onto a CD.

Digital audiotape is often used in sound recording. The most popular format for the consumer market is the rotary-head digital audiotape, sometimes called RDAT or DAT, which uses a recording system similar to that of the videotape recorder. As in the CD, the recording is digital, and thus offers the advantages of a large dynamic range and a virtually infinite signal-to-noise ratio. These tapes are smaller than the standard audiocassette, but have a longer playing time, around two hours. Several sampling rates are used, depending on the application. The DAT format provides excellent sound, as does the CD. However, like standard audiocassettes, DAT does not provide the random access available with the CD but must also be wound to the desired position.

Because the waveform is encoded digitally it is possible to duplicate digital audiotapes indefinitely without loss of fidelity or increase in noise. Thus, when digital audiotape was introduced, many record manufacturers had great reservations regarding possible illegal duplication of their recordings. To avoid this possibility, most CDs are coded so that they cannot be duplicated onto digital tapes without first converting to an analog signal, which degrades the quality of the recording.

7.10 Compression and MP3

Virtually as soon as CDs started to roll off the production lines, it was realized that the CD format is much better than necessary to produce high-fidelity sound reproduction. CDs have fidelity from 20 Hz to 22 kHz, beyond even what is necessary to obtain excellent music reproduction, while audiotape extends only to about 17 kHz, FM radio to about 15 kHz, and AM radio to about 5 kHz. The dynamic range of the CD is over 90 dB, way beyond the useful dynamic range of music. All of this would be irrelevant, except that to store the digital waveform for music in such an inclusive manner limits severely the amount of music that can be stored on a CD.

Several techniques exist for data storage so that it will not take up so much space or require so much time and effort to transmit. They fall into two categories: those that transmit all of the data with no loss and those that are "lossy." Zip files and other compression techniques for scientific data or for written documents are *non-lossy*, because every character in the data is significant. Digital waveforms on a CD and WAV files are non-lossy as well. Compression techniques used to store video and digital images, such as jpeg, gif, and mpeg, for use with computers and on the Internet involve some degree of loss of data. However, to a large extent it is either not noticeable or not bothersome.

MP3 is a system for compression of audio data while retaining as much of the fidelity as necessary to provide high-quality reproduction. It originated as part of a system for compressing video developed by the Motion Picture Experts Group (MPEG), an industry organization—hence the title MPEG for videos used on the Web. In contrast to audio, virtually all digitized video requires some degree of compression because of the enormous amount of data required. If a picture is digitized at 600 dots per inch, it will likely be very clear and look nearly as good as the original, but at the expense of consuming a lot of memory. On the other hand, if it is digitized at 50 dots per inch sharp edges will become a bit fuzzy and the picture will not have the quality that most people desire. Thus you may choose to maximize your picture quality or the number of pictures stored, or to find some compromise that satisfies both requirements. MP3 is a system that lets the user choose how much he or she would like to compress the data, and therefore how much music can be stored per unit of storage space. For example, CD-quality audio requires about 10 megabytes of memory per minute of audio, so a music enthusiast can easily use up large amounts of memory.

MP3 compression begins with the observation that much of the "space" or information available on a CD encodes sounds that we cannot hear, or that are outside the audible regions on the Fletcher-Munson diagram (Figure 6-4) that are normally covered by music. By simply removing very high or very low frequencies, and very soft sounds, the amount of data can be significantly reduced. For example, eliminating frequencies in the upper octave (out of nine) of the audio spectrum would cut the size of the file in half. However, MP3 goes well beyond this minimal level of reduction.

First, the music is divided into variable time segments, each a small fraction of a second, and a spectral analysis is performed on each segment in order. Any components that are outside the range of frequencies or intensities the human ear can detect can be eliminated. Knowledge of psychoacoustics, such as that studied in Chap. 6, can then be used to eliminate any Fourier components that are masked by other larger sounds: frequency components that will be masked by other larger nearby components may also be eliminated with little or no loss of fidelity. Temporal masking—masking of softer sounds that occur very close in *time* to louder sounds—can also be used to eliminate some wave data.

CDs use a constant bit rate code—that is, the wave file consisting of 16 bits per point is read at a constant rate of 44.1 kHz throughout the music. Often it is not necessary to provide a signal continuously at this rate. For example, the system could be told to remain silent for some period during a rest, or continue to play the same sound for some period of time, using fewer kilobytes per second than in a very dense musical passage. By using a variable bit rate, the repetition time at which the data is collected is adjusted, depending on the nature of the music; typically the bit rate increases whenever the sound contains lots of high frequencies.

Using some combination of all of these compression techniques, it is possible to reduce the amount of data by at least a factor of ten, so more than ten times the music can be stored in the same amount of memory. A single CD holds about 660 megabytes of memory, about 75 minutes of music in CD format, but over twelve hours of music compressed into MP3 format.

Each MP3 user can determine the fidelity of his or her music by selecting the bit rate at which it is stored. For example, telephone seems adequate at 16 kBs (kilobytes per second, where a *byte* is eight bits), FM radio at about 112 kBs, CDs at about 192 kBs and studio-quality sound at about 256 kBs. For most use, a bit rate of 128 kBs would be considered adequate, 160 kBs good, and 192 kBs excellent. The user selects the desired bit rate and either constant or (usually) variable bit rate, and records the music. He or she can then reduce the bit rate until the quality of the music is no longer acceptable.

Because the conversion of a music file to MP3 is very computation intensive, it requires a very fast computer or a commercial MP3 converter to carry out the coding in real time—that is, as the music is playing. With less powerful computers it is necessary to first store the music in memory as a WAV file, then let the computer convert the WAV file to MP3 at its leisure.

Using the MP3 format, musical files can be readily shared over the Internet, an act that is generally illegal, as is making direct copies of CDs or commercial videotapes of copyrighted movies. It *is* legal for you to convert your own CDs to MP3 format for your own personal use.

SUMMARY

(***Section 7.1***) Electrical circuits can be viewed as analogous to hydraulic circuits, and are governed by **Ohm's law of circuits:** $I = V/R$, where I is electrical **current**, V is **voltage**, and R is electrical **resistance.** The **power** delivered to a circuit component is given by the equation: $P = I^2R$. Audio systems use **ac** rather than **dc** signals, so the resistance of a circuit, called the **impedance**, is more complex. The typical signal level, known as **high level** or **line level**, has about 1.5 V maximum amplitude. The output and input impedance, typically about 50,000 ohms for most audio components, must be matched in order to prevent wave reflections and to transfer the signal efficiently. (***Section 7.2***) In an audio system, signals from the various sources, such as CD players, DVD and videotape audio, radio tuners, tape decks, and microphones, are sent into a **preamplifier.** The preamplifier acts as the control center for the audio system. The power amplifier must have the proper voltage, current, and impedance so that it can properly drive the loudspeaker. (***Section 7.3***) A **dynamic microphone** uses **Faraday's law** of induction to convert sound waves into electrical vibrations. The **condenser microphone**, or **electrostatic microphone**, uses the principle of the voltage across a charged capacitor to create its signal. Microphones can be **omnidirectional**, or directional, such as the **cardioid** or **shotgun**, or have a reflector, like the **parabolic** microphone. Microphones must have good **frequency response** and **transient response.** (***Section 7.4***) The weakest part of the reproduction chain, the loudspeaker, is usually electromagnetic, or **dynamic.** They may be packaged in **infinite-baffle**, **acoustic-suspension**, or **tuned-port (bass reflex)** enclosures. To get good frequency response, you must have at least two speakers: a **woofer** and a **tweeter**; many systems also have one or more **midrange speakers.** A system with multiple speakers requires an electrical **crossover network** to route the frequency components in the signal to the appropriate component speaker. Stereo headphones have more linear frequency response than loudspeakers. (***Section 7.5***) Preamplifiers are the control center for the audio system, containing controls such as input selection; **volume**, **bass**, and **treble** boost or attenuation; **loudness**; **filters**; and **balance.** (***Section 7.6***) The power amplifier is simply a black box that takes the signals from the preamplifier and prepares them to drive the loudspeakers. (***Section 7.7***) Tuners accept the **AM** or **FM** radio signal, **demodulate** it to obtain the audio signal, and send it to the preamplifier. When FM radio is broadcast in stereo, the signal contains a 19-kHz **pilot signal**, indicating that a second stereo channel is being simultaneously broadcast on a 38-kHz **subcarrier.** (***Section 7.8***) One-eighth inch stereo tape decks, used in both home and auto systems, generally hold either 30 or 45 minutes of music per side. The magnetic tape replays using the principle of **Faraday's law of magnetic induction**, with the tape deck sensing magnetic fields formed by alignment of **magnetic domains** on the tape. Because of **hysteresis**, the tape must have a **bias** signal. **Tape hiss** and nonlinearities inherent in the recording and playback processes require **preemphasis** and **equalization** for low-noise audio signals. **Dolby noise reduction** is commonly used with high-quality tape decks. (***Section 7.9***) **CDs** employ **digital** electronics to achieve high fidelity, low noise, and great **dynamic range.** A CD player uses a laser beam to control the player and sense the signal, which in turn uses several types of coding to avoid problems in data reading and handling such as **burst errors.** The frequency range of a CD extends to over 22 kHz in order to avoid high-frequency phase distortion and **aliasing. Digital audiotape (DAT)** is used in record production but has not become popular in the consumer market. (***Section 7.10***) To reduce the need for large digital memories of CDs, **digital compression** techniques such as **MP3** have become popular for storage of digital music. MP3 is a **lossy compression** technique, but it allows a user to compress the music to the greatest extent that still sounds good to the user, generally at least a factor of 10.

QUESTIONS

1. **a.** What is Ohm's law of electrical circuits? Explain each symbol and give its unit.
   b. If a 100-Ω resistor is connected across the terminals of a 1.5-volt battery, what electrical current will flow?
   c. What electrical power does the resistor in (b) consume?
   d. What form does the energy take, and where does it go?

2. **a.** If a 100-Ω light bulb is connected across a 10-volt battery, how much electrical current will flow?
   b. What electrical power will be consumed by the light bulb?
   c. What form does the energy take and where does it go?

3. Describe the principal functions of the following components in an audio system:

   a. CD player
   b. AM-FM tuner
   c. tape recorder
   d. microphone
   e. preamplifier
   f. power amplifier
   g. loudspeaker

   Draw a block diagram indicating how they are usually connected together.

4. **a.** State and explain Faraday's law of magnetic induction in your own words.
   b. Relate this law of physics to dynamic microphones.

5. **a.** How does the voltage between two parallel, oppositely charged, electrically conducting plates vary as the distance between them increases?
   b. Relate this law of physics to the operation of an electret condenser microphone.

6. What is tape bias and why is it needed?

7. Name and describe the three types of microphones as related to directionality. Which is best suited for

   a. recording an orchestra concert with a live audience
   b. recording a large crowd scene for a film
   c. recording a birdcall among rustling trees

8. **a.** What are the names of the component speakers that comprise a "two-way" and a "three-way" speaker system?
   b. What are the approximate frequency ranges of each?
   c. What is a crossover network and why is it needed?

9. Listen to music from a two-way or three-way speaker system. Closely observe the sound that comes from each speaker and describe what you hear.

10. Describe the primary features of the following loudspeaker systems:

a. infinite-baffle

b. acoustic-suspension

c. tuned-port

If appropriate, use graphs of the loudspeaker response versus frequency to clarify any points you make.

11. Obtain an MP3 coding device (either on-line or from an audio store) and make MP3 compressed recordings at several bit rates. Play them in succession and discuss which bit rate seems optimal for you. Do other listeners agree?

PROBLEMS

1. **a.** Calculate the power transferred from a microphone to a preamplifier if the microphone voltage is 2 mV and the impedance of the preamplifier is 300 Ω.

b. Calculate the power transferred from a preamplifier to a power amplifier if the voltage is 1.5 volts and the input impedance of the preamplifier is 50,000 Ω.

c. Calculate the power transferred from a power amplifier to a loudspeaker if the amplifier output voltage is 10 volts and the speaker impedance is 8 Ω.

2. Convert the following decimal (base 10) numbers to binary: 1, 12, 15, 25, 33, 64, and 127.

3. Convert the following binary numbers to base 10: 0001, 0110, 00000001, 00101101, 11111111, and 110110111011.

4. Suppose that the smallest amplitude a CD could create were 1 binary bit and the largest amplitude 2^{15} bits, or 32,768 times the smallest amplitude. What would the dynamic range be in this ideal situation? As a basis, assume that the SIL corresponding to the smallest amplitude is 0 dB, and use the formula $SIL = SIL_0 + 20 \log(A/A_0)$.

5. An RCA Victor CD with Van Cliburn playing Beethoven piano concertos #4 and #5 has continuous uniform pit tracks between radii of 2.3 and 5.7 cm.

a. If the spacing between tracks is 1.6 microns, how many tracks are there on the CD?

b. Determine the average radius of the tracks and, using that number, the total length of the spiral CD track.

c. If the total playing time of the recording is 72 min 6 sec, what length of track does the laser reader view each second of the recording?

d. If the data on the disc consists of 16-bit binary numbers read at a rate of 44,100 numbers per second for each of the two stereo channels, how many binary bits per second are being read? Assume that an additional equal number of bits is necessary for bookkeeping and other error checks.

e. Using the results of (c) and (d), determine the average space along the track for one binary bit.

REFERENCES

BACKUS, JOHN. *The Acoustical Foundations of Music.* 2nd ed. New York: W. W. Norton & Company, Inc., 1977. Contains a good reference list.

CLIFFORD, MARTIN. *The Complete Compact Disc Player.* Englewood Cliffs, N.J.: Prentice-Hall, Inc., 1987. An outstanding book covering virtually every area of compact disc technology and use, written in an unusually readable style.

DAVIS, DON, and CAROLYN DAVIS. *Sound System Engineering.* Indianapolis, Ind.: Howard W. Sams & Co., Inc., 1975. Good but relatively straightforward book dealing with technical and practical aspects of audio systems.

EARGLE, JOHN. *Sound Recording*, 2nd edition. New York: Van Nostrand Reinhold, 1980.

———. *Handbook of Recording Engineering*, 2nd edition. New York: Van Nostrand Reinhold, 1992.

This pair of books contains an enormous amount of technical information regarding recording and audio electronics. The first volume includes older technologies such as LP discs and quadraphonic sound; the latter covers digital audio equipment.

GREINER, R. A., and DOUGLAS E. MELTON. "A Quest for the Audibility of Polarity." *Audio Magazine*, December 1993. This article discusses the effects of speaker polarity and phase on both acoustical and electronic sounds and furnishes a number of recent references.

JOHNSON, KENNETH W., and WILLARD C. WALKER. *The Science of High Fidelity.* Dubuque, Iowa: Kendall/Hunt Publishing Company, 1977. About the same technical level as this chapter but more complete.

Understanding High Fidelity. Tokyo, Japan: Pioneer Electronic Corporation, 1972. This booklet contains a collection of articles on various components of the typical high-fidelity sound system and discusses their basic features and operation.

WATKINSON, JOHN. *RDAT.* Boston: Focal Press, 1991. An excellent survey of the technical aspects of rotary-head digital audio tape. Other good sources of technical literature are pamphlets and specification sheets available from audio-equipment stores and manufacturers.

Acoustical Reference

MAZER, ELLIOT, and LOREN RUSH. *The Digital Domain: A Demonstration.* Electra/Asylum Records, 1983. An early compact disc recording illustrating many of the features of compact discs and containing several types of noise and signals that are of use in the teaching of acoustics.

Chapter 8

Room and Auditorium Acoustics

In this chapter we shall survey the criteria used to describe the acoustical properties of rooms and auditoriums and relate these properties to the physical properties of such rooms. Wave phenomena, such as the inverse square law, Huygens's principle, reflection, refraction, absorption, and diffraction, as well as our knowledge of logarithms, the decibel intensity scale, and noise and spectral characteristics of musical sounds, will be important here. After a brief look at open-air theaters, we shall see how certain improvements in the acoustical characteristics of this simple structure can be made, resulting in the basic structure of indoor auditoriums. Auditoriums, concert halls, opera houses, and theaters differ from this basic structure in a number of ways.

Attaining proper acoustical characteristics in an auditorium is no accident; good acoustics must be designed into the building, just as are good visibility and adequate ingress and egress. The size, shape, wall and ceiling material, structure, and type of seating must be consistent with the purpose for which the room is designed. Different characteristics are required for optimum listening for speech, chamber music, or large performing groups; furthermore, somewhat different characteristics are required for music from different periods in music history. Home listening rooms differ considerably from performance rooms and will be discussed separately. Suggestions will be made concerning speaker and furniture placement, construction materials, and so forth, to aid in obtaining the greatest listening enjoyment from your home audio system.

8.1 Criteria in Acoustical Design

Several important acoustical criteria arise from the reverberation time of a room and the characteristics of the reverberant sound. Consider the room shown in Figure 8-1; the direct sound from the source arrives at the listener first, followed shortly thereafter by the reflected sound. The reflected sound is composed of the original wave reflected

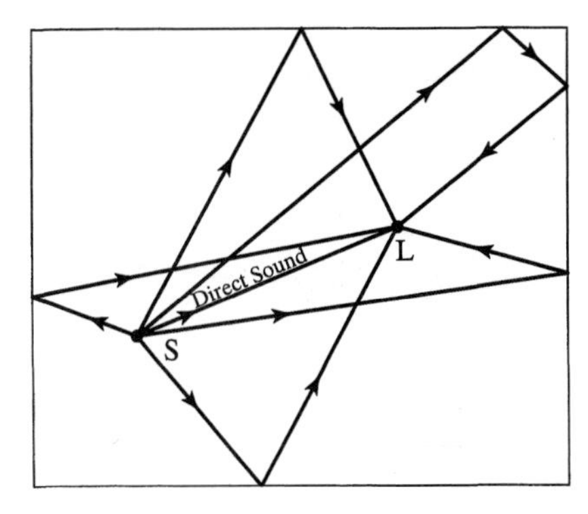

Figure 8-1 Some paths of direct and reflected sound from source to listener in a room.

from the four walls, the ceiling, and floor, plus the successive reflections of these waves. Because of the decrease in the intensity of the wave as it propagates (the inverse square law) and the absorption of some of its energy each time the wave reflects off a surface, the later reflected waves a listener receives are lower in intensity. If we create a sharp sound, such as a snap or a handclap, at time $t = 0$, the sound heard by the listener might typically appear as shown in Figure 8-2. Figure 8-2 is an idealization; the successive pulses would be wider and the entire intensity curve smeared into a more continuous curve because of the continuous range of distances over which the sound wave can pass from the source to the observer.

When a musical tone is attacked and held, the listener will hear something else. If the intensity of the initial sound is I_0 at the listener, the listener will hear the direct sound at this intensity, shortly thereafter reinforced by the reflected waves, with the sum reaching a peak intensity greater than I_0. When the steady tone is released, its intensity will decrease exponentially toward zero, as shown in Figure 8-3. The time required for the sound to decrease from its maximum intensity to 1/1,000,000 of its intensity is defined to be the *reverberation time*, T_R. Other ways of stating this are that the reverberation time is the time required for the sound intensity to decrease to 10^{-6} of its original value, or for the SIL to decrease by 60 dB. The reverberation time of a room is determined

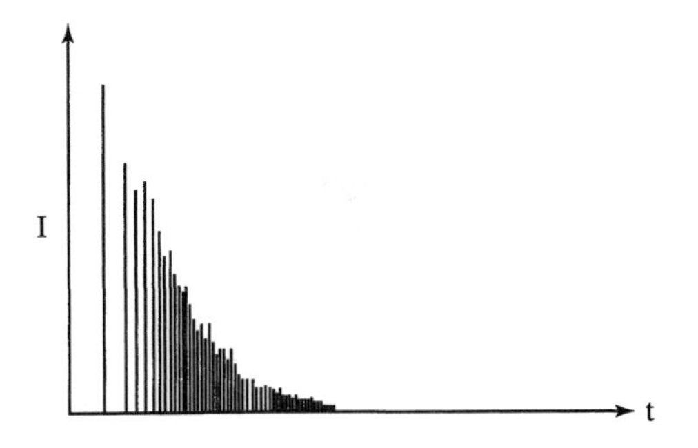

Figure 8-2 Intensity versus time for a sharp sound pulse created at point S and observed at point L of Figure 8-1. The direct sound is the first pulse; reflected pulses follow.

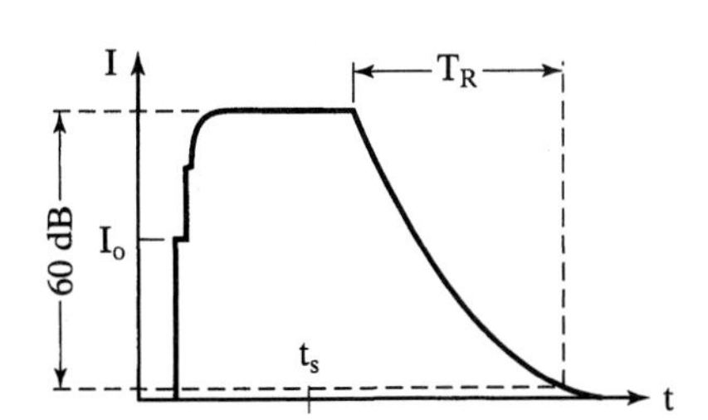

Figure 8-3 Intensity versus time for a steady-state tone in the room shown in Figure 8-1; t_S denotes the time when the source stops, and T_R is the reverberation time.

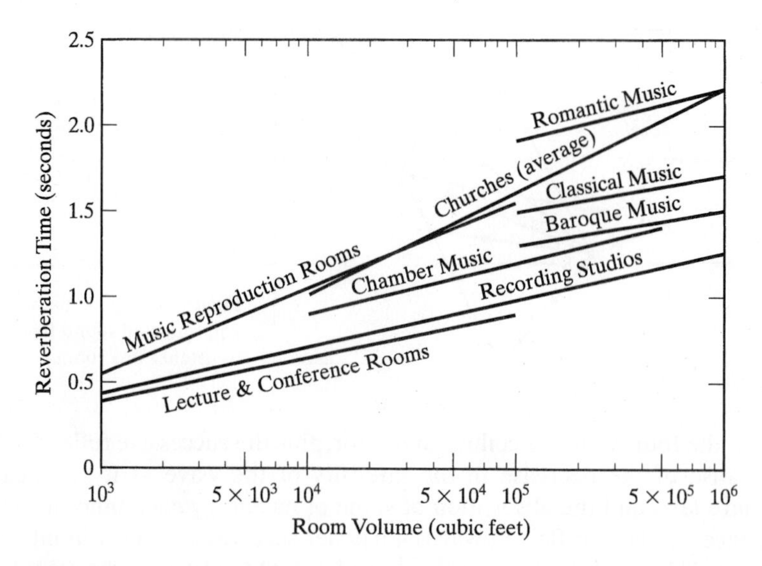

Figure 8-4 "Ideal" average reverberation time versus room volume for several types of rooms. Also shown are the optimum average reverberation times for several important types of music.

by its volume (in cubic feet), the total area of the surfaces both within and enclosing the room (including floor and ceiling), and the absorptive characteristics of the surfaces. The reverberation time may be different for different frequencies.

The reverberation time most suited to a particular application (e.g., speech, chamber music) in a specific room will depend on the nature of that application. Shown in Figure 8-4 are lines indicating the approximate ranges of reverberation times appropriate for the various types of rooms and applications. There is no universally accepted "ideal"; this figure is merely a general guide.

Some of the more important acoustical characteristics, along with the physical properties of sound from which they arise, are discussed next.

Liveness

Liveness is a qualitative term that expresses the physically measurable reverberation time. The longer the reverberation time, the more "live" the room is. The most appropriate reverberation time depends on the nature of the sound or music and on the size of the room. Furthermore, even subtle differences in the mood or period of the music when played by the same ensemble or orchestra can ideally be suited to rooms with different reverberant qualities, as we shall see. A particular hall may be most suited to the music of certain periods, and may thus influence the type of music for which the group performing in that hall becomes renowned.

Intimacy

Intimacy refers to how close the performing group sounds to the listener. Intimacy is achieved whenever the first reflected sound reaches the listener less than about

20 ms after the direct sound, a condition easily met in small halls. For large halls, this goal can be achieved by placing a reflecting canopy above the performers, tilted so as to direct the reflected sound toward the audience. Such a canopy is sometimes used in large cathedrals above the pulpit to achieve intimacy of the speaker's voice to the congregation. A canopy also forms the basic part of the structure of a bandshell, helping to increase the sound level by reflecting the upward sound out toward the audience.

Fullness

Fullness refers to the amount of reflected sound intensity relative to the intensity of the direct sound; the more the reflected sound, the more "full" the hall will be. For slow, romantic music performed by very large groups, fullness is required, whereas chamber music or music from the classical or baroque periods does not require great fullness. In general, greater fullness implies a longer reverberation time.

Clarity

Clarity, the acoustical opposite of fullness, is obtained when the intensity of the reflected sound is low relative to the intensity of the direct sound. Great clarity is required for optimum listening to speech and is particularly important when performing early orchestral music, such as the fast movements of the Mozart symphonies. In designing a symphony orchestra hall, it is necessary to strike a compromise between fullness and clarity to allow adequate performance of both early and later music. In general, greater clarity implies a shorter reverberation time.

Warmth

Warmth is obtained when the reverberation time for low-frequency sounds is somewhat greater than the reverberation time for high frequencies. The "ideal" reverberation time, as a function of frequency, required to obtain the feeling of warmth is shown in Figure 8-5. Above 500 Hz the reverberation time should be roughly constant, but below that frequency it should increase to about 1.5 times the reverberation time for high frequencies. This is accomplished by the proper choice of materials for the walls and ceiling, which will be discussed in a later section. If the reverberation time is too long at low frequencies, however, the sound can become muddy and will lack clarity.

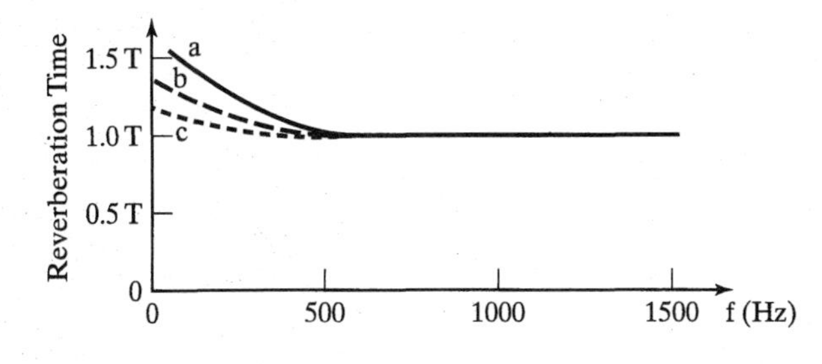

Figure 8-5 Relative reverberation time versus frequency (a) in a "warm" room, (b) in a normal room, and (c) in a "brilliant" room. T is the nominal reverberation time.

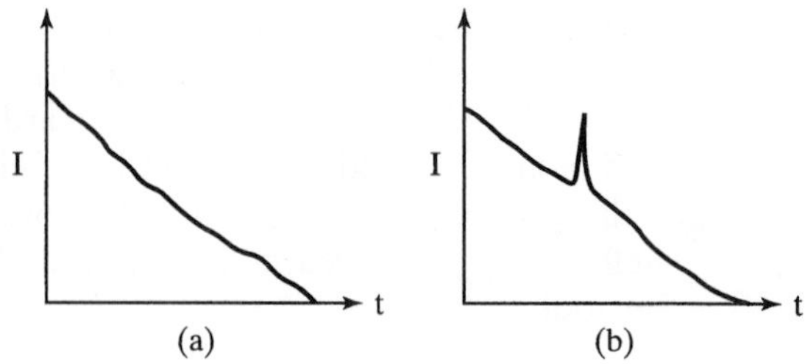

Figure 8-6 Intensity versus time for reverberations in a room with (a) good texture, and (b) poor texture resulting from a large, late group of reflections.

Brilliance

Brilliance is the opposite of warmth and exists if the reverberation time for high frequencies is larger relative to that of the low frequencies. For an average room, the reverberation time for the low frequencies is slightly greater than that for the high frequencies because of the absorption of higher frequencies by the walls, floor, and ceiling. In a more brilliant room, the difference is less pronounced; the reverberation time for high frequencies is more nearly equal to that of the low frequencies. If the reverberation time for high frequencies is excessively long, a continual high-pitched ringing sound may result.

Texture

Texture refers to the temporal pattern of reflections reaching the listener. The first reflection should quickly follow the direct sound to achieve intimacy, with successive reflections following quickly thereafter. To achieve good texture, it is generally necessary to have at least five reflections within about 60 ms after arrival of the direct sound. An important related requirement is that the overall sum of the reflected waves combined should, on average, decrease uniformly in intensity. Shown in Figure 8-6 are two curves of sound intensity versus time, as in Figure 8-2. Figure 8-6(a) shows good texture, whereas Figure 8-6(b) shows poor texture owing to a late-arriving intense reflection. The structure of the hall should not create any single unusually large echo or have a tendency to focus the sound; otherwise, poor texture may result.

Blend

Blend refers to the mixing of the sound from all the instruments of the orchestra or ensemble throughout the audience. In a concert hall with poor blend, a member of the audience in some specific location may hear one section or player louder than he or she hears the others. The simplest technique for achieving proper blend is to mix the sound from the various instruments and voices on the stage before distributing the sound to the audience. This need not imply using electronic mixing equipment but rather using appropriate reflecting surfaces surrounding the performing area. It is also necessary that there be no surfaces that focus or direct the sound from some part of the performing group to some small section of the audience. All the sound must be diffused over the entire audience by appropriately shaped reflectors.

Ensemble

Ensemble refers to the ability of the members of the performing group to hear each other during the performance, enhancing the ability of the players to play together effectively. To achieve good ensemble there must not be strong reflections that are delayed longer than the duration of the fast notes in the music being performed. This puts a practical limit on the size of the stage. The lateral distance between the nearly parallel sides of the stage should be less than about 65 feet for use with a classical orchestra, one of the larger performing musical ensembles. A more serious problem is the sense of ensemble in the performance of an opera between the members of the orchestra in an orchestra pit and the singers (soloists or choir) on the stage. One method of improving the ensemble is to put reflectors in front of the orchestra, to reflect some of the orchestral sound back to the singers. Doing this, however, reduces the sound of the orchestra to the audience and results in part of the audience hearing only reflected sound with no direct sound, an undesirable situation. Another way to increase the sound level of the orchestra to the singers is to use small loudspeakers aimed at the stage from the wings. Ensemble remains one of the most problematic areas in opera production.

Another situation in which the sense of ensemble leads to problems is in the performance of a marching band, particularly on a football field, where the band is spread out over a large area. The time delays for the sound of instruments at one end of the band to reach the other end are large in comparison with the rhythmic pattern. This can be compensated for only by having a single reference by which every player sets his timing. By watching a director and playing in time with his or her conducting, rather than with the sounds of the other instruments, the band can be kept synchronous. The director's hand signals arrive at all the members of the band essentially simultaneously, because the speed of light is extremely great. Someone in the stands may not hear the different instruments as synchronized, however, because of the varying distances.

8.2 Problems in Acoustical Design

Several important acoustical problems must be avoided in the design of auditoriums and other rooms. Many of these problems result from improper architectural designs. The following are some of the more serious problems, the architectural features from which they arise, and their basis in the fundamental properties of waves.

Focusing of sound

Domes, curved walls, or ceilings approximating the shape of spheres, ellipsoids, or paraboloids, or series of small straight panels approximating these shapes, should generally be avoided. Such surfaces may produce focal points at which the sound is much louder than that in the surrounding region, or can lead to the whispering-chamber effect, as shown in Figure 2-17. In the extreme, if part of the orchestra is at one focus of an ellipse and part of the audience at the other focus, that section of the audience could hear some instruments of an orchestra louder than other instruments. Under these conditions, audience noise could also become inordinately loud for the performers sitting near a focal point. Most large, gently curving walls at the rear of an auditorium do not focus the sound unacceptably.

Echoes

To obtain good texture it is necessary to avoid any particularly large single echoes; the direct sound followed by a large echo gives the illusion of a performance inside a large can. One way to obtain a very large echo is to make the stage in the shape of a parabola with highly reflective walls, with the performers at the focus. The direct sound will then be followed by a large echo obtained when the sound reflecting off the rear of the stage reaches the audience as a plane wave (see Sec. 2.2).

Another important echo effect is the flutter echo, which can be obtained when the two side walls of the auditorium are parallel to each other and made of highly reflecting materials. In this case the sound reflects back and forth between the two sides, creating a rapid sequence of echoes. These reflections can even lead to destructive interference between the various reflected waves, as in the discussion of standing waves in Chap. 3.

Shadows

Acoustical shadows, or quiet regions, can be produced when there are long overhanging balconies or other structures jutting into the hall. There should be a large open space extending over a large angle in front of each person in the audience. When some protruding object, such as a balcony, wall, or column blocks any observer's view, the sound will undergo diffraction around that structure. As we have seen, the diffraction is different for different frequencies, leading to distortion of the sound passing around the obstruction.

Another place where diffraction can become important is in the design of reflecting surfaces, sometimes called clouds, which can be lowered into place, say in a hall with very high ceilings, to create intimate acoustics. Small reflecting surfaces with spaces between them can reflect the higher frequencies adequately, but may allow the lower frequencies to diffract through the spaces and be lost, again resulting in a distorted reflected sound and inadequate warmth. On the other hand, this may be desirable in certain situations to compensate for other defects of the room.

Resonances

Resonances arise primarily in smaller rooms in which the dimensions of the room may only be up to a few wavelengths long for the lower audio frequencies. One place where resonances almost always appear is in a bathroom, particularly in a shower or stall. This is because the walls are parallel and made of highly reflecting materials such as metal, tile, or glass. A resonance can occur whenever any dimension of the room is any multiple of one half-wavelength long for some particular frequency. There will be a node at each wall, and the wave will resonate as though it were in a tube closed at both ends. There can be a large number of resonances in each of the three directions (up-down, left-right, and backward-forward).

The following equation relates the various resonant frequencies of the room to its physical dimensions:

$$f_{xyz} = \frac{S}{2}\sqrt{\left(\frac{N_x}{x}\right)^2 + \left(\frac{N_y}{y}\right)^2 + \left(\frac{N_z}{z}\right)^2} \tag{8.1}$$

where S is the speed of sound in air; x, y, and z are the three dimensions of the room, which is assumed to be in a regular rectangular shape; and N_x, N_y, and N_z are any integers. The subscripts x, y, and z refer to the values of N_x, N_y, and N_z, respectively. Changing any of these three numbers results in a different resonant frequency. For $N_x = 1$ and $N_y = N_z = 0$, the formula becomes

$$f_{100} = \frac{S}{2x}, \tag{8.2}$$

which is the resonance occurring when the wavelength of the wave of frequency f_{100} is twice the x dimension of the room. The room is one half-wavelength long in this dimension. For a small room, such as a bathroom or a practice room with dimensions of 1.75 m by 3.5 m by 2.8 m high, some of the resonant frequencies are shown in Table 8-1.

Not only is there a set of resonant frequencies associated with each of the dimensions of the room (waves reflecting back and forth between two walls to produce standing waves), but there is also a new set of complex standing waves occurring when the wave reflects obliquely between two or more of the pairs of walls. At high frequencies, the resonances get more closely spaced, leading to good reinforcement of higher harmonics, whereas at low frequencies the resonances are more widely spaced, giving a less uniform reinforcement. These high-frequency resonances lead to the nice sounds of singing in the shower or practicing your instrument in the bathroom. Practicing in a bathroom may be all right at times, but it clearly gives one a distorted perspective on tone quality.

Table 8-1 Resonances for a Room

With $x = 3.5$ m, $y = 2.8$ m, and $z = 1.75$ m

N_x	N_y	N_z	f_{xyz}(Hz)
1	0	0	50.0
0	1	0	62.5
0	0	1	100.0
1	1	0	80.0
1	0	1	111.8
0	1	1	117.9
1	1	1	128.1
2	0	0	100.0
0	2	0	125.0
0	0	2	200.0

You might try to find a few resonances by singing in a shower stall and measuring the dimensions of the stall to verify that the formula works. There should be general agreement between your theoretical and experimental results.

The formula for resonances in rooms applies equally to resonances in smaller boxes like speaker enclosures and must be employed in the design of some types of enclosures. It is necessary in such design to separate the resonances of the speaker and the box enclosure of a tuned-port system to provide the most linear response possible.

One way to avoid some room resonances is to avoid parallel walls. In many cases, small music practice rooms are constructed with oblique opposing walls.

External noise

Careful design of the architectural features of an auditorium is of no avail if either external or internal noise is too loud. If there are neighboring airport flight patterns, railroads, or highways, special efforts must be made to isolate the auditorium acoustically from the surrounding area. This can be accomplished by building a box within a box, with acoustical insulation between the boxes, as in the design of the John F. Kennedy Center for the Performing Arts in Washington, D.C., which is close to the flight path of a nearby airport. Adjacent rooms must also be insulated from one another, again as is the case in the Kennedy Center, which has three major performing halls.

Another important source of noise is air conditioners or blowers, which are often sources of white noise or noise rich in lower frequencies. These low frequencies can, in extreme cases, mask out the higher-frequency sounds of the music.

The maximum tolerable noise level depends on the type of environment and the room's purpose. In general, the noise requirement is most stringent in studios and theaters, where words must be understood, but is less critical in homes and offices. A partial list of noise requirements is given in Table 8-2; the sound-intensity levels are given in decibels, as defined in Chap. 6.

One way to control the noise level is to match the size of the room to the size and type of performing group. It is also important to provide for sufficient absorption of sounds created by the audience, which, if not immediately absorbed, may become loud enough to mask out the music.

Shielding outdoor auditoriums from external sound is particularly important. This can be done in part by putting the auditorium in a valley surrounded by high hills.

Table 8-2 Average Normal Noise Levels for Several Acoustical Environments

	dB
Studios (recording, radio, TV)	25
Auditoriums and theaters	30
Classrooms or lecture halls	30
Hospitals	30
Homes	40
Offices	45
Restaurants	50

Such noise shielding is especially important at night owing to the very long distances over which sounds will travel as a result of refraction in the atmosphere during times of temperature inversion. The focusing of sound by air currents must also be considered in outdoor open-air auditoriums.

Double-valued reverberation time

The reverberation time in rooms used for recording playback can be different from the reverberation time of the hall in which the music was recorded. If the reverberation time of the playback room is very long, this can lead to a rather confusing and unpleasant sound. Decreasing the reverberation time or using headphones in the listening room can combat this problem. In the case of speech reproduction, a double-valued reverberation time can actually render the speech incomprehensible.

Unfortunately, the requirements for use of a room for listening are somewhat different from those when the room is used for performance or even rehearsal, so some compromise must often be accepted.

8.3 Control of Reverberation Time

We have seen that the reverberation time is the most important characteristic of an auditorium. We shall now study how to control the reverberation time by using sound-absorptive materials.

Recall that the reverberation time is the time required for the intensity of a sound to drop to 1 part in 1,000,000 of its original intensity. An approximate formula for the reverberation time in seconds is

$$T_R = 55.2\frac{V}{SA} \tag{8.3}$$

where V is the volume of the room, S is the speed of sound, and A is the total absorption of the surface of the room. The reverberation time is proportional to the volume of the room, because the bigger the room is, the longer the time required for all waves to reach the absorbing surfaces. In this formula we will use the volume of the room in cubic feet, because architects and engineers generally work in these units. If we use room volume V in cubic feet and speed of sound $S = 1100$ feet per second, then the absorption A is in units of the sabin. The reverberation time formula in this case can be written as

$$T_R = 0.050\frac{V}{A}. \tag{8.4}$$

The unit of sound absorption, the sabin, was named after Wallace C. Sabine (1868–1919), the physicist generally acknowledged as the founder of the science of architectural acoustics. The sabin is an absorption equivalent to 1 square foot (ft^2) of perfectly absorbing surface, say a 1-ft^2 opening in a wall. A 1-ft^2 opening, 5 ft^2 of material with an absorption coefficient of 0.2, or 10 ft^2 of material with an absorption coefficient of 0.1 will each have a total absorption of 1 sabin. The total absorption of the surfaces of a room is

$$A = a_1A_1 + a_2A_2 + a_3A_3 + a_4A_4 + \cdots \tag{8.5}$$

where the $A_1, A_2, \ldots$ are the areas of the various types of absorbing surfaces and the $a_1, a_2, \ldots$ are the absorption coefficients of the respective surfaces. The formula for reverberation time then becomes

$$T_R = \frac{0.050V}{a_1A_1 + a_2A_2 + a_3A_3 + a_4A_4 + \cdots}. \tag{8.6}$$

This formula has omitted the absorption of sound by the air. Although this effect cannot be neglected in the design of large auditoriums, it is insignificant for small halls, and its effect is only significant at high frequencies. This effect would appear as another term in the series in the denominator of the equation.

As an example of reverberation time calculation, consider a room 13 ft by 20 ft with an 8-ft-high ceiling. The volume $V = 13 \times 20 \times 8 = 2080\ \text{ft}^3$. If the four walls are of plaster with an absorption coefficient of 0.1, the total absorption of the walls would be $A_w = 0.1 \times 8 \times (13 + 20 + 13 + 20) = 52.8$ sabins. If the floor is covered by carpet with an absorption coefficient of 0.3, it adds an additional absorption of $A_f = 0.3 \times 13 \times 20 = 78$ sabins. An absorptive tile ceiling with absorption coefficient of 0.6 would provide an additional absorption of $A_c = 0.6 \times 13 \times 20 = 156$ sabins. The total absorption of all the surfaces of the room is then the sum

$$A = A_w + A_f + A_c = 286.8\ \text{sabins}. \tag{8.7}$$

The reverberation time then becomes

$$T_R = \frac{0.050 \times 2080\ \text{ft}^3}{286.8\ \text{sabins}} = 0.36\ \text{sec}. \tag{8.8}$$

A room of the same size with all surfaces, including floor and ceiling, of marble with an absorption coefficient of 0.1 would have a reverberation time of about 1 sec. If an audience were present, the total absorption of the people plus the seats would have to be added to the denominator.

It is usually desirable to provide warmth by making the reverberation time longer for low frequencies than for high frequencies. Fortunately, the reverberation time can be adjusted independently over a broad frequency range by choosing the proper materials to cover the surfaces of the room. Shown in Table 8-3 are the absorption coefficients for a number of building materials at several frequencies. The absorption coefficients are usually given at frequency intervals of one octave, rather than in equal frequency intervals. Using the formula for reverberation time and Table 8-3, the reverberation time can be calculated at a number of frequencies to determine whether the required warmth or brilliance will be obtained.

It is also necessary to take into consideration the absorptive effect of the audience. Shown in Table 8-4 is the average total absorption, in sabins, at a number of frequencies for some types of seats with and without people. This is only an approximation, as can be

Table 8-3 Average Absorption Coefficients for Several Types of Building Materials at Octave Frequency Intervals

Material	Frequency (Hz)					
	125	250	500	1000	2000	4000
Concrete, bricks	0.01	0.01	0.02	0.02	0.02	0.03
Glass	0.19	0.08	0.06	0.04	0.03	0.02
Plasterboard	0.20	0.15	0.10	0.08	0.04	0.02
Plywood	0.45	0.25	0.13	0.11	0.10	0.09
Carpet	0.10	0.20	0.30	0.35	0.50	0.60
Curtains	0.05	0.12	0.25	0.35	0.40	0.45
Acoustical board	0.25	0.45	0.80	0.90	0.90	0.90

Table 8-4 Average Absorption in Sabins at Octave Frequency Intervals of Seats and a Person

	Frequency (Hz)				
	125	250	500	1000	2000
Unupholstered seat	0.15	0.22	0.25	0.28	0.50
Upholstered seat	3.0	3.1	3.1	3.2	3.4
Adult person	2.5	3.5	4.2	4.6	5.0
Adult in upholstered seat	3.0	3.8	4.5	5.0	5.2

inferred from the rather large range of variation of absorption coefficients for each case. In "tuning" the acoustics of an auditorium before its general use, it is often possible to use blocks of absorbing material in place of people to obtain their effect, as was done in tuning New York's Avery Fisher Hall with artificial "instant people," who were quiet as well as absorbent.

Ultimately, after the room has been designed and built, the listener will judge the acoustics of the room. A combination of listening in concert situations and actual physical measurement of the reverberation times may reveal inadequacies, which then can be corrected by adding to or changing the surfaces.

Anechoic chambers, often used in acoustics research, are rooms whose walls consist of absorbent wedges that project in toward the center of the room. A sound wave reflects off the side of the wedge and then strikes nearby wedges. The multiple reflections off the extremely absorbent material make the room virtually echo free and the reverberation time extremely short.

8.4 Design of Auditoriums

We shall now "design" an auditorium using the basic acoustical laws and properties discussed previously. We start with an open-air auditorium and end by covering the sides and top to obtain the design of an indoor hall.

The most elementary type of "auditorium" would be an audience seated in front of the speaker or performer out in the middle of a flat field. This has several problems: no shielding from external noise, the problem of rapid attenuation of the sound as the wave progresses near the ground level, and the lack of any reflected sound. We can correct the first situation by locating our auditorium in a small valley between large hills or, in the case of an indoor auditorium, by providing walls and a ceiling to keep out outside sounds.

Away from any walls, ceiling, or floor (for instance, up in the sky) a sound signal diminishes 6 dB as it doubles its distance from the source. This is simply the inverse square law; the intensity decreases by a factor of 4 each time the distance from the source doubles (see Exercise 14). However, the calculation for waves near the surface of the ground, or near any other highly absorbent surface, is not so simple. In Chap. 2 we discussed how the addition of Huygens's wavelets from the surface of a wave front produces the new wave front at a later time. If the surface is absorbent—a grassy field or an audience—resulting in propagation of the wave over only one hemisphere, one-half of the wave that normally acts as a source of wavelets for the succeeding wave will then be missing, having been absorbed. The amplitude of the new wave will then be only about half that of its value with all the point sources included. The intensity will therefore drop off near the ground almost as the inverse fourth power of distance, or at a rate of about 12 dB as it doubles its distance from the source. To keep the sound level as high as possible, it is necessary to provide each member of the audience with a large open field of view, so that the effect of absorption of the waves by the ground or other members of the audience will be minimized. This can be done by raising the stage, sloping the audience seating area, or a combination of these. In early Greek open-air amphitheaters the slope increased with distance from the stage.

To obtain fullness, with a good strong reflected sound, it is necessary to provide a shell over the performers. The shell must be low enough to give a reasonably rapid reflected sound to ensure an adequate sense of intimacy. Adding sides to the shell provides additional fullness and texture, acts to diffuse the reflected sound, and assures a better sense of ensemble for the performers. The shell should be made of only a few straight sections, without curved shapes, to avoid any focusing or other undesirable reflections.

Additional large flat reflecting surfaces above, behind, and to the sides of the audience (walls and ceiling) can be added to obtain additional fullness and to preserve the sound level. The two sides of the hall should fan out and be of absorbent material to avoid flutter echo and any standing-wave effects. We now have an indoor auditorium.

If a balcony is desired, it should be strongly sloped and should not extend out far enough to cause any diffraction of waves passing underneath the balcony. Pillars or support columns that interfere with the audience either in or underneath the balcony should be avoided.

The rear wall of the auditorium can be lined with absorbent material to prevent strong reflection of the direct sound, if necessary. Materials for the other walls, the seats, and the floor can be chosen to provide the proper reverberation time for the desired application.

An additional possibility is to provide removable or retractable absorbers, which can be used to reduce the reverberation time when the hall is to be used for speech and removed during musical performances. Smaller shells could be provided for small groups. Movable ceiling sections could be lowered and the balcony left unused during chamber-music performances.

Our last design step is to listen to musicians on stage from various places in the audience section to test for blend and then measure the reverberation time at several frequencies. Different construction materials could then be substituted or appropriate absorbent panels added, if necessary.

An additional consideration in some large halls is the use of electronic amplification. Two basic loudspeaker configurations are popular. In one, speakers are simply installed in the front of the room; in the other, the speakers are spread out more or less uniformly over the ceiling. Calculations must be made to ensure that adequate loudness can be obtained while ringing, feedback squealing, and other distortions are avoided. It is important for some types of musical performance to use the speakers to simply support the direct sound of the music and to avoid having excessive amplification. It is also necessary to delay the sound from the speakers so that it reaches the listener after the direct sound has reached the listener. The audience will then perceive the electronic sound as part of the reverberation. The situation in which the amplified sound from above reaches the listener before the direct sound does can create the illusion that the performers are on the ceiling. This is because of the precedence effect, according to which listeners perceive the location of the sound source as the location from which the earliest sound arrives.

The Elsie and Marvin Dekelboum Concert Hall of the University of Maryland Clarice Smith Performing Arts Center is a new concert facility with both outstanding visual appeal and excellent acoustics. Architects Moore, Ruble, Yudell, in association with Ayers, Saint, Gross and acoustical consultant Kirkegaard and Associates, created the design, which has several interesting features.

As can be seen in Figure 8-7 and Figure 8-8, it has unique visual appeal, with its expansive stages, balcony, and unusual structure. As seen in Figure 8-8, looking from the rear of the stage into the audience, the seating is gently sloped, with a total of 1,000 seats

Figure 8-7 View from the back of the Elsie and Marvin Dekelboum Concert Hall at the University of Maryland Clarice Smith Center for the Performing Arts.

Figure 8-8 View from the front of the Elsie and Marvin Dekelboum Concert Hall at the University of Maryland Clarice Smith Center for the Performing Arts.

on the main floor, the balcony, and boxes along the sides. On a balcony behind the stage are an additional 170 seats that can be used for audience or for a large choir performing with an orchestra on the stage.

Reverberation time is adjusted by lowering heavy curtains that unroll from the center of the ceiling, as seen in Figure 8-9. In the figure, the set of three curtains at the upper right of the picture has been raised about halfway from its fully extended position to its fully retracted position. In general, the curtains remain in their low positions for concerts but are raised to obtain a longer reverberation time for recording. Further small adjustment of reverberation time can be made using curtains located behind wooden louvers at the rear of the stage. Curtains behind the box seats on the balcony level can be opened or retracted. The Concert Hall has been excellent for large ensembles and combined orchestral and choral groups. To accommodate smaller ensembles and soloists, the concert hall stage area is to be fitted with retractable clouds. This will improve intimacy in listening to smaller performing groups and increase the sense of ensemble for the performers.

As technical improvements are made in the design and installation of electronics and speaker systems, and as more is learned about the use of speakers in large auditoriums, more complex and innovative loudspeaker systems are being used in large concert halls. However, most critics agree that direct unamplified sound in a well-designed auditorium often provides the most enjoyable listening experience for classical concert music.

Figure 8-9 Acoustical curtains that can be lowered from the ceiling to control reverberation time in the Dekelboum Concert Hall at the University of Maryland Clarice Smith Center for the Performing Arts.

8.5 Home Listening Rooms

Although very few people will ever become involved in the design or construction of auditoriums, many at some time or another will assemble their own audio reproduction systems. Further, although most of us are not wealthy enough to completely design or rebuild a music room to conform to the optimum specifications for music listening, there are certain things you can do readily and at little cost to ensure maximum enjoyment of your own installation. Several factors will influence the sound ultimately obtained in your listening room: (1) the type of materials used in the walls, the floor, and the ceiling of the room, (2) the size and shape of the room, (3) types of rugs and furnishings used, (4) loudspeaker placement, and (5) quality of audio components.

In most cases, without undue expense there is not much one can do about the room size and shape or the building materials. If deadening of the room is required, it may be possible to install a suspended ceiling with absorbent tiles or hang rugs from the walls. If it is desirable to retain as much liveness as possible, furnishings can be obtained that offer little absorption of sound. If it is necessary to reduce the reverberation time, overstuffed chairs and couches and other absorbent furnishings can be used. Fortunately, as a usual rule, a normally furnished living room has a reverberation time that is within the range acceptable for use as an audio reproduction and listening room.

Figure 8-10 Areas of good listening for stereophonic music.

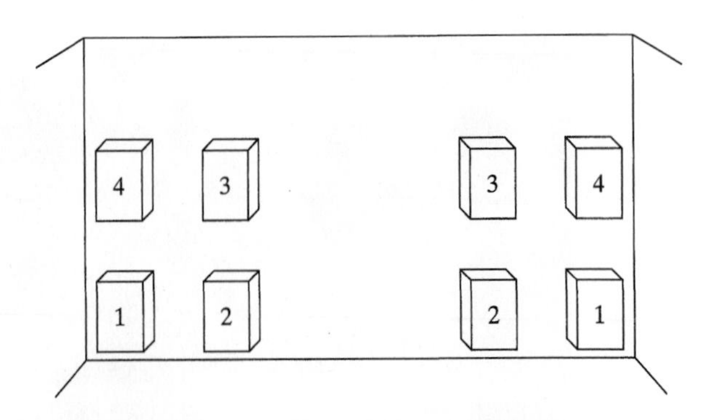

Figure 8-11 Good positions for stereo loudspeakers along a wall of a listening room.

The two factors over which one has most effective control, and which probably affect your listening pleasure the most, are the quality of the audio components and the placement of the speakers relative to the sitting areas. In the previous chapter we discussed some of the more important criteria in the choice of components for an audio system. One additional consideration should be mentioned here: The output power from the audio system should be matched to the size and quality of reverberation for the room in which it is to be used. Large rooms and rooms with shorter reverberation times require greater power output from the speakers. However, almost any loudspeaker of reasonable quality will produce sufficient power output, when driven by a sufficiently powerful amplifier, to fill a normal home listening room with sound; normally less than 1 watt of acoustic power is required. Use of high-power components is only necessary for unusually large rooms or when it is desired to raise the volume of the music above the normal level.

The placement of loudspeakers can be of considerable importance, particularly in listening to stereophonic or quadraphonic programs. Figure 8-10 shows the approximate extent of the listening area for stereo music with normal positioning of speakers. For optimum listening to stereo sound, the listening area should be confined to a region between the two speakers, and not much farther from one than from the other, to assure balance of intensity. The listening area should begin back some distance from the speakers; most speakers are somewhat directional, and some space is required for the sound to spread out uniformly, owing to diffraction effects.

Figure 8-11 shows possible placement of speakers against one wall of a rectangular room. The speakers should in general be placed symmetrically in the room, and can be placed either against the wall or out from the wall. The closer the speakers are to the wall and corner (positions 1 and 4), the greater will be the support of the bass frequencies by the coherence of reflected waves off the corner or wall. However, if the speakers themselves provide sufficient bass sound, it might be desirable to move them out from the wall (out of the plane of the paper) and to position 2 to obtain the greatest uniformity of sound. Raising the speakers to position 3 allows exposure to direct sound from the speakers unhindered by either reflections from the corners or by absorption from the floor. Placing high-quality speakers at position 3 and slightly out from the rear wall is possibly optimal. More recent home theater loudspeaker systems include two additional speakers that are positioned on either side at the rear of the room.

One additional recent innovation should be mentioned at this point: The development of reasonably inexpensive equalizer circuits has provided us with a way of "tuning" our listening rooms, in combination with the audio system, to remove any strong resonances that may emphasize certain frequencies. The average power per unit frequency of orchestral music, and to a good approximation of other types of music, falls off at about 3 dB per octave. In the audio-engineering field, noise that falls off in power per unit frequency at the rate of 3 dB per octave is called pink noise, as discussed in Sec. 4.4. In tuning a room, pink noise is played through the audio system and picked up by a high-quality microphone that is connected to a sound-level meter to measure the intensity of the pink noise at the position of the microphone. The microphone is placed at the optimum position for listening to the speakers. If room resonances or resonances in the speakers exist, the sound level will be higher than that of the pink noise at those frequencies.

On the other hand, there may be frequencies at which the room absorbs sound particularly strongly or at which the speaker responds less well. In either case, *equalizers* consisting of special filters can be used to modify the power-amplifier signals sent to the speaker to provide pink noise, which falls off with frequency at exactly the right rate, in the listening area. In most cases, it is only necessary to use about four to six separate frequency bands spanning the frequency region up to a few thousand Hertz to provide a response that is within about 1 dB of the ideal pink noise. This procedure is limited; in suppressing a very strong resonance it may also suppress frequencies on either side of the resonance to below the desired intensity level.

Other techniques can be used to remove or limit the most serious room resonances. Because these resonances are sometimes the result of reflections between parallel surfaces forming standing waves, it might be possible to install a wall or an extension of the existing wall at an angle. Alternatively, it might be possible to cover adjacent walls with a rug or some other absorbent wall hanging to reduce reflections.

SUMMARY

(***Section 8.1***) The acoustical quality of a room is determined largely by its **reverberation time (T_R)**. Terms used to describe the quality of a room resulting from features of the reverberation time include **liveness**, **intimacy**, **fullness**, **clarity**, **warmth**, **brilliance**, **texture**, **blend**, and **ensemble**. (***Section 8.2***) Acoustical problems to be avoided in a concert hall include **focusing of sound**, **echoes**, **shadows**, **resonances**, **external noise**, and a **double-valued reverberation time**. (***Section 8.3***) The design of a good concert hall is based on calculating the reverberation time, adjusting it over a range of frequencies through the geometry of the room, and using various sound-absorbing materials. The unit of sound absorption in architectural acoustics is the **sabin**. (***Section 8.4***) Good auditoriums provide visual appeal and allow audience members to see the entire performing group and to hear the sound without acoustical problems. Reverberation-time adjustments can often be made in the hall to provide the best acoustics for various types of performances. (***Section 8.5***) The acoustical performance of home listening rooms can be optimized by using high-quality equipment with the power level matched to the room, placing the speakers in the best locations, reducing reverberations in the room, and possibly equalizing the sound for the particular room.

QUESTIONS

1. Define each of the following acoustical characteristics of rooms and relate them to the features of the reverberation time:

   a. liveness
   b. intimacy
   c. fullness
   d. clarity
   e. warmth
   f. brilliance
   g. texture
   h. blend
   i. ensemble

   How might the presence or absence of an audience affect these properties and the sound of a musical performance?

2. Describe the following problems arising from improper acoustical design, and discuss how these problems could be avoided or corrected:

   a. flutter echo
   b. excessive liveness
   c. poor texture
   d. problems relating to the precedence effect
   e. focusing

3. A room is 25 ft × 40 ft × 30 ft. The front (25 ft) is covered by glass, the back (25 ft) by curtains, and both sides (40 ft) by plasterboard. The ceiling is of acoustical board, and the floor is made of concrete.

   a. What is the reverberation time of this room at 125 Hz? at 1000 Hz?
   b. How could these reverberation times be increased? decreased?
   c. Discuss the acoustical properties of this room.
   d. For which purposes might this room be suited?
   e. For which purposes might it not be suited?

4. Design a 200-seat lecture hall with emphasis on clear visibility and good acoustics. Compute the reverberation time for the room you design for 125 and 1000 Hz, including the absorption effects of a full audience. What changes might you make if the room were to be used for chamber music?

5. **a.** Design a home listening room suitable for stereo system listening and piano and small ensemble rehearsal.
   b. Compute its reverberation time.
   c. What might you do to increase intimacy? warmth? liveness?

6. Intimacy is achieved when the first reflected sound arrives less than 20 ms after the direct sound.

   a. What extra path length must the reflected wave travel to exceed this delay?

   b. Could a cathedral have intimate acoustics?

   c. A lecture hall?

   d. A large concert hall?

7. What can be done to improve listening in your dormitory, apartment, or home music-listening area? How can transmission of sound to neighboring areas be reduced?

8. Why do some music schools construct their small practice rooms with unusual shapes, such as nonparallel walls and tilted ceilings?

9. You sit in the center of your music room listening to your favorite music emanating from the expensive new stereo speakers that you just installed. Although the speakers seem to be in a proper position and you are in the center of the good listening area, the bass seems to be very weak. What might be the problem, and how would you check it?

10. Why might you suffer from a significant decrease in sound when you sit under a balcony at the rear of the concert hall?

11. If the floor of a concert hall is nearly horizontal and the stage is at nearly the same level as the seats, the attenuation of the sound with distance from the stage is very great. What is the mathematical form of the attenuation? Why is it greater than when the seating slopes up toward the rear of the auditorium?

12. Your church sanctuary is exceptionally noisy, and the reverberation is so bad that it is often difficult to understand the words of a speaker. On the basis of your knowledge of acoustics, make suggestions as to how to correct this problem.

13. Suppose that you are on the church committee to plan and build a new sanctuary. The room must be flat and hold about 300 people. How would you suggest that the committee proceed? How would you relate to the architect to obtain the best possible facility?

14. With the appropriate calculation, show that an inverse square decrease in intensity is equivalent to a fall-off of 6 dB for each factor of 2 in distance from the source.

PROBLEMS

1. Calculate the reverberation time at 500 Hz for a conference room that is 12 feet wide, 20 feet long, and 12 feet high. It has a carpeted floor, curtains on both long sides, plasterboard (or equivalent) short sides, and a plasterboard ceiling. What will the reverberation time be if it is occupied by twenty people in comfortable upholstered seats?

2. Calculate the frequencies of the first few standing waves in a bathroom stall that is 1.5 m wide, 2.5 m long, and 3.0 m high. (Can you demonstrate their existence by singing any of them?)

REFERENCES

BACKUS, JOHN. *The Acoustical Foundation of Music*. 2nd ed. New York: W. W. Norton & Company, Inc., 1977.

BENADE, ARTHUR H. *Fundamentals of Musical Acoustics*. New York: Oxford University Press, Inc., 1976. See their reference lists.

BERANEK, LEO L. *Music, Acoustics, and Architecture*. New York: John Wiley & Sons, Inc., 1962. Perhaps the most complete work on architectural acoustics. Fairly advanced.

GEERDES, HAROLD P. *Planning and Equipping Educational Music Facilities*. Reston, Va.: Music Education National Conference, 1975. A summary booklet describing some of the concerns when designing a practice or rehearsal facility, as well as auditoriums. Includes an excellent recording, *Acoustics for the Music Educator*, illustrating reverberation and other effects.

KNUDSEN, VERN O. "Architectural Acoustics." *Scientific American*, November 1963. A good survey article for the nontechnical student.

SABINE, WALLACE C. *Collected Papers on Acoustics*. New York: Dover Publications, Inc., 1964. Contains reprints of the papers on architectural acoustics by the founding father in the field. Some are advanced.

SCHROEDER, MANFRED R. "Toward Better Acoustics for Concert Halls." *Physics Today*, 33, no. 10 (October 1980): 24. The article discusses recent innovations, with some math, and gives references dealing with improvements in concert hall acoustics.

TALASKE, R. H., E. A. WETHERILL, and W. C. CAVANAUGH. *Halls for Music Performance: Two Decades of Experience*. Woodbury, N.Y.: Acoustical Society of America, 1982. An introductory-level book containing data and photographs of 87 concert halls.

Chapter 9

Musical Temperament and Pitch

Unless the reader has listened attentively to musical passages in different temperaments, he or she is unlikely to appreciate the concern regarding these differences. To experience them, one should hear an organ tuned in a temperament other than the usual equal-tempered scale or a recording (see References) that illustrates these differences. Despite their subtleties, temperaments involve several interesting and important principles of mathematics, physics, and psychology, and so are well worth our study.

After a brief historical background, we will turn to a discussion of the historical change in pitch standard and the effects of this change. Finally, we shall describe several types of musical temperaments, along with their strengths and weaknesses, following roughly historical order. We shall also give procedures by which each temperament can be set.

9.1 Background and Historical Perspective

In tuning a keyboard instrument, such as the piano, organ, or harpsichord, two major decisions must be made. First, one must decide the pitch level to be set, say A = 440 Hz (U.S. standard), A = 415 Hz (the approximate value used during the baroque era), C = 256 Hz (the "physics standard," which is now obsolete), and so on. Second, one must decide on a *temperament*—that is, the exact frequencies of all 12 notes in the chromatic scale. After the pitches of the 12 notes within one octave have been set, the remaining notes on the instrument can be tuned by octaves.

Before we discuss temperaments, we shall review the basics of keys, scales, and harmonic relationships. Figure 9-1 is a graphic illustration of the *circle of fifths*. The 12 letters around the perimeter represent the 12 major keys. Notes in adjacent positions are separated by a musical interval of a fifth—hence the name "circle of fifths." The shaded inner wedge can be rotated to show the notes that make up the major scale in

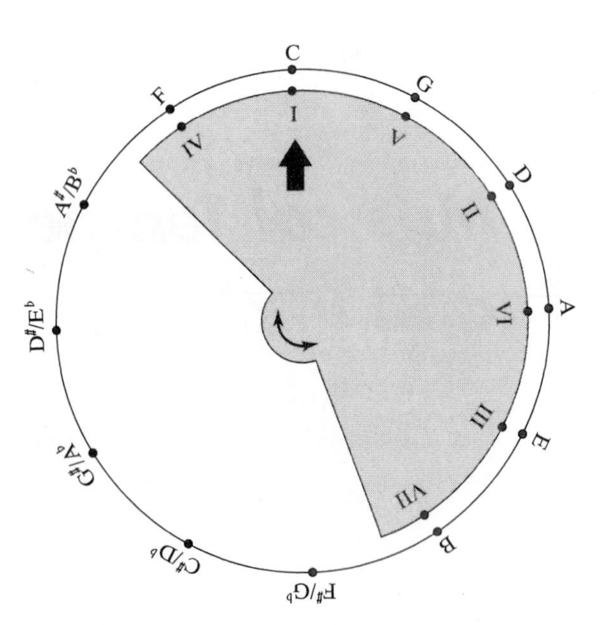

Figure 9-1 The "circle of fifths." The notes along the outer ring are separated by a musical interval of a fifth, and therefore repeat after twelve steps. Numbers on the inner wedge (shaded) show steps in the major scale. The inner wedge can be rotated to show the notes of a major scale with the chosen note as its tonic (I). For the position shown, the tonic is C, and the subsequent notes in the scale are D, E, F, G, A, and B. If the wedge were rotated to have G as its tonic (I), the scale would be G, A, B, C, D, E, and F♯.

any key. For instance, to find the seven notes that comprise the C major scale, the inner wedge is rotated such that I is placed in the C position. The note C is called the *tonic* in the key of C major. The Roman numerals on the inner wedge indicate the sequence of notes in the major scale for that tonic. Thus the Roman numerals also indicate the musical interval between the corresponding notes on the perimeter and the tonic. For example, F is the fourth note in the C scale and an interval of a fourth above C, G is a fifth above C, and so on.

You can see that the key of G, adjacent around the circle, shares six of its seven notes with the key of C. The key of G is said to be closely related harmonically to the key of C. Likewise, F is closely related to C. The key of D♭ shares only two notes (F and C) with the key of G and is therefore not closely related to C. Keep in mind the relationships summarized in Figure 9-1 as the discussion of temperament proceeds.

There is no such thing as an "ideal" or "best" temperament; the particular choice of temperament should reflect the requirements of the music to be performed and the instruments to be used. Although the variations between different temperaments may seem minute or insignificant to the untrained listener, to the trained musician they can be very substantial, creating vast differences in the sound of the music. For instance, one can choose a temperament that will have most intervals of a fifth well in tune; that is, notes a fifth apart will have a ratio of frequencies of exactly 3:2. However, this demands that the intervals of a third will not have the exact frequency ratio of 5:4 and thus they will sound somewhat unpleasant. On the other hand, one can choose a temperament based on many exact thirds, but then the fifths will deviate from the frequency ratio of 3:2. These two temperaments will sound different. Other temperaments strike a compromise and have all intervals slightly out of tune. In each temperament that we shall study the octave is tuned to the exact frequency ratio of 2:1. The single exception to this rule is the tuning of the piano, which will be discussed in Chap. 13.

Although the problem of temperament dates back to antiquity, it only emerged as a practical problem with the advent of well-tuned keyboard instruments and string and wind instruments whose pitch could be controlled well enough that a temperament could be discerned. This occurred in the late medieval period, when music in the church began to use two different parts, rather than the traditional monophonic structure. The Pythagorean temperament, based on exact intervals of fifths, was used throughout antiquity because of its simplicity and basic symmetry, which was considered a philosophical matter of beauty and order. The very earliest two-part music used parallel fifths predominantly and was perfectly in tune in the Pythagorean system.

A more sophisticated temperament was needed in the early Renaissance when music with several parts became commonplace and the interval of a third became more popular. As we shall see later, a serious deficiency in the Pythagorean system became obvious: Major thirds are grossly out of tune. (Other deficiencies of that temperament, such as the "wolf fifth," became important several centuries later as musical tastes and styles changed further.) The *just scale*, or temperament, improved the sound of the thirds but had exact tuning for important chords (I-IV-V) in only one key, and acceptable tuning for nearby, harmonically related keys.

The limited number of properly tuned chords in the just scale was not a serious drawback in the Renaissance (about 1430–1650), because the music of that period used primarily these few chords. As the music deviates further harmonically from its good key, the intonation becomes worse, and some keys have abominable intonation in just temperament. For instance, a harpsichord tuned justly in the key of C would sound good in the key of C major but bad in the key of F♯, which is harmonically distant from C. Just temperament is an example of an *open* temperament. In open temperaments, the intonation problems are worse as the instrument is played in keys harmonically distant from its good key, and there are one or more intervals in which all or most of the tuning error is concentrated. A *closed* temperament, on the other hand, involves tuning compromises that result in the instrument being at least playable in all keys, even though they might not all be equivalent or sound equally good.

Later in the Renaissance it became obvious that the just scale was inadequate to handle the increasingly complex harmonies. Around 1500, there was considerable experimentation with a set of temperaments called *mean-tone* tuning, the most popular of which was the quarter-comma mean-tone temperament. (We shall discuss tuning errors and commas shortly.) This scale was based on improved tuning of thirds at the expense of well-tuned fifths, which did not seem as harsh as mistuned thirds. Because the quarter-comma mean-tone temperament is an open system, there are certain intervals that remain uncorrected in which a disproportionate amount of the tuning error is concentrated.

Closed temperaments, in which all keys could be played with some success, were first seriously used during the seventeenth century, although they had been discussed for some time. Harmonic developments of the late Renaissance and baroque (about 1600–1750) required keyboard instruments to be played in many keys and to be easily tuned. The term "well-tempered," made famous by J. S. Bach, originated with the music theorist Andreas Werckmeister (1645–1706) in his treatise, *Musicalische Temperatur*, published when Bach was six years old, and was applied to any closed temperament. At that time, keys containing many sharps and flats were not used often, and the tuning of these keys was compromised in favor of the keys more often used. Nevertheless, as

Bach demonstrated in his *Well-Tempered Clavier*, all keys could be used within a single temperament. It appears now that Bach was not necessarily advocating the use of equal temperament, and although the temperament that he preferred is unclear, some historians believe that he favored a closed temperament developed by Werckmeister. An interesting feature of the *Well-Tempered Clavier* provides evidence on this matter. The pieces in keys such as C, G, and F (closely related to the key of C) tend to be slower than those in more distant keys such as F♯, in which slight deviations from pure intervals are not as noticeable because of their shorter duration.

Many closed temperaments were developed during the seventeenth and eighteenth centuries and a few became popular. Although the use of the equal-tempered scale was first documented in the early seventeenth century, it was not widely used at first. Most organs were tuned to unequal closed temperaments during the eighteenth century, and regular use of such tuning systems continued into the middle of the nineteenth century.

Since the nineteenth century, virtually all instruments have been tuned in equal temperament. In this closed temperament there is a single frequency ratio between any two notes separated by the same musical interval. In equal temperament, corresponding intervals in all keys have the same frequency ratio and therefore sound the same, except for absolute pitch level. An instrument tuned in this temperament can be played in any key with equal success, but because the tuning error is spread over all keys there are no keys that have the pure sound of the better keys in unequal closed temperaments. This "equality" was considered a disadvantage in the baroque period; it took nearly two centuries for the equal temperament to achieve regular use. Romantic and modern music uses a wide variety of keys and modulations, and equal temperament is the only practical temperament for such music. However, the persistent beating that occurs in any equal-tempered organ or piano can be bothersome.

There is a movement now among some organists to return to unequal closed temperaments. This has been done most effectively where eighteenth-century churches have been restored and baroque-style organs installed or renovated. The baroque organ literature sounds very good in these temperaments, especially in the acoustical environment designed for such music. Many scholars and historians believe scales like that of Werckmeister are considerably better than the equal-tempered scale in many keys. Although Werckmeister's scale is slightly worse in a few keys, this is more than compensated for by the beauty of the sound in its best keys.

Other closed, unequal temperaments were developed by Kirnberger, Valotti, and Neidhardt. Each temperament has good and bad features and may be best suited to slightly different circumstances. These scales are easier to tune than the equal-tempered scale, which requires some subtle judgments or modern electronic equipment for proper tuning.

Several points in musical history and tradition are related to problems associated with temperaments. For instance, some types of musical ornamentation arose in part from the desire to avoid inherent intonation problems by using a trill, passing tone, or some other simple ornament. This permitted the performer or composer to avoid lengthy use of an ill-tuned note. Such practice was recommended by music theorists of the Renaissance. Organist and theorist Arnolt Schlick, in his book *Spiegel der Orgelmacher und Organisten* (1511), as quoted by Franssen and van der Peet (1970), stated, "the descant may be disguised or hidden with a little pause at its start or a sequence of short notes, a grace, run, shake or ornament—name it as thou wilt—so that the harshness of the close shall not be noticed...."

Table 9-1 Frequency Ratios for Notes of the Pythagorean, Just Major, Quarter-Comma Mean-Tone, Werckmeister No. 1, and Equal-Tempered Scales

	Pythagorean scale		Just major scale		Mean-tone scale		Werckmeister scale		Equal tempered scale
C	2.0000	C	2.0000	C	2.0000	C	2.0000	C	2.0000
B	1.8984	B	1.8750	B	1.8692	B	1.8793	B	1.8877
B♭	1.7778	B♭	1.8000	B♭	1.7889	B♭/A♯	1.7778	B♭/A♯	1.7818
A	1.6875	A	1.6667	A	1.6719	A	1.6705	A	1.6818
A♭	1.5802	A♭	1.6000	A♭	1.6000	A♭/G♯	1.5802	A♭/G♯	1.5874
G	1.5000	G	1.5000	G	1.4953	G	1.4951	G	1.4983
F♯	1.4238	F♯	1.4063	F♯	1.3975	G♭/F♯	1.4047	G♭/F♯	1.4142
F	1.3333	F	1.3333	F	1.3375	F	1.3333	F	1.3348
E	1.2656	E	1.2500	E	1.2500	E	1.2528	E	1.2599
E♭	1.1852	E♭	1.2000	E♭	1.1963	E♭/D♯	1.1852	E♭/D♯	1.1892
D	1.1250	D	1.1250	D	1.1180	D	1.1175	D	1.1225
C♯	1.0679	C♯	1.0417	C♯	1.0449	D♭/C♯	1.0535	D♭/C♯	1.0595
C	1.0000	C	1.0000	C	1.0000	C	1.0000	C	1.0000

Finally, the idea that each key possessed its own harmonic flavor or emotion arose from the use of the unequal closed temperaments. Because of the nature of the compromises, each key is slightly different from the others in an unequal temperament. Perhaps this variety is one of the subtleties and beauties of earlier musical tradition that has been lost in the contemporary era of "progress."

Table 9-1 shows the ratio of frequencies between notes within an octave tuned in the Pythagorean, just, mean-tone, and Werckmeister temperaments (all based in the key of C), as well as equal temperament. The exact frequencies of these notes, based on the frequency standard $C_4 = 261.63$ Hz, are shown in Table 9-2. We shall refer to these tables regularly.

Table 9-2 Frequencies of Notes Between C_4 and C_5 in the Pythagorean, Just Major, Quarter-Comma Mean-Tone, Werckmeister No. 1, and Equal-Tempered Scales Based on $C_4 = 261.626$ Hz

	Pythagorean scale		Just major scale		Mean-tone scale		Werckmeister scale		Equal-tempered scale
C	523.25	C	523.25	C	523.25	C	523.25	C	523.25
B	496.67	B	490.55	B	489.03	B	491.67	B	493.88
B♭	465.12	B♭	470.93	B♭	468.02	B♭/A♯	465.12	B♭/A♯	466.16
A	441.49	A	436.05	A	437.41	A	437.05	A	440.00
A♭	413.42	A♭	418.60	A♭	418.60	A♭/G♯	413.42	A♭/G♯	415.30
G	392.44	G	392.44	G	391.21	G	391.16	G	392.00
F♯	372.50	F♯	367.92	F♯	365.62	G♭/F♯	367.51	G♭/F♯	369.99
F	348.83	F	348.83	F	349.92	F	348.83	F	349.23
E	331.11	E	327.03	E	327.03	E	327.76	E	329.63
E♭	310.08	E♭	313.96	E♭	312.98	E♭/D♯	310.08	E♭/D♯	311.13
D	294.33	D	294.33	D	292.50	D	292.37	D	293.66
C♯	279.39	C♯	272.54	C♯	273.37	D♭/C♯	275.62	D♭/C♯	277.18
C	261.63	C	261.63	C	261.63	C	261.63	C	261.63

9.2 Historical Development of Pitch Level

The first thing that is done when a piano is tuned or an orchestra tunes up is to sound a tuning fork (or perhaps an electronic oscillator) at A = 440 Hz, tune the A to that frequency, and then tune all the other notes relative to that A. This has not always been the standard for pitch level; in fact, a strong case could be made for the statement that the pitch level used over the course of the last several centuries has been in a state of almost total anarchy. In general, though, over the past 500 years the pitch level has risen approximately a minor third, but during this time frame there have been many variations. There are important implications of this substantial rise in pitch, both for instruments and voices. We shall survey many of these problems here and discuss some in more detail in later chapters.

Although there was no systematic choice of pitch level in the medieval and early Renaissance periods, it was generally about a minor third lower than the level today, particularly in the church. The implication of this for modern performance practice is clear: If the editor has not transposed the music down about a minor third from its original notation, the performer must do it or suffer the consequences of voice strain and/or uncontrolled sounds resulting from the unnaturally high range, or *tessitura*.

Some editors of medieval music transpose the notes down about a fourth as a standard part of the publication preparation. The music of Palestrina and his contemporaries should probably be performed about a minor third lower than written in the original. This transposition can lead to problems if instruments are used. In some cases the notes will no longer be in range of the instruments, and in many situations transposition will result in a key with many chromatic notes, which will be difficult to play on historical instruments, especially wind instruments. This problem is one of the factors motivating construction of early wind and string instruments at their original pitch level, rather than the A = 440 Hz now used.

Although during the Renaissance and early baroque most churches used a pitch level that was at least one semitone below A = 440 Hz, there were some churches and areas that tuned considerably higher, and the level of pitch common in much secular music was often higher than that common in the church. Isolated areas in northern Europe used a pitch level almost one-fourth higher than that used today, and some of the music of late Renaissance and early baroque German composers should therefore be performed about a fourth higher than its original notation. Most editions of this music are transposed by the editors to make the ranges appropriate to the standard voice ranges. Bach composed music for churches and performers in which the pitch level varied from about a semitone below today's level to almost a whole step higher, depending on where he was at the time or where he intended the music to be performed. In the early seventeenth century, Mersenne documented pitch levels regularly used in his region that included a range of almost a musical fifth.

The rise and variation in pitch level has created some problems for musicians performing the music at contemporary pitch level, particularly when the composer used the entire range of instruments in combination with voices. Lowering the pitch to make the music comfortable for the singers makes transposition necessary for the instruments, again putting them in awkward keys or extending the pitch level below the range of the instrument. Although the larger corrections of earlier Renaissance music could often be accomplished by changing to another instrument in the family, more subtle changes were involved in the later baroque music. By this time, composers were writing for specific instruments and combinations, and it is necessary to use these instruments at the correct pitch in the intended combinations to achieve the tonal color desired by the composer.

Although there remained considerable fluctuation in pitch, by the late baroque the general level of pitch had generally settled down to about a half-step below that of today. Bach wrote much of his music at a frequency of about 415 Hz for A, and Handel used 422.5 Hz as his standard. The invention of the tuning fork by John Shore (*c.* 1662–1752) aided the documentation and standardization of pitch level. The standard baroque low pitch of A = 415 Hz is often used by contemporary instrument makers seeking to return to more authentic baroque performance.

Within a limited range, adjustments in pitch level can be obtained by increasing or decreasing the instrument length using a tuning slide on wind instruments and string tension in stringed instruments. On modern wind instruments one can simply pull out the slide between the mouthpiece and finger holes or valves to elongate the instrument, thus lowering the pitch level, or, conversely, push in the slide to raise the pitch of the instrument. Baroque flute players often carry several barrel sections of different lengths, and choose the one that produces the desired pitch level. This can even be done to change the pitch between baroque low pitch and contemporary pitch, about a half-step, but the relative intonation between the notes suffers considerably in such a large change.

By the end of the nineteenth century the pitch level commonly used increased to about A = 435 Hz, although there was considerable variation documented between major music centers and instrument builders. The pitch used by such prominent composers as Mozart and Beethoven in the late eighteenth and early nineteenth centuries was still almost one half-step below contemporary pitch. During this time, much of the motivation for the rising pitch level came from the strings, which tended to sound more brilliant when tuned higher. Although this created severe problems for wind instruments, generally necessitating major redesign of the finger-hole positions on woodwinds and tube lengths on brasses, the strings were also affected by the changes they themselves wrought. The increase in tension of the strings and the concomitant increase in forces on the back and belly of the instruments made additional reinforcement of the wood necessary and demanded use of a steel string, rather than the traditional gut, for the highest string of the violin. A violin built in the seventeenth or early eighteenth century, such as those by Stradivari or Amati, and still in use today, is in many respects not the same instrument as it was when it was built, owing to the modifications required by the rise in pitch, as well as other modifications. A Stradivari violin performed at today's pitch level differs in tone quality from one performed at the pitch level of Stradivari's day. This is also a factor in the return to baroque pitch and performance practice.

By the beginning of the twentieth century, as the greater exchange of musicians between orchestras all over Europe and the United States became commonplace, a pitch standard became necessary. The response to this pressure was to establish A = 440 Hz as the international standard in about 1920. Most modern instrument makers now use this as their standard.

However, there remains pressure to raise this level even further among some of the major symphony orchestras, with pitch levels at 442 to 444 Hz being used occasionally, and even a level as high as 448 Hz being used by one major American orchestra. Some European orchestras use an even higher standard of pitch level. These changes are usually resisted by the wind players. Increases of 2 to 3 Hz require some modification of blowing technique to obtain proper intonation, while changes of 4 to 5 Hz or more require major modifications of the woodwind finger hole design, or tube lengths for brasses, to make the instrument comfortable to play in tune. A pitch level of about A = 444 Hz is often used in Europe, and some European instruments, including many recorders, must be pulled out at their barrels in order to play in tune properly at A = 440 Hz.

Since the late baroque and the invention of the tuning fork, the data on pitch level have been documented as an exact science. Previous to this, pitch level had to be determined by less direct techniques. Some of the sources are existing historical wind instruments, detailed designs of wind instruments, and the notes of scientists and artists containing certain details about the instruments and their design. Bell-tuned organ pipes are also an important source of historical pitch data and will be discussed in Chap. 10. Other details concerning the effects of the changes in pitch level on instruments will be discussed in the sections describing the instruments.

9.3 Pythagorean Temperament

In all of the temperaments we shall discuss, octaves are required to be tuned exact; that is, they must have a frequency ratio of 2:1. The Pythagorean scale is based on the musical interval of a fifth, which is tuned exact for every fifth within an octave except one.

It is impossible to tune *all* octaves and fifths perfectly; there will necessarily be notes separated by the interval of a fifth that will not be true. To see this, consider the note C_4. Starting with that note, we can tune up and down the keyboard by fifths, as indicated in Figure 9-2. Going up, we can continue to $C_8^\sharp$, and going down we can tune $D_1^\flat$, which is the enharmonic equivalent of $C_1^\sharp$—that is, it is the same note on the keyboard. We can then tune the other notes on the keyboard by setting octaves properly. Now we can compare the frequency ratio between $C_8^\sharp$ ($D_8^\flat$) and $C_1^\sharp$ ($D_1^\flat$) using two different methods: one based on octaves (ratio of frequencies 2:1) and the other based on fifths (ratio of frequencies 3:2). Because there are 12 jumps of a fifth between these two notes, their ratio of frequencies is $3/2 \times 3/2 \ldots$ (12 times) $= (3/2)^{12} = 129.75$. On the other hand, there are 7 jumps of an octave between the notes, so their ratio of frequencies should be $2 \times 2 \times \ldots$ (7 times) $= 2^7 = 128$. This discrepancy ($129.75/128 = 1.0136$) corresponds to a significant part of a half-step and is clearly audible.

Thus not all the octaves and fifths can be tuned exactly. As the octaves must remain perfect, there must be some mistuned fifths. In the Pythagorean temperament, all fifths except one are tuned true, or beatless. Starting with C_4, successive fifths are tuned true, going up to $C_8^\sharp$ and down to $A_1^\flat$, then the octaves are properly set. Every interval $C^\sharp$ to $A^\flat$ will be less than a true fifth by the amount determined previously, called the *comma of Pythagoras*. This final fifth is very badly out of tune and is called the *wolf fifth* by musicians, owing to its growling character. Its use in music is considered unacceptable.

The Pythagorean scale is the most easily tuned scale among those we shall discuss. After choosing the frequency of the note C_4, true fifths are obtained by eliminating

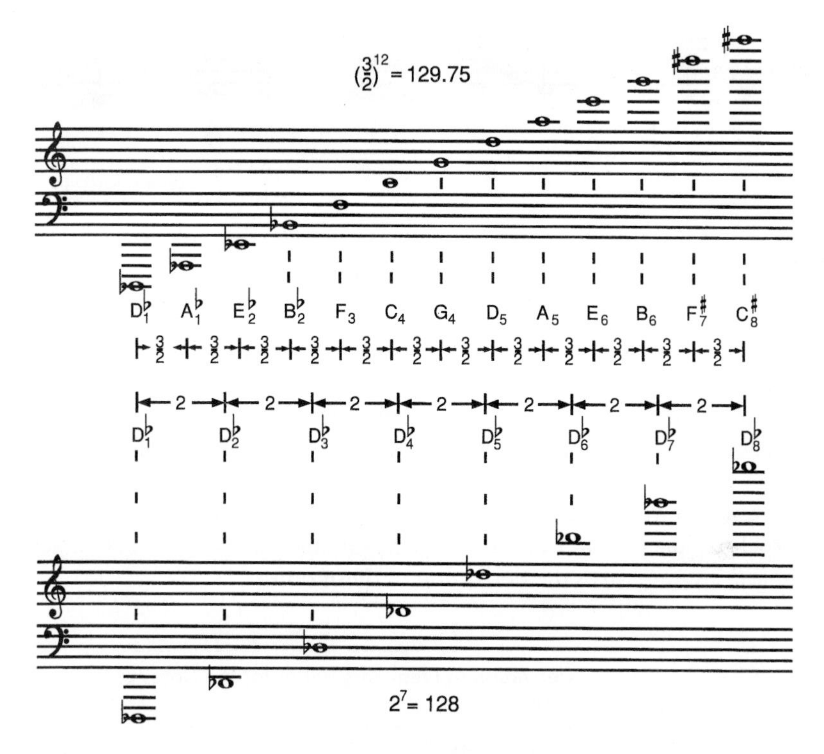

Figure 9-2 Origin of the comma of Pythagoras. Tuning upward from $D_1^\flat$ to $D_8^\flat$ by octaves (factors of two in frequency) produces a note that is different in frequency from its enharmonic equivalent $C_8^\sharp$, obtained by tuning upward from $D_1^\flat$ by fifths (factors of 3/2 in frequency).

beats between the third harmonic of the lower note and the second harmonic of the upper note. This is indicated in Figure 9-3 for the particular fifth between the notes G_2 and D_3. When the fifth is exactly in tune—a pure, or true, fifth—no beats will be heard between the third harmonic of the lower note and the second harmonic of the upper note. The frequency ratio between the notes will be exactly 3:2.

Fifths are used in succession to set one note of each letter name of the chromatic scale; after that, octaves are tuned beatless to set all the notes on the keyboard to their proper frequencies. Setting the correct value for the first note of each letter name is called *setting the temperament* and is much harder than fixing the frequencies of the other notes by tuning octaves.

One need not go up and down the keyboard to set the temperament, as outlined; one can set the temperament within slightly more than a one-octave range. In practice,

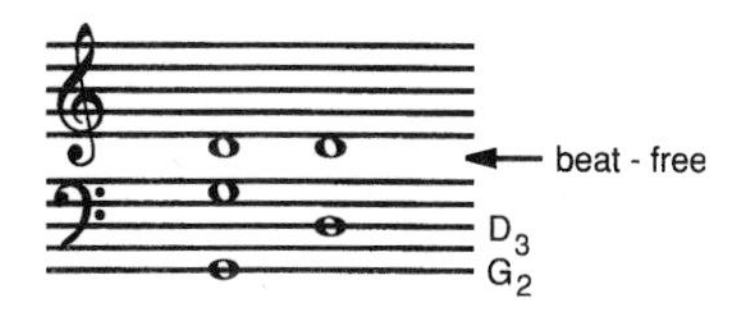

Figure 9-3 G_2 and D_3 are tuned to an interval of a perfect fifth by eliminating beats between the third harmonic of G_2 and the second harmonic of D_3 when these two notes are played simultaneously.

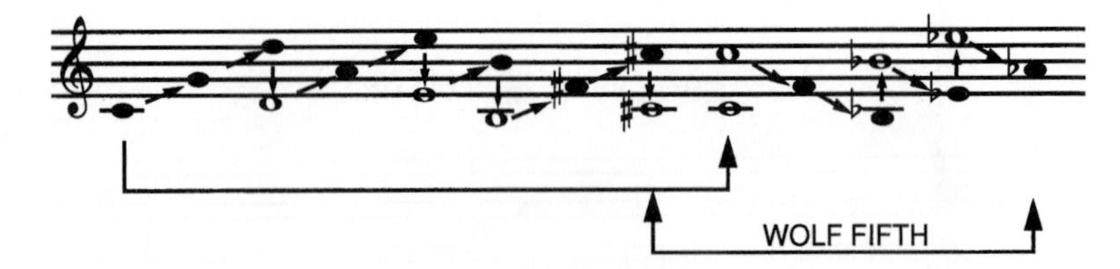

Figure 9-4 Sequence for setting the Pythagorean temperament. The solid notes are the first tuning of a note with that letter name.

the temperament is generally set in the octave around C_4 to C_5, perhaps including a few notes lower. Figure 9-4 outlines a procedure for setting the Pythagorean temperament that involves correct tuning of a sequence of true fifths and octaves, starting from C_4 and proceeding by tuning the intervals shown. The blackened notes indicate the first tuning of the note of that letter name. The remaining notes on the keyboard are then tuned in octaves. The wolf fifth between all $C^\sharp$ to $A^\flat$ intervals results from tuning by this procedure. Frequencies and ratios for a Pythagorean scale are given in Tables 9-1 and 9-2, and Table 9-3 summarizes the derivation of these ratios.

Table 9-3 Derivation of Frequency Ratios for the Notes of the Pythagorean Scale

C_4	$= 1.000$ (start)			$= 1.0000$
$C_4^\sharp$	$= \left(\frac{1}{2}\right)^4 C_8^\sharp$	$= \left(\frac{1}{2}\right)^4 \left(\frac{3}{2}\right)^7 C_4$	$= \frac{3^7}{2^{11}} C_4$	$= 1.0679$
D_4	$= \frac{1}{2} D_5$	$= \frac{1}{2} \times \left(\frac{3}{2}\right)^2 C_4$	$= \frac{3^2}{2^3} C_4$	$= 1.1250$
$E_4^\flat$	$= 2^2 E_2^\flat$	$= 2^2 \left(\frac{2}{3}\right)^3 C_4$	$= \frac{2^5}{3^3} C_4$	$= 1.1852$
E_4	$= \left(\frac{1}{2}\right)^2 E_6$	$= \left(\frac{1}{2}\right)^2 \left(\frac{3}{2}\right)^4 C_4$	$= \frac{3^4}{2^6} C_4$	$= 1.2656$
F_4	$= 2F_3$	$= 2 \times \frac{2}{3} C_4$	$= \frac{2^2}{3} C_4$	$= 1.3333$
$F_4^\sharp$	$= \left(\frac{1}{2}\right)^3 F_7^\sharp$	$= \left(\frac{1}{2}\right)^3 \left(\frac{3}{2}\right)^6 C_4$	$= \frac{3^6}{2^9} C_4$	$= 1.4238$
G_4	$= \frac{3}{2} C_4$			$= 1.5000$
$A_4^\flat$	$= 2^3 A_1^\flat$	$= 2^3 \times \left(\frac{2}{3}\right)^4 C_4$	$= \frac{2^7}{3^4} C_4$	$= 1.5802$
A_4	$= \frac{1}{2} A_5$	$= \frac{1}{2} \times \left(\frac{3}{2}\right)^3 \times C_4$	$= \frac{3^3}{2^4} C_4$	$= 1.6875$
$B_4^\flat$	$= 2^2 B_2^\flat$	$= 2^2 \times \left(\frac{2}{3}\right)^2 C_4$	$= \frac{2^4}{3^2} C_4$	$= 1.7778$
B_4	$= \left(\frac{1}{2}\right)^2 B_6$	$= \left(\frac{1}{2}\right)^2 \times \left(\frac{3}{2}\right)^5 C_4$	$= \frac{3^5}{2^7} C_4$	$= 1.8984$
C_5	$= 2C_4$			$= 2.0000$

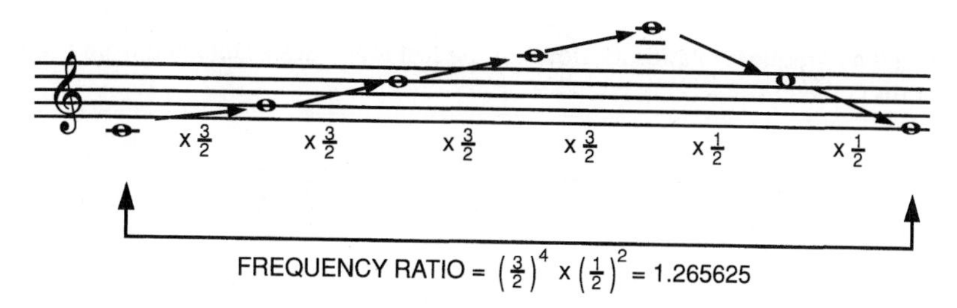

Figure 9-5 The Pythagorean major third.

The wolf fifth became problematic when keys using many chromatic notes (sharps and flats) came into common use. A more serious problem historically was the use of the interval of the major third. A true beatless major third, formed as the fourth and fifth harmonics in an overtone series, has a ratio of frequencies of 5:4, or 1.250. Thirds do not have this ratio in the Pythagorean temperament, as can be seen by considering E_4 and C_4, as diagrammed in Figure 9-5. Going up four intervals of a true fifth from C_4 to E_6 gives a frequency of $(3/2)^4$ times the frequency of C_4. Going down two octaves from E_6 to E_4 gives a frequency of $(1/2)^2$ times the frequency of E_6. Thus the frequency of E_4 in the Pythagorean temperament is $(3/2)^4 \times (1/2)^2 = 81/64 = 1.265625$ times the frequency of C_4. This ratio differs considerably from the true major third ratio of 1.25 obtained from the notes in an overtone series. The difference between these two thirds, known as the *comma of Didymus*, is a significant fraction of a half-step, causing the Pythagorean thirds to sound out of tune, just as the wolf fifth is out of tune.

9.4 Just Temperament

If some of the thirds are to be tuned true—that is, without beats—then all of the fifths cannot be perfectly in tune. With the historical development of more musical emphasis on thirds, the *just* scale was devised, in which the thirds and fifths of three major chords—the tonic (I), the dominant (V), and the subdominant (IV)—of some key were tuned true. The tonic chosen was usually C major. Because music of that period used these chords primarily, the deficiencies of the other keys were little noticed. In addition to those three chords, two more chords could be tuned beatless before all 12 notes of the temperament were determined, as shown in Figure 9-6. The twelfth note in the

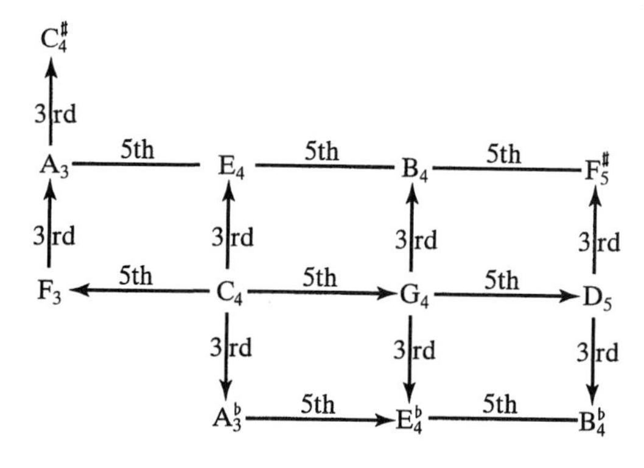

Figure 9-6 Setting the temperament for the just major scale. Arrows indicate the sequence in which the notes can be tuned, starting at C_4.

Table 9-4 Frequency Ratios for Notes of the Just Major Scale Obtained from the Three Major Chords C, F, and G

C	D	E	F	G	A	B	C	D
4		5		6				
			4		5		6	
				4		5		6
1.000	1.125	1.250	1.333	1.500	1.667	1.875	2.000	2.250

chromatic scale could be chosen as B^b, in tune with D (as a major third). Note that the interval of a fifth, F$^\sharp$ to D^b (C$^\sharp$), is not in tune, because those notes are obtained via two different routes. We saw earlier that following two different routes of successive intervals results in a discrepancy between the frequencies to be assigned to a single note. In the Pythagorean temperament, one route consisted of sequential octaves, the other route consisted of sequential true fifths, and the discrepancy was called the comma of Pythagoras. In other temperaments, the same principle applies, although the routes are more complicated than in setting the Pythagorean temperament. In this case, the fifth from F$^\sharp$ to D^b is in the ratio of 1.517, considerably off from the ideal ratio of 1.500.

The derivation of the frequencies in the just diatonic major scale, in terms of the three major chords, is shown in Table 9-4. The ratios of the frequency of any note to the tonic C can be obtained by referring to the ratios in the three chords shown. Notes with ratios of 4:5:6 form a major chord, which contains a major third (ratio 4:5) and a minor third (ratio 5:6) in that order.

If the scale is to be tuned for a minor key, a different set of ratios must be used to make the I, IV, and V chords in tune in the minor key. This just diatonic natural minor scale is shown in Table 9-5. The frequency ratios between the notes in a pure minor chord are 10:12:15, formed from a minor third (5:6 or 10:12) and a major third (4:5 or 12:15) in that order. The fifth is true, as it is in the major key.

The frequency ratios for the notes in the just major scale can be calculated using the procedure shown in Table 9-6; for notes in a minor scale (such as E^b, A^b, and B^b), the minor-chord ratios can be used. C$^\sharp$ and F$^\sharp$ do not appear in any of the three chords used to set the temperament, and these two notes can be set to fit in best with any desired chord. The procedure for setting the temperament is outlined in Figure 9-6, starting from the note C_4 and proceeding by tuning the intervals indicated by the arrows.

Table 9-5 Frequency Ratios for Notes of the Just Minor Scale Obtained from the Three Minor Chords C, F, and G

C	D	E^b	F	G	A^b	B^b	C	D
10		12		15				
			10		12		15	
				10		12		15
1.000	1.125	1.200	1.333	1.500	1.600	1.800	2.000	2.250

Table 9-6 Derivation of Frequency Ratios for the Notes of the Just Major Scale

C_4	$= 1.000$ (start)				$= 1.0000$
$C_4^{\sharp}$	$= \frac{5}{4}A_3$	$= \frac{5}{4} \times \left(\frac{5}{4}F_3\right)$	$= \frac{5}{4} \times \frac{5}{4} \times \left(\frac{2}{3}C_4\right)$	$= \frac{25}{24}C_4$	$= 1.0417$
D_4	$= \frac{1}{2}D_5$	$= \frac{1}{2} \times \left(\frac{3}{2}G_4\right)$	$= \frac{1}{2} \times \frac{3}{2} \times \left(\frac{3}{2}C_4\right)$	$= \frac{9}{8}C_4$	$= 1.1250$
$E_4^{\flat}$	$= \frac{4}{5}G_4$	$= \frac{4}{5} \times \left(\frac{3}{2}C_4\right)$	$= \frac{6}{5}C_4$		$= 1.2000$
E_4	$= \frac{5}{4}C_4$				$= 1.2500$
F_4	$= 2$ F3	$= 2 \times \left(\frac{2}{3}C_4\right)$	$= \frac{4}{3}C_4$		$= 1.3333$
$F_4^{\sharp}$	$= \frac{5}{4}D_4$	$= \frac{5}{4} \times \left(\frac{9}{8}C_4\right)$	$= \frac{45}{32}C_4$		$= 1.4063$
G_4	$= \frac{3}{2}C_4$				$= 1.5000$
$A_4^{\flat}$	$= \frac{4}{5}C_5$	$= \frac{4}{5} \times (2C_4)$	$= \frac{8}{5}C_4$		$= 1.6000$
A_4	$= \frac{5}{4}F_4$	$= \frac{5}{4} \times \left(\frac{4}{3}C_4\right)$	$= \frac{5}{3}C_4$		$= 1.6667$
$B_4^{\flat}$	$= \frac{4}{5}D_5$	$= \frac{4}{5} \times \left(\frac{3}{2}G_4\right)$	$= \frac{4}{5} \times \frac{3}{2} \times \left(\frac{3}{2}C_4\right)$	$= \frac{9}{5}C_4$	$= 1.8000$
B_4	$= \frac{5}{4}G_4$	$= \frac{5}{4} \times \left(\frac{3}{2}C_4\right)$	$= \frac{15}{8}C_4$		$= 1.8750$
C_5	$= 2\,C_4$				$= 2.0000$

9.5 Mean-Tone Temperament

An important observation made during the early Renaissance was that compromising the intonation of the thirds creates more harshness in chords than does compromising the fifths. A number of temperaments were therefore developed that are based on using as many true thirds as possible, at the expense of the fifths. The most widely accepted of these was the quarter-comma mean-tone temperament. Referring again to Figure 9-5, we see that the Pythagorean third, obtained by sequential tuning of fifths, is too large by the amount of the comma of Didymus. The frequency of the note E can be brought down to a true third (a ratio of 1.250) by reducing the four intervals of a fifth (C to G, G to D, D to A, and A to E) by one-quarter of that amount, dividing the comma into four parts—hence the term *quarter-comma mean-tone temperament.* The scheme for setting the quarter-comma mean-tone temperament is shown in Figure 9-7. Each interval marked 5th* is shortened by a quarter-comma. For the interval of a fifth between $B_2^{\flat}$ and F_3, this results in a beat rate of 1.10 Hz. Similarly, the interval of a fifth between F_3 and C_4 has a beat rate of 1.64 Hz, and the fifth between C_4 and G_4 has a beat rate of 2.45 Hz.

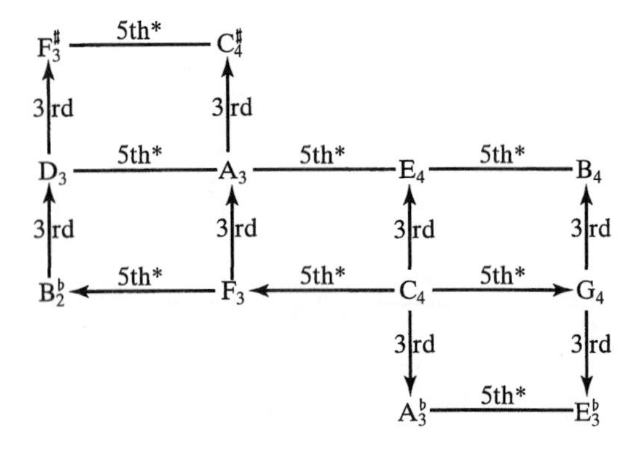

Figure 9-7 Setting the temperament of the quarter-comma mean-tone scale. Arrows indicate the sequence in which the notes can be tuned, starting at C_4. Intervals marked 5th* are narrowed from true fifths by one-quarter of the comma of Didymus.

All the fifths except one can then be tuned one quarter-comma flat, just as they were tuned true in the Pythagorean scale. Referring to Figure 9-2, the ratios from $D^{\flat}_1$ to F_3, from F_3 to A_5, and from A_5 to $C^{\sharp}_8$ will each be exactly 5.000, giving a ratio of 125.000 between the frequencies $D^{\flat}$ and $C^{\sharp}_8$. This factor should be exactly 128 for true octaves, and is 129.75 in the Pythagorean scale, so there will be an even greater wolf tone where the last fifth fails to close. Thus the quarter-comma mean-tone temperament is also an open temperament.

The Pythagorean scale produced chords with thirds that were badly out of tune, and the just scale produced five well-tuned chords with all the others unsatisfactory. The quarter-comma mean-tone scale as presented here yields five chords, which, although they are not as true as in the just scale, are nevertheless still quite adequate. The main advantage of this mean-tone scale is that it has an additional chord that is good, and several of the other chords are more usable than with the just scale.

One of the proponents of the mean-tone scale around the beginning of the sixteenth century, Arnolt Schlick, suggested that the frequency of the note $G^{\sharp}/A^{\flat}$ be set halfway between the good $G^{\sharp}$ and the good $A^{\flat}$. However, if $A^{\flat}$ were tuned a true third below C, the intonation defect resulting when $G^{\sharp}$ was used as the third of the E major triad could be minimized by ornamentation rather than by holding the note, while the other notes in the chord could be held because they are in tune.

Despite these advantages, because the mean-tone scale is an open temperament, there are some unsatisfactory keys that simply cannot be used. Therefore, additional compromises were required in order that more keys could be successfully used, leading eventually to the development of closed temperaments. The frequency ratios and frequencies for the quarter-comma mean-tone temperament are given in Tables 9-1 and 9-2.

9.6 Closed Unequal Temperaments

The temperaments previously described are all open temperaments, characterized by certain chords and keys that are unplayable, because the tuning error is concentrated in a few chords. The necessity for greater harmonic modulation increased steadily through the seventeenth century, and temperaments that allow unlimited modulation came into regular use by the late seventeenth century. Because the tuning error of these temperaments was distributed over more, or all, of the triads, all keys were usable, and the temperaments were known as closed temperaments. There are two types of closed temperaments: equal temperament, in which the tuning error is distributed equally over all possible triads, making all triads identical; and unequal temperaments, in which the greater proportion of the tuning error was put into the less frequently used keys, whereas the more frequently used keys, whereas tuned more nearly to true intervals.

Many closed temperaments were popular at various times and places throughout the eighteenth century. Among the more popular were those suggested by Andreas Werckmeister (1645–1706), Philipp Kirnberger (1721–1783), Johann George Neidhardt (1685–1739), and Francesco Antonio Valotti (1697–1780). The criteria for the "good" temperaments they developed were that the circle of fifths be "closed"—that is, wolf-free; that the thirds and sixths, both major and minor, vary in size systematically so that they could be readily tuned; and that the temperament be playable in any key, favoring, however, the more common keys.

In most closed temperaments the choice of which intervals deviate from the true intervals, and by what proportion, is relatively unrestricted, determined by which errors the artist deems most tolerable. In all cases, tempered fifths are tuned smaller than true fifths, while tempered fourths, major thirds, and minor sixths are larger than the true intervals.

In the Valotti temperament, the thirds in keys farther removed from C major become sharper symmetrically, with the thirds on F, C, and G the best, and those on D♭, G♭, and B tuned farther from true. In addition, the fifths on F, C, G, D, A, and E are slightly tempered, with the remaining fifths being true. This temperament was considered excellent both because of its pleasing variation of tonality between keys and its basic symmetry. Most unequal temperaments are not symmetric and do not vary uniformly in tuning error, as does the Valotti temperament.

Probably the most famous of the closed unequal temperaments was that of Werckmeister, the Werckmeister Correct Temperament No. 1 (sometimes erroneously called Werckmeister III). In the Werckmeister scale, the thirds on F and C are very close to true, those on B♭, G, and D are somewhat farther off, and those on A♭, D♭, and G♭ are off by almost one-quarter of a semitone. The fifths on F, B♭, E♭, A♭, D♭, and G♭ are tuned true, as are the fifths on A and E. The fifths on D, G, B, and C are tempered to differing degrees. In addition to being one of the earliest proposed closed systems, the Werckmeister system is one of the simplest to tune.

The tuning procedure, proceeding in steps from $A_4 = 440$ Hz, is shown in Figure 9-8. A simplifying feature of this temperament, common to most unequal closed temperaments, is that the notes are largely obtained by either tuning to true intervals or comparing beat rates between two different intervals. Thus it is not necessary to listen to complex beat patterns, as is the case in the equal-tempered scale, as we shall see. Another simplifying feature is that a list of intervals is given that can be compared for beat rate to indicate tuning errors in previous steps, as indicated in Figure 9-8.

In the Werckmeister temperament, eight of the fifths are true, as in the Pythagorean temperament, and the entire Pythagorean comma is distributed among the four remaining fifths, C to G, G to D, D to A, and B to F♯. The character of the scale therefore changes with key, being close to mean-tone tuning in the key of C and more

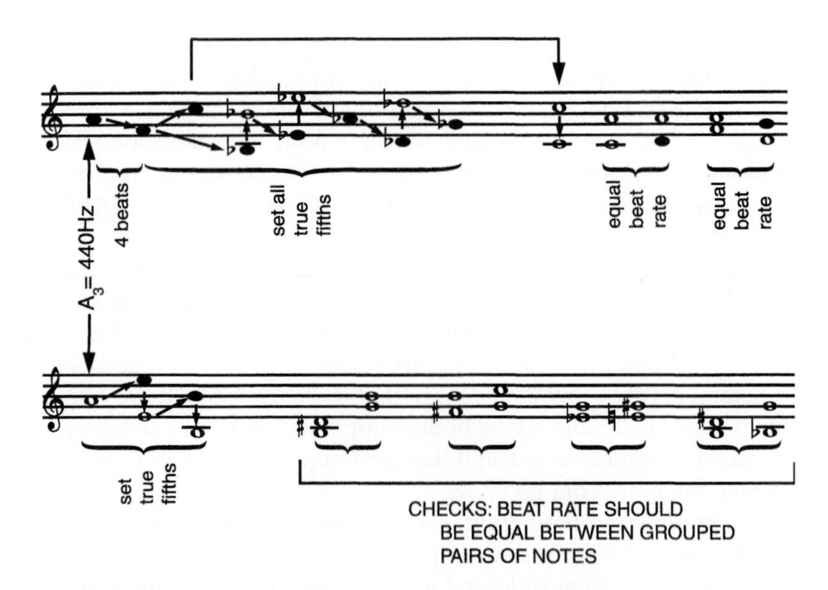

Figure 9-8 Setting the temperament for the Werckmeister Correct Temperament No. 1. The solid notes are the first tuning of a note with that letter name.

nearly Pythagorean in the key of F♯. Thus, although all keys are usable, the character of the tuning varies from key to key. This type of temperament is one source of the "difference" between the overall character of various keys that still persists to this day in the minds of some musicians.

9.7 Equal Temperament

The natural progression of scale development has taken us through a sequence of open temperaments and unequal closed temperaments. The final step in this progression is an equal closed temperament, the *equal-tempered scale*. Equal temperament is based on equal intervals between all notes one semitone apart on the keyboard. The only true interval in the equal-tempered scale is the octave, the notes on the opposite ends of which vary in frequency by a ratio of exactly 2:1. Noting that it requires 12 half-steps of equal frequency ratio R to go from one note to the note one octave higher, we can write the frequency relationships shown in Table 9-7. Applying the ratio 12 times to the original frequency raises the frequency to twice its original value:

$$f_{12} = Rf_{11} = R(Rf_{10}) = R(R(Rf_9)) = \ldots = R^{12}f_0 = 2f_0 \tag{9.1}$$

so

$$R^{12} = 2 \text{ or } R = \sqrt[12]{2} = 1.05946\ldots. \tag{9.2}$$

an irrational number whose first five decimal places are given. Thus the frequency of any note of the equal-tempered chromatic scale is equal to 1.0595 times the frequency of the note a half-tone lower.

Table 9-7 Derivation of Frequency Ratios Within One Octave of the Equal-Tempered Scale

Octave	$f_{12} = Rf_{11} = R^{12}f_0$	2.00000
Major seventh	$f_{11} = Rf_{10} = R^{11}f_0$	1.88775
Minor seventh	$f_{10} = Rf_9 = R^{10}f_0$	1.78180
Major sixth	$f_9 = Rf_8 = R^9f_0$	1.68179
Minor sixth	$f_8 = Rf_7 = R^8f_0$	1.58740
Perfect fifth	$f_7 = Rf_6 = R^7f_0$	1.49831
Augmented fourth Diminished fifth	$f_6 = Rf_5 = R^6f_0$	1.41421
Perfect fourth	$f_5 = Rf_4 = R^5f_0$	1.33483
Major third	$f_4 = Rf_3 = R^4f_0$	1.25992
Minor third	$f_3 = Rf_2 = R^3f_0$	1.18921
Major second	$f_2 = Rf_1 = R^2f_0$	1.12246
Minor second	$f_1 = Rf_0$	1.05946
Unison (start)	f_0	1.00000

The frequencies of the notes of the equal-tempered scale are given in Appendix A, Figure A-1, for all the notes of the piano keyboard. A comparison of the frequency ratios and frequencies of the equal tempered and other scales is given in Tables 9-1 and 9-2. Figure 9-9 compares the frequencies of the notes of the overtone series, as we have written it throughout the book, and the closest notes of the equal-tempered scale. Recall that the notes of the overtone series shown form the true, or beatless, chords. The fifths (D) are off a slight amount, but not enough to be upsetting to most listeners. However, the major thirds (B) are off a considerable amount from the true major third, and the minor seventh (F) is grossly out of tune, which is why it was put in parentheses in Figure 3-12. In each case, simultaneously playing the exact harmonic and the nearest equal-tempered note will produce beats at a rate equal to the difference between their frequencies.

This system has one important basic feature: all the scales and triads in any key will sound the same, because the frequency ratios from which they are formed are all identical. However, although this compromise improves the worst keys of earlier temperaments, it simultaneously degrades their better keys. The positive result of this temperament is that all keys, being exactly the same in sound except for pitch level, can be used with equal success.

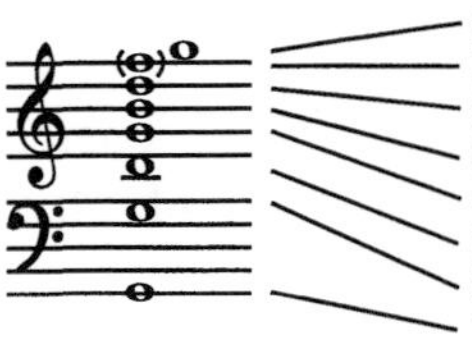

HARMONIC NUMBER	EXACT HARMONIC FREQUENCY	EQUAL-TEMPERED FREQUENCY
8	784.0	784.0
7	686.0	698.5
6	588.0	587.3
5	490.0	493.9
4	392.0	392.0
3	294.0	293.7
2	196.0	196.0
1	98.0	98.0

Figure 9-9 Comparison of the frequencies of notes of the overtone series with nearest notes of the equal-tempered scale.

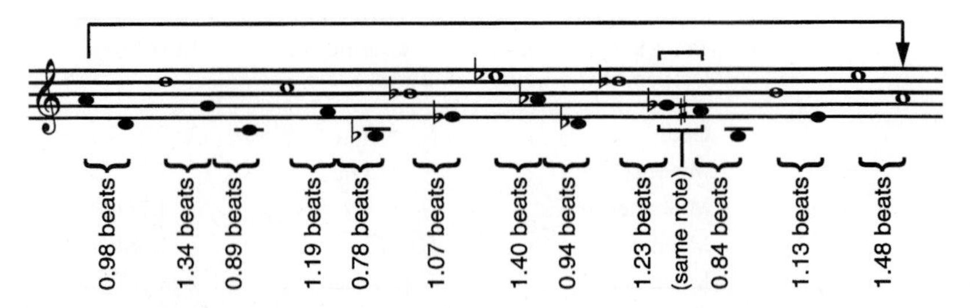

Figure 9-10 Setting the temperament for the equal-tempered scale. The solid note at the right of each bracketed pair is tuned to make the tempered fifth small by the indicated beat rate, in beats per second.

Some musicians, especially historians whose emphasis is on the baroque period or earlier, believe that the equal-tempered scale is too much of a good thing and prefer to suffer a little every now and then by listening to a few less true chords to obtain the generally better intonation available in closed unequal temperaments for the often-used keys. Apparently, the musicians of that era also had considerable reservations about the equal-tempered scale, for although it came into universal use during the latter half of the eighteenth century, it had been seriously proposed and thoroughly documented almost 200 years earlier and had been known for several centuries before that. The reason for this lack of acceptance was the considerable beating of all chords in the equal-tempered scale, particularly the thirds. Most people with even a moderate ear for music can hear the difference between the beating of the equal-tempered scale and the purer sound of the Werckmeister scale in its good keys, and many musicians believe that the beating that occurs in even a properly tuned piano tuned to equal temperament is very distracting, or even annoying.

The prescription for setting the temperament of the equal-tempered scale involves listening to beats between notes a fifth apart, as shown in Figure 9-3. Applying this prescription to a sequence of fifths, one begins with a beatless interval and narrows the fifth by raising the lower note or by lowering the upper note until the proper beat rate is reached, as shown in Figure 9-10. Table 9-8 lists the frequencies of both notes of the fifths used in Figure 9-10, along with the third harmonics of the lower notes and the second harmonics of the upper notes. The beat rate, the frequency difference between these frequencies, is also shown. To set the temperament of the equal-tempered scale properly, these beat rates must be set very close to the given values. Notice that the beat rate is proportional to the frequencies of the notes of the particular interval of a fifth involved. There are few exact-integer beat rates, and no intervals that can be used as a comparison or check, as in the case of the Werckmeister and other unequal temperaments. This feature makes the equal-tempered scale the most difficult temperament to set properly. However, the systematic nature of the changing beat rate makes it somewhat easier to cope with than is implied here. Also, modern electronic equipment can be used to set this temperament, although many piano and organ tuners do not rely on it.

Despite the problems in tuning and listening to music in the equal-tempered scale, it is destined to remain the most important temperament, as it has for almost two centuries. The demands of unlimited modulation in most romantic and modern music render any other temperament inadequate; equal temperament is probably the only practical choice.

Table 9-8 Calculation of Beat Rates Used in Setting the Temperament by Tuning Perfect-Fifth Intervals of the Equal-Tempered Scale

Interval	f_{lower}	$3f_{\text{lower}}$	F_{upper}	$2f_{\text{upper}}$	Beat rate
$B^{\flat}_3$-F_4	33.08	699.24	349.23	698.46	0.78
B_3-$F^{\sharp}_4$	246.94	740.82	369.99	739.98	0.84
C_4-G_4	261.63	784.89	392.00	784.00	0.89
$D^{\flat}_4$-$A^{\flat}_4$	277.18	831.54	415.30	830.60	0.94
D_4-A_4	293.66	880.98	440.00	880.00	0.98
$E^{\flat}_4$-$B^{\flat}_4$	311.13	933.39	466.16	932.32	1.07
E_4-B_4	329.63	988.89	493.88	987.76	1.13
F_4-C_5	349.23	1047.69	523.25	1046.50	1.19
$G^{\flat}_4$-$D^{\flat}_5$	369.99	1109.97	554.37	1108.74	1.23
G_4-D_5	392.00	1176.00	587.33	1174.66	1.34
$A^{\flat}_4$-$E^{\flat}_5$	415.30	1245.90	622.25	1244.50	1.40
A_4-E_5	440.00	1320.00	659.26	1318.52	1.48

9.8 Recent Innovations in Tuning and Temperament

An important contribution of modern digital electronics to the area of tuning is the electronic tuner. Inexpensive and easy-to-use electronic tuners have not only simplified tuning for many amateur instrumentalists, but have also spawned a whole generation of do-it-yourself piano tuners. Many of these extremely flexible tuning devices, an example of which is shown in Figure 9-11, can be used to tune in several of the historical temperaments, thus helping amateur musicians in performing baroque musical literature in a more historically correct manner.

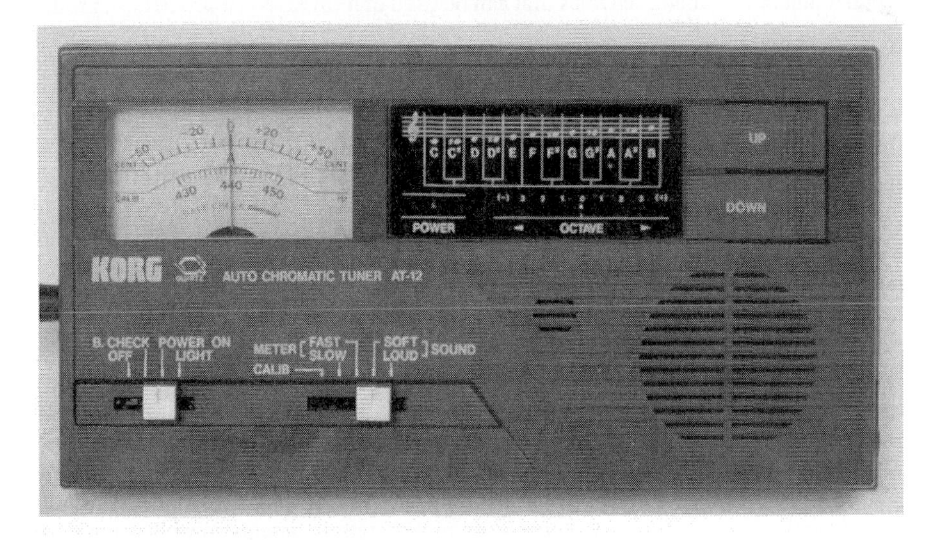

Figure 9-11 A Korg multitemperament electronic tuner. Notice that the scale is marked in *cents*, where 100 cents corresponds to one half step. (Courtesy Korg USA, Inc.)

Many piano tuners believe that the use of an electronic tuner will not yield the high quality of tuning that can be obtained by a well-trained craftsman using the ear alone. Although it is possible to set the tuning as close to perfect as patience allows, good piano technicians adjust the frequency relationship between the three strings of a single note to compensate for possible irregularities in the voicing or other tonal defects. Although it is possible to tune the "stretched" octave using an electronic tuner, some tuners believe that it is both faster and more accurate to do this by ear. (We shall discuss the tuning of the piano in more depth in Chapter 13.)

Many modern digital synthesizers can be tuned in the popular baroque temperaments, and played using a baroque organ or harpsichord sound, to obtain a good feel for the differences in sound that occur between various temperaments. Some examples are suggested in the exercises in this chapter. Because many synthesizers can be tuned to quarter-tone temperament, this temperament has become used more often in contemporary electronic music. Some synthesizers can even be tuned so that their entire keyboard covers only a single octave.

SUMMARY

(***Section 9.1***) Tuning any instrument involves establishing a **pitch level** and setting a **temperament**. Temperament is dependent on which chords will be used in the music. Deviation from the primary key in which the music is written can be described in terms of the **circle of fifths**; use of modulation further from the key in which the piece is written requires greater uniformity of all chords. Earlier music remained closer to its primary key than does more modern music, so temperament has become more uniform over time. Western music is based on the modern keyboard, which has eight notes in each scale and twelve half-steps per octave. **Open temperament** refers to tuning in which keys with more sharps and/or flats become progressively further out of tune, so that only a limited number of keys are usable. **Closed temperament** refers to a number of tuning schemes that can be used in most keys but are better in tune for keys with fewer sharps and flats. (***Section 9.2***) Pitch level has risen approximately one half step since the baroque era. Much of the influence to raise pitch level has resulted from the desire of performers to obtain more brilliance from the stringed instruments. This has necessitated substantial modification of the violin as well as correction of intonation defects in other instruments, and has resulted in much early vocal music having an uncomfortably high **tessitura**. (***Section 9.3***) The **Pythagorean scale**, an example of open temperament, is based on beatless fifths, resulting in a serious tuning problem known as the **comma of Pythagoras**. It is fine for monophonic music, but cannot be used for music employing harmony other than perfect fifths. (***Section 9.4***) **Just temperament**, also an open temperament, involves beatless tuning of the primary chords but results in tuning errors that become worse with further deviation from those chords and is therefore not usable for music with more than the nearest modulation. (***Section 9.5***) **Mean-tone temperament** uses well-tuned thirds, but as an open temperament it is also not usable for distant modulation. (***Section 9.6***) **Closed unequal temperaments**, developed as music expanded its use of modulation in the baroque, include one proposed by Werckmeister that Bach employed for his *Well-Tempered Clavier*. (***Section 9.7***) **Equal temperament** is equally in tune and therefore equally usable in all keys. (***Section 9.8***) The electronic tuner has proven to be a very important aid in helping to tune pianos as well as to help amateur performing groups tune their instruments. Many modern digital synthesizers can be used to perform in early temperaments as well as modern quarter-tone and other exotic temperaments.

QUESTIONS

1. Why are temperaments necessary?

2. a. List and discuss the important features of each temperament discussed in the text:

   i. Pythagorean
   ii. just
   iii. mean-tone
   iv. Werckmeister
   v. equal temperament

   b. Contrast open and closed temperaments.
   c. Which temperaments are open and which are closed?

3. For each of the intervals listed, draw on the staff two overtone series, one for each of the notes that make up the interval. Use the notes you have drawn to explain how the following intervals would be tuned true:

   a. octave
   b. fifth
   c. fourth
   d. major third
   e. minor third

4. Look at the score for a Palestrina mass or other major piece of the same period. Examine the tessitura of each voice and describe the effect of lowering the pitch level by about a minor third.

5. Listen to a recording of the Bach Brandenburg concerti played on modern instruments at contemporary pitch and then listen to a recording of the concerti played on historically authentic instruments at low pitch. Describe differences in the tone, blend, overall sound, and performance styles.

6. Try tuning a harpsichord in the Werckmeister temperament using the procedure given in Section 9.6. Listen to the basic chords in various keys to get a feel for the differences from equal temperament. Play, or have someone play, pieces in different keys, such as selections from Bach's *Well-Tempered Clavier*. Is this temperament usable in all keys? How would you characterize the sound? Play a simple melody transposed to another key and compare the sounds.

7. Listen to an organ (or a recording of one) in an unequal closed temperament. How does it differ from an organ in equal temperament? Can you hear the differences between the keys?

8. Obtain a digital synthesizer that can play in several temperaments. Try the following:

   a. Play a simple hymn or song in C major and C$^{\#}$ major using Pythagorean temperament, just temperament, mean-tone temperament, Werckmeister temperament (or another baroque unequal closed temperament), and equal temperament. Notice how the sound difference between the two keys progresses as you get closer to equal temperament.

b. Play some intervals in these temperaments. Which intervals sound good and which sound bad?

9. Some historically accurate wind instruments, such as certain baroque flutes, are tuned in mean-tone temperament. Obtain such an instrument, and play music for that instrument and harpsichord using a synthesizer with its harpsichord stop tuned in

a. mean-tone temperament

b. equal temperament

Describe the intonation problems and the ease with which the music can be performed using each temperament.

PROBLEMS

1. If the frequency ratio between two notes one half-step apart in the equal-tempered scale is 1.05946, calculate the frequency ratio for the following musical intervals in the equal-tempered scale: a major third, a perfect fourth, a perfect fifth, a major sixth, and an octave. Using the notes of the overtone series, determine the frequency ratios for these same intervals. Using $C_4 = 261.63$ Hz (middle C), determine the frequencies of the notes E, F, G, A, and C in the octave above middle C using each of these two techniques. Make a table including these data.

2. If $A_4 = 440$ Hz for standard tuning in the equal-tempered scale, use the frequency ratio of 1.05946 between half-step intervals to determine the frequency of E_5.

3. If $C_4 = 261.63$ Hz (middle C) for the equal-tempered scale, use the frequency ratio of 1.05946 between half-step intervals to determine the frequency of E_4. Compare this with the frequency of a true major third above $C_4 = 261.63$ Hz.

4. If the frequency ratio between two notes one half-step apart in the equal-tempered scale is 1.05946, what would be the frequency ratio between two notes one quarter-step apart in the "equal-tempered quarter-tone scale"?

5. If you were to invent a "pentatonic" scale, with five equally spaced notes, what would be the frequency ratio between two adjacent notes?

6. Starting at $C_4 = 261.63$ Hz (middle C), use the ratio of 3:2 for a true perfect fifth and 2:1 for an octave to determine the frequency of the note $F_4^\sharp$ by going upward by fifths and downward by octaves. Starting at $C_4 = 261.63$ Hz (middle C), use the ratio of 3:2 for a true perfect fifth and 2:1 for an octave to determine the frequency of the note $D_5^\flat(C_5^\sharp)$ by going downward by fifths and upward by octaves. What is the ratio between these two frequencies? Is this noticeably different from an equal-tempered perfect fifth?

REFERENCES

BARBOUR, JAMES MURRAY. *Tuning and Temperament: A Historical Survey*. East Lansing, Mich.: Michigan State College Press, 1951. The early magnum opus on temperament, covering a large number of temperaments and variants.

BLOOD, WILLIAM. "'Well-Tempering' the Clavier: Five Methods." *Early Music*, October 1979, pp. 491–495. A concise article detailing tuning techniques for five closed temperaments popular in the baroque.

FRANSSEN, N. V., and C. J. VAN DER PEET. "Digital Tone Generation for a Transposing Keyboard Instrument." *Philips Technical Review*, 31, no. 11/12 (1970): 354–365. A summary article dealing with the main concepts in temperament. Included with the article is an excellent record, showing some of the intonation defects occurring in certain temperaments and a comparison of Pythagorean, just, mean-tone, Werckmeister, and equal temperaments.

JORGENSEN, OWEN. *Tuning the Historical Temperaments by Ear*. Marquette, Mich.: Northern Michigan University Press, 1977. By far the most modern and complete book on temperament; literally all you ever wanted to know about any temperament.

Recording

BACH, JOHANN SEBASTIAN. *Brandenburgische Konzerte 1–6*, Erstaufnahme mit Original Instrumenten in Originalbesetzung, Concentus Musicus Wien. Lechtung: Nikolaus Harnoncourt, TELDEC, 9031-77611-1 and 9031-77611-2. This two-CD set is perhaps the premiere example of baroque music played at low pitch on replica instruments.

Chapter 10 Woodwind Instruments

Second only to the percussion instruments, the woodwind family, which includes the pipe organ, has the richest variety in both sound and acoustical properties. In this chapter we shall trace a brief history of the development of the contemporary woodwind instruments, summarize the important elements in their tone production, and discuss some of the acoustical details relevant to performance technique.

10.1 History of Woodwind Instruments

Although the earliest woodwind instruments were primitive bamboo and pipe flutes, it is the earliest organs that are best considered the forerunners of the modern reed woodwind instruments. Wind pressure for these organs was supplied first by water columns and later by hand-driven bellows. The keys were linked directly to valves to allow airflow to the pipes. Each key controlled one pipe, which was usually made of reed. The *tracker action* organ, common in the baroque era, evolved from these early reed organs.

The original hydraulic organs were bulky, difficult to control, and usually out of tune, problems that were largely responsible for the exclusion of the organ from the church service. Several people were required to operate the organ, often including two keyboard players; the keys were large and difficult to depress. In the Middle Ages, bellows and other technical improvements in the organ mechanism helped it to become useful in church settings. Bellows were also used in smaller portable or *portative organs*, which were popular in the Renaissance. Electrical blowers and switches today replace the bellows and direct mechanical linkage on earlier organs, and progress in materials science and acoustics has enabled modern builders to construct organs that stay in tune for a longer time and behave more reliably.

A direct precursor to the reed instruments was the bagpipe, which was developed in medieval Spain as a modification of a Middle Eastern ancestor. The later Scottish bagpipe was developed from this original Spanish model. The drones—pipes that produced the same note all the time—were simply reed-organ pipes attached to the air sack, a modified bellows. A typical medieval bagpipe might consist of two drone pipes and a chanter, a reed pipe with finger holes providing a scale extending about an octave.

The *capped-reed* instruments were direct descendants of the early bagpipe; they achieved their pinnacle of use during the high Renaissance, and were ultimately replaced by the oboe and the bassoon. In capped-reed instruments like the krummhorn, kortholt, rauschpfeife, and cornemuse families, the double reed is enclosed in a box or cylindrical enclosure that is connected to the resonant pipe, which in turn is fingered like a recorder or similar woodwind instrument. This allowed the player to blow the instrument directly through a hole in the box enclosing the reed but without touching the reed with the lips, a technique that permitted considerably greater control of the tone, articulation, and intonation than in the bagpipe.

An even greater degree of tone control was obtained by removing the cap and blowing the instrument directly with the lips on the reed, as in the Renaissance shawm and rackett families, and later in the dolcian, a predecessor to the baroque bassoon. Modifications of the shawm in the late Renaissance and early baroque periods led to the baroque oboe. Few acoustical innovations over the baroque oboe and bassoon are found in their contemporary counterparts other than the addition of a good bit of machinery and additional holes to provide more adequate intonation of the chromatic scale throughout the full range of the instrument.

The recorder and transverse flute families developed alongside the double-reed instruments, beginning many centuries ago with primitive flutes made of bone, bamboo, or other materials. By the late medieval era, the wooden flute had developed into the basic form used in the Renaissance, and during the Renaissance the flute developed into a family of straight cylindrical instruments.

The continuing acoustical evolution of the flute is very interesting with respect to bore shape—that is, the change of diameter of the tube along its length. During the Renaissance, the flute was a simple cylindrical tube with embouchure hole and finger holes, stopped (i.e., closed) at the end above the embouchure hole. To achieve a greater range, the bore of the baroque flute was modified to a conical shape, with the larger radius at the embouchure hole and the smaller radius at the bell end. As in the cylindrical Renaissance flute, the short end adjacent to the embouchure hole was stopped, and the bore shape in this end section was modified to give the best intonation throughout the entire range. The modern flute reverted back to a cylindrical bore and achieved the desired range and acceptable intonation by elongating the end section above the embouchure hole and modifying the sizes and positions of the finger holes. The addition of the considerable machinery used in fingering allowed the flute to accommodate the demands of modern music, particularly the range, intonation, expression, and volume of sound.

A very important improvement in woodwind design was a system of fingering introduced by Theobald Böhm (1794–1881), a German watchmaker, goldsmith, master court flutist, and acoustician. The Böhm fingering system, introduced in the middle of the nineteenth century, was a method of organizing the fingering machinery for the flute. It was later adapted to the clarinet and saxophone families but has not yet achieved acceptance on the oboe and bassoon families, although significant technical improvement could be obtained by its adoption. Some musicians believe that modification of the tone holes of the oboe and bassoon degrades their unique tone quality.

Recorders are similar to the flutes acoustically, differing primarily in the way in which the initial noise source is generated. Recorders are *fipple flutes*, in which the air from the mouth of the player is blown through a windway and impinges on a sharp edge, producing an edge tone. One of the earliest fipple flutes is the medieval tabor flute, or one-handed flute, which has only two finger holes positioned near the end of the tube so that both the holes and the end of the flute can be covered. The performer could play the tabor flute with one hand and the tabor, a type of drum, with the other, perhaps dancing at the same time. Such minstrels were very popular during the Middle Ages, and although virtually no written music of this type remains (it was probably never written down but only passed on informally from performer to performer), paintings and descriptions of such performances abound in medieval history. The bore of the tabor flute is so narrow that it is very difficult or impossible to play the fundamental; in any fingering configuration the octave and the fifth above that could be obtained by proper breath control. Because the higher harmonics are produced, the notes playable with any fingering are closely spaced. With three or four sets of closely spaced overtones, one for each fingering, a full scale of notes over a range of more than one octave could be easily obtained.

The earliest Renaissance recorders were cylindrical instruments with eight finger holes. As in the case of the flute, the baroque recorder developed with a conical bore having the larger diameter at the fipple and the smaller diameter at the bell end. The recorder was the most advanced of the baroque woodwinds, possessing the best intonation and voicing, the result of an incredible number of man-years of development during the seventeenth and eighteenth centuries. This was summarized in extremely detailed formulas relating the bore diameter, hole sizes and positions, and other physical dimensions. Although the recorder required centuries to develop into its final form, in the classical orchestra it was supplanted relatively quickly by the flute because of the superiority of the flute in range, volume, and expression.

The clarinet is the youngest of the standard modern orchestral woodwinds; it was developed during the middle of the seventeenth century from the chalumeau, a very soft, single-reed Renaissance woodwind with a narrow cylindrical bore. The saxophone, a single-reed instrument with a very large conical bore, is an even more recent development, having been invented in the middle of the nineteenth century. Although the saxophone is seldom used in the standard symphony orchestra, it has achieved immense popularity in jazz and dance music and is a member in good standing of the contemporary concert band.

Figure 10-1 shows some of the woodwind instruments that acted as important stepping stones in the development of contemporary woodwinds. As in the contemporary instruments, they can be classified by their bore shape (cylindrical or conical), the type of reed (single or double), or the type of noise source (fipple or embouchure hole).

Virtually all Renaissance instruments came in families of different sizes and pitch ranges, the basic sizes being the soprano, alto, tenor, and bass. When appropriate, this basic family was extended in either direction, and when size became a critical factor, the larger members were simply left out or fell rapidly into a state of benign neglect.

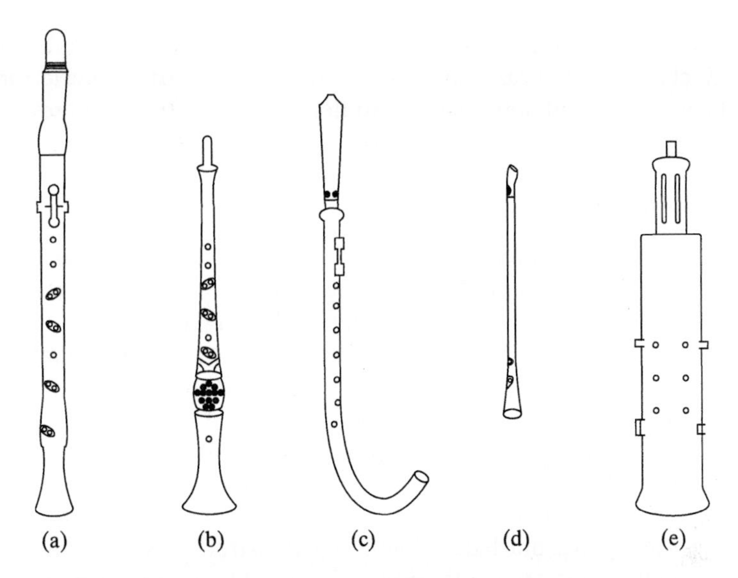

Figure 10-1 Some ancestors of the contemporary woodwind instruments: (a) the chalumeau, a single-reed cylindrical Renaissance precursor to the clarinet; (b) the shawm, a double-reed direct-blow conical Renaissance precursor to the oboe; (c) the krummhorn, a cylindrical Renaissance instrument with a capped or enclosed double reed; (d) the tabor flute, a medieval narrow-bore fipple flute played with one hand; and (e) the rackett, a direct-blow, double-reed Renaissance instrument with a long cylindrical bore, which has been folded back on itself nine times, a precursor to the bassoon. (Drawings are not to scale.)

Many modern woodwinds come in a family of sizes, as did their Renaissance ancestors. The members of the family that survive as members of the modern symphony orchestra are based on tradition and on their usefulness in the orchestra.

Some instruments, such as the piano, guitar, and voice, can be played in most keys with equal ease. There is nothing inherent in the human voice, for instance, that "prefers" the key of C major to the key of A♭ or F♯. This is not true of some other instruments, such as the clarinet family. One often-used scale on a B♭ clarinet sounds B♭ major. It is convenient, then, to write music for this instrument in which the key that is simplest notationally, C major (no sharps or flats), corresponds to the key most often played, sounding B♭ major. When the clarinet player sees the note C written in music, he or she fingers that note on the clarinet, but B♭ sounds. Composers must allow for this change between written note and sounded note and write accordingly, or *transpose*. For instance, if the composer wants the B♭ clarinet to sound a C (called "concert C"), a D must be written. The clarinet is thus a *transposing instrument.*

In practice, this system has another benefit. Once a clarinetist learns to play one instrument, say the B♭ soprano clarinet, it is a simple matter to play a different member of the clarinet family, for instance, the E♭ alto clarinet. The musician learns one fingering for each written note. When reading the note C, the same finger combination is used on each instrument. A B^b will sound on the B^b clarinet, while an E^b will sound on the E^b clarinet. The musician need not worry: The written music for each instrument has been transposed appropriately so that the desired note will sound.

10.2 Woodwind Tone Quality

For woodwind instruments, there are three primary elements that determine the tone quality and overall sound: (1) the source of noise or vibration, (2) the size and shape of the bore, and (3) the sizes and positions of the finger holes. In addition, details in construction, such as the inside finish of the bore, reed and body materials and construction, and others, contribute to the tone quality. Although some of these effects are well understood and their importance generally agreed upon, others are controversial and appear to have little if any basis in physics. We shall try to discuss the elements that affect the quality of woodwind tones and put some of the more controversial items in perspective using a combination of physics and musical experience. Unfortunately, as we shall see, there is much emotionalism and extremely strong tradition involved in the choice of musical instruments and the details of their construction.

The most obvious factor determining the tone quality of an instrument is the source of the vibration. Perhaps the simplest in appearance, but not the simplest acoustically, is the *edge tone*, which is used in the flute and recorder families. In the fipple flute shown in Figure 10-2, a jet of air is blown through the fipple, strikes a sharp edge, the *lip*, and creates acoustical noise. Components of this noise that are at resonant frequencies of the air column will be strengthened, and after a short time the oscillations of the air column begin to strongly affect the motion of the air stream about the lip. This feedback of the motion of the standing wave in the tube strengthens the

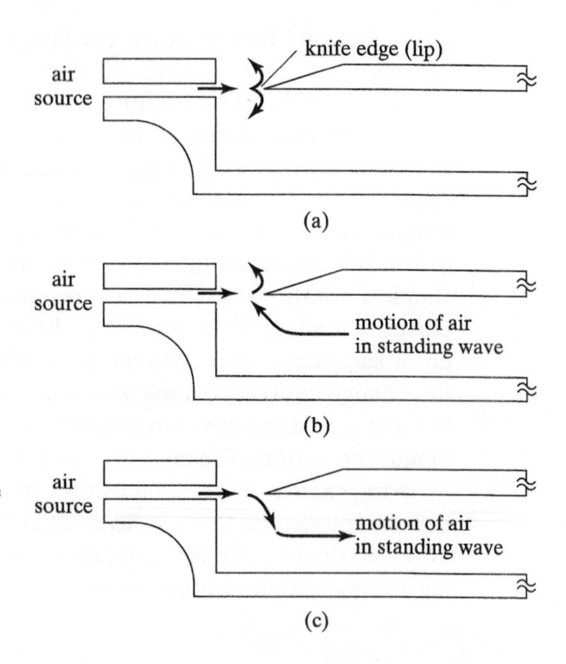

Figure 10-2 Action of sound source of a fipple flute. After the initial attack transient period (a), the air jet from the source is forced to oscillate back and forth across the knife edge, or lip, by the motion of the standing wave, as illustrated in (b) and (c).

frequency components in the noise spectrum at the resonant frequencies of the tube and thereby strengthens the tone of the instrument. The changing harmonic structure at the beginning of the tone is described by the general term *attack transient*. After the initial transient period, the oscillating standing wave in the recorder drives the air stream back and forth across the lip at the sounding frequency of the instrument.

Because there is rapid motion of the air in and out of the opening by the lip, as well as at the other end (at the first open hole or at the bell, if all the holes are covered), both ends are velocity (displacement) antinodes or pressure nodes; the instrument will be one half-wavelength long and will behave acoustically like an open tube. The antinodes are not exactly at the end of the bell and the edge of the fipple opening but will be strongly affected by details of the geometry. As all recorder players know, blowing too hard will raise the pitch of the tone. This effectively pushes the antinode farther into the bore while leaving the bell end almost unaffected, thus shortening the effective length of the tube. Alternatively, reducing breath pressure allows the antinode to move farther out from the fipple opening, thus increasing the effective length of the air column and lowering the pitch.

The sound-production mechanism in a flute is similar to that in the recorder, with a few significant differences. The lips must be carefully shaped when blowing onto the edge of the embouchure hole to produce the initial sound. Although the feedback mechanism between the standing wave in the tube and the air stream oscillating across the edge of the embouchure hole is similar to that of the recorder, two additional adjustments can be made on the flute.

First, the flute has a section above the embouchure hole whose bore can be modified and in which a cork end can be inserted and adjusted in position to correct some tuning problems. This was very important in the development of the modern cylindrical flute. The second and possibly the most important element in the flute, which allowed it to supplant the recorder and become a member of the modern symphony orchestra, was the greater dynamic expression made possible by the control the player has over the air stream. The player can compensate for intonation defects in the baroque flute by "rolling" the mouthpiece slightly about the lip. This allows a greater range of dynamics with good pitch compensation, leading to considerably greater "expression."

In a reed instrument, the oscillating air column drives the reed at the resonant frequency of the standing wave in the tube. Although each reed has a resonant frequency at which it vibrates freely, this resonance is usually very broad and does not play an important role in determining the sounding frequency of the instrument. On the contrary, undesirable resonances in the reed are responsible for the assorted high-pitched squeaks and squawks typical of poorly played reed instruments. Such effects are increased when the reed is dry, is shaped improperly, has cuts or cracks, or otherwise cannot vibrate uniformly. In most woodwind instruments the natural vibration frequency of the reed is considerably higher than the frequency range of the instrument as it is normally played, even with the damping obtained when the lips are in contact with the reed. One exception to this general rule is that of reed-organ pipes. Because each pipe and reed sounds only at one frequency, the reed is generally "tuned," by a technique that we shall discuss later, to sound at a frequency very close to the resonant frequency of the pipe.

A reed end acts as a closed end and thus is the acoustical opposite of a fipple or embouchure hole. Consider what happens in a cylindrical tube with a reed on one end.

A large burst of air flowing rapidly through the reed opening will cause it, per the Bernoulli principle, to close. The compressional air burst travels down the length of the tube, reflects off the open end, and returns to the reed end as a rarefaction. The rarefaction in turn reflects off the reed and travels down the length of the tube, reflecting off the bell end back to the reed end as a compression, forcing the reed open and allowing another air burst to enter the reed from the player's mouth. Because a burst of air entering the reed from the instrument causes the reed to open, allowing another burst of air to enter from the player's mouth moving in the opposite direction, there is a large pressure variation at the end of the reed. There is therefore a pressure antinode or a velocity node at the reed end. In the clarinet, a cylindrical reed instrument, the reed end is a velocity node and the bell end a velocity antinode, so the clarinet behaves acoustically like a closed tube. This behavior of the reed is modified considerably if the bore shape is conical rather than cylindrical, as we shall discuss later. In reality, for soft tones the reed does not close entirely, but closes only when the volume of sound is at a relatively high level. This means that there is some change in tone quality as the volume increases, when the reed closes.

Double reeds function acoustically in the same manner as single reeds; the tone quality of most reed instruments depends much more strongly on the size and shape of the bore. There are some differences in detail that will be discussed as each of the instruments is surveyed.

Although the intensity of tone is largely determined by the bore diameter, the bore shape affects the tone quality. Shaping the bore of the instrument can control two features of the sound produced by a woodwind instrument: (1) The sequence of overtones that resonate in the instrument is adjusted to be exact harmonics. (2) The bore is shaped to preserve this harmonic relation as the finger holes are opened; in other words, the harmonicity of the overtones must remain stable as the effective length of the instrument is changed.

There are many shapes for bores; it is convenient to limit the discussions to the family of exponential horns, in which the cross-sectional area of the bore is proportional to some power of the distance from one end. Of all the possible bore shapes, only two satisfy the harmonicity requirements and are physically possible to produce and play. These are the cylinder, in which the area is constant, and the cone, in which the area is proportional to the square of the distance from the pointed end. A complete cone need not be used; any section of a cone will resonate in the same basic manner as an open tube of approximately the same length, even if one end of the cone is a reed end. These shapes can be determined theoretically by computing the details of standing waves in these instruments. They were discovered experimentally by trial and error during the Renaissance; the task of Renaissance wind-instrument makers was to produce playable families of instruments with good sounds. The instruments developed during that era evolved into the orchestral woodwinds, which have either cylindrical or conical bores. Considerable modification of the bore shape is used to correct intonation problems and improve the voicing or ease with which some instruments can be played, but the basic bore shape must be either cylindrical or a straight conical section. Figure 10-3 shows examples of cylindrical and straight conical tubes that might be useful shapes for woodwind-instrument bores.

A great deal of study and care is involved in designing the finger holes. Basically, the finger hole opens to allow the air to come in and out, effectively moving the antinode from near the bell end to near the position of the first open hole.

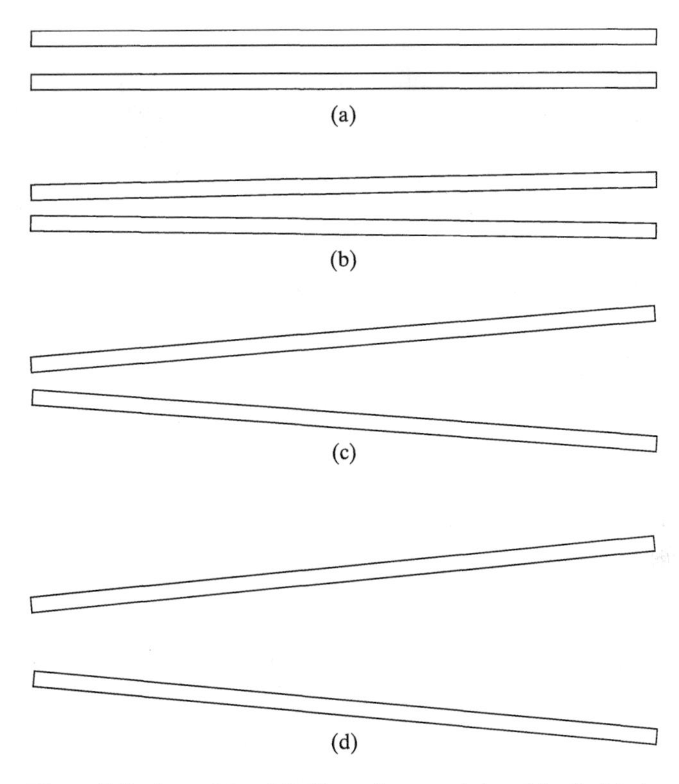

Figure 10-3 Several woodwind bore shapes and sizes: (a) cylindrical; (b) narrow cone with small flare; (c) narrow cone with larger flare; (d) wide cone with large flare.

Figure 10-4 shows a hypothetical cylindrical bore fipple flute with a row of finger holes designed to produce a chromatic scale. The instrument behaves acoustically like an open tube, with one end at the fipple opening and one end at either the bell or the open hole closest to the fipple. Each hole is opened in order to play the chromatic scale, with no "fork" fingerings or other complexities. (Fork fingering is when one finger is up while both neighboring fingers are down.) The length of the instrument is such that one half-wavelength of the note C is the distance L between the fipple opening and the bell. When all holes are open, the effective length is $L/2$, one-half the original wavelength, producing twice the original frequency, or a musical interval of one octave above the lowest note. Notice that the hole spacing increases for holes farther down the tube from the fipple. Closing each hole successively, beginning from the top, must increase the length of the standing wave by a factor of 1.0595 times the previous length to lower the frequency by that factor, or one half-step.

Another way to obtain the note C′ is to keep all the holes closed except hole C′, allowing air to move in and out through hole C′ and thereby forcing a velocity antinode at that position. The instrument will then be two loops (one full wavelength) long, and produce the note C′. If both hole C′ and hole D♭ are opened, the standing wave will adjust itself so that the antinode in the middle will move slightly upward toward

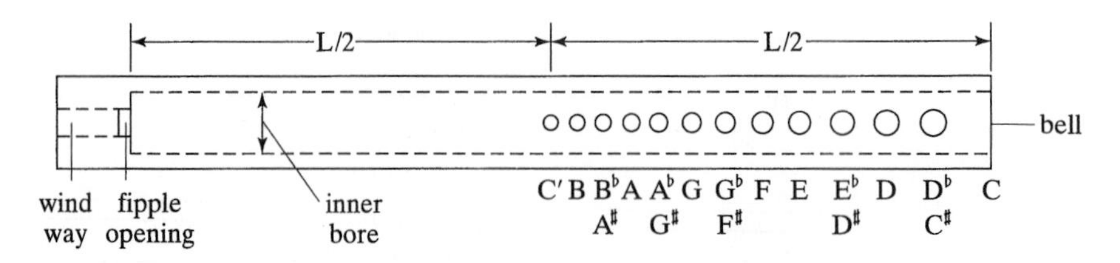

Figure 10-4 A hypothetical cylindrical fipple flute with finger holes that, if covered successively, would provide a chromatic scale of one octave.

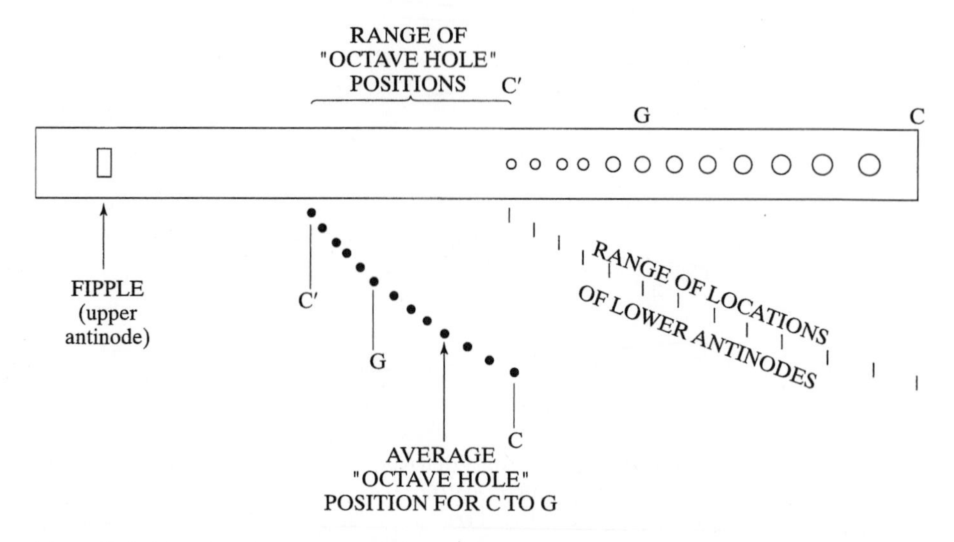

Figure 10-5 Location of average octave hole position for a single hole servicing the notes C′ through G on the hypothetical fipple flute.

the fipple, while the antinode at the end will be pushed slightly from the D^b hole toward the bell. This produces two loops of a standing wave whose wavelength is slightly greater than that of the note one octave above D^b; this D^b is therefore slightly flat. By opening the remaining holes successively, notes about one octave above D, E^b, E, F, and so on, can be produced, each becoming successively flatter in pitch relative to the notes one octave above the original chromatic scale. By moving the position of hole C′ up toward the fipple, additional notes in the second octave can be played more nearly in tune, because that hole will be nearer to the position of an antinode for more notes, as illustrated in Figure 10-5. The antinode is not exactly at the position of the hole for all notes, so there remains an intonation problem of varying degrees for notes farther from the average. Also, moving the C′ hole up the tube causes C′ to be out of tune when played with all holes open and renders this note unusable. The hole must therefore be made smaller if it is to play C′ more nearly in tune. This repositioning of the C′ hole allows it to function as an octave hole, a feature of contemporary woodwinds.

One other minor problem for this hypothetical instrument is that it could only be played by someone with 12 fingers and no thumbs. Additional modifications required

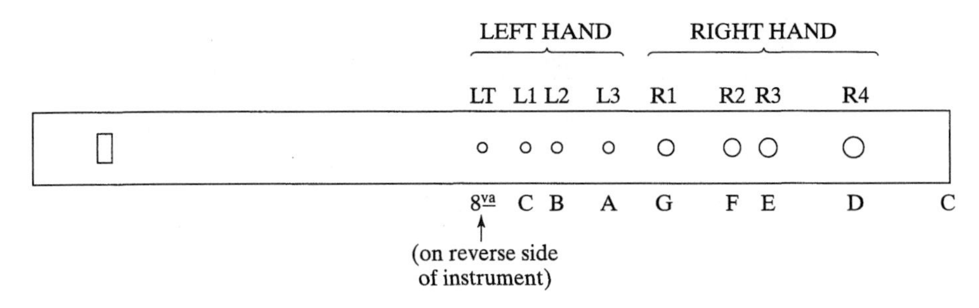

Figure 10-6 Modifications to the finger holes of hypothetical fipple flute to make it playable by a human.

to make the instrument playable by a normal human are shown in Figure 10-6. First, some of the holes must be eliminated. The normal choice is to eliminate the chromatic notes, leaving an instrument capable of playing one complete octave of a diatonic scale. Even with this modification, the spacing of the holes is not comfortable, because it is not the normal positioning for fingers. Therefore, in addition to placement of the thumb hole on the side opposite the finger holes, the positions of the holes are adjusted so that the hand feels comfortable. However, changing the hole positions introduces additional acoustical problems in intonation and voicing.

Moving a hole up the tube toward the mouthpiece decreases the length of the standing wave and increases the pitch of the note controlled by that hole; moving the hole toward the bell lowers the pitch of that note. Making a hole larger allows the air to flow in and out more easily, moving the antinode up, and thereby shortening the wavelength and raising the pitch; making a hole smaller lowers the pitch of the note. In the Renaissance these factors were used of necessity; the positions of the holes were chosen so that the instruments were playable without mechanical keys, and the sizes were adjusted to provide the best intonation. Today, without this constraint, it is best acoustically for the holes to become progressively larger as the notes go lower, perhaps requiring keys. This not only allows proper intonation, but also creates better *voicing*—that is, more uniform tone quality of all the notes. Listening to chromatic notes played on early instruments gives one an appreciation for the improvements in intonation and voicing that have been achieved by the adjustment of hole size and the addition of rings and keys to modern instruments. The added machinery permits large holes at large spacings to be used comfortably.

Modification of hole positions for comfort (and out of necessity for some of the bigger instruments) and appropriate adjustment of their sizes to retain the proper intonation for the diatonic scale results in a fipple flute very closely resembling the Renaissance recorder, shown in Figure 10-7. To play the note with the right-hand little finger (R4), that hole is usually placed toward the right side of the line in which the other holes are positioned. Certain other modifications in the hole positions and sizes are then made to provide for best intonation and voicing of the fork fingerings used for some of the more common chromatic notes. The basic baroque recorder finger-hole layout is shown in Figure 10-8, and the typical fingering chart for a baroque alto recorder is given in Figure 10-9.

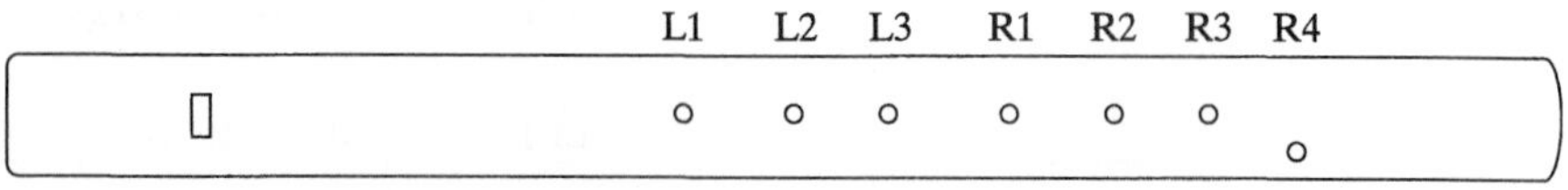

L1 L2 L3 R1 R2 R3 R4

Figure 10-7 Basic design of a typical cylindrical Renaissance recorder.

L1 L2 L3 R1 R2 R3 R4

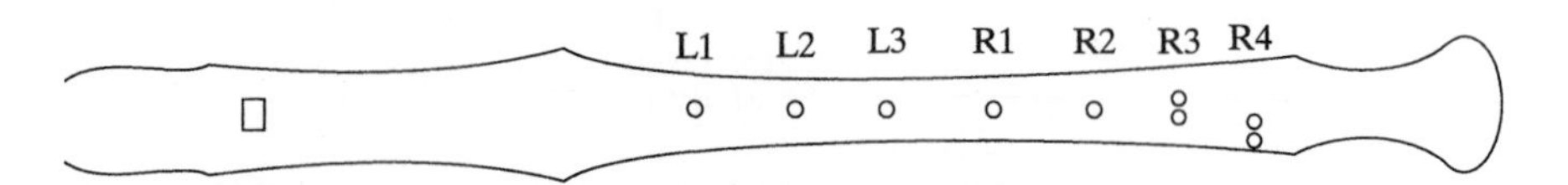

Figure 10-8 Basic design of a typical conical baroque recorder.

Perhaps the most important and exotic secondary effect of the tone holes is related to the *tone-hole cutoff frequency*. A resonance curve for a woodwind instrument like the hypothetical fipple flute, say for a note with only about half of the holes covered, shows that above a certain frequency the resonances diminish rapidly in amplitude and then disappear. The frequency at which the resonances disappear is called the tone-hole cutoff frequency. A higher-frequency standing wave is capable of extending its oscillation much of the way along the tube, despite the existence of the array of open holes. At lower frequencies, the first open hole will act as a strong cutoff point for the standing wave and effectively preclude strong oscillations beyond that point. Well below the cutoff frequency, the first open hole has by far the strongest influence on the frequency of the note. However, toward the higher ranges of some instruments, particularly the smaller woodwinds, the standing wave extends well into the array of open

F F♯ G♭ G G♯ A♭ A A♯ B♭ B C C♯ D♭ D D♯ E♭ E F F♯ G♭ G G♯ A♭ A A♯ B♭ B C C♯ D♭ D D♯ E♭ E F G
LEFT HAND
LT L1 L2 L3
RIGHT HAND
R1 R2 R3 R4

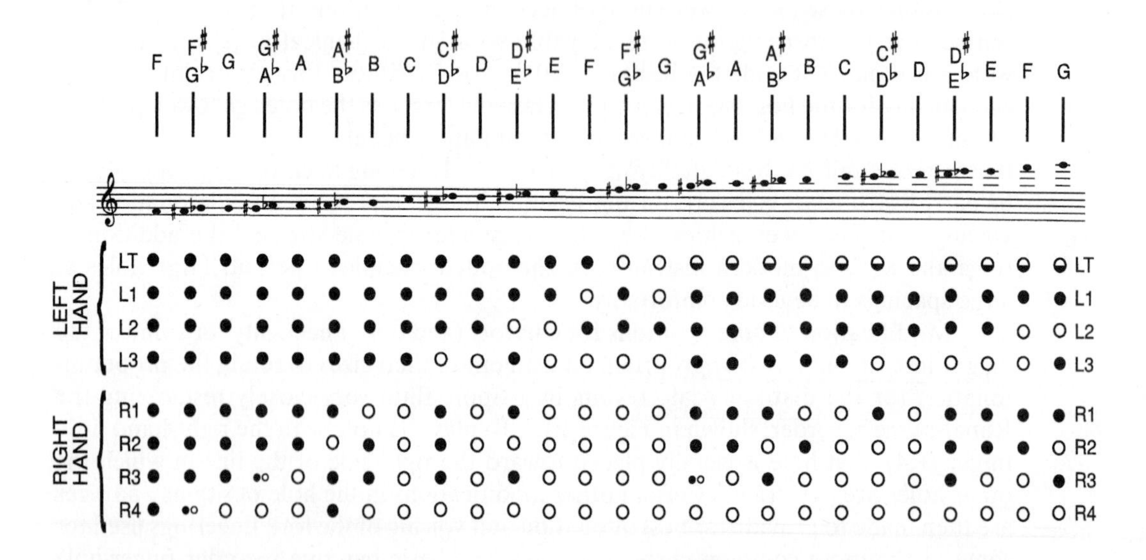

Figure 10-9 Fingering chart for a typical baroque alto recorder. The symbol ○ denotes an open finger hole; the symbol ● denotes a closed finger hole; ◒ denotes a half-closed finger hole (left thumb only for most baroque recorders), and ●○ denotes a half-closed double hole (third and fourth fingers of right hand only for most baroque recorders). This fingering chart will work acceptably for most F recorders, but only alto recorders sound at the pitch written here.

tone holes, and closure of holes at some distance down the tube can have a dramatic effect. This feature of woodwind instruments makes possible the use of fork fingerings. A very dramatic and surprising fork fingering involves the note D at the top of the range of some Renaissance alto recorders. The note C is played with L1, L2, and L3 covered and LT half-covered, as an octave hole. The note D is obtained by *closing* holes R2 and R3 while leaving R1 open and the left hand as it was! Fork fingering is the only way to obtain any of the chromatic notes, the notes other than the original seven notes in the basic scale of the hypothetical fipple flute that we designed.

Another important effect of the open-hole array is to increase the effective size of the first open hole; to provide the proper tuning, the holes nearer the upper end of the instrument must therefore be slightly smaller than if they were opened individually.

The holes have other, more subtle effects. Particularly in instruments with thick walls, the volume of the holes is not negligibly small compared with the volume of the tube itself, and the frequencies of the notes will be affected by the existence of the covered holes. Furthermore, as higher notes are played, fewer holes remain along the resonating column. The standing wave is partially broken up by the irregularity of the holes along the tube. This leads to a change in tone quality between various notes of the scale and is thought to be partially responsible for the unique quality of the "throat" tones of the clarinet (the highest notes in the lower register, just below the register break). Another feature that changes the effect of the holes is chamfering of the inside of the holes, that is, tapering the inside edge so that there is not a right-angle corner where the finger hole enters the bore. Chamfering is sometimes used in more expensive instruments as a final very subtle step in adjusting the tone and voicing.

Important improvements can be obtained in woodwind intonation and voicing by modifying the bore diameter from the basic cylinder or straight-cone geometry. This is analogous to the adjustment of frequencies of vocal formants by changing the size and shape of the vocal tract. If the size of the bore is increased at the position of a velocity node (pressure antinode), the frequency of that mode of oscillation will be lowered; if the size of the bore is increased at the position of a velocity antinode (pressure node), the frequency of that mode of oscillation will be increased. This effect is involved in fixing the amount of flare of the clarinet bell, which actually extends about halfway up the instrument, to produce overtones closest to exact harmonics. The bore of the clarinet near the mouthpiece end must also be adjusted from the cylindrical shape, an effect usually accomplished in the mouthpiece and adjacent barrel sections.

Such modifications are particularly significant in conical instruments: Small changes in the volume of the small end of the cone (near the reed) can create very large intonation deviations. As a result, the shape and size of the saxophone mouthpiece can have a very dramatic effect on its intonation, and a mouthpiece must be chosen very carefully. This problem is more acute in the oboe and particularly in the bassoon because of their small bores. It is necessary to control very carefully the volume of air that is contained within the reed and to minimize any changes in the reed shape as breath and lip pressure are changed in the normal course of playing over the entire range of the instrument. Some notes on the bassoon can be varied up and down by more than a semitone by changing breath pressure and lip pressure on the reed.

A long-standing conflict involves the effect of choice of material on tone quality. Although there is apparently a difference between the wooden flute of the early 1900s and the modern cylindrical metal flute, this can be largely explained by differences in the geometry of the bore and mouthpiece. On the other hand, it is more difficult to explain the difference between otherwise identical flutes that are plated with, say, gold, silver, platinum, or aluminum. Calculations have been made that indicate that there should be no noticeable difference between these metals with respect to their vibrations or damping of the oscillations of the air column. Yet many flutists claim substantial differences in tone even among these metals.

Another controversy involves the choice of wood, hard rubber, or other "artificial" materials in constructing clarinets. Physically, there should be negligible differences between instruments made from these materials, yet strong claims of differences are made, and expert clarinetists often express strong preferences. A partial explanation of this certainly lies with tradition. As long as clarinetists do not accept instruments of materials other than wood, makers will not apply the highest quality craftsmanship to other materials. As long as the substitute materials are finished with less adequate workmanship, they will continue to receive less than total acceptance. Thus the cycle continues. Perhaps we shall some day see "artificial" materials with excellent acoustical properties and superior physical properties, like resistance to temperature and moisture and superior strength, finished with exquisite workmanship. If this occurs, acceptance may come.

10.3 Recorders

We shall look in some detail at the acoustical properties and problems of the recorder, because the recorder is a very basic instrument and is extremely popular. Most of the acoustical problems discussed can be applied in a more sophisticated way to modern woodwind instruments.

The recorder derived its English name from its use during the Renaissance to teach songs to birds. An early form of the sopranino recorder called the *bird flageolet* was used to play short songs repeatedly for birds, thus teaching them the songs by rote. This process, called *recording*, was extremely popular for several centuries, and hundreds of short songs were composed and taught to dozens of different types of birds.

The recorder, being a fipple flute, achieves its sound by resonance of the noise created by the airstream from the windway striking the lip and by feedback of the vibrating air column to drive the oscillating air jet. Because both the fipple end and the bell end are acoustically open ends, the recorder behaves like an open tube, with one end near the fipple opening and the other end located either near the bell or near the open hole closest to the fipple opening.

The earliest Renaissance recorders were simple cylindrical tubes with eight finger holes; during the baroque period the bore developed a slight conical taper with the larger diameter at the fipple end. In addition, the two lowest holes became double holes, used to produce the chromatic notes one half step above each of the two lowest diatonic notes. The range of the Renaissance recorder was about one octave plus a sixth; the improved bore and finger-hole geometry of the baroque recorder extended that range to about two octaves plus one note. Figure 10-8 shows the typical fingering layout of a Baroque recorder with English fingering, and Figure 10-9 is a standard fingering chart for such an instrument.

To achieve the most uniform tone down to the lowest note, the holes should become larger toward the bell end of the tube, and the spacing between holes should become greater. To adapt these requirements to the shape of the human hand, a compromise is necessary: the holes are smaller and slightly closer together near the bell end. By making the last few holes gradually smaller while concomitantly tapering the bore to a smaller radius at the bell end, the most uniform voicing of the notes (quality and loudness) is obtained, although the notes near the bottom of the range are somewhat weaker. This effect is not so pronounced in large recorders in which keys are used for the last few notes, so the holes can remain relatively large.

Ideally, it is best to have one hole for each note on the instrument, in which the size and position of each hole is chosen to give the correct intonation and uniform tone quality, or voicing, over the entire range. The most obvious limitation of the recorder, because it has no keys, is that one finger can only cover one hole, limiting the instrument to about eight useful holes (two of which are often made as double holes). Obviously, the first acoustical compromise that must be made is that each hole must have a substantial effect on more than one note, and fork fingerings must be employed for all chromatic notes as well as some diatonic notes.

Another necessary compromise is to use the thumb as an octave hole. Opening the thumb hole allows air to flow in and out of the tube, creating an antinode at or near that point. This single thumb hole must serve as an octave hole for a range of notes, as was shown in the fingering chart. This is an additional compromise, because the position of the vent hole can be the exact position of an antinode for only one of the notes in the upper octave. In fact, it is used to produce the octave for an interval of over a fourth, from A♭ to D on the alto recorder. As can be seen on the fingering chart, the recorder uses fork fingerings a great deal, particularly for the higher notes. In practice, this fingering chart will serve only as a guide for many of these fork-fingered notes, because they are extremely sensitive to fingering changes, sometimes requiring half-holes at the bottom of the instrument. Many of these fork fingerings will differ in detail among similar recorders, even those by the same maker, owing to extremely small variations in dimensions when they were manufactured or changes in dimensions with the environment. A significant improvement in modern instruments is greater resistance to the effects of humidity and different kinds of weather, so that modern instruments are not critically affected by these factors.

Some recorders are made with small variations in the bore diameter to provide small final adjustments in the pitch of certain notes. One of the major problems with inexpensive recorders and replicas scaled to high pitch is that the thumb hole must be very carefully half-covered for the upper notes; small changes in the thumb opening will drastically affect both the pitch of the upper notes and how well they "speak." A slight constriction of the bore between the left thumb hole and the fipple opening has been suggested for some recorders as a way to improve the ease with which the last few notes in the upper range of the instrument will speak. This might allow these notes to be played in tune more easily and permit greater leeway in the thumb-hole opening to obtain good high notes.

The correct hole positions for the recorder were obtained historically in a sequence of "choices," as indicated earlier in this section. First, the sizes and positions of the holes were adjusted to provide a diatonic scale that is in tune using the fingering shown in Figure 10-9 for the low octave. Then, using the thumb hole half opened as an "octave" hole, the hole sizes and positions were adjusted to tune the upper octave.

Chromatic notes are played by finding the most accurately tuned fork fingerings, as illustrated in Figure 10-9. The use of fork fingerings to "force" the positions of the antinodes to the proper location results in a tone that is different in quality from the diatonic notes. This results in considerable variation in voicing over the entire range of the instrument and increases the technical complexity of performance. It was necessary to further modify the hole sizes and positions to obtain the best compromise between making the upper octave in tune, making the chromatic notes in tune, and obtaining the best voicing.

Several problems remain in the recorder because of the acoustical compromises in its design. Some of the most important compromises deal with intonation, simply because one hole must be a strong determinant for more than one note. One such tuning compromise is between the notes D and F♯ in the middle range of the alto recorder. Referring to the fingering chart, we see that the middle F♯ is obtained by fingering D and opening the left thumb hole. However, the thumb hole position has already been (over) prescribed by the dual requirements that it make the note G in tune and that it properly act as an octave vent over a considerable frequency range. As a result, if D is in tune, F♯ will be flat, or if F♯ is in tune, then D will be sharp. On most baroque replica recorders, D is made exactly in tune, because that note is played much more often than F♯. Thus F♯ must be either fingered differently or blown harder than normal to raise its pitch. In a recorder designed to play more chromatic modern music, a compromise is reached in which D is slightly sharp and F♯ slightly flat. It is then necessary to underblow D and to overblow F♯ so that they will be in tune. All recorders are built with this intonation compromise, and an accomplished player learns how to correct the pitch by adjusting breath pressure.

Possibly the most limiting feature of the recorder is that the player has very little flexibility in control of the airstream. Changes in air pressure strongly affect the intonation, leaving very little possible adjustment in volume, and thus limited possibility for dynamic expression in music.

Another complication in recorder design results from the overall rise in pitch standard over the centuries since the baroque period. Unfortunately, despite the great detail involved in specifying the proper dimensions for recorder bore, holes, fipple, and bell geometry, these dimensions do not scale well with pitch. Slight adjustments in hole positions can correct for minor intonation problems resulting from shortening the design length of the recorder so that it plays at A = 440 Hz rather than A = 415 Hz, the "standard" baroque pitch, which is about one half-step below A = 440 Hz. Although the intonation can be readily adjusted, the voicing and the ease with which the notes can be played over the entire range are very strongly affected by scaling. It is extremely difficult and time-consuming to correct these problems. In this era of high labor costs, very few craftsmen are willing to take the time required to modify instrument design at high (A = 440 Hz) pitch to achieve a response similar to that obtained in the baroque. On the other hand, by reproducing instruments according to baroque specifications, at low pitch, the problem of redesign is avoided. This is a strong motivation for performing baroque music at low pitch on low-pitch instruments, as is now done by several leading performing ensembles in Europe and the United States. An additional factor is the more authentic tone quality produced by instruments played at low pitch. Although the problems arising from the pitch change since the baroque appear in many instruments and in vocal music as well, it is possible that the effect is greatest in recorders, because they were probably the most highly developed wind instruments in that period, despite their severe limitations. Many makers of baroque instruments now produce recorders, and other instruments as well, at low pitch, generally A = 415 Hz, and performance at low pitch is becoming more commonplace.

The recorder comes in a family of sizes, typical of Renaissance instruments. The soprano in C, alto in F, tenor in C, and bass in F form the basic set, as with most other families. This is supplemented by the great bass in C and the double great bass in F (very rare) on the low end and the sopranino in F on the upper end. An even smaller recorder also appears irregularly in a variety of sizes. The ranges are given in Table 10-1 for various sizes of baroque recorders.

The recorder is not a transposing instrument; all present-day recorders sound at the written pitch. Two sets of fingerings are used, one set for the F instruments, and one set for the C instruments. The F fingering is similar to the fingering of the low register of the clarinet; the C fingering is similar to the fingering in the clarinet middle register. For all the C instruments, some octave of C is the note produced with all fingers down; on the F instruments, some octave of the F is produced with all the holes closed.

Table 10-1 Ranges of Selected Woodwind Instruments

Instrument	Lowest note	Approximate highest note
Soprano recorder in C	C_5	D_7
Alto recorder in F	F_4	G_6
Tenor recorder in C	C_4	D_6
Bass recorder in F	F_3	G_5
Piccolo in C	D_5	C_8
Flute	C_4	D_7
Bass flute	G_3	G_6
Sopranino clarinet in $E^\flat$	G_3	$B_6^\flat$
Soprano clarinet in $B^\flat$	D_3	$B_6^\flat$
Soprano clarinet in A	$C_3^\sharp$	A_6
Alto clarinet in $E^\flat$	G_2	$B_5^\flat$
Basset horn in F	F_2	C_6
Bass clarinet in $B^\flat$	D_2	F_5
Soprano saxophone in $B^\flat$	$A_3^\flat$	$D_6^\flat$
Alto saxophone in $E^\flat$	$D_3^\flat$	$A_5^\flat$
Tenor saxophone in $B^\flat$	$A_2^\flat$	$E_5^\flat$
Baritone saxophone in $E^\flat$	$D_2^\flat$	$A_4^\flat$
Bass saxophone in $B^\flat$	$A_1^\flat$	$D_4^\flat$
Oboe	$B_3^\flat$	G_6
English horn	E_3	A_5
Bassoon	$B_1^\flat$	$E_5^\flat$
Contrabassoon	$B_0^\flat$	$E_3^\flat$

10.4 Flutes

The basic acoustical principles of the flute are very similar to those of the recorder. The single most important acoustical difference is that the player is in direct control of the angle at which the air from the lips strikes the embouchure hole. This flexibility, along with control of the rate of air flow, provides the flute player with a greater range of volume and expression. Like the recorder, the flute was a straight cylindrical wooden tube during the Renaissance. The range was extended by making the baroque flute in the shape of a slightly tapered cone, with the small end at the bell. This modification, along with some improvements in the tone-hole design and cork joint, also allowed somewhat greater expression and improved voicing in the baroque flute.

During the nineteenth century, several additional modifications were made. In the early nineteenth century, more holes and keys were added to the baroque two-key and five-key flutes, improving the intonation of the flute and vastly simplifying the fingering of the chromatic notes. Eliminating many fork fingerings for chromatic notes also made the voicing and tone quality more uniform. One result of these additional keys was a rather extensive array of complicated machinery. The flute bore again became cylindrical in the middle of the nineteenth century, and metal began to replace the traditional wood around the beginning of the twentieth century. Modifications of the head joint and the movable cork improved the intonation in the third octave, thus increasing the range by almost one octave. One of the most important improvements in the flute was the Böhm fingering system, developed by Theobald Böhm over the first half of the nineteenth century. Böhm also revised the basic design of the contemporary flute, including the hole layout, design of the head joint and cork position, and details of which keys and rings are attached to each other and controlled with each finger.

The Böhm fingering system organized the machinery to provide a relatively straightforward and systematic set of fingerings by specifying how the keys and rings should be interconnected to provide the simplest basic fingering plan. By providing the appropriate normally open keys, large tone holes could be employed, thus allowing greater volume.

Figure 10-10 shows the cross section of a modern Böhm flute. The primary reason for the extension of the usable range of the flute to its modern limits lies in the details of the construction and placement of the cork. Although a movable cork was used in

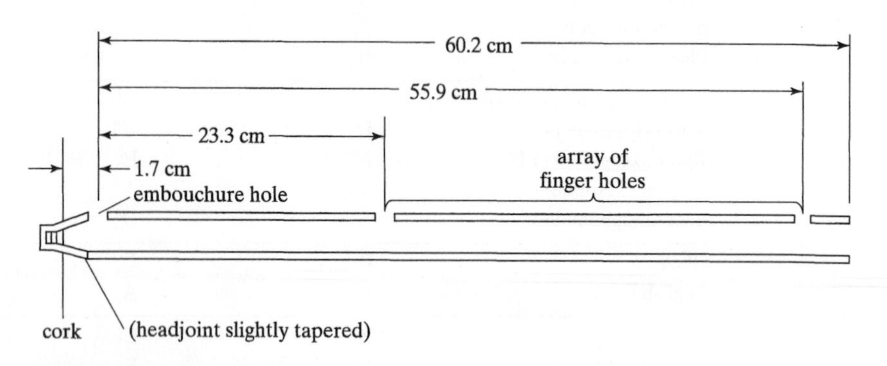

Figure 10-10 Cross section and basic dimensions of a Böhm flute.

the baroque flute, its normal position was changed dramatically when the flute bore was returned to its original cylindrical shape by Böhm. Böhm added a slight parabolic taper to the section of pipe above the embouchure hole that holds the cork, with the length of the head joint from the embouchure hole to cork equal to the bore diameter at the embouchure hole. The existence of the additional tone holes and the machinery with which they could be closed allowed the entire chromatic scale to be played with uniform voicing and without using complicated fork fingerings. Böhm used large holes to obtain greater volume from the flute and covered these holes with keys when they became too big for fingers to cover.

The contributions of Böhm to the musical world are important to more than just the flute, as the Böhm fingering system has been extended to the clarinet and saxophone families as well. It also appears that some of the problems of the oboe and bassoon might be helped by introducing a modified Böhm fingering system. One desirable feature of the Böhm system is that after learning one Böhm instrument any other is relatively easy to learn, because of the similarity in fingering and key placement.

Very few acoustical changes were made in the flute from the time Böhm designed his original model in the middle of the nineteenth century until late in the twentieth century. In fact, some makers advertised that they used a design as close as possible to the Böhm original. However, during the intervening century the standard of pitch rose from about A = 435 Hz, the standard at which Böhm designed his flute, to A = 440 Hz. As a result, certain intonation defects developed. Measurements on contemporary flutes by John Coltman in the mid-1960s showed that relative to the equal-tempered scale, the notes at the bottom of the flute, with most fingers down, were too low in pitch, whereas those with most fingers up were too high in pitch, even in high-quality flutes. This error repeated in the second octave, as shown in Figure 10-11 for several modern Böhm flutes tuned to A = 440 Hz. Performers normally compensate for this error by

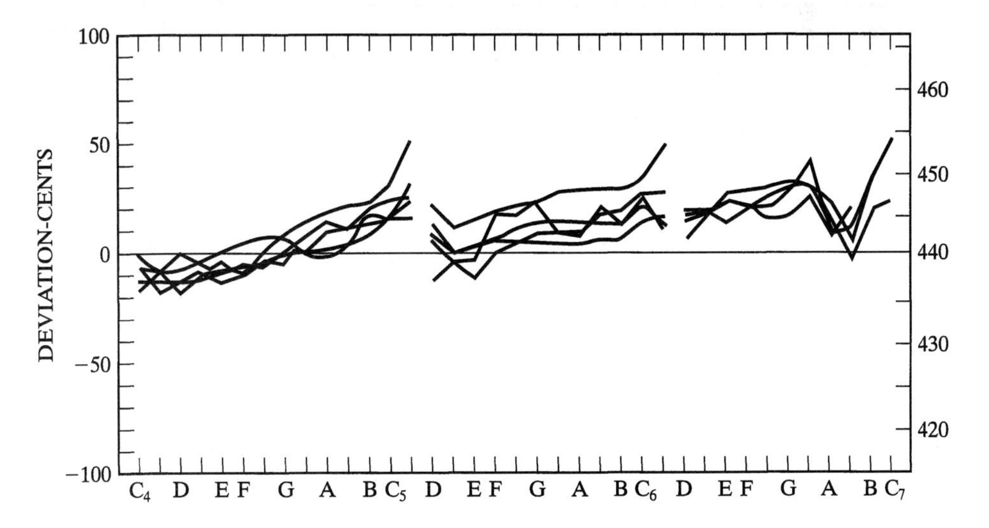

Figure 10-11 Tuning errors obtained with several modern Böhm flutes. A deviation of 100 cents corresponds to one half-step. (Reprinted from *The Instrumentalist* [March 1975, pp. 78, 80]. © 1975 by The Instrumentalist Company. Used by permission of *The Instrumentalist.*)

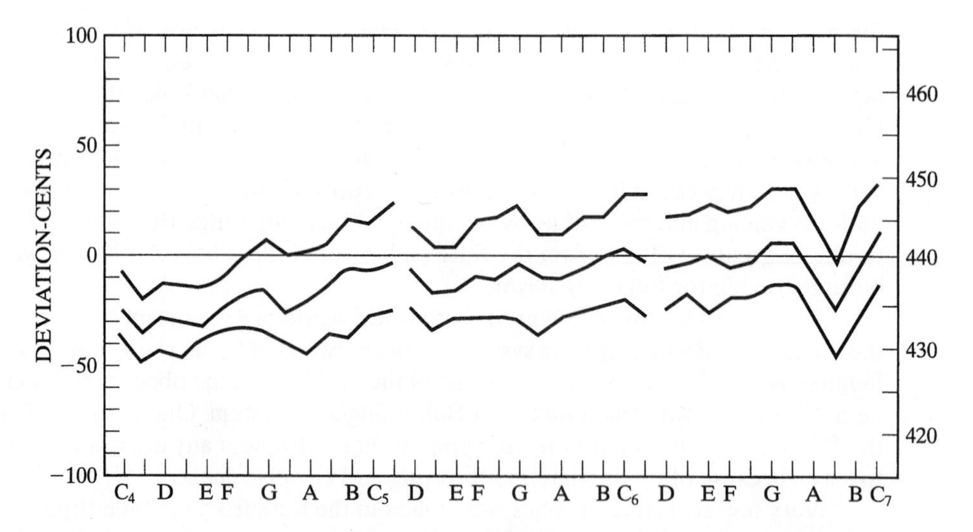

Figure 10-12 Tuning errors in a Haynes Böhm (top), and the errors on the same flute extended between the embouchure hole and the first finger hole by 8 mm (middle) and 13 mm (bottom). The flute is better in tune when it is extended to play at a lower pitch. (Reprinted from *The Instrumentalist* [March 1975, pp. 78, 80]. © 1975 by The Instrumentalist Company. Used by permission of *The Instrumentalist*.)

rolling the instrument about the lip to adjust the angle at which the air strikes the embouchure hole, and perhaps blowing louder or softer. Such performer compensation was not included in Coltman's measurements.

If a longer tuning slide is used and the flute is extended between the embouchure hole and the first finger hole to play at a pitch level about A = 430–435 Hz, the tuning error disappears and the intonation across the octave break is more continuous. This requires an additional length of some 8 to 13 millimeters (up to about 1/2 inch). As the standard pitch level rose, the flute was "corrected" by simply shortening its length by removing some tubing between the embouchure hole and the finger holes while the spacings between the finger holes remained unchanged. If this simple shortening were the only adjustment made, it would overcorrect the high notes in each octave, leaving them too high in pitch, and undercorrect the low notes in each octave, leaving them too low in pitch, as seen in Figure 10-11. The effect of extending the tuning slide on one such contemporary flute is shown in Figure 10-12. The notes become in tune for all octaves, and the break between octaves is dramatically reduced.

The tone-hole spacing can be adjusted to compensate for the tuning errors obtained when the flute is shortened by simply removing tubing from the tuning slide. The upper holes must be moved slightly farther down the tube, while the lower holes must be moved slightly up the tube, both corrections being a few millimeters. This has been done by the twentieth-century English flute maker Albert Cooper. Figure 10-13 shows the deviations in pitch of a flute manufactured to the Cooper scale standard; it is obviously much better over its entire range. The Cooper scale flute has now become much more acceptable, and virtually all flute players and manufacturers have now adopted it. Lamentably, some do not recognize Cooper as the innovator of their upgraded designs.

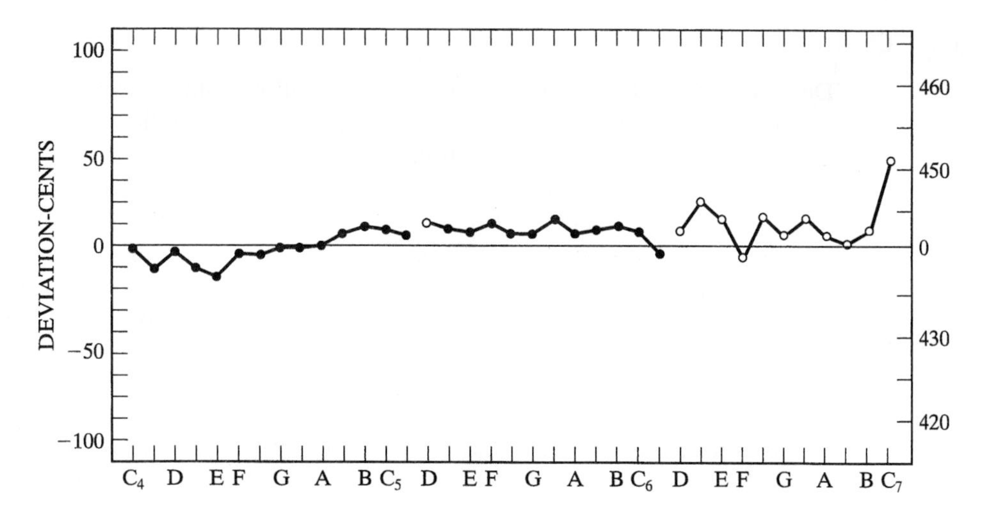

Figure 10-13 Tuning errors obtained on a Cooper scale flute, tuned to $A_4 = 440$ Hz, but with the Böhm finger hole spacing modified for the increase in pitch level. (Reprinted from *The Instrumentalist* [March 1975, pp. 78, 80]. © 1975 by The Instrumentalist Company. Used by permission of *The Instrumentalist.*)

The history of the development of the Cooper scale flute illustrates how much tradition means in musical instrument design. Tradition dictated that the Böhm design, having been in use for over 100 years, was the only acceptable design, and any deviation therefrom must be bad. To the contrary, in this case the design change was a dramatic improvement. Flute players learned to compensate for this defect automatically and thus played out of tune on "properly" constructed flutes, or at least the flute had a different "feel" to the player.

Any further rise in the standard of pitch would again increase the magnitude of this problem, and make the flute harder to play in tune, unless it is accompanied by an appropriate additional design modification.

Two additional members of the flute family are used in the contemporary band and orchestra. The piccolo plays one octave higher than the flute and is in common use. Less common is the flute a fourth lower than the orchestral flute, called either the alto or bass flute in various scores. Their ranges are given in Table 10-1. Two members of the flute family are shown in Figure 10-14.

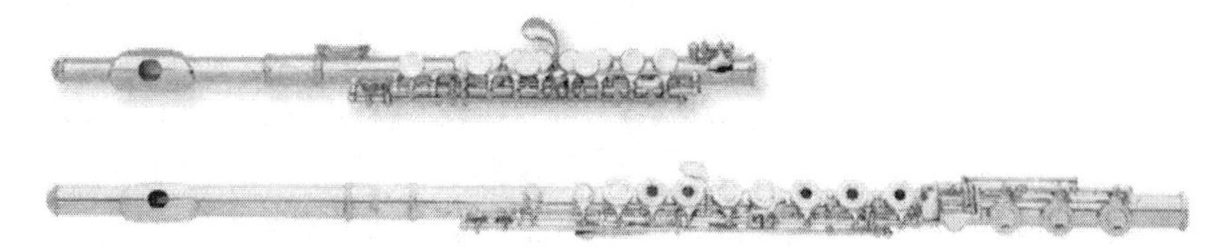

Figure 10-14 Two members of the modern flute family: the C piccolo and the orchestral flute. (Courtesy of Conn-Selmer, Inc.)

10.5 Clarinets

J. C. Denner (1655–1707) made significant modifications and improvements in the narrow-bore single-reed woodwind instrument, the chalumeau, thus developing the precursor to the modern clarinet. The tone of the chalumeau was very soft and possessed the typical "woody" character of the lower register of the clarinet, called the chalumeau register. The middle register of the clarinet is often called the clarion register, in recognition of the clear, more bell-like characteristic of its tone quality, which in its early stages was likened to that of the trumpet. The clarinet emerged as a distinct instrument in the late seventeenth century and took its place in the symphony orchestra in the late eighteenth century.

The clarinet is unique among the contemporary woodwinds, in that it is the only instrument that behaves acoustically like a closed tube, the reed end being the closed end. A reed instrument can behave like a closed tube only if the bore is predominantly cylindrical. The clarinet bore is very nearly cylindrical from the barrel, just below the mouthpiece, to just below the middle joint, about halfway down the instrument.

There are several important implications of the clarinet's acoustical behavior as a closed tube. As in the closed tubes studied in Chapter 4, the resonance curve has its largest peaks at the odd harmonics of the fundamental frequency. The Fourier spectrum of typical clarinet tones is dominated by the odd harmonics, especially in the first several overtones. This is not true for higher harmonics, say above the fifth or sixth, when the even harmonics become as large as the odd harmonics. The clarinet shares this emphasis on odd harmonics with the square wave, which has only odd harmonics, and thus the clarinet tone resembles that of a square wave.

Because the clarinet is effectively a closed tube, its fundamental wavelength will be four times the distance from the reed, or node, end to the open end, either the bell or the highest open tone hole. The wavelength of the lowest note on the instrument is about four times the length of the clarinet. The flute, on the other hand, is an open tube, so its lowest note has a wavelength of twice the length of the flute. Although the flute and clarinet are approximately the same length, this means that the lowest note on the clarinet is about one octave below that of the flute.

The clarinet overblows at the third harmonic, a musical interval of a twelfth, rather than the octave, as in all other woodwind instruments. This can be understood by referring to Figure 10-15. The fundamental mode in a closed tube has a velocity (or displacement) node at the closed end and an antinode at the open end, as shown. By venting the tube at one-third of its length from the closed end, an antinode is forced at that position. The overblown standing wave is just that of the third harmonic, or a musical interval of a twelfth, the next possible resonant mode of the closed tube. This hole is

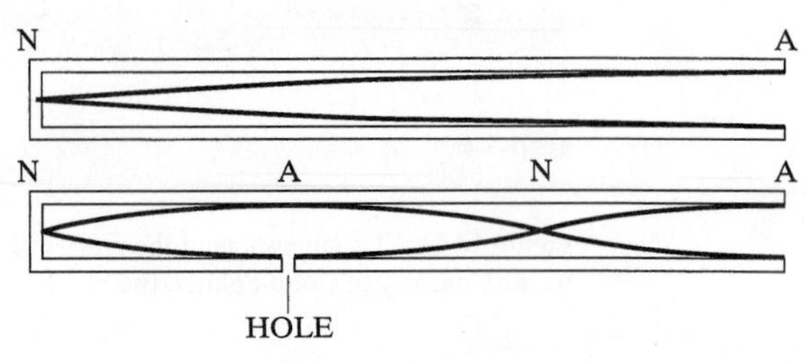

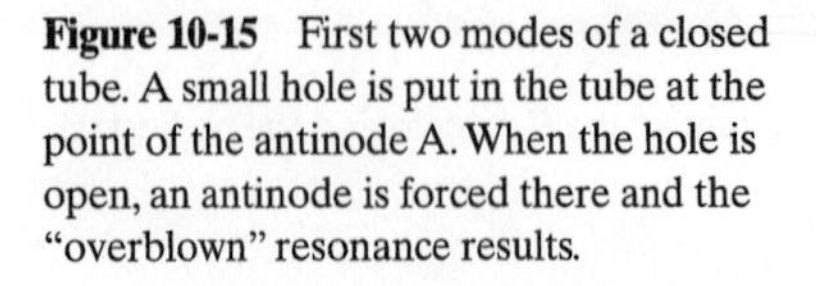
Figure 10-15 First two modes of a closed tube. A small hole is put in the tube at the point of the antinode A. When the hole is open, an antinode is forced there and the "overblown" resonance results.

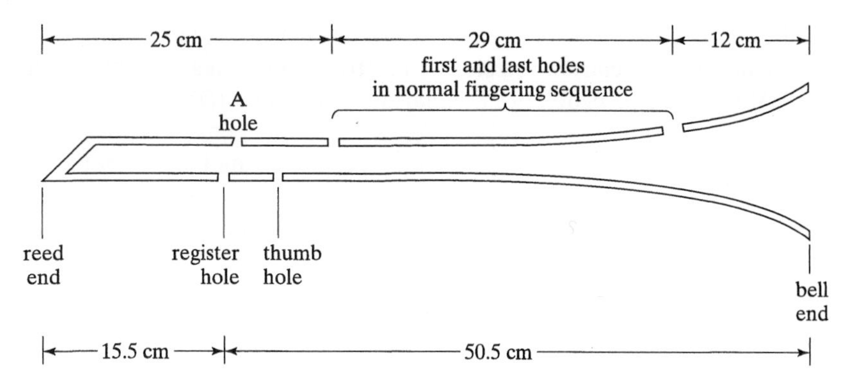

Figure 10-16 Cross section of a clarinet.

called the register hole on a clarinet, because the clarion (middle) register is up one twelfth, not an octave from the chalumeau (lower) register. The altissimo (upper) register consists entirely of forked fingerings.

Figure 10-16 shows, roughly to scale, the positions of the tone holes and register hole of a clarinet. Notice that, as in the recorder, the register key provides the venting to force an antinode over a wide range of notes, where the effective tube length varies by over a factor of 2. The total length of the clarinet from the reed end to the bell is about 66 cm, so for the lowest note the register key should be about 22 cm from the tip of the reed. The highest note on the clarion register has the hole open 25 cm from the reed end, so the register key should be about 8 cm from the tip of the reed. The total change of position of the antinode is almost 17 cm (over $6\frac{1}{2}$ inches) in this simplified approximation.

Another result of the fact that a clarinet overblows at a twelfth (one octave plus a fifth) is that 11 notes in the normal fingering sequence must be provided before the register is changed by opening the register key. This is more complicated than in other woodwinds, such as the recorder or flute, which overblow at the octave and thus require only seven notes. It is necessary, therefore, to use the left thumbhole and a special key (the A hole) operated by the left index finger to extend upward the range of the lower register. The range of both registers is also extended downward one half-step by adding a key controlled by the left little finger to close the lowest hole on the instrument, providing the note B. The B^b between these two notes is obtained by opening the register hole while the left thumb hole and the A key are simultaneously open. The notes A and B^b on the clarinet, called the *throat tones*, have a tone quality somewhat different from either of the other registers. This is partly because for most notes on the clarinet there are closed finger holes along the tube that affect the tone, but there are fewer or none remaining above these holes to affect the tone of the throat tones. In addition, the register key is really at the wrong position and is the wrong size to produce B^b with the same voicing as the other notes, and it comes out very muffled. The clarinet has an alternate key to produce B^b with better voicing, but it is inconvenient to use in even moderately fast passages and is generally used only as a trill key. A clarinet would be more adequately voiced and easier to play if two register keys were used to provide for the throat tone B^b and the venting of the entire upper register. Such a double key could be made to automatically open the proper hole, depending on which note was being played in its clarion register, or if throat-tone notes were being played.

Despite these problems, the clarinet is in some respects the easiest reed instrument to learn to play acceptably, even with an inexpensive instrument. The fingering has been standardized using the Böhm system, making the clarinet easier to play than the oboe or bassoon. In contrast to the conical instruments, as the player changes pressure on the clarinet reed there is relatively little change in the effective length of the instrument, so it is easier for the beginner to play in tune. On the other hand, the fact that the clarinet overblows at the musical interval of a twelfth means that notes one octave apart are fingered differently.

The two lower registers of the clarinet encompass the two sets of fingerings used in most woodwind instruments. Virtually all woodwinds, including all transposing instruments, are fingered in the same way as the clarion register, whereas the F recorders and some other Renaissance woodwinds are fingered like the chalumeau register of the clarinet. As in the recorders and flutes, there is a unique set of fork fingerings required to obtain the notes in the highest register, above the highest note in the clarion register.

Although the clarinet is acoustically a cylindrical instrument, it begins to flare toward the bell just below its midpoint. This flare is necessary to keep the overtones of all notes harmonic and to keep the frequencies of the overblown third harmonics of the clarion register in the correct ratio to the frequencies of the fundamentals in the chalumeau register. These problems arise in part from tuning errors introduced by the single register key, because it cannot be properly positioned for all the notes in the clarion register. Double register keys would significantly reduce this intonation problem, but such a dramatic change would deviate from the existing clarinet tradition and change the sound of some notes, which might be deemed undesirable.

The clarinet comes in a family of transposing instruments, the E^b sopranino, the B^b and A sopranos, the E^b alto, the basset horn in F, and the B^b bass being the most common. The B^b soprano, E^b alto, and B^b bass clarinets are used as standard members of the concert band; the B^b and A clarinets are the standard members of the symphony orchestra; the A clarinet sounds one half-step below the B^b clarinet. Ranges of the members of the clarinet family are given in Table 10-1, and several members of the clarinet family are shown in Figure 10-17.

10.6 Saxophones

The saxophone, invented by the French musician Antoine Joseph (known as Adolphe) Sax (1814–1894) during the middle of the nineteenth century, is a relative newcomer to the musical scene and is sometimes scorned by musical purists as a "hybrid" instrument. Because of its recent development, it has not achieved general acceptance in the symphony orchestra, but it is used in the orchestra on occasion and has achieved popularity in the concert band and especially in dance and jazz bands. The saxophone is a conical, single-reed instrument with a large bore diameter and very large tone holes, all covered by pads. The large bore diameter allows it to produce loud sounds appropriate for jazz and dance bands. The saxophone's conical bore shape yields resonances at all harmonics, which distingushes it from the cylindrical bore of the clarinet, which

Figure 10-17 Members of the clarinet family: the E^b sopranino, the B^b soprano, the E^b alto, and the B^b bass clarinets. (Courtesy of Conn-Selmer, Inc.)

emphasized the odd harmonics. The saxophone overblows at the octave. It has also been standardized to the Böhm system, making its fingering somewhat simpler than that of the clarinet.

One important feature of the saxophone, an improvement over the clarinet, is that the saxophone is equipped with two octave holes. Figure 10-18 shows the approximate dimensions of the saxophone, along with the rough positions of some of the tone holes and the octave holes. The effective length of the saxophone to produce its lowest note, D_3^b (about 140 Hz), is slightly longer than the instrument. Ideally, the distance of the octave key from the effective end should be one-half the distance from the

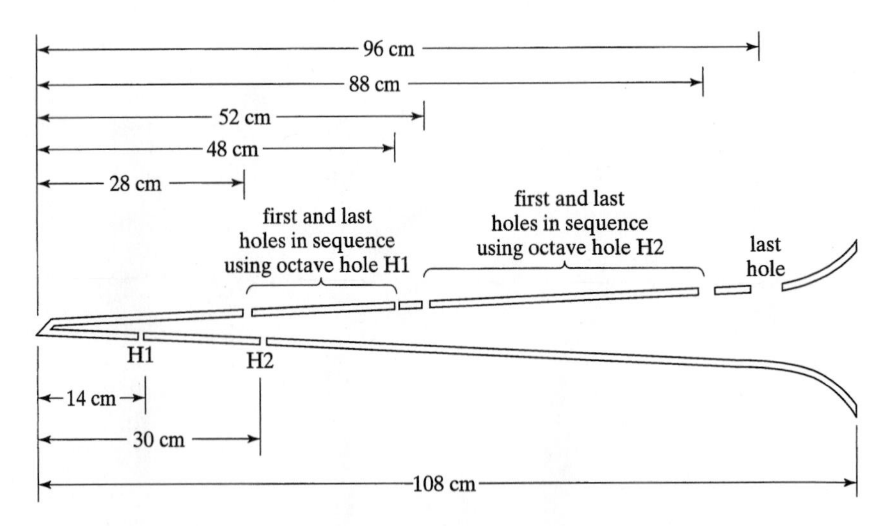

Figure 10-18 Cross-sectional view of a straightened out E^b alto saxophone.

effective end to the first open hole. Octave hole H2, about 30 cm from the end, serves for a range of lengths from 52 to about 88 cm. Octave hole H1, about 14 cm from the end, serves for a range of lengths from about 28 to about 48 cm. However, the effective end extends the instrument several centimeters beyond the mouthpiece back toward the apex of the cone, placing the effective position of the hole H1 at the halfway point for a hole lying somewhere between the tone holes 28 and 48 cm from the actual reed tip. There is only one octave key; the appropriate octave hole is opened by rings and levers attached to the tone holes whenever the octave key is pressed. Thus, when the instrument is fingered up its scale, opening the lowest finger hole requiring use of octave hole H1 automatically opens that octave hole and closes H2. This automatic, double, octave-hole system is a decided improvement over the system used in the clarinet and improves the saxophone intonation significantly.

The saxophone is probably the easiest reed instrument on which to obtain a tone, but some acoustical design problems arise for the beginner that are not present in the flute or clarinet. One characteristic of conical instruments is that small changes in the bore of the instrument near its narrow end can result in unexpectedly serious variations in the effective length of the instrument, changing its pitch considerably. Because the saxophone has a rather large mouthpiece, changes in embouchure—that is, the position and pressure of the mouth and lips—can cause rather large changes in the volume of the air space near the apex of the cone, which can detrimentally affect the pitch. Although this sensitivity of pitch to embouchure may hinder the beginner, accomplished jazz saxophonists use it for expressive purposes by "bending" notes. Tuning the instrument by sliding the mouthpiece in or out changes the effective length by a greater amount than normally would be expected, and the intonation could easily vary throughout the range of the instrument more than in the case of either the flute or the clarinet. These problems are substantially exacerbated by any dimensional errors in the mouthpiece or narrow part of the bore of the saxophone, which occur in poorly made instruments.

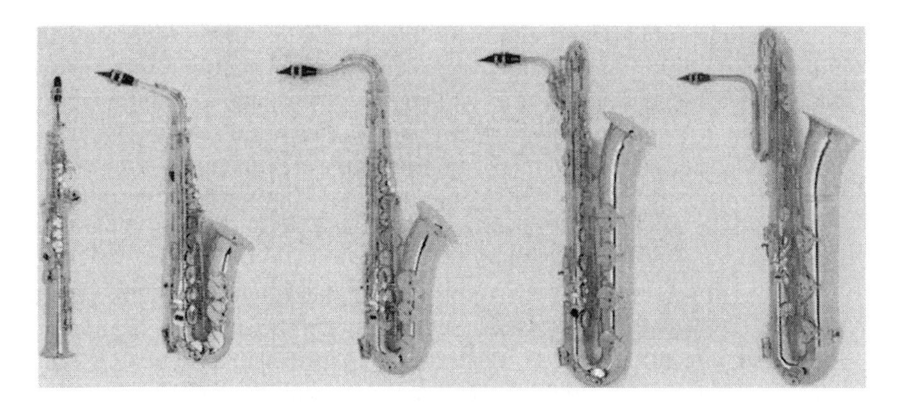

Figure 10-19 Members of the saxophone family: the B^b soprano, the E^b alto, the B^b tenor, the E^b baritone, and the B^b bass saxophones. (Courtesy of Conn-Selmer, Inc.)

Because the saxophone is relatively easy to play for beginners, and the sounds they are likely to get are often very loud and out of tune, the saxophone has perhaps unfairly achieved a bad reputation among serious musicians. An expert player with a good saxophone and a properly matched mouthpiece can achieve a remarkable variety of sounds, volumes, and expressions.

Saxophones are transposing instruments. As in the case of the clarinet, one basic set of fingerings covers all instruments. Common sizes of saxophones found in the full concert band are the E^b alto, B^b tenor, E^b baritone, and the B^b bass. The B^b soprano and several larger sizes are used primarily in jazz ensembles and large dance bands. The ranges of some members of the saxophone family are given in Table 10-1. Figure 10-19 shows several members of the saxophone family.

10.7 Oboes

The oboe is a double-reed instrument with a narrow conical bore. It overblows at an octave, as do the flute, recorder, and saxophone. All harmonics are present in its resonant curve, and therefore all harmonics are typically present in the Fourier spectra of its steady-state tone.

Because it is conical, all the acoustical problems arising from mismatching the reed cavity to the instrument are present, as in the saxophone. However, because the bore is narrow, intonation problems arising from changes in the effective length are severe. Furthermore, although the double reed is acoustically very similar in its operation to the single reed, the oboe is inherently more susceptible to variations in its volume as the embouchure of the player is changed. It is probably these inherent acoustical problems that make the oboe a very difficult instrument to play in tune with good tone, giving origin to its description as an "ill wind that no one blows good."

The performance problems of the oboe are significantly exacerbated by the fact that the fingering system has never been converted to the Böhm system. The fingering system remains more difficult to learn and use compared with that of Böhm-fingered instruments.

An interesting experiment with the oboe is to substitute a small clarinet mouthpiece for the traditional double reed. The sound of the oboe remains very nearly the same, and it becomes easier to blow because of the simpler embouchure technique for single-reed instruments. This verifies the hypothesis that the physical behavior of single and double reeds is very similar, and that the tone quality of a reed instrument is determined primarily by the bore shape and size. Clarinet-type mouthpieces have been used on oboes to simplify the process of learning to finger the oboe while a student is converting from clarinet to oboe.

The alto member of the oboe family, the English horn, is unique among the woodwinds, in that it has a region near the bell end where the diameter of the instrument is significantly enlarged. This roughly spherical enlargement acts as a type of resonator and is probably responsible for the formants in the English-horn tone, which are believed by some physicists to be the origin of its typical plaintive sound. Actually, the English horn is neither English nor a horn. Its name probably derives from the old French words for "angle," because the instrument has a sharp bend near the mouthpiece, and for "horn," because it originally looked somewhat like an animal's horn.

The oboe is not a transposing instrument, but the English horn sounds a fifth below its written notes. The ranges of the oboe and English horn are given in Table 10-1. The contemporary oboe is shown, along with the bassoon, in Figure 10-20.

10.8 Bassoons

If the oboe is a bit antiquated, the bassoon is a true relic of the past, having been the subject of very few significant acoustical improvements since it emerged as a member of the orchestra in the eighteenth century. The bassoon, like the oboe, is a double-reed, conical-bore instrument. The ratio of its length to its bore diameter is greater than that of any other woodwind instrument. Its very narrow bore makes the bassoon an acoustical nightmare, its intonation problems dwarfing even those of the oboe. Many notes on the bassoon can be easily lipped up and down by more than one half-tone, some as much as a minor third.

The bassoon bore is over 8 feet long, and its tube must be bent twice in order to be playable. It is fitted with a metal tube called the *bocal* on which the reed is mounted, which increases the length of the instrument and makes it more comfortable to play. The bocal is part of the narrow end of the resonant conical tube, as opposed to the bocals of some large recorders, which simply transmit the air stream from the player's mouth to the windway and which do not have the standing waves within them. The small diameter of the bocal at the reed end, typically a few millimeters, only serves to increase the sensitivity of the instrument to variations in the effective volume of the reed cavity and can result in considerable pitch variation for a single note.

Due to the extreme length of the bassoon, it must be equipped with three octave keys, the uppermost of which, called the *whisper key*, is mounted on the bocal near its entry to the wooden tube. Because the bassoon has not been modernized to the Böhm fingering system, the three octave keys must be independently operated by the left thumb. A significant improvement would be to interconnect the octave keys by means of levers and rings, as in the saxophone. Several of the holes in the bassoon are not

Figure 10-20 Double-reed orchestral instruments: the oboe and the bassoon. (Courtesy of Conn-Selmer, Inc.)

covered by rings and pads but are closed directly by the player's fingers. To make the bassoon playable, some holes must be located at compromise positions, modified from the ideal diameter, and drilled into the bore at large angles. Owing to the lack of a modern system of interconnected rings and keys, fork fingerings are regularly required on the bassoon. As a result of these tone-hole limitations, the bassoon is very cumbersome to play and suffers from considerable intonation and voicing defects.

The contrabassoon, which is twice the length of the normal bassoon and therefore sounds one octave lower, is the only other member of the bassoon family in use in the modern symphony orchestra. An interesting experiment in tone projection involves use of the contrabassoon doubling the part of the bass viol section of an orchestra. Even though the contrabassoon cannot be heard as a distinct entity when it plays with the basses, it gives their tone a substantial solidity, and a great difference may be noticed when the contrabassoon drops out. Perhaps this effect would be an appropriate subject for further investigation. The bassoon is not a transposing instrument, but the contrabassoon sounds one octave below its written pitch; their ranges are given in Table 10-1. The contemporary bassoon is shown with the oboe in Figure 10-20.

10.9 Pipe Organs

Although several large volumes have been written about the organ, we shall confine ourselves to a brief summary of some of its basic features, especially the sound-producing parts, as they relate to the woodwind instruments.

A typical pipe organ has two or more keyboards, called *manuals*, and a single set of pedals played by the feet. A pipe organ is usually equipped with several complete sets, or *ranks*, of pipes; some large organs have 50 or 60 ranks. A single rank includes one pipe for each note on one of the manuals and, in general, each rank will be a different type of pipe. A set of levers called *stops* controls which ranks are actuated by each manual. Various stops will usually allow most of the ranks of pipes to be played by any of the manuals, and two or more manuals can be set to control the same rank of pipes.

Organ pipes operate acoustically as either fipple flutes or Renaissance capped-reed instruments, and because the air pressure is kept constant to keep the pitch level constant, there is no possibility for change in volume of any single rank of pipes. Changes of volume are obtained by two techniques. One of the manuals, called the *great* manual, adds additional ranks of pipes as a volume pedal is pushed by the foot. The *swell* manual has its pipes in a large enclosure, and pushing the swell foot pedal opens a set of baffles or dampers in front of the box, allowing more of the sound to escape and increasing the volume.

Two types of action, or techniques for controlling the pipes by the keys, are in common use. In *tracker action*, the key mechanism extends directly to a valve that opens the air to the pipe to be sounded, allowing direct control of the attack transients by the player. In an *electrical action*, pressing the key closes an electrical switch that operates valves to allow air flow to the pipes. Tracker action was used exclusively until the twentieth century, and many organists even today prefer the closer control and more rapid response that can be obtained in a good tracker-action organ.

There are two basic types of pipes, flue pipes and reed pipes. A *flue pipe* operates by the same basic technique as a fipple flute and may be either open or stopped (closed) at the end opposite the fipple. The basic parts of the typical flue pipe are shown in Figure 10-21. Air enters the pipe through the foot, usually positioned at the bottom of the vertical pipe. The air stream is then directed between the lower lip and the *languid* in the case of a metal pipe, or between the cap and block in the case of a wooden pipe, onto the top lip.

The top lip is sharp, as is the lip on a recorder; the noise created by the air stream striking the upper lip and the feedback of the oscillating air in the standing wave within

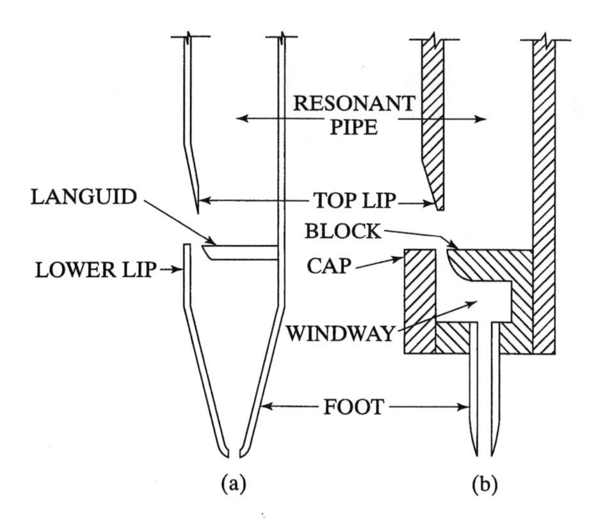

Figure 10-21 Basic sound-producing mechanism for the (a) metal and (b) wooden flue pipes of a pipe organ.

the pipe then create the sound of that pipe. Air is stored under pressure in the wind chest, a large reservoir that is required to keep the pressure steady while the various combinations of notes being played are changed.

Figure 10-22 shows several types of common flue pipes. Ranks of pipes are often labeled by a length, such as 4 feet, 8 feet, 32 feet, and so on. This length refers to the approximate length of the open pipe in the rank that sounds when the key C_2 is depressed. When the 8-foot rank is engaged and the C_2 key depressed, an 8-foot long open pipe sounds. The wavelength of the note emitted is twice the length of the open tube, about 16 feet, and has a frequency equal to that of C_2 on the piano. Thus the 8-foot rank, or unison rank, plays at the written pitch. If, instead, the 16-foot rank is engaged and C_2 depressed, a 16-foot open pipe sounds. This is an octave lower than the note written. The same principle applies to all ranks: The name of the rank corresponds to the length of the open tube sounded when the C_2 key is depressed.

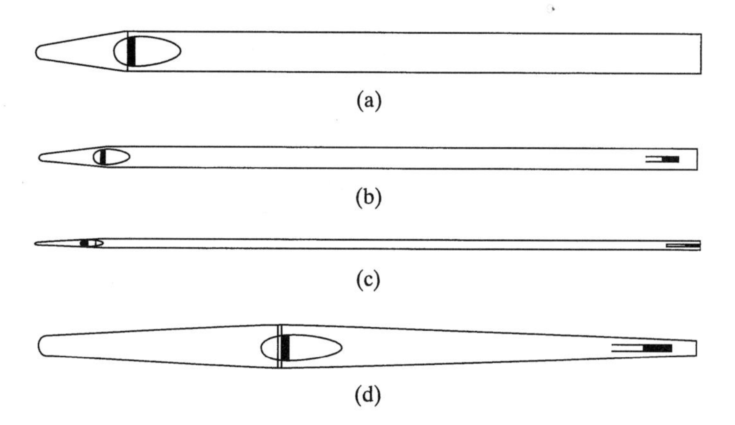

Figure 10-22 Four examples of flue pipes: (a) diapason; (b) dulciana; (c) viole d'orchestre; and (d) spitz flute. (From Audsley, *The Art of Organ Building*, 1965.)

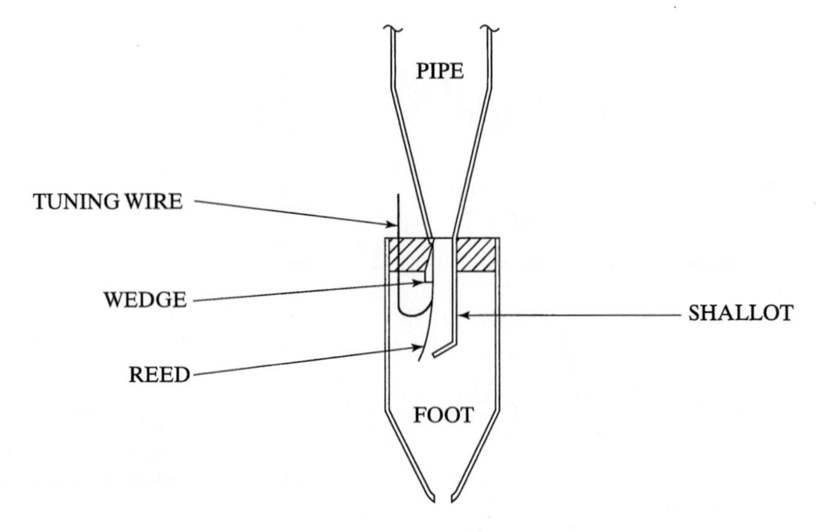

Figure 10-23 Sound-producing mechanism of a reed organ pipe.

Figure 10-23 shows the basic parts of a typical reed pipe. Air again collects in the foot and passes through the opening between the reed tongue and the shallot. This causes the reed to vibrate as a result of a combination of Bernoulli's principle and feedback of pressure pulses in the standing wave, as in the case of any reed instrument. The enclosure of the reed in the air box is very similar to the situation in Renaissance capped-reed and other reed instruments, with one important exception. Because the reed is fixed to one pipe of a specific length and always sounds at the same pitch, it is common to tune the *reed* by sliding the tuning wire up or down its surface. The reed is tuned so that it will sound the proper pitch even if the pipe is not connected. The "tuning" mechanism of the pipe is generally adjusted to provide for uniform voicing. Figure 10-24 shows several types of common reed pipes.

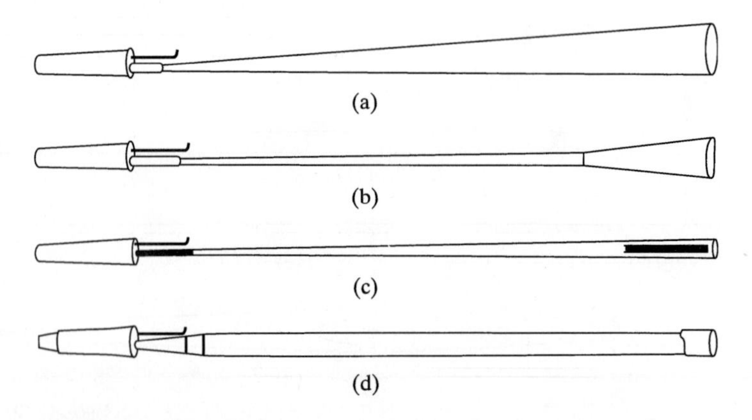

Figure 10-24 Four examples of reed pipes: (a) one type of chorus reed; (b) organ oboe; (c) orchestral oboe; and (d) clarinet. (From Audsley, *The Art of Organ Building*, 1965.)

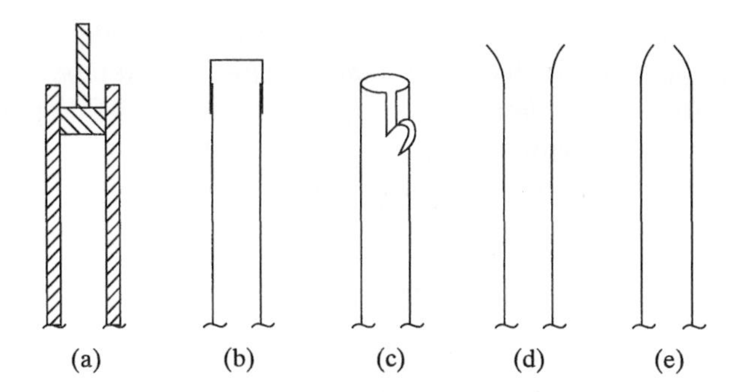

Figure 10-25 Tuning techniques for flue pipes: (a) wooden and (b) metal stopped pipes; (c) standard modern tuning for metal open pipes; and bell tuning to (d) raise and (e) lower pitch of pipe.

Several types of tuning techniques are used, depending on the nature of the pipe, as illustrated in Figure 10-25. Stopped pipes are generally tuned by adjusting the position of a metal cap in the case of a metal pipe or a wooden plunger in the case of a wooden pipe. For open metal pipes, tuning is usually accomplished by cutting a strip out of the side of the pipe and rolling the strip down to raise the pitch of the pipe to the proper level, as shown in Figure 10-25(c). In some open pipes a movable section, added to the top of the pipe, can be slid in or out to vary the pipe length, just as the cap does for a closed pipe. Another tuning device used on open pipes is a small flap that partially covers the end of the pipe, as seen in Figure 10-26(e) and (f). Such a flap controls the position of the pressure node at the open end.

Many metal organ pipes were tuned during the baroque by a technique known as *bell tuning*. The entire end of the pipe was tapered in to lower the pitch or tapered out to raise the pitch. This type of tuning provides the most stability over a long time, but requires exceptional care in order that the end of the pipe not be damaged. Significant information about early pitch level comes from analysis of bell-tuned organ pipes, because in the absence of physical abuse they can remain relatively well in tune for centuries.

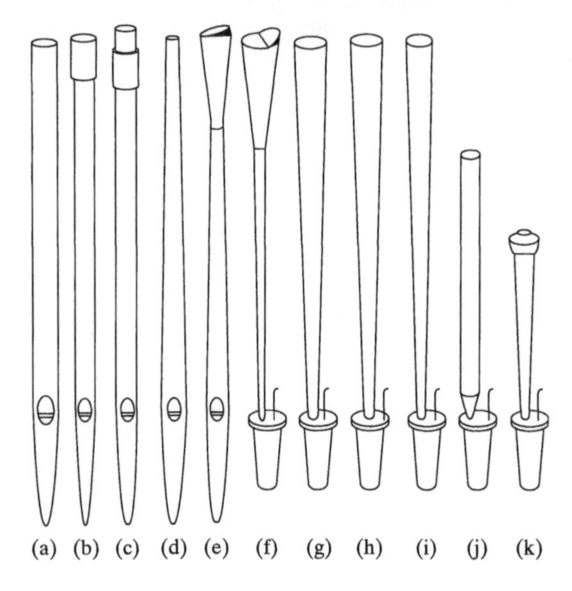

Figure 10-26 Examples of different types of pipes that sound at the same pitch: (a) open diapason; (b) salicional diapason; (c) dulciana; (d) gemshorn; (e) bell gamba; (f) oboe; (g) trumpet; (h) tuba; (i) cornopean; (j) clarinet; and (k) vox humana. The covers on (e) and (f) are used to tune the pipes. (From Audsley, *The Art of Organ Building*, 1965.)

Because the reed in a reed pipe, and not the pipe itself, is tuned, reed pipes react somewhat differently from flue pipes to changes in temperature. A reed pipe is more able to remain in tune despite small variations in temperature than is a flue pipe, so that reed pipes will sound out of tune with flue pipes when the temperature changes. A change of about 5°C (9°F) will change the pitch of flue pipes by about one-sixth of a semitone, sufficient to sound badly out of tune to even most musical amateurs. Even in an age of energy conservation, churches and concert halls are being heated somewhat differently than they might have been otherwise. Because it can take a long period to uniformly heat or cool an entire set of organ pipes, they must sometimes be tuned at some temperature different from "normal." Many good organ tuners will adjust the pitch level at which they tune an organ on any given day so that the pitch will be exactly A = 440 at, say, 70°F.

The variety of sound obtained with different ranks of organ pipes depends on several physical features of the pipes. Flue pipes can be open, to produce all harmonics, or stopped, to emphasize the odd harmonics. In addition, the shape of the languid and lips, the pressure and volume of air, and the material, size, and shape of the pipe are important. For reed pipes, the details of reed material and shape, tuning technique, and type of shallot all affect pitch and tone quality. Some of the different types of organ pipes that sound at the same pitch are shown in Figure 10-26.

SUMMARY

(***Section 10.1***) The **organ** is considered the ancestor of the woodwind instruments; the first truly successful organs used **bellows** to achieve a constant air supply. The medieval Spanish **bagpipe** is the direct parent of the reed instruments. The earliest reed instruments, such as the krummhorn family, used **capped reeds**; an improvement in instruments such as the shawm family allowed the player to obtain greater control of the instrument by placing the reed directly in the player's mouth. Primitive bamboo instruments were the likely ancestors of the recorder and flute families. The recorder is a **fipple flute**, in which a **windway** directs air onto a **knife edge** to create **edge tones**; the flute requires the player to control the air flow onto the edge of an **embouchure hole**, and thus achieves greater volume and expression. The modern flute developed from its Renaissance and baroque parentage; the oboe and the bassoon developed from Renaissance reed instruments, such as the shawm. The clarinet developed during the seventeenth century from the chalumeau. The **Böhm fingering system** of connected rings and keys is used on the modern flute, clarinet, and saxophone, but not on the oboe or bassoon. Many woodwinds are **transposing instruments**; the note that sounds is different from the one that they play on their music. (***Section 10.2***) The tone of woodwind instruments is a function of the source of sound, the size and shape of the **bore**, and the sizes and positions of the **finger holes**. Renaissance and even baroque woodwinds were difficult to play with expression because they lacked the number of holes required for correct **intonation** and **voicing**. It was necessary to compromise on the proper acoustical properties of an instrument in order to build one that was playable. (***Section 10.3***) The recorder derived its English name from "**recording**," the use of a sopranino version called the **bird flageolet** to teach songs to birds. Recorders were simple to make, so they were very popular in the Renaissance and the baroque, at which time they were supplanted by the transverse flute. Renaissance recorders, cylindrical in shape, were typically played in families of three to six sizes. In the baroque, the alto recorder, conical in shape with the small end at the bell, dominated. (***Section 10.4***) The

flute was cylindrical in the Renaissance, became conical in the baroque, and reverted back to a cylindrical bore in the nineteenth century. The Böhm fingering system was developed for the flute, and the development of the flute suffered because of strict adherence, until the 1960s, to the design that Böhm created. The **Cooper scale flute** allowed the performer to play relatively in tune without adjusting the angle at which the air stream from the performer's lips strikes the edge of the embouchure hole. (***Section 10.5***) The **clarinet**, a single reed, cylindrical-bore instrument, developed from the chalumeau by J. C. Denner during the middle of the eighteenth century, is the only instrument that behaves acoustically like a closed tube, so it overblows at a musical interval of a twelfth. Clarinets come in a family of transposing instruments. (***Section 10.6***) The **saxophone**, a single-reed, wide-bore instrument, was invented by Adolphe Sax in the middle of the nineteenth century and has no historical precedent. Because the intonation is very sensitive to the geometry of the mouthpiece, it requires two octave keys, and still remains difficult to play in tune. The saxophone comes in a large family of transposing instruments, and is used more in popular music than in the classical orchestral environment. (***Section 10.7***) The **oboe**, a double-reed instrument with a narrow, conical bore, has never been modified for Böhm fingering, and suffers from the a host of acoustical problems. (***Section 10.8***) The **bassoon**, a very narrow, conical-bore, double-reed instrument, has three octave keys to improve its response and intonation. (***Section 10.9***) The **pipe organ** may have several **manuals** and a number of **ranks** of pipes. The **flue pipe**, which works like a fipple flute, can be either open or stopped; the **reed pipe** works basically like a capped-reed Renaissance wind instrument. A variety of geometries provide a wide range of sounds. The tuning mechanism of some pipes is extremely stable, and has provided contemporary musicologists with much information about early temperaments.

QUESTIONS

1. Examine a flute, a clarinet, and a saxophone. Describe the similarities in their actions and fingerings, which result from the Böhm fingering system they all use.

2. Compare the fingering systems of the clarinet (Böhm) with that of an oboe (not Böhm). Describe particular notes or passages that might be simpler to play on a clarinet than on an oboe.

3. Design and construct a simple transverse "flute" using a piece of plastic tube with a cork or rubber stopper in one end. Calculate the positions of the holes before you start making your instrument, and comment on the accuracy of your calculations after you have finished tuning your instrument.

4. Design and construct a simple "clarinet," using a piece of plastic tubing and a standard clarinet mouthpiece. Be sure to consider the register key. What problems do you observe when playing your instruments, and how do standard flutes and clarinets avoid them?

5. Play a chromatic scale on a Renaissance instrument like a krummhorn, shawm, or recorder. Notice and describe the problems in fingering and voicing. What other problems do you notice?

6. List the basic acoustical features of the various woodwind instruments discussed in the text. Which of these features do you think are most important in determining the sound quality of each instrument?

7. Examine the pipes from a pipe organ. Listen to the sounds of various ranks of pipes and relate them to the different sound-production mechanisms of the different pipes. Listen to and compare electric and tracker organs.

8. Attach a clarinet mouthpiece to a flute, and play it. Describe the tone obtained and explain how this tone quality arises. Discuss intonation problems that may appear.

9. Attach a flute embouchure joint to a clarinet, and play it. Describe the tone obtained and explain how this tone quality arises. Discuss intonation problems that may arise.

10. Draw the standing waves for a typical overtone series in a clarinet with the register key closed, and with the register key open.

   a. Which harmonics will be present in the sound of the lower note (register key closed)?

   b. Which harmonics might be present in the sound of the upper note (register key open)?

   c. If theory were correct, would a clarinet have primarily odd harmonics for its entire range, or would it have all harmonics in the upper register?

11. The oboe and the clarinet are about the same length, but the lowest note on a clarinet is about half the frequency of that on an oboe. Explain why this is so.

12. The flute and the clarinet are about the same length, but the lowest note on a clarinet is about half the frequency of that on an flute. Explain why this is so.

13. Use the lowest note on an alto recorder, F = 349.23 Hz, to determine the speed of sound. *Hint*: Measure the length of the instrument from the bell end to the fipple opening to determine the wavelength. What error might affect this value?

14. Use the notes C = 523.25 Hz and E = 659.26 Hz on your alto recorder to determine the speed of sound in air. *Hints*: Measure the distance from the fipple opening to the finger holes responsible for the notes C and E, and then use the ratio of the frequencies to determine how much longer the effective length of the standing wave in the recorder is than the actual distance between the two openings. Use the corrected wavelength to determine the speed of sound with the equation $s = f\lambda$.

PROBLEMS

1. If the lowest note on a flute is C = 261.63 Hz, how far should its embouchure hole be to its end? Measure a flute and check your calculated value. Explain any difference.

2. If the lowest note on a clarinet is concert D = 146.83 Hz, how far should the tip of its reed be from the end of the bell? Measure a clarinet and check your calculated value. Explain any difference.

3. The range of a flute is C_4 = 261.63 Hz to D_7 = 2349.32 Hz. How many octaves does this encompass?

4. The range of an alto saxophone is $D_3^\flat$ = 138.59 Hz to $A_5^\flat$ = 830.61 Hz. How many octaves does this encompass?

5. If the lowest note on a soprano saxophone is $A_3^\flat$ and the lowest note on a bass saxophone is $A_1^\flat$, what is the ratio of the length of the bass saxophone to that of a soprano saxophone?

6. If the lowest note on a bassoon is $B_1^\flat$ = 58.27 Hz, what would you expect its length to be?

7. What is the length of an open-tube organ pipe that sounds at the pitch C_3 = 130.81 Hz?

8. What is the length of a closed-tube organ pipe that sounds at the pitch G_3 = 196.00 Hz?

REFERENCES

ANDERSON, PAUL-GEHARD. *Organ Building and Design.* Translated from the Danish by Joanne Curnutt. New York: Oxford University Press, Inc. 1969. One of the more recent books on the organ; it contains some good information on recent innovations.

AUDSLEY, GEORGE A. *The Art of Organ-Building.* Volumes I and II. New York: Dover Publications, Inc., 1965. Contains a wealth of detailed information on organ construction and operation.

BACKUS, JOHN. *The Acoustical Foundations of Music.* 2nd ed. New York: W. W. Norton & Co., Inc., 1977. See his reference list.

BAINES, ANTHONY. *Woodwind Instruments and Their History.* New York: W. W. Norton & Company, Inc., 1963. A classic book on woodwind instruments written for musicians.

BARNES, WILLIAM. *The Contemporary American Organ: Its Evolution, Design and Construction.* 3rd ed. New York: J. Fischer and Bros., 1937. The organ classic, containing a vast wealth of excellent information; easy to read and contains many excellent drawings and photographs.

BATE, PHILIP. *The Flute: A Study of Its History, Development and Construction.* New York: W. W. Norton & Company, Inc., 1969. One in a set on the instruments of the orchestra, written for the musician.

BENADE, ARTHUR H. *Fundamentals of Musical Acoustics.* New York: Oxford University Press, Inc., 1976. See his reference list.

———. "The Physics of Woodwinds." *Scientific American,* October 1960. An excellent article dealing with the basic acoustical features of woodwind instruments, written by one of the foremost authors in the field.

BÖHM, THEOBALD. *The Flute and Flute Playing.* New York: Dover Publications, Inc., 1964. The magnum opus for flutes, in which the design of the flute dimensions and the Böhm ring and key system is explained.

COLTMAN, JOHN W. "Acoustics of the Flute," *Physics Today,* November 1968, p. 25. A good general article on flute acoustics.

———. "Effect of Material on Tone Quality." *Journal of the Acoustical Society of America* 49, no. 2, part 2 (1971): 520–523. This excellent article deals with the physics of the effect of various types of materials on the sound of the flute, including some experimental data.

———. "Flute Scales: Pitch and Intonation (Part I)." *The Instrumentalist,* May 1976, p. 39. Discusses the early history of the flute and gives background for three additional articles in the set discussing flute acoustics.

———. "The Intonation of Antique and Modern Flutes (Part IV)." *The Instrumentalist,* March 1975, p. 77. A discussion of tuning problems in the flute leading up to the proposal by Murray on how to improve the intonation of modern flutes by modifying the Böhm finger-hole geometry.

———. "Sounding Mechanism of the Flute and Organ Pipe." *Journal of the Acoustical Society of America* 44, no. 4 (October 1968): 983–992. Deals with the edge tone mechanism, the source of sound in the flute, recorder, and organ flue pipe.

FLETCHER, N. H., and T. D. ROSSING. *The Physics of Musical Instruments*. New York: Springer-Verlag, 1990 (Hardcover), 1993 (Softcover). An excellent, up-to-date book containing a wealth of detailed information about virtually all types of musical instruments.

LANGWILL, LYNDESAY G. *The Bassoon and Contrabassoon*. New York: W. W. Norton & Company, Inc., 1965. The basic volume on the bassoon, written for the musician as part of the Benn/Norton series on musical instruments.

RENDALL, F. GEOFFREY. *The Clarinet: Some Notes upon Its History and Construction*. 2nd ed. London: Ernest Benn Ltd. New York: W. W. Norton & Company, Inc., 1957. Contains the basic historical and contemporary information on the clarinet, written for musicians.

TOFF, NANCY. *The Development of the Modern Flute*. New York: Taplinger Publishing Company, Inc., 1979. Discusses recent developments in the flute, including the Cooper scale modification to the Böhm flute.

Chapter 11

Brass Instruments

The instruments of the contemporary brass family are very similar in their acoustical properties and problems. We shall look briefly at the history of brass instruments, which contains considerably more variety than remains in the contemporary brasses, and then discuss some of the basic acoustics of brasses. Finally, we shall consider some specific modern brass instruments and survey some of the features of the remaining members of the family.

11.1 History of Brass Instruments

The earliest brass instruments were probably created by primitive people buzzing their lips into the end of animal horns. Many centuries ago this simple "horn" was improved by adding a specially shaped mouthpiece to one end of a tube and a flared "bell" to the other end. These ancient horns have been depicted and discussed throughout antiquity, and are, along with the harp, the instruments most often pictured with angels. Figure 11-1 shows some ancestors of the contemporary brass instruments.

In the Middle Ages and early Renaissance the vast growth in the varieties of brass instruments began. Two developments of that time were particularly important: the use of finger holes and the invention of the slide.

Instruments of the *cornetto* family consist of conical tubes, usually curved, blown at the narrow end with a very small mouthpiece. These instruments had finger holes and were fingered similarly to the recorder and other woodwinds, so they are sometimes classified with the Renaissance woodwinds. Paintings of the period often show the cornetto played out of the side of the mouth, not the center as with contemporary brasses. Although this type of instrument was known as early as the eighth century, it reached its pinnacle in the late Renaissance and achieved a reputation for incredible technical capacity and flourish in the performance of typical Renaissance divisions and

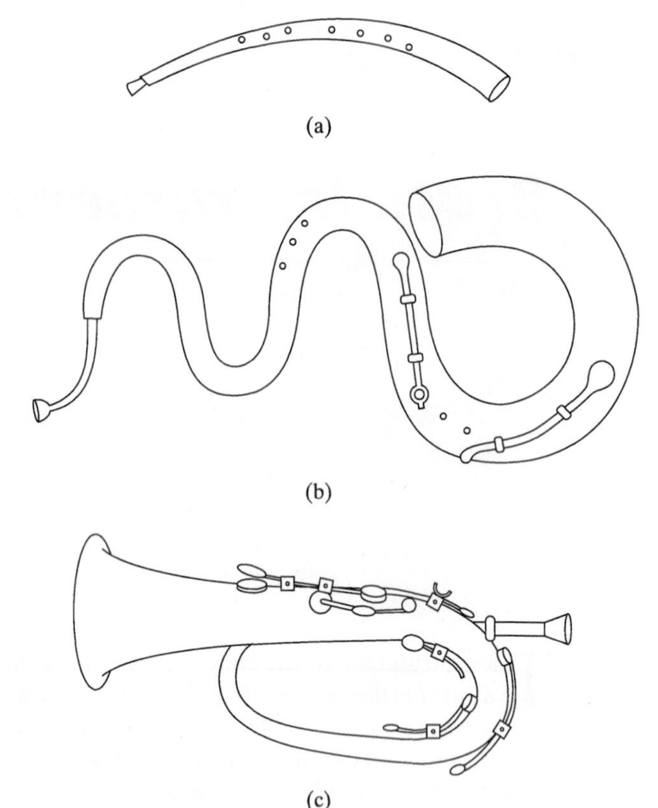

Figure 11-1 Some early "brass" instruments: (a) the cornetto, and (b) the serpent, two members of the fingered Renaissance "brasses," and (c) the keyed bugle. (Part (c) from *The History of Musical Instruments* by Curt Sachs. Copyright 1940 by W. W. Norton & Company, Inc., renewed © 1968 by Irene Sachs. Used by permission of W. W. Norton & Company, Inc.)

ornamentation. The cornetto came in a family of sizes, as did the Renaissance woodwinds. The bass instrument in the cornetto family is the serpent, which has a large bore with several curves and a broad bell. The serpent is particularly difficult to play, more so even than the small cornetto, owing to the variation in pitch that can occur when any note is fingered. These difficulties are also present to a lesser extent in the cornetto and are probably why there are very few excellent cornetto players at present.

The slide, first introduced in the fifteenth century, was used on both the trumpet-size and trombone-size instruments. The early slide trombone was called the *sackbut*, derived from the French "pull-push," had a narrow bore and a small bell, and came in several sizes. The sackbut provided a more reliable bass than any of the woodwinds or the serpent, and consorts of cornettos (sometimes called *cornets*, but not the same as the modern band cornet) and sackbuts were very common in the late Renaissance. The slide trumpet was less popular than the sackbut, probably because of the immense popularity of the cornetto. It was used into the eighteenth century and even appeared in some of Bach's scores. The trombone, as we know it today, developed directly from the sackbut, with an increase in bore diameter and a larger bell, to provide a somewhat louder and more brilliant sound.

The modern trumpet, however, went through several additional steps in its evolution. Much of the trumpet music in the baroque involved a highly developed technique of playing by which an almost complete diatonic scale could be obtained without using valves or a slide. The baroque trumpet was roughly twice the length of the contemporary trumpet, so its overtones have about half the spacing of its modern counterpart when played in their normal ranges. Furthermore, the typical baroque trumpet technique called for the instrument to be played primarily in its upper octave, where the overtones are spaced as closely as a whole step over much of the range. This very high tessitura trumpet playing became a fine art in the high baroque, but was almost immediately lost with the advent of the classical orchestra, which either called for no trumpets or had the trumpets simply playing the notes of the overtone series in their lower range. The orchestral literature for the brass instruments remained about the same in style and technique until early in the nineteenth century, when the trumpet and the horn were fitted with valves. The trumpet used three valves, which lowered the pitch by two, one, and three semitones, respectively, and which could be operated together to obtain the cumulative effect of lowering the pitch by six semitones. An entire chromatic scale could be obtained in this way, even in the lower register. This development dramatically expanded the capability of the trumpet and its role in the orchestra, beginning in the early romantic era.

The trumpet and trombone both use basically cylindrical tubes. As in the woodwinds, conical tubes are also used in brasses, although the acoustical difference between conical and cylindrical tubes is not as dramatic for brass instruments as it is for reed instruments. Both conical and cylindrical brass instruments support all harmonics of a fundamental. Conical reed instruments support all harmonics, whereas cylindrical reed instruments, such as the clarinet, support predominantly odd harmonics. The most important early brass instrument using a conical tube was the French horn, which was given its name by the British when it was introduced to England during the seventeenth century. The French horn (or simply "horn") is actually a modified conical hunting horn, which is, like the baroque trumpet, played in its upper range where the harmonics are closely spaced. Because the horn is over twice the length of the baroque trumpet, it plays more than one octave lower in its normal range. Horn technique developed during the eighteenth century, and involved inserting the hand into the bell to change the frequencies of the notes; this is the source of the characteristic muffled tone quality of the horn. This hand-in-bell stopping technique greatly extended the useful diatonic range of the French horn, and it could even be played chromatically over much of its range. The hand-stopping technique was not entirely satisfactory, however, and when the valve was invented it was immediately applied to the horn as well as the trumpet.

An additional family of conical brass instruments arose in the nineteenth century, consisting of the modern cornet (not the Renaissance cornetto), the flügelhorn, the euphonium, and the tuba. These instruments, played normally in the lower part of their ranges, use valves to achieve complete scales and differ primarily in their length. The flügelhorn is sometimes used today in marching bands, and the euphonium is used occasionally in the symphony orchestra. The baritone, a conical horn that plays at the same pitch as the euphonium, but is slightly smaller in diameter and therefore has a slightly softer sound, is used almost solely in the concert band. The tuba is used in the symphony orchestra and the concert band, and the sousaphone, which is acoustically similar to the tuba but easier to carry, is used in concert and marching bands.

Figure 11-2 The trumpet (top), the cornet (center), and the flügelhorn (bottom). (Courtesy of Conn-Selmer, Inc.)

Some of the instruments in the cylindrical and the conical brass families are shown in Figures 11-2, 11-3, 11-4, and 11-5. Figure 11-2 shows the conical cornet and the flügelhorn compared with the cylindrical trumpet. The trombone and bass trombone (see Section 11-4), and the double horn (see Section 11-5), are shown in Figure 11-3. The euphonium and the tuba are shown in Figure 11-4, and the sousaphone is shown in Figure 11-5.

Another nineteenth-century innovation, now obsolete, was the use of keys to cover holes, particularly on the bugle. The bugle is a conical instrument, without valves, that is constructed in different sizes to form a family. Whereas the cornet and its relatives used three valves, the keyed bugle used about five keys to obtain the chromatic scale. Like the Renaissance cornetto, the keyed bugle was somewhat more difficult to play than its valved relatives, and the sound was not of the same quality and loudness. The keyed bugle became obsolete by about the end of the nineteenth century, leaving valves as the primary mechanism for obtaining the chromatic scale, along with the slide in the case of the trombone. The modern-day bugle, used in the military and drum-and-bugle marching bands, is an unvalved or a single-valved cone with a mouthpiece and wide-flaring bell. Several different sizes are used in some large drum-and-bugle corps.

There are many types of valves, though they can generally be classified under two categories, *rotary* or *piston*, as illustrated in Figure 11-6. In a piston valve a straight piston is slid up and down so that at its depressed position an additional section of tubing extends the overall acoustical length of the instrument. When the piston is released, that section of tubing is disconnected. In a rotary valve, the motion of the finger is converted into a rotation of a valve that similarly introduces an additional length of tubing. Three main valves are used on all valved brass instruments. The first valve lowers

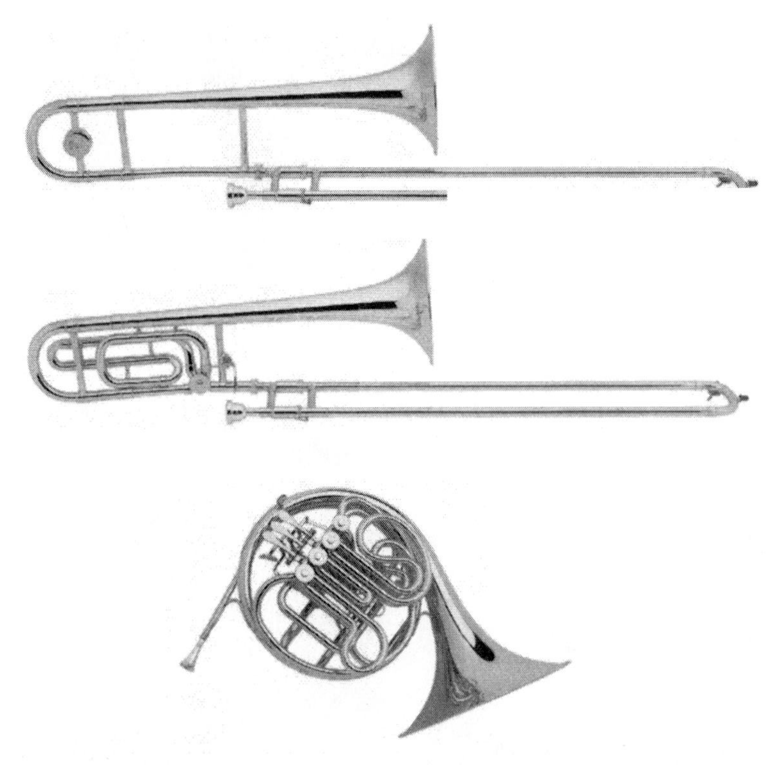

Figure 11-3 The trombone (top), the bass trombone (center), and the double horn (bottom). (Courtesy of Conn-Selmer, Inc.)

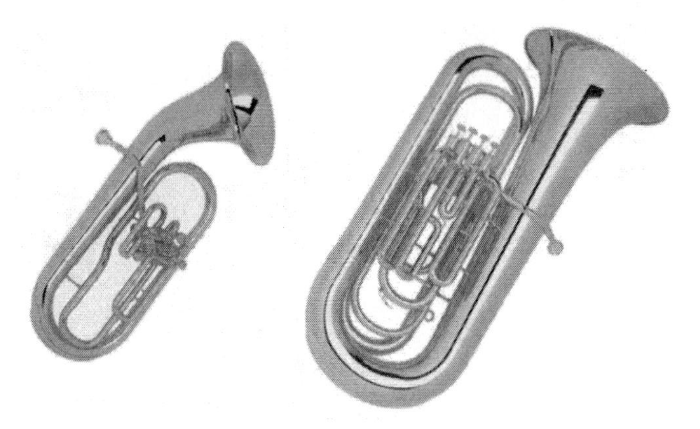

Figure 11-4 The euphonium (left) and the tuba (right). (Courtesy of Conn-Selmer, Inc.)

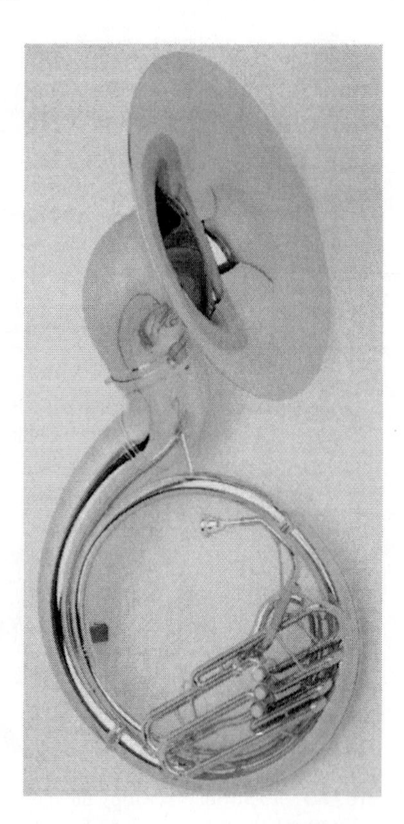

Figure 11-5 The sousaphone. (Courtesy of Conn-Selmer, Inc.)

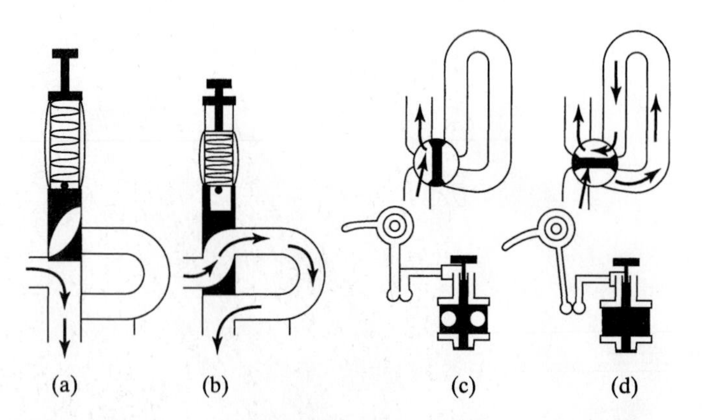

Figure 11-6 Piston valve, (a) open and (b) pressed; and rotary valve, (c) open and (d) pressed. (From *The History of Musical Instruments* by Curt Sachs. Copyright 1940 by W. W. Norton & Company, Inc., renewed © 1968 by Irene Sachs. Used by permission of W. W. Norton & Company, Inc.)

Table 11-1 Ranges of Selected Brass Instruments

Instrument	Lowest note*	Approximate highest note
D trumpet	$G_3^\sharp$	D_6
C trumpet	$F_3^\sharp$	C_6
B^b trumpet (cornet)	E_3	B_6^b
Flügelhorn in B^b	A_2	E_6^b
French horn in F	B_1	F_5
French horn in E^b	A_1	E_5^b
Trombone in B^b	$E_2(E_1)$	D_5
Bass trombone in F	$B_1(B_0)$	F_4
Euphonium (baritone)	B_1^b	B_4^b
Tuba in BB^b (sousaphone)	B_0^b	B_3^b

* Lower limits of pedal tones are given in parentheses.

all the notes of the open (unvalved) instrument by two semitones, the second by one semitone, and the third by three semitones. Today most French horns and some euphoniums and tubas are made with rotary valves, while the other common brass instruments usually use piston valves. Although the valve apparently has won the war with keys as the most acceptable method for obtaining the chromatic scale on brass instruments, valves have a basic intonation defect that will be discussed in the next section.

By the middle of the twentieth century, families of brass instruments had largely disappeared, whereas woodwind families remained. The trumpet is used in the modern symphony orchestra, along with the French horn, trombone, and tuba, and the concert band also uses the cornet, baritone or euphonium, and sousaphone. The ranges of some of these instruments are shown in Table 11-1.

Many of the brass instruments are, like some woodwinds, transposing instruments. The brasses, also like the woodwinds, can be classified by their shape: cylindrical (the trumpet and trombone) or conical (the cornet, flügelhorn, baritone, euphonium, and tuba). The French horn is theoretically conical, but the modern French horn includes a long section of cylindrical tubing.

11.2 Sound Production in Brass Instruments

The woodwinds have a wide variety of acoustical properties and sounds; the brass instruments, on the other hand, have similar tones and tone production mechanisms. The only modern brass instrument that has somewhat unusual acoustics is the French horn. We shall begin with a discussion of how the buzzing of the lips into the end of a cylindrical tube produces sound and then summarize the effects of the mouthpiece and bell. Finally, we shall consider the acoustical properties of the valves and problems that occur when the valves are used in combination.

The simplest "brass" instrument, a length of straight tube, can be played by buzzing the lips into its end. The standing-wave oscillation in the tube drives the lips just as the oscillating air column drives the reed in the woodwinds. When a burst of air is emitted from the lips into the tube, there is a decrease in pressure in the space between the lips

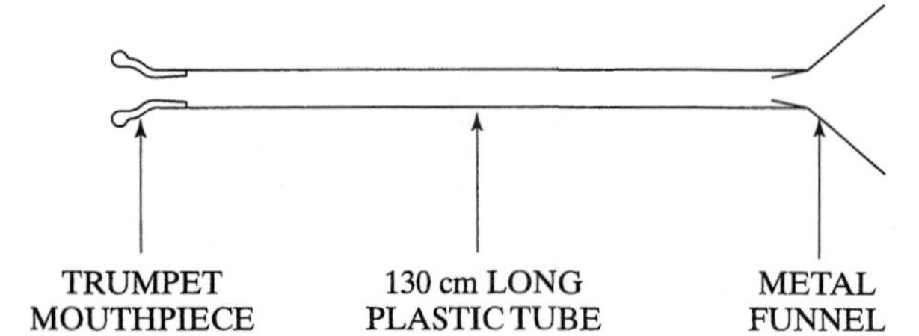

Figure 11-7 Simple experimental "trumpet."

where the air flow is most rapid, and the lips are drawn together owing to Bernoulli's principle and lip tension. Eventually, the burst reflects back to the lips and forces them open again, allowing another air burst to be emitted from the lips. Then, the process repeats. The lip end behaves acoustically like a closed end of an air column, as a reed end does, so a velocity node (or a pressure antinode) appears there, whereas the bell end behaves acoustically like an open end and has a velocity antinode (or a pressure node). The tube behaves acoustically like a closed tube, with the fundamental mode having a wavelength approximately four times the length of the tube.

We can perform a series of experiments in which we first blow into a length of plastic tube alone, and then repeat the experiment after inserting a trumpet mouthpiece into one end of the tube and a metal funnel into the other end, as shown in Figure 11-7. Here we shall use a tube about 130 cm long, producing a fundamental frequency of about 66 Hz. The possible resonant modes in the unmodified tube are the odd harmonics, as shown in Figure 11-8. Adding the mouthpiece to one end and the funnel to the other end adjusts the frequencies of the harmonics, as shown in Figure 11-9. The result is to increase slightly the effective length of the tube and to cause the instrument to behave acoustically approximately like an open tube of the new length. It then produces a set of resonances including *all* the notes of the overtone series. The fundamental mode of this new overtone series is slightly lower in frequency than twice the fundamental frequency of the original tube.

The size and shape of the mouthpiece predominantly affects the frequencies of the upper resonances, causing these resonances to decrease in frequency and thus reducing their spacing. The bell has two important effects: (1) it provides the proper coupling, or impedance "match," between the instrument and the outside air to transfer the sound most efficiently, making the sound considerably louder, and (2) it modifies the frequencies and stability of the resonances, especially the lower ones. The fundamental is extremely difficult to play, and its frequency is too low relative to the other notes of the overtone series produced by the "tube plus mouthpiece plus bell" system. The effect of the bell on the lowest playable note is to raise its pitch from that of the third harmonic of the original closed tube (G) to that of the second harmonic of the final overtone series (B^b) in Figure 11-9. The third harmonic note (F) in Figure 11-9 is

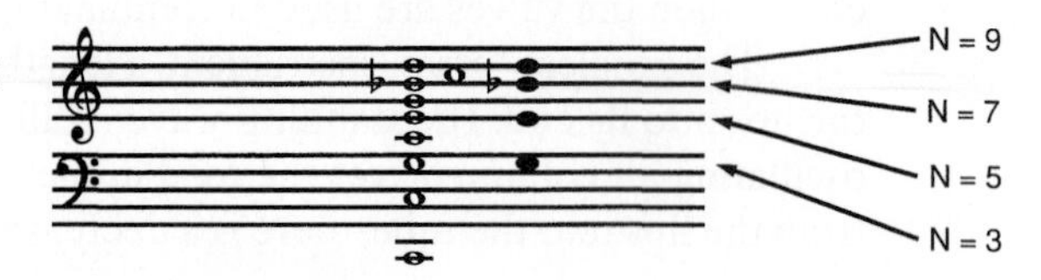

Figure 11-8 Solid notes are those obtained by buzzing the lips into one end of a 130-cm tube. All the notes of the complete overtone series of the fundamental of a 130-cm long closed tube are shown to the left.

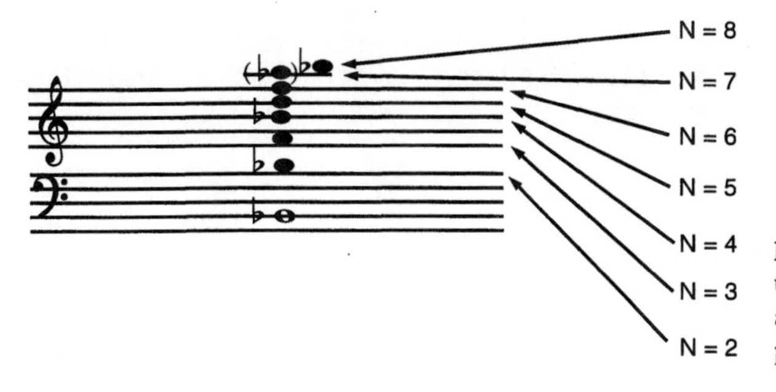

Figure 11-9 Notes obtained when a trumpet mouthpiece and a metal funnel are inserted into the ends of the 130-cm plastic tube of Figure 11-3.

also moved slightly upward by the bell. By varying the size and shape of the mouthpiece and the flare and length of the bell, the frequencies of the resonances can be adjusted to most nearly match the harmonics of the overtone series, except for the fundamental. The player can adjust the frequencies of the notes by varying the lip tension to tune the notes.

The most obvious effect of the bell is to increase the volume of sound from the instrument. The bell provides a good impedance match for conducting the sound vibrations in the horn to the surrounding space most efficiently. This is very clear in the case of our plastic tube described previously. In addition, the bell helps to project the sound in a single direction.

The frequencies of the notes produced by a high-quality trumpet are given in Table 11-2. Although they are neither perfect harmonics nor notes of the equal-tempered scale, they are considerably closer to exact harmonicity than the resonant frequencies of our simple plastic tube with mouthpiece and funnel. This is because the size and shape of the mouthpiece and bell of the trumpet have been carefully chosen to provide the closest approximation to the notes of the harmonic series. It is clear that to play a brass instrument in tune (even without using the valves) it is necessary to listen carefully and adjust the pitch of the notes during performance.

We now have developed a brass instrument that can play only the notes of the overtone series. To play the complete chromatic scale, valves must be added. Pressing a valve increases the effective length of the instrument by introducing an extra section of tube. Each valve introduces a different length of tubing and changes the pitch accordingly.

Table 11-2 Frequencies Produced by a Trumpet with No Valves Pressed*

Harmonic number N	f_n (Hz)	f_n/N (Hz)
2	230	115.0
3	344	114.7
4	458	114.5
5	578	115.6
6	695	115.8
7	814	116.3
8	931	116.4

*The deviations from harmonicity are less than those in the plastic tube trumpet.

Table 11-3 Frequency Ratios on Ideal Valved Brass Instruments in Which (a) All Three Valves Are Individually Made to the Correct Length, (b) Valve 1 + 3 Combination is Made Correct, and (c) Valve 2 + 3 Combination is Correct

			Valve configurations tuned exactly*		
			(a)	(b)	(c)
Note	Ratio (equal-tempered scale)	Valve configuration	1, 2, 3	1, 2, and 1 + 3	1, 2, and 2 + 3
C	2.0000	0	2.0000	2.0000	2.0000
B	1.8877	2	1.8877	1.8877	1.8877
B♭/A♯	1.7818	1	1.7818	1.7818	1.7818
A	1.6818	1 + 2	1.6922	1.6922	1.6922
		3	1.6818	1.6946	1.6660
A♭/G♯	1.5874	2 + 3	1.6017	1.5725	1.5874
G	1.4983	0	1.4983	1.4983	1.4983
		1 + 3	1.5248	1.4983	1.5118
G♭/F♯	1.4142	2	1.4142	1.4142	1.4142
F	1.3348	1	1.3348	1.3348	1.3348
E	1.2599	1 + 2	1.2677	1.2677	1.2677
		3	1.2599	1.2358	1.2481
E♭/D♯	1.1892	2 + 3	1.1999	1.1780	1.1892
D	1.1225	1 + 3	1.1423	1.1225	1.1326
D♭/C♯	1.0595	1 + 2 + 3	1.0927	1.0746	1.0839
C	1.0000	0	1.0000	1.0000	1.0000

*Boxes indicate notes that are out of tune.

First, consider the effect of each valve individually, assuming that the open notes (no valves pressed) have perfect intonation in the equal-tempered scale. (They do not, as Table 11-2 shows.) Table 11-3 shows the frequency ratios of the notes of a valved instrument if the open notes are perfectly in tune in the equal-tempered scale. Also shown in the table are the relative frequencies of the notes produced by pressing each valve individually. Table 11-4 gives the number of half-steps the pitch is lowered for

Table 11-4 Change of Pitch of Overtone Series Due to Trombone Slide Position and Trumpet Valve Configuration

Trombone slide Position	Trumpet valve combination	Pitch reduction in half-steps	Musical interval
1	0	0	Unison
2	2	1	Half step
3	1	2	Whole step
4	1 + 2	3	Minor third
5	2 + 3	4	Major third
6	1 + 3	5	Perfect fourth
7	1 + 2 + 3	6	Augmented fourth

various valve configurations. To lower the pitch by one half-step, a length of tube equal to 5.95 percent of the length of the open tube must be added; this is done with the second valve. The first valve lowers the pitch by one full step or two half-steps by adding 12.2 percent to the length of the open tube. The third valve adds 18.9 percent to the length of the tube to obtain a pitch reduction of three half-steps or a minor third. If it were only necessary to use one valve at a time, each could be made the proper length to produce its notes exactly in tune with respect to the open notes by choosing the proper length of tube. However, there is a problem when the valves are used together.

If the first valve is pressed, lowering the pitch of an open note by two half-steps, the second valve is used to obtain the note an additional half-step lower. To lower the pitch by this additional half-step, however, the valve must add 5.95 percent to the length of the open horn *plus* the length of the first valve. Because the second valve is already made equal in length to exactly 5.95 percent of the open horn, it is therefore short by 5.95 percent of the length of the first valve, or about 0.7 percent too short. Thus, with valves one and two pressed, the instrument would produce notes about 0.7 percent high in frequency, about one-eighth of a half-step, a mistuning easily detected by a good musician. The notes obtained by valves 1 and 2 in combination, which are high in pitch, could be produced exactly by using valve 3 alone if it were made to the exact length. If valves 2 and 3 were used together, the notes would be 5.95 percent of 18.9 percent high, or about 1.1 percent high, over one-sixth of a semitone. If valves 1 and 3 were used together, the notes would be over one-third of a semitone high; if all three valves were used simultaneously, the notes would be over one-half a semitone high in pitch.

Table 11-3 shows the frequency ratios obtained for different valve-tuning compromises. For instance, the column for valves 1, 2, 3 all tuned exactly shows the ratios of frequencies for an ideal instrument with valves that individually lower the pitch from the unvalved or open configuration by exactly a whole step, a half-step, and a minor third. As we have seen, any valve combination will then deviate from the equal-tempered scale. The column for 1, 2, and 1 + 3 shows the ratios of frequencies obtained when the first two valves are tuned individually, but the third valve is tuned so that, in combination with valve 1, the interval of a tempered perfect fourth is obtained. The various tuning schemes produce different notes in tune.

This problem could be approached in several other ways. First, the brass player must adjust the pitch of the notes by changing lip tension. This will be necessary, anyway, because even the open notes (that is, those that don't require pressing any valves) on the instrument are not exactly in tune. However, the errors introduced by the valve-length discrepancies become so large that "lipping" the note into tune is very difficult or impossible and cannot be used. The tube sections inserted by the individual valves could be made longer so that each is low in pitch when used alone but more nearly in tune when used in combination, but this is usually not done. A third alternative involves moving an additional slide out when the valves are used in certain combinations to lower the pitch of those notes, particularly when the third valve is used. Another possibility is to provide a fourth valve, which is actually an elongated third valve, to be used in lieu of the third valve whenever it is to be used in combinations, or to provide an additional valve that lowers the pitch of the instrument by a perfect fourth. The corrections most often used will be discussed in the sections dealing with the various instruments.

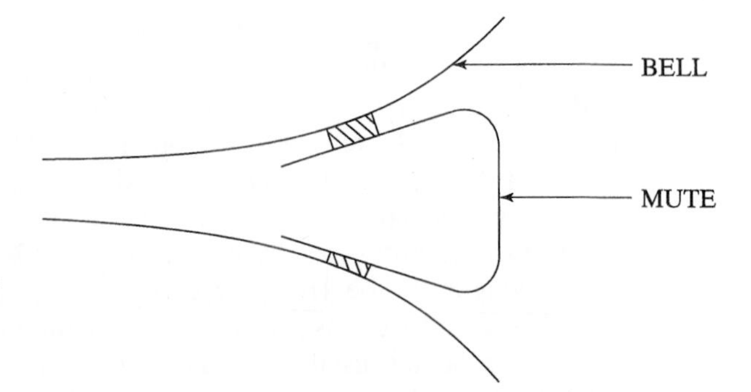

Figure 11-10 A mute for a brass instrument, in position in the bell. The cross hatches indicate the positions of the small cork retainers, used to hold the mute in place.

Brass instruments, like woodwinds, are made in two basic bore shapes. The trumpet and trombone are cylindrical throughout much of their length, with the end section gradually flaring to the bell. The other common brass instruments are flared throughout and can be considered approximately conical. One effect of the flare, in the bell region for all brasses and throughout the instrument for the conical brasses, is that the higher-frequency harmonics are not as readily reflected back inside the horn. The standing waves for higher harmonics extend further out than those for the lower harmonics, because of the more nearly uniform change in impedance of the flared bore, and a greater fraction of the energy in the wave is radiated before forming a standing wave for the higher frequencies. Therefore, the standing wave within the instrument does not build up to as great an amplitude for these higher harmonics. This yields a more mellow tone for the conical instruments, such as the cornet and the euphonium, relative to their cylindrical counterparts, the trumpet and the trombone.

A mute, shown in Figure 11-10, is used to muffle the sound of many brass instruments in a way similar to a French horn player inserting his hand into the bell. Because the mute leaves only a small opening at the end of the instrument, the higher harmonics pass out of the instrument more readily than do the lower harmonics. Different types of mutes can produce somewhat different sounds.

11.3 Trumpets

The trumpet is the treble instrument in the contemporary brasses, and is usually a transposing instrument. Whereas the trumpets in C and D were popular in the baroque, the B^b trumpet is used in bands and orchestras today, with the C trumpet used less often. In a B^b trumpet the notes sound one whole step lower than written; that is, a C played on a B^b trumpet sounds at B^b on the piano, and so on. Pushing down the second valve lowers the entire harmonic series by one half-step; the first valve lowers the notes by one full step; the third valve alone or the first and second valves together, a minor third; the second and third valves together, a major third; the first plus third, a perfect fourth; and all three together, an augmented fourth. The notes obtained for each of these valve configurations are given in Figure 11-11.

If the three valves were made the appropriate length to be in tune individually, the combination notes would be very sharp, as we have seen. Therefore, only the first

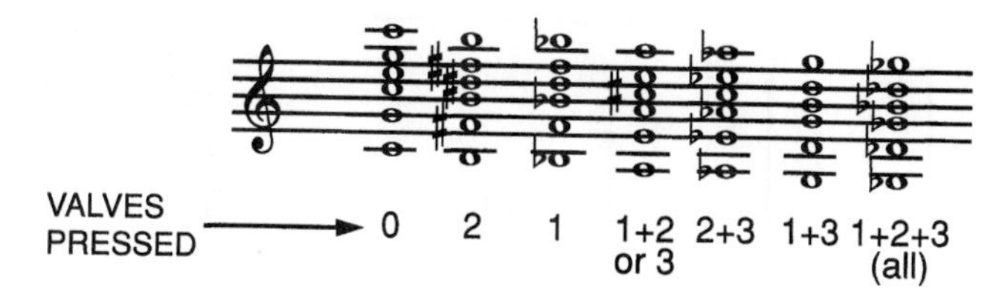

Figure 11-11 Notes that can be played using each of the seven trumpet valve configurations.

and second valves are made the correct length to be exactly in tune when used individually, and the third valve is often made of such a length that the notes played with the second and third valves in combination are exactly in tune. Thus the third valve used alone will produce notes that are too low in pitch, so the third valve is not used as an alternative for the first and second valves in combination. The notes produced by the first and second valves in combination are still sharp, however, and must be adjusted in performance. The intonation of this configuration of valve lengths is shown in the right column of Table 11-3.

The notes produced when the first and third valves are used together, and especially the notes produced when all three valves are used together, are still quite sharp and require special treatment. Most trumpets are equipped with a tuning slide for each valve, which can be used to adjust the length of tube added to the horn when that valve is pressed. In general, these tuning slides are not adjusted after they are set to their proper positions. However, many trumpet players will have their third valve tuning slide fixed so that it moves very easily with one of the fingers on the left hand, which is normally used only to hold the instrument. This slide can then be pushed out for notes using either the combination of first and third valves or all three valves, when these notes are to be held for a long time and the intonation defect is likely to be noticed. Most nonprofessional trumpets will not have a third valve tuning slide that can be moved easily, so it is usually necessary to align the two tubes of the slide so that it can be moved easily during performance. Repair shops are usually familiar with this standard trumpet modification. The frequency ratios of notes played in the lower octave of a trumpet with the third valve length chosen in the preceding manner are shown in Table 11-3.

The assumption in Table 11-3 that the open notes (the overtone series) are in tune is not true. The third is flat (low) compared with the note of the equal-tempered scale, whereas the fifth is slightly sharp (high). Each of the series of notes for a given valve configuration deviates from the equal-tempered scale in this way. A trumpet player must compensate for these intonation defects, using a combination of lip pressure and alternate fingerings, during performance.

The resonance curve of the trumpet, shown in Figure 11-12, provides insight into the stability of the notes. Any particular note on the trumpet becomes stable when all of the harmonics that exist for that note are very closely in tune. The group of overtones for each note, called its *regime of oscillation*, provides stability if more of them exist, if they are strong, and if they are closely aligned with the overtone series of the note sounded. The figure shows the resonances of the trumpet with no valves pressed;

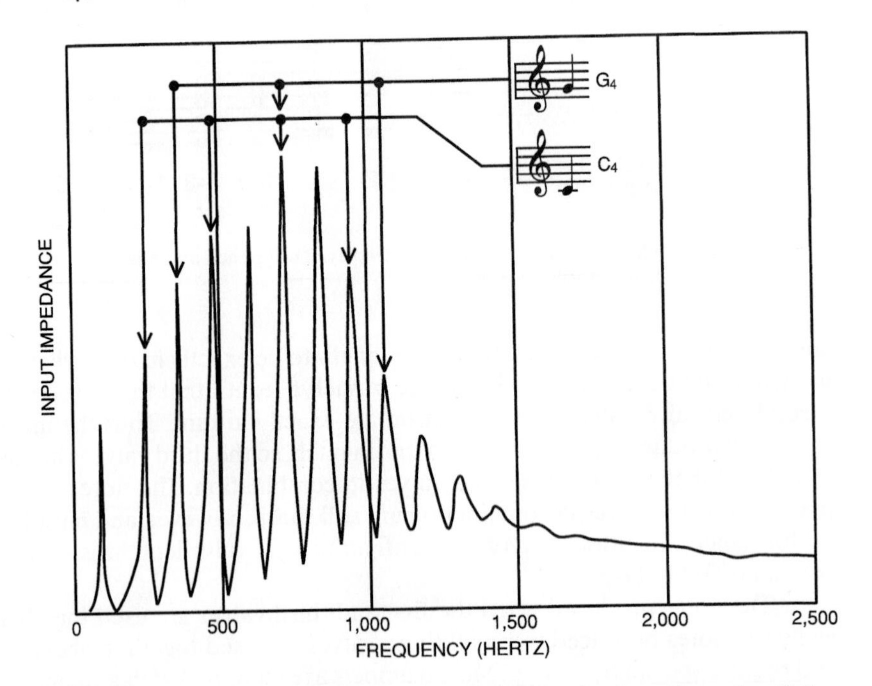

Figure 11-12 Resonance curve for a modern trumpet, showing the regimes of oscillation for the notes C_4 and G_4. (From "The Physics of Brasses" by Arthur H. Benade, *Scientific American*, July, 1973. Graphics by Dan Todd, used with permission.)

at the top, the members of that overtone series that comprise each note's regime of oscillation are identified. The notes C_4 and G_4 are strong, because they have several strong, well-tuned overtones. The pedal note, C_3, is weak, because its fundamental is weak and low in frequency compared with the other members of its overtone series. On the other hand, the fundamental of the harmonic series is stronger and much better in tune for the trombone, so the pedal note can be readily played and is used in considerable music literature.

The range of the orchestral B♭ trumpet is given in Table 11-1; the trumpet and the cornet are shown in Figure 11-2.

11.4 Trombones

The trombone is one of the simplest brass instruments, acoustically and mechanically, and the only brass or woodwind instrument that can be played perfectly in tune without requiring adjustment of the player's embouchure. The slide of the trombone replaces the valves of the trumpet, giving infinitely adjustable pitch control. Continuous sliding notes or glissandos can be played on the trombone. There are seven approximate slide "positions" on the trombone, shown in Figure 11-13. The correspondence between slide position and valves pressed on a valved instrument is shown in Table 11-4. The trombone, like the trumpet, is a cylindrical instrument, and because the trombone has an effective length twice that of the trumpet, it sounds one octave lower in first position than the trumpet with no valves pressed.

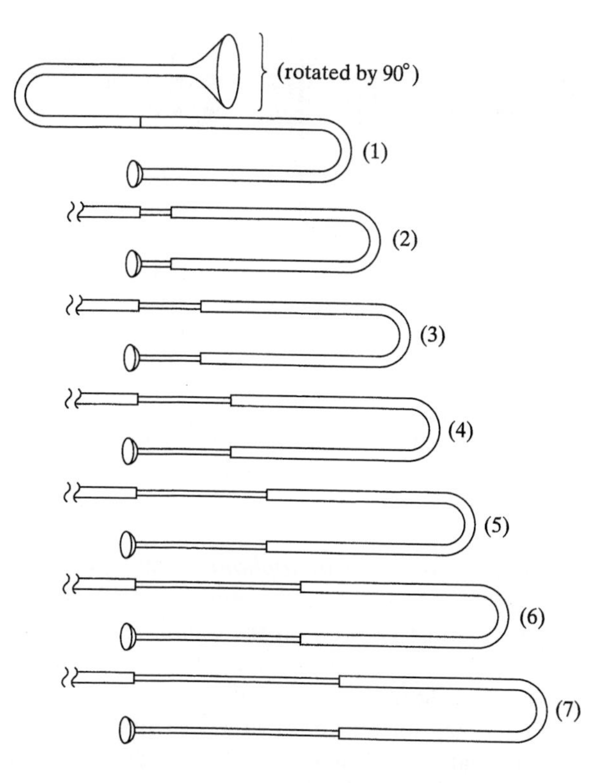

Figure 11-13 The seven slide positions on the trombone. The increase in slide length between positions 6 and 7 is approximately one-third longer than that between positions 1 and 2.

Whereas the trumpet has a usable range down to the second harmonic for each valve configuration, the larger size of the trombone mouthpiece permits the lips to vibrate at the fundamental frequency of the trombone. The very low notes, known as *pedal tones*, are quite stable and usable. The notes playable on a trombone in each of the seven positions, including the pedal tones, are shown in Figure 11-14.

Some modern professional-quality trombones can be shortened slightly above first position during performance by pulling the slide in against the force of a spring. This allows the player to obtain a fifth harmonic (the D above middle C), which is sharp with respect to the exact harmonic of the overtone series but in tune with the equal-tempered scale.

The trombone is not a transposing instrument. The standard orchestral and band instrument is called the tenor or B^b trombone, because the fundamental note of the

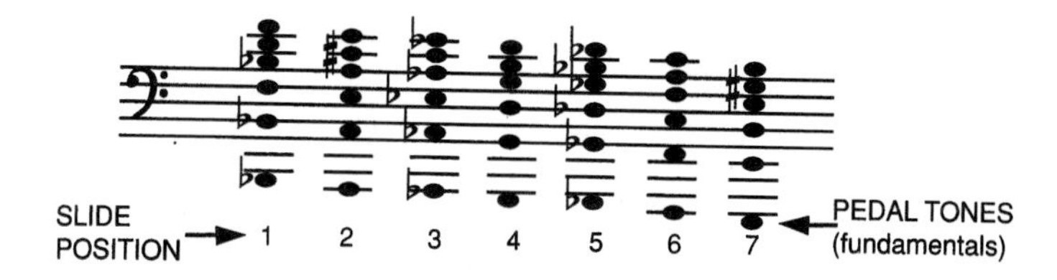

Figure 11-14 Notes that can be obtained in the seven trombone slide positions.

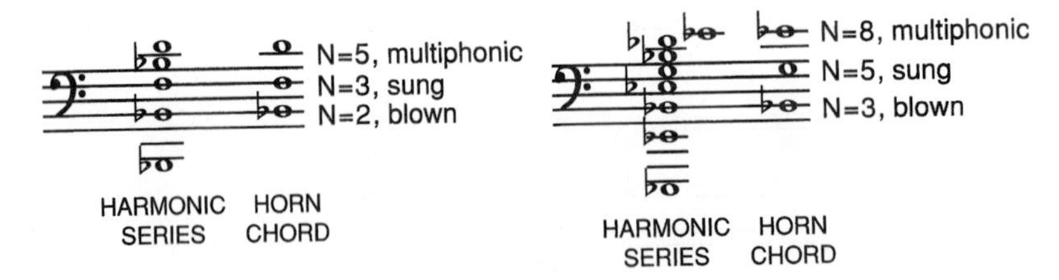

Figure 11-15 Examples of horn chords, or multiphonics. The notes of the relevant harmonic series are shown to the left of each chord. Blowing the lowest note of the chord while singing the middle note produces the multiphonic shown for each case.

instrument in first position is a B$^{\flat}$. The bass trombone, playing a fourth lower, is either built larger by a factor of one-third or equipped with a *valve*, operated with the left hand, that introduces a section of tube to lower the pitch by a musical fourth. When the valve is engaged, the slide positions must all be lengthened to account for the additional length of the tube in first position.

An interesting effect, called *horn chords* or *multiphonics*, can be obtained by simultaneously playing and singing into a low brass instrument. Although it can be done on several of the brasses, it has been most popular on the trombone. By playing one note and simultaneously singing another note that is a member of the same overtone series, a third note is produced, the sum of the two original frequencies. The third note is nearly as loud as the two original notes, forming a chord whose quality is determined by the tone quality of the two original notes. Some typical horn chords are shown in Figure 11-15. Additional chords can be obtained by using two notes forming the same intervals, so the harmonics obtained by this technique are virtually limitless.

Horn chords are produced as a type of combination tone created when the two original notes and their harmonics mix in the standing wave in the instrument. They are not due to a nonlinearity of the ear, as were the combination tones of Chapter 6. The combination of the two complex waves creates an additional set of harmonics belonging to the sum frequency of the two additional waves. If the two original notes are members of a common overtone series, the combination wave will also be a member of that overtone series, and a chord will be formed.

Some well-known traveling concert bands of the late nineteenth and early twentieth centuries, like the John Philip Sousa band, had players who could perform remarkable harmonies using horn chords. One show-stopping performance involved a trumpet player doing sets of very complex variations, accompanied by a single trombone player playing horn chords. Horn chords or multiphonics have been used occasionally in serious contemporary music for the trombone.

The tenor and bass trombones shown in Figure 11-3, and their ranges are given in Table 11-1.

11.5 French Horns

The French horn is the member of the brass family whose physics is somewhat different from that of the other members. The horn bore is cylindrical over much of its length, but the flare of the bell is quite large. Relative to the other brass instruments, the normal range of the horn is higher in its overtone series. During the eighteenth century, before the invention of the valve, a player could change the pitch of the French horn by over a half-step by inserting a hand into the bell. This created new notes and allowed accurate tuning of all the notes in the normal range of the instrument, In addition, this hand-stopping technique increased the stability of some of the notes. Horn players of today still follow this tradition of inserting a hand into the bell to affect the timbre.

Although nearly a complete diatonic scale can be obtained on the French horn without using valves, its rotary valves improve intonation and permit easier changes from one note to another. Compared with other brass instruments, the small mouthpiece and closely spaced harmonics of the French horn make it difficult to play any desired note. Large leaps to high notes are particularly treacherous. Because of these difficulties, as well as the problem of exactly how far into the bell the hand should penetrate, it is very important for the horn player to listen carefully to pitch and tone quality and adjust carefully.

The modern double horn is equipped with an additional valve, to insert additional tubing and put the horn in either the key of F or the key of B♭, a fourth higher. This flexibility would introduce additional intonation problems, because the valves cannot be the correct lengths for both horns. Therefore, two separate sets of valve tubing are provided, one for each horn (F or B♭). The set of three valve tubes to be used is automatically determined by whether the B♭ horn valve has been pressed. Performers still must compensate by adjusting their embouchure.

The double horn is shown in Figure 11-3, and the range of the French horn is given in Table 11-1.

11.6 Other Contemporary Brass Instruments

Several other members of the modern brass family merit additional discussion; there is great variety in the details of how brass instruments are constructed and used. The trumpet is the principal high voice for the brass choir. However, the modern cornet, a conical instrument similar in almost every other respect to the trumpet, is often used in bands.

Because of its flare, the cornet has relatively less amplitude in its upper harmonics and thus a somewhat more mellow tone. The cornet and the flügelhorn, the alto member of the cornet family, are shown in Figure 11-2.

Additional intermediate low brasses are the baritone and the euphonium. Whereas the trombone is basically cylindrical, the baritone and euphonium are conical and must use valves, rather than slide positions, to change their overtone series. All three of these instruments have virtually the same range, but the trombone has a more brilliant tone due to its cylindrical shape. The euphonium has a somewhat greater flare and larger bore than the baritone, and therefore produces a bigger, fuller tone. Because a large bore exacerbates the valve tubing length problem, the euphonium suffers from this problem more than does the baritone. To compensate, most modern euphoniums are equipped with a fourth valve, which lowers the pitch of the instruments by a perfect fourth and provides some alternate fingerings to improve the intonation of notes requiring valve combinations including the third valve.

The tuba is generally pitched one octave below the trombone and euphonium; it has a very large bore and flare. The intonation problems of tubas are much greater than for the higher brasses, and some tubas are therefore equipped with two additional valves, the fourth valve being similar to the fourth valve on a euphonium. To play in tune, most tuba players find it necessary not only to use all valves in exotic combinations, but also to be continually adjusting several of the individual valve tuning slides.

The ranges of some of these instruments are given in Table 11-1, and the euphonium and the tuba are shown in Figure 11-4.

SUMMARY

(***Section 11.1***) Brass instruments likely originated when someone buzzed his lips into an animal horn; improvements such as a mouthpiece and a bell were then added. The **cornetto**, a hybrid instrument with a trumpetlike mouthpiece but played like a recorder, appeared as early as the eighth century. The **sackbut**, precursor to the trombone, developed with the invention of the **slide** in the fifteenth century. The **baroque trumpet** is a valveless instrument played in its upper range, where the overtone series includes a complete, if somewhat out of tune, diatonic scale. The modern **trumpet** and the **trombone** have cylindrical bores. The **French horn** became popular during the seventeenth century; it used a longer tube but played in its upper register to obtain a complete diatonic scale. The technique of **hand stopping** evolved, in which the performer inserted one hand into the bell of the instrument to improve the stability of the notes and to produce a complete chromatic scale. The invention of the **valve** in the early eighteenth century allowed the trumpet to produce the complete chromatic scale. Types of valves in common use are the **rotary valve** (used in the French horn and some low brasses) and the **piston valve** (used in the trumpet and most other brasses). During the eighteenth century, the family of conical-bore instruments was developed. (***Section 11.2***) A cylindrical tube blown by buzzing your lips into one end acts as an acoustical closed tube. Adding a **mouthpiece** to one end and a **bell** to the other end adjusts the frequencies of the overtones so that the instrument functions as an open tube, including all overtones except perhaps the fundamental. The bell also improves the **impedance match**, allowing the instrument to produce more sound. (***Section 11.3***) The B^b trumpet is commonly used in the contemporary band and orchestra, with the C and D trumpets less often seen. The first, the second, and the second and third valves together are usually made in tune, but the player must adjust for other valve combinations. The third valve tuning slide is usually made so that it can be adjusted during performance to mitigate intonation problems. The **regime of oscillation** of a trumpet note largely determines its stability. (***Section 11.4***) The trombone changes notes by adjusting its slide position and, unlike the trumpet, has usable pedal tones. **Horn chords**, or **multiphonics**, can be performed on a trombone and other brasses with a similar range by playing one note and singing another into the instrument to produce a third note as a sum tone. Trombones typically come in two versions: the regular B^b trombone and the F trombone, which uses a valve to introduce additional tubing, lowering the sound by a fourth. (***Section 11.5***) French horns are unique, in that they are very long and play higher in their overtone series than other brasses. The horn is a transposing instrument in the key of F or the key of B^b; the double horn includes an extra valve to allow the performer to play using either the F or the key of B^b. (***Section 11.6***) Other brass instruments form a conical brass family, including the (modern) **cornet, flügelhorn, baritone, euphonium, tuba**, and **sousaphone**. The lower instruments are often equipped with a fourth valve to help with intonation problems.

QUESTIONS

1. Classify the brass instruments commonly used in the band and orchestra by range and by bore shape (cylindrical or conical). Discuss the differences in tone between instruments of the same range that can be attributed to bore shape and size.

2. Attach a trumpet mouthpiece to a section of rigid plastic tube about 40 cm long. Drill finger holes, as if making a flute, and play it like a Renaissance cornetto. Describe how the instrument plays, and discuss some of its acoustical problems.

3. Blow a length of rigid plastic tubing by buzzing your lips into one end without a mouthpiece.

   a. Measure the resulting frequencies with a frequency counter or by comparing them to the notes on the piano (read the frequencies from Appendix A).

   b. Calculate the frequencies expected for a closed tube of this length.

   c. Why does this tube behave acoustically like a closed tube?

   d. Do the calculated and experimental frequencies agree?

   e. What frequencies would be expected if a trumpet mouthpiece were added to one end and blown in the same manner?

4. Make a "trumpet" by attaching a trumpet mouthpiece to a long section of tubing or hose.

   a. Add a bell made from a metal funnel. Measure the frequencies of the harmonics, and discuss the intonation problems that would be met when playing this instrument with an organ tuned in equal temperament.

   b. Play the trumpet with and without the funnel, and discuss the effects of the bell.

   c. Calculate the expected frequencies of your trumpet, based on its measured length.

   d. How well do the calculated and experimental frequencies agree?

5. Discuss the valve/tubing-length problem. What compromises can be made with regard to it?

6. Examine the horn chords given in Figure 11-10. Using your knowledge of the overtone series, determine the notes that you must sing and play to produce other horn chords. Describe the difficulties in producing horn chords on a trombone or a baritone.

7. Compare the sounds of the trumpet and the cornet. How are they different? Use a spectrum analyzer to compare their spectra for several notes. How are they different? Does the difference between their spectra explain at least in part the difference between their timbres?

8. In the baroque era the trumpet did not have valves but was played in its upper octave, much like the contemporary French horn. The music was often performed in mean tone or true temperament. Would you expect that more or fewer intonation problems would occur if the instrument played mainly the notes of a major chord? What about when the music required more notes of the scale?

9. **a.** Measure the increase in length L_{12} of the trombone when the slide is moved from first to second position. Measure the increase in length of the trombone L_{17} when the slide is moved from first to seventh position. What is the ratio of L_{17} to L_{12}?

b. If each half-step of pitch reduction increases the length of the trombone by a factor of 1.05946 (the twelfth root of 2), then what would be the theoretical ratio for the length increases that you measured in part (a)?

c. Measure the length of the trombone and calculate the actual length increases for L_{12} and L_{17}. Comment on the accuracy of the results that you obtained. (See problem 1 following.)

10. The range of a trumpet is $E_3 = 164.81$ Hz to $B_6^\flat = 932.33$ Hz; the range of the trombone is $E_1 = 41.20$ Hz to $D_5 = 587.33$. If the trombone is twice as long as the trumpet and both are cylindrical tubes with a mouthpiece and a bell, how much lower than the trumpet would you expect the trombone to play? Why does the trombone go one octave lower than this?

PROBLEMS

1. The pedal tone for a trombone in first position is $B^\flat = 58.27$ Hz. If the temperature of the air in the trombone rises to about 90°F, so that the speed of sound in the air in the trombone bore is about 350 m/s, then what is the wavelength for the pedal tone in the trombone? What would the effective length of the trombone be?

2. The fundamental frequency of a trumpet with no valves depressed (if it were playable) would be $B^\flat = 116.54$ Hz. If the temperature of the air in the trumpet rises to about 90°F, so that the speed of sound in the air in the trumpet bore is about 350 m/s, then what is the wavelength for the pedal tone in the trumpet? What would the effective length of the trumpet be?

3. If a tuba sounds one octave below a trombone, what would the frequency of its pedal tone be (if playable)? What would the effective length of the tuba be if this note were the fundamental?

4. A bass trombone uses a valve to insert extra tube length, so that the notes are lowered by a perfect fourth—for example, from $B_3^\flat$ to F_3 in first position. If the effective length of the trombone is about 2.85 meters, what is the length of the additional tube?

5. When the bass trombone valve is used, it lowers the notes for first position by a perfect fourth—for example, from $B_3^\flat$ to F_3 in first position. However, the slide positions must all be made longer (i.e., the slide extended farther out) to compensate for this extra starting length. What length of additional tubing must be added in going from first position to second position with the initial tubing (going from F_3 to E_3 rather than from $B_3^\flat$ to A_3)? If the effective length of the trombone is about 2.85 meters, then how much further out must the slide be pushed in order to use second position with the F trombone?

6. The range of a trumpet is $E_3 = 164.81$ Hz to $B_6^\flat = 932.33$ Hz. How many octaves does this encompass?

7. The range of the trombone is $E_1 = 41.20$ Hz to $D_5 = 587.33$. How many octaves does this encompass?

REFERENCES

BACKUS, JOHN. *The Acoustical Foundations of Music.* 2nd ed. New York: W. W. Norton & Co., Inc., 1977. See his reference list.

———, and T. C. HUNDLEY. "Harmonic Generation in the Trumpet." *Journal of the Acoustical Society of America* 49, no. 509 (1971): 309. A somewhat detailed description of sound production in a trumpet.

BAINES, ANTHONY. *Brass Instruments: Their History and Development.* London: Faber & Faber, Ltd., 1976. An important reference on brass instruments, written for musicians.

BATE, PHILIP. *The Trumpet and the Trombone.* New York: W. W. Norton & Company, Inc., 1966. Describes the basic brass instruments in terms easily readable by the musician.

BENADE, ARTHUR H. *Fundamentals of Musical Acoustics.* New York: Oxford University Press, Inc., 1976. See his reference list.

———. *Horns, Strings, and Harmony.* Garden City, N.Y.: Doubleday & Company, Inc., 1960. Contains some elementary but excellent qualitative concepts on how the trumpet works.

———. "The Physics of Brasses," *Scientific American,* July 1973. Contains considerable detailed information on the nature of the trumpet and how its sounds are produced, as well as some of their characteristics.

FLETCHER, N. H., and T. D. ROSSING. *The Physics of Musical Instruments.* New York: Springer-Verlag, 1990 (Hardcover), 1993 (Softcover). An excellent, up-to-date book containing a wealth of detailed information about virtually all types of musical instruments.

MORLEY-PEGOE, R. *The French Horn: Some Notes on the Evolution of the Instrument and of Its Techniques.* New York: Philosophical Library, Inc., 1960. Describes the basic features of the French horn and its history; written for the horn player.

Chapter 12

Stringed Instruments

The history of stringed instruments is as diverse as the history of music itself. Surprisingly, out of the incredible variety of historical stringed instruments, only one family of four sizes of violins and a few varieties of plucked strings are all that can be called "contemporary."

The violin family is the sole surviving family of bowed strings, and the guitar (acoustical and electronic) and harp are the main plucked strings in modern use, although the banjo, ukulele, and dulcimer are often used for folk and country music. In addition to developing in their own right, the historical string families laid the foundation for the clavichord and harpsichord, and their modern descendant, the piano.

In this chapter we shall first summarize the history of some stringed instruments. We shall then discuss the basic theory of bowed strings and apply this to the contemporary violin family. Finally, we shall briefly look at the acoustics of some plucked stringed instruments.

12.1 History of Stringed Instruments

Plucking a stretched fiber to obtain a musical tone is probably as old as music. As early as several centuries B.C., artists made drawings of stringed instruments, many similar to the lyre and the harp. Pythagoras and other Greek philosophers discussed the sounds of a stretched string, as in the monochord, and the effects of stopping the string (i.e., artificially shortening it temporarily to obtain a higher-frequency note by pressing it against the instrument frame) to obtain musical intervals. These intervals could be consonant or dissonant, depending on the ratio of the lengths of the two strings; the very concept of consonance was largely seen as a matter of philosophy and mathematical relationships.

By the medieval era, the harp and lyre had become associated with the music of heaven and were regularly depicted as being held or played by angels. Additional plucked stringed instruments had evolved, using primarily animal shells, skulls, or other naturally occurring objects as the resonators, over which the wires or gut strings were stretched. A modern-day descendant of such an instrument is the charango, a Peruvian folk instrument that uses an armadillo shell as its resonator.

During the medieval era, two new techniques evolved for creating the sound in a stringed instrument: hammering and bowing. In the hammered dulcimer, small hammers with heads of leather or a similar material were used to strike the strings. This technique was later used in the clavichord and finally adapted to the piano. The technique of bowing or scraping the strings took two directions. The hurdy-gurdy, or organistrum, used a disc that was turned by a hand crank so that its edge scraped the strings. The hurdy-gurdy often had three strings, two of which might be drones, while the third could be stopped, using a set of keys, to play a melody; its sound was similar to that of the bagpipe. Early hurdy-gurdy playing required two people, because the instrument was some 5 feet long, with the crank at one end and keys along one side.

A group of bowed stringed instruments, similar to the later violin family, also evolved around the latter part of the twelfth century. They generally had from one to six strings, with four being a popular compromise; most were unfretted. Among the more popular of these were the rebec and the medieval fiddle or vielle.

An important technical development during the medieval period involved improved woodworking capabilities. The new woodworking facilities allowed construction of a wide-ranging variety of new stringed instruments, like the rebec and fiedel, with superior physical properties. The lute was probably the most important of the new families of string instruments. The lute was a loose descendant of the Arabic ud, a plucked stringed instrument with many strings. The word *ud*, and its relative, *lute*, come from the Arabic word for wood; the instruments are so named because they were among the earliest with wooden resonating boxes, in contrast to those made with natural resonators like animal shells. The lute had between 6 and 24 strings, which, with the exception of the highest string, were tuned with adjacent pairs in unison. Despite its limitations, the lute was destined to become by far the most important stringed instrument of the European Renaissance, retaining its vast popularity for almost 500 years.

No authentic medieval stringed instruments still exist; the primary source of information on them is works of art in which they are depicted. One type of medieval fiddle is known largely from a single painting done by Hans Memling (ca. 1480) on the case of a Spanish organ. One of about ten figures in the painting depicts the playing of a fiedel in some detail, including the bow and bowing technique, and the way in which the instrument was held, as well as some construction details of the instrument and the bow. The same paintings show the psaltery, tromba marina or nun's fiddle, lute, and harp, as well as several reed and brass instruments and the portative organ.

During the Renaissance several new stringed instruments were developed, usually appearing in families of three or four sizes, and often in many varieties. Most of the new varieties of bowed strings were fretted, as is the modern-day guitar. The viola da gamba, perhaps the most popular family of Renaissance bowed string instruments, was held between the knees and bowed in the style of the contemporary French bass violin. Gambas had six strings, came in three sizes, and were often played in groups of up to six voices. The viola da braccio was played in a manner similar to that of our modern violin. The viola d'amore was unique in possessing a set of strings that were not bowed but instead vibrated sympathetically as the bowed strings sounded.

In addition to the lute, several other varieties of plucked strings arose. The guitar took roughly its contemporary form. The cittern, which had an oval belly and back, and the pandora (or bandora), which had a three-lobed, scallop-shaped body, were prominent throughout Europe, and the vihuela became popular in Spain for a lengthy period.

During the Renaissance the keyboard was developed to its present form and applied to two types of stringed instruments. In the clavichord, the key controls a tangent, which strikes the string. In the harpsichord the key controls a plectrum, which plucks the string. Use of the harp was expanded, while many other stringed instruments fell into oblivion when they could not meet the increasing technical demands or requirements for expression.

During the late Renaissance and early baroque, many changes occurred in the popularity of various stringed instruments. For centuries, the lute had been popular as both a melody and accompaniment instrument and could be called upon to play or double any instrumental or vocal line. Despite its immense popularity, it had overwhelming technical problems, particularly in tuning. It used gut strings, which are very sensitive to temperature and humidity variations, and which tend to stretch easily and therefore to become flat in pitch rather quickly after they are tuned. A saying of the time stated, "If the lutenist shall attain the age of ninety, he shall have spent seventy of those years tuning his lute." With the advent of reliable harpsichord tuning and action, the harpsichord often replaced the lute by playing chords to accompanying solo singers, instruments, or small ensembles. The greater technical capacity of keyboard stringed instruments also led them to supplant other plucked strings in popularity.

A similar change occurred in bowed strings. The desire for greater expression in early baroque music required a rather more versatile instrument than the viol family. In response, the violin family was developed. The new instruments were played under the chin, and, being unfretted, could be played somewhat more expressively than the viols. They also had generally greater technical capacity; for instance, they could use vibrato, glissando, and harmonics. An attempt was made to prolong the life of the viols by removing the frets and modifying their bowing technique, but they were nevertheless rapidly supplanted by the violin family. One remnant of the viol family is the string bass, which is in many respects more like a member of the viol family than the violin family. It is tuned in fourths, as are the viols, rather than fifths, as are the members of the violin family, so that scales can be played with minimal shifts of hand position. It has sloped shoulders and is generally shaped like a viol, and its bowing technique is more like that of the gamba than that of the cello. In fact, the string bass is sometimes called the *bass viol*.

During the baroque era most plucked instruments, including the lute, decreased in popularity. The guitar and, later, several other plucked instruments, such as the banjo, ukulele, and dulcimer, increased in popularity, and to this day are prominent in folk, dance, and country music. The only plucked instrument to be used in the contemporary orchestra is the harp, which has grown to massive proportions in comparison with its medieval ancestor, the lute. Use of the guitar in serious music is growing, especially in chamber and solo compositions.

Two additional developments affected the acoustics of the violin family: (1) The orchestra grew substantially from the baroque to the present, increasing the sound of each of the string sections. (2) The instruments required significant modifications as a result of the rise in pitch over the past three centuries, which gave them a more brilliant sound. These changes will be discussed in the section on the violin family.

A significant recent development is the use of electronic pickup and amplification in instruments such as the guitar. The *acoustic* guitar simply uses the vibration of the instrument as its source of sound, whereas the *electric* guitar makes use of special pickup devices called *transducers* that convert the sound into electronic signals, which are amplified to drive a loudspeaker.

Figure 12-1 shows some of the early stringed instruments that preceded our contemporary bowed and plucked string families. Table 12-1 gives the normal ranges for the instruments in the violin family and the notes of their open strings, along with the ranges of selected stringed instruments, both bowed and plucked.

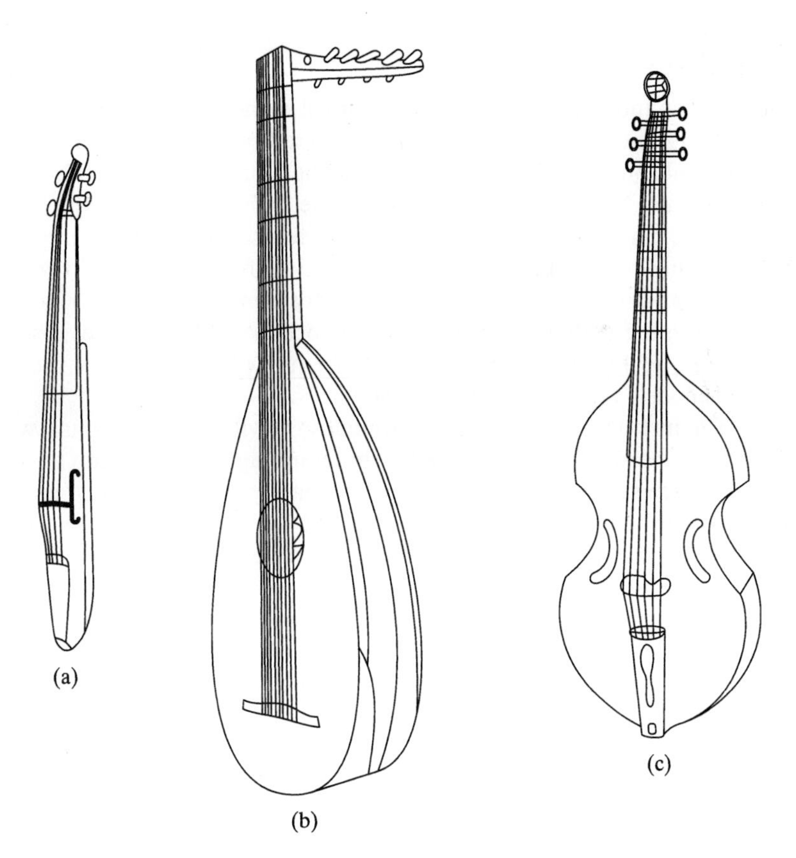

Figure 12-1 Some early stringed instruments: (a) the rebec, an unfretted medieval bowed instrument with four strings; (b) the lute, a fretted Renaissance plucked instrument with a large number of strings; and (c) the viola da gamba, a six-stringed, fretted, bowed Renaissance instrument that came in three sizes. (Drawings not to scale.)

Table 12-1 Ranges of Some Bowed and Plucked Stringed Instruments

Instrument	Open strings	Lowest note	Approximate highest note
Violin	$G_3D_4A_4E_5$	G_3	$E_7(A_7)^b$
Viola	$C_3G_3D_4A_4$	C_3	$A_6(D_7)^b$
Cello	$C_2G_2D_3A_3$	C_2	$E_5(G_6)^b$
Double bass	$E_1A_1D_2G_2$	$E_1(C_1)^a$	$D_4(F_5)^b$
Treble viol	$D_3G_3C_4E_4A_4D_5$	D_3	D_6
Tenor viol	$G_2C_3F_3A_3D_4G_4$	G_2	G_5
Viola da gamba	$D_2G_2C_3E_3A_3D_4$	D_2	D_5
Harp	(All)	$C_1^\flat$	$G_7^\sharp$
Guitar	$E_2A_2D_3G_3B_3E_4$	E_2	A_5

[a] Denotes with use of extender.

( )[b] Denotes upper ranges using harmonics.

12.2 Theory of Bowed Instruments

The basic sound-producing mechanism in all stringed instruments is virtually the same; the only significant difference is that some are bowed and others are plucked to obtain the original source of the vibration. After the vibrations in a string are produced, these vibrations are transmitted by a bridge to the belly and back of the instrument, causing these large plates, generally made of wood, to oscillate with the frequencies present in the vibrating string. The bridge must act as an efficient coupler between the string and the body, and if the body is to resonate effectively, there must be actual resonant frequencies in the plate similar to the resonances in Chladni plates discussed in Chap. 3. Furthermore, it is desirable to have low-frequency resonances in the wooden plates to help produce a full tone in the low range.

The instrument transfers the sound of the vibrating string and wood to the air, so there must be an effective means for this coupling. The front and back plates of the instrument are set in vibration so they alternately force air out of and into the hole or holes cut in the front plate. Each instrument has its own geometry for its hole or holes. The hole is circular in the case of the guitar, whereas in the violin family there are two holes, called *f*-holes because of their shape. The sizes and shapes of the holes and body determine a resonance similar to that of a Helmholtz resonator. For most stringed instruments, this resonance is rather broad, although it is isolated, as is the primary resonance of a Helmholtz resonator. The frequency of this Helmholtz-type air resonance is usually in the lower range of the instrument, so it will act to improve the tone and intensity of the instrument in its lower range. This effect can be viewed as similar to the low-frequency air resonance in a tuned-port loudspeaker.

In general, the stringed instruments of today have developed with these characteristics. Instruments lacking any of those characteristics produced a less-than-adequate sound for the music of the time and thus became obsolete. For example, the medieval rebec has a very small air volume and small wooden plates. As a result, the air and wood resonances are too high relative to the normal range of pitches at which the instruments play, and the tone is too nasal and squeaky for modern taste. The violin family, which has survived as the bowed string family of the orchestra, is the family with air and wood resonances best suited to producing good tone over their entire range.

Let us look in more detail at the mechanism by which the drawn bow causes vibrations in a string. The strands of the bow are generally made of hair from the tail of a horse. The bow hair is coated with rosin, increasing the friction between bow and string. As the bow is drawn across the string it tends to grab the string, pulling the string along at the speed of the bow. At some point the tension in the string, because of its displacement, will become too much and will overcome this frictional force. At this point the string will snap rapidly back to a point on the other side of its equilibrium position equal to the distance it was drawn by the bow. The bow then begins to pull the string back in the original direction again. The displacement of the point on the string at which the bowing occurs is thus approximately the shape of a sawtooth wave. The frequency with which this alternating motion of the string occurs is controlled by the length, tension, and size of the string, according to Mersenne's laws, as discussed in Chap. 3; it is exactly the fundamental frequency of the string, for normal bowing pressure and bow velocity. If too light a bowing pressure is used, the tension in the displaced string will overcome the frictional force more quickly, and the string will snap back sooner. Its frequency, again controlled by the resonance of the string, becomes that of the second harmonic. This is why inexperienced string players, playing timidly and without sufficient bow pressure, often produce squeaky notes an octave too high.

The string actually vibrates in a rather complex manner, more complex than our simple idea of the sum of simultaneous motion at the frequencies of the harmonics, as described in Chapters 3 and 4. Instead, the shape of the string remains roughly triangular, with the vertex of the triangle tracing out an envelope that looks like the string vibrating in its fundamental mode. This is illustrated in Figure 12-2. This is a type of standing wave, which, in the case of violin string bowed with normal pressure at the normal bowing point, contains all the harmonics with amplitudes that decrease with harmonic number.

The position along the string at which the bow is drawn affects the harmonics produced. If the bow were very thin and if it produced the exact pull-and-release motion of the string as described, the number of harmonics present in the vibration of the string would be limited as follows. If the point at which the string was bowed were $1/N$ of the length of string, it would produce an antinode in the vibration at that point. Therefore, the Nth harmonic would be missing, because it must have a node at that point. For example, bowing a string at its midpoint would fail to produce the second harmonic, and bowing a string at one-third of the distance from the end would fail to produce the third harmonic. Bowing one-tenth of the distance from one end would produce no tenth harmonic. Furthermore, the bowing point would be close to a node for the ninth harmonic, so the amplitude of the ninth harmonic would be small. The eighth harmonic would be slightly greater in amplitude, the seventh greater yet, and so forth, down to the fundamental.

In reality, the situation is not so simple. Through an additional interaction between the bow and the string, the Nth harmonic is actually produced with moderate amplitude even when the bow is at $1/N$ of the distance from the end. However, when a string is bowed normally, the amplitudes of the harmonics generally decrease until the Nth harmonic is reached.

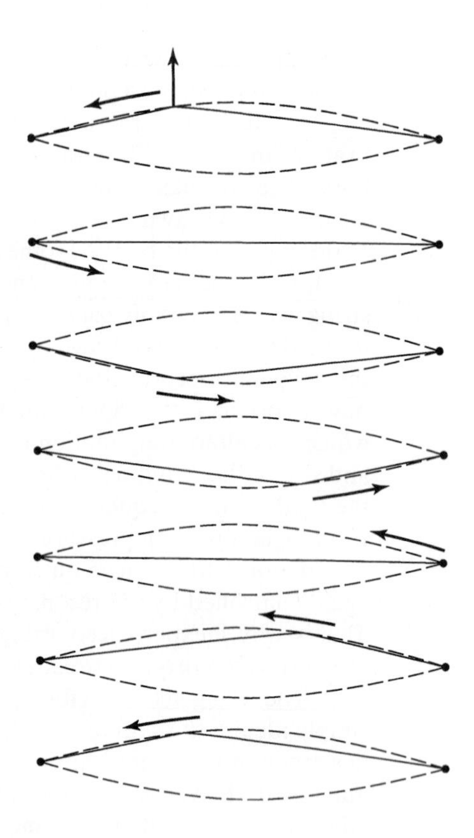

Figure 12-2 A string bowed at the point shown at the top vibrates in the unusual "standing wave" pattern shown here. The bow creates a triangular displacement of the string. During one full period, the vertex of the triangle traces out a figure that has the shape of the envelope of the normal fundamental mode of a freely vibrating string. (After Schelling, 1973)

If the bow is moved closer to the bridge (nearer the end of the string) than normal, the N value will be greater and the tone will have more harmonics. This bowing technique is called *sul ponticello* (on the bridge). If, on the other hand, the string is bowed farther from the bridge (or closer to the middle of the string), the N value will be less, and fewer harmonics will be produced. This bowing technique is called *sul tasto* (on the fingerboard), because it usually involves moving the bow above the fingerboard. Thus, by moving the bowing point along the string, the player can produce either a richer or a more mellow tone.

One additional effect is important in determining the amplitudes of harmonics in the vibrating string: the damping of the string by the fingers when stopped notes (as opposed to open strings) are played. Finger damping decreases the amplitudes of the higher harmonics and causes the tone to be less brilliant. By avoiding open string notes, the violin tone can be made more uniform note to note; when a violin player has the option of playing a note either on an open string or stopped with a finger on a lower string, the stopped note is generally considered preferable. Stopped notes also permit vibrato, whereas open notes do not. On fretted, bowed instruments like the Renaissance viol family, the frets were designed to produce a sound more nearly like that of the open strings for *all* notes, giving the viols a unique nasal quality.

When the string is bowed with great pressure by the bow, the very-large-amplitude oscillations elongate the string and cause an increase in the tension of the string. This increased tension leads to a rise in the frequency of the string, according to Mersenne's laws. Thus bowing a string with great pressure causes a rise in pitch.

12.3 The Violin Family

Figure 12-3 shows some views of the interior and exterior of the violin. The bow is held perpendicular to the strings and drawn across the string at a point between the bridge and the fingerboard. This causes the string to vibrate primarily in the plane defined by the string and the bow. These vibrations are first transmitted to the bridge, causing it to rock rapidly back and forth in the plane of the bridge, and then to the wooden back and belly.

The sound post is a rigid wooden dowel wedged between the two plates of the violin. As the bridge rocks back and forth, it flexes the front surface of the violin, causing small changes in the volume of the air space within the instrument. The bass bar has two functions. First, it reinforces the belly of the instrument so that the force placed on the strings and bridge will not crack or shatter the belly. Second, its most important acoustical function is to extend the vibration of the front surface over a wider area, thus producing greater oscillatory change in the volume of the air cavity. As the front surface oscillates in and out at the frequency of the fundamental being sounded, it causes a change in the volume of the air space and thereby alternately forces air out of the instrument and draws the air back in, both through the *f*-holes. This motion of the string–bridge–front plate system creates motion of the air at the sounding frequency of the violin.

The bridge (its size, shape, mass, elasticity, and placement) may have more of an effect on the violin tone quality than any other single component. The motion of the bridge as described earlier is a bit simplified; the motion also includes some longitudinal vibration at twice the string frequency. This vibration is prevented from coupling to the belly of the violin by exact matching of the bridge feet to the plate. If the bridge

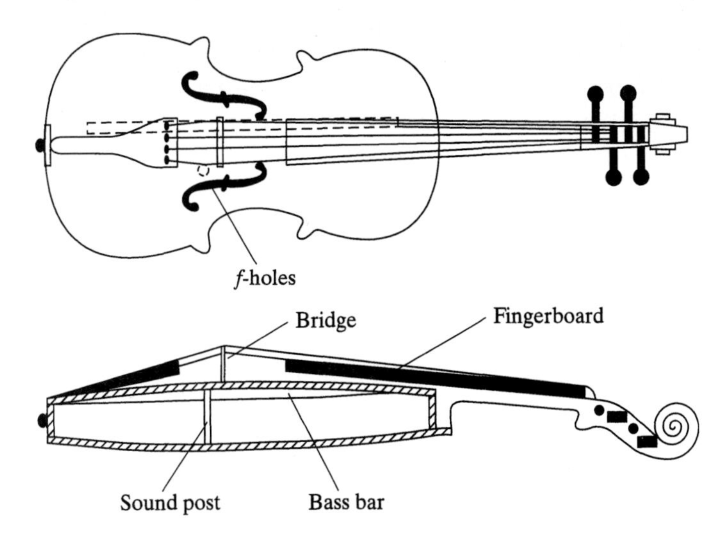

Figure 12-3 Sectional views of the violin.

were not held down by the other strings, it would rattle, as in the case of the tromba marina, a bowed medieval instrument with a single string mounted asymmetrically on the bridge.

Bridge height is clearly a factor in coupling the string motion to the belly; the height of the bridge was increased when the baroque violin was modified to its modern form, to help increase the volume and brilliance of the violin tone. The bridge can affect the wolf tone in the cello (see discussion later in this section), and certain changes of the bridge have been proposed to help in eliminating the wolf tone. The position of the bridge with respect to the sound post is crucial in minimizing the wolf tone, as it is in obtaining the best tone from a violin. The exotic shape of the violin bridge probably helps to suppress undesired modes of oscillation while maximizing coupling of the desired modes.

The air resonance of a violin is very important in reinforcing the tone of the instrument in its lower range. On most quality violins the air resonance is rather broad, with its peak at a frequency just below that of the D string. It therefore substantially reinforces the tone of the two lowest strings. The frequency of the air resonance is determined primarily by the size and shape of the air cavity within the violin and the size, shape, and position of the *f*-holes. The frequency of the air resonance can be determined by blowing sharply across the *f*-holes, just as the resonant frequency of a jug can be determined by blowing sharply across its mouth.

Resonances in the wood play an important part in the reinforcement of the tone of the violin as well. As the foot of the bridge alternately presses on the belly and then releases the pressure, the entire front of the instrument moves in and out. At its main resonant frequency, the entire plate along the bass bar moves back and forth as a long antinodal region. This is called the *wood resonance;* for most violins the frequency of this resonance is just below the frequency of the A string. This resonance is largely controlled by the type of wood, the thickness and shape of the wood plate, and variations in the thickness of the plate, which can be chosen to make the standing-wave vibrations assume the proper shape. The frequency of the wood resonance is determined by the violinmaker through the use of tap tones. The maker holds the wood lightly or lays it on a foam rubber sheet or other soft surface, taps it gently, and listens for tap tones. As the plate is made thinner, the tap tone will become quite bell-like and attain the correct frequency. Often this measurement is made using electronic equipment to produce Chladni-type patterns in the plate, which can then be observed using a type of interference pattern of laser light. The resonant frequency of the wood plate changes significantly when it is glued to the sides of the instrument, as the nodal lines move toward the edges where the plate is glued. The wood resonance just below the frequency of the A string is with the instrument assembled. The main wood resonance also reinforces the notes within a few half-steps of the low note of the instrument; the second harmonics of these notes are resonated by the main wood resonance. This resonance effect has been called the *wood prime* resonance. Additional resonances of the wood exist in the form of Chladni-type resonances that are closely spaced throughout the audible frequency range. This type of resonance is illustrated in Figure 12-4 for the top and back plates of a violin.

The sum effect of all these resonances can be seen in the *loudness curve.* If the violin is bowed up the scale chromatically from its lowest note with equal bowing intensity,

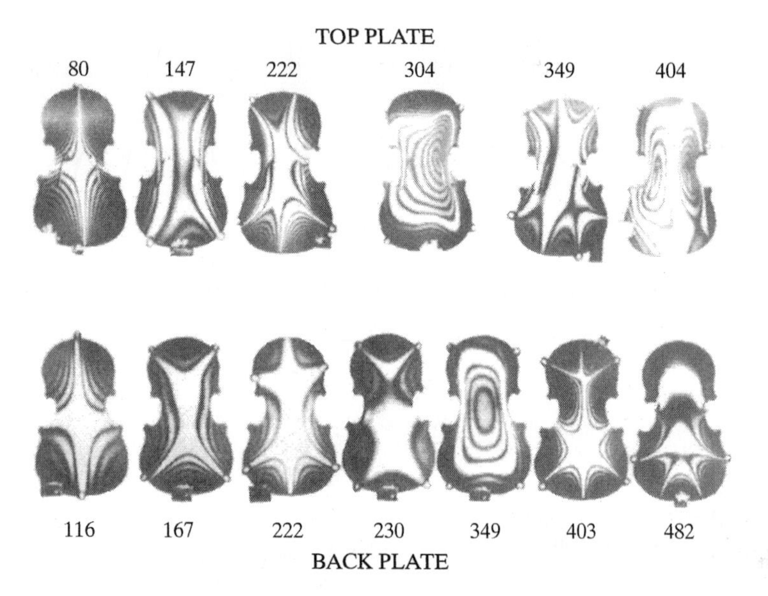

Figure 12-4 Holographic interferograms of oscillations of the top and the back plates of a violin. The frequency of each oscillation is given adjacent to its interferogram. (Courtesy of the Catgut Acoustical Society)

the notes will vary in volume owing to the differing intensities of the resonances. In a good violin this loudness should not vary much; nonetheless, the air and wood resonances will be clearly observable. A violin with a good tone will have a relatively flat loudness curve with a gradual drop in level to about one octave above the E string; this drop in the loudness curve apparently gives a more mellow and less squeaky tone quality to the notes around the open E. Figure 12-5 shows the loudness curves for a good violin and a poor one. Good violins differ from poor ones in that the loudness curve is more uniform for a good violin, and more sound is generally available on a better instrument.

When string instruments are simply scaled down (half-size, quarter-size) for use by small children, their tone usually suffers drastically. Because the instruments are smaller, their wood and air resonances are higher in frequency; as a result, the low notes are weak. Furthermore, the higher harmonics will be overemphasized, and the tone will therefore be too brittle or squeaky. Current work in this area is apparently paving the way for significant improvements in the tone quality of reduced-sized violins.

The choice of wood type and the way in which the wood is cut and trimmed are clearly important to violin tone. Norway spruce has been very popular because of its density and elastic properties. Although there has been some recent experimentation with substitute materials like plastics and resins, the usual choice for the back, front, and sides of the violin is limited to the types of wood that have been traditional for centuries. The strings of the violin are usually wire-wound gut for the lower three strings and thin solid steel for the highest (E) string. Wound Perlon, a type of plastic fiber, has recently emerged as a high-quality substitute for gut, owing to its superior ability to remain in tune. The use of steel E strings causes the sound to be extremely brilliant in the upper range of

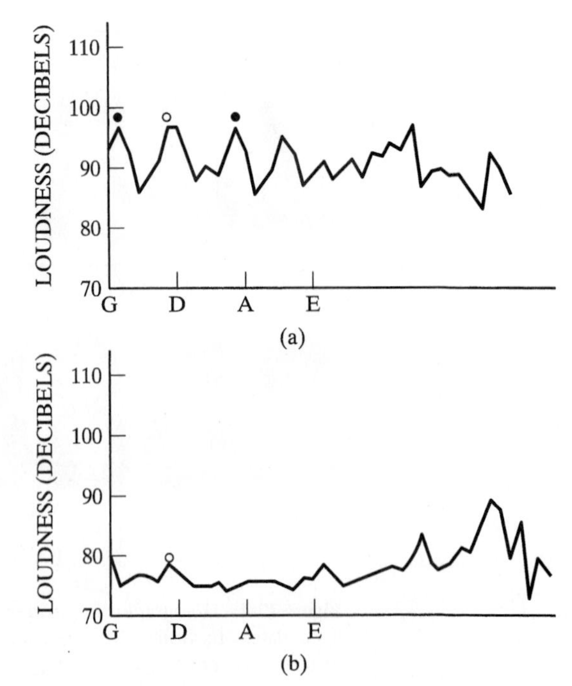

Figure 12-5 Loudness curves for (a) a good violin and (b) a poor violin. The black dots mark the wood (upper) and wood prime (lower) resonances; the circle marks the air resonance. (From "The Physics of Violins" by Carleen Maley Hutchins, *Scientific American*, November 1962. Graphics by Hatti Sauer Prentice, used with permission.)

the violin and is one reason for the requirement that the loudness curve decrease in amplitude for notes going up the scale above the frequency of the open E string.

Several other factors have been suggested as being important to the tone of the modern violin. For example, the type of varnish, the type of glue, and certain techniques for finishing the surface are believed to be important by many experts. Whether the finishing oil and varnish penetrate the surface of the wood is believed to be important; oils used in the eighteenth century were significantly less penetrating than some oils used today.

Regular playing throughout its life improves the tone quality of a violin. It is of course necessary to construct the instrument of high-quality wood of the appropriate type that has been properly aged. The vibrations of the wood produced when it is regularly played apparently increase its resiliency, and therefore improve the response of the resonances in the wood. Experiments have been performed in which one of two identical instruments is played while the other is allowed to sit idle. The one that is played develops a tone quality and responsiveness that are better than in the unplayed instrument.

The changes in the construction and sound of the violin over the last three centuries have important acoustical implications. The early violins by Antonio Stradivari (*c.*1644–1737) and his contemporaries were pitched about a half-step lower (approximately $A_4 = 415$ Hz, the baroque average), had much shorter fingerboards that were set straight into the body, and used gut strings exclusively. The tension in the strings was considerably less than that of modern violins, with important consequences; the violin tone was much more mellow and less brilliant. The bass bar was also shorter and less massive.

As the orchestra grew in size, more wind instruments were added, and the tone of the winds became louder and more brilliant. In response, more brilliance was required from the strings. This was achieved by lengthening the fingerboard and the violin strings and considerably increasing the string tension. The fingerboard was also angled more sharply to the body. These modifications result in a substantial increase in the force that the bridge exerts on the belly. To prevent the leg of the bridge from pushing through the belly and to increase the amplitude of the vibration of the wood, it was necessary to increase the size and strength of the bass bar. Further increases in string tension were used to obtain additional rises in pitch level throughout the nineteenth and early twentieth centuries. This also introduced increasing brilliance into the tone quality of the violin. Because gut strings, particularly the E (highest) string, could not withstand this additional tension, it was necessary to use a thin steel string in place of gut. To match the increased brilliance of the steel E string, the other strings are now generally made of wire-wound gut or a synthetic substitute.

The air and wood resonances in old violins remained at about the same frequency over the centuries, while the pitch level at which the instrument is played rose slightly over one half-step. Therefore, the frequencies of the resonances, with respect to the frequencies of the open strings, have changed by about one half-step from their positions in the seventeenth and eighteenth centuries. This seems to indicate that the positions of these resonances is not critical, as long as they are close to the open strings and have all the desirable physical characteristics such as the correct width and intensity. This is a point of considerable interest, though, and many modern violinmakers attempt to duplicate the resonant properties of Stradivarius violins, because they sound so good.

The other members of the violin family are the viola, the violoncello or cello, and the string bass. Although the viola can generally be made as a larger-scaled version of the violin, the cello and bass would be too large to play if they were properly scaled. Because the cello is smaller than a scaled-up violin of equal pitch, the air and wood resonances appear in different frequencies relative to the frequencies of the open strings. For the cello, the main wood resonance generally appears very close to the note $F^\sharp_2$, often with serious consequences. When the cellist plays the note $F^\sharp_2$, the main wood resonance vibrates at its frequency as the cello sounds the frequency of the note $F^\sharp_2$. A loud beating sound results between these nearby frequencies; this is known as the *wolf note,* because it is a growling tone. In fact, the mechanism by which the cello wolf note is produced is rather complicated. The wood resonance appears to be split into two frequencies by the driving force of the sounding string. These two periodic resonances beat with each other. This wolf note must be eliminated or significantly reduced in amplitude for the cello to play the nearby notes. This can often be accomplished by modifying the cello front plate or moving the sound post. Some experimentation is being done using other techniques to either reduce the wolf note or move its frequency farther from one of the chromatic notes so that it will not normally be strongly driven.

The string bass is, in many respects, more like a member of the Renaissance viol family than the violin family (see Section 12.1). In the modern orchestra, the basses often double the cello parts, sounding one octave lower. The string bass is sometimes equipped with an extender on its lowest (E) string to allow a low C to be played. By using this extender, it is possible for the basses to double the cellos down to the lowest note on the cello. If a bass were merely a scaled-up version of a violin, it would be enormous. Instead, the bass has had to be made somewhat smaller, and as a result the bass tone can be rather weak in its lower range.

An interesting effect used in music for the modern violin family is *harmonics*. There are two types, *natural* (or open-string) *harmonics* and *artificial* (or stopped-string) *harmonics*. To produce a natural harmonic, the string is lightly touched at a distance of about $1/N$ from the bridge, and the string is bowed very lightly between the bridge and the point being touched. This produces the Nth harmonic of that string. The tone obtained is very pure and soft, and generally well tuned, because it only depends on the string having been tuned properly; vibrato is generally not used on these harmonics. It is often possible to play a rapid arpeggio ending with a natural harmonic more easily than a regular finger-stopped note, because to produce these notes the finger need only barely touch the string in an extremely high position, perhaps beyond the fingerboard. This increases the effective range of the instrument slightly and gives some variation to the sound.

Artificial harmonics are produced by playing a normal finger-stopped note and simultaneously touching the string with another finger at $1/N$ of the length of the stopped string; one advantage of artificial harmonics is that they can be played with the hand in a relatively low position. An interesting effect, called the "seagull," can be obtained by using artificial harmonics on a string bass or cello. An artificial harmonic is produced in a high position. Keeping the thumb (stopping finger) and little finger (lightly touching the string) with the same relative spacing, the hand is slid up along the string. This results in a sequence of short downward glissandos, each lower in pitch, characteristic of the sound of the seagull.

12.4 Plucked Stringed Instruments

After the onset of the plucked tone, the tone of a plucked stringed instrument consists entirely of transients. Usually the string must be plucked rather strongly to retain an audible sound for a reasonable time. Therefore, as in the case of the violin bowed with great pressure, the string tension is increased substantially above that of the string at rest or vibrating with small amplitude. Thus the fundamental of the initial tone will be somewhat too high in frequency, as is the fundamental of a violin string when bowed with above-normal pressure. Because the higher harmonics are relatively unaffected by this apparent increase in tension, they will remain nearly constant in frequency as the tone dies out. In most plucked instruments, therefore, the decay transient consists of a general damping out of the tone accompanied by continuous changes in the wave shape as the overtones decay at different rates and as the overtones shift in phase with respect to the fundamental. The tone of a guitar is an excellent example of this. Although the guitar tone lasts a relatively long time, the wave form continually changes, because the overtones are continually changing in phase and amplitude with respect to the fundamental.

The details of how the string motion couples through the bridge to the wooden plate are somewhat different for plucked strings. The bridge on plucked instruments is lower, and the instruments usually do not have sound posts. The coupling mechanism is more like that of a piano than that of the violin family. The bridge of plucked instruments couples oscillations in the plane perpendicular to the plate more efficiently than those parallel to the plate, giving prompt and sustained components to the decaying sound. This can be heard by listening carefully to the tone of the guitar.

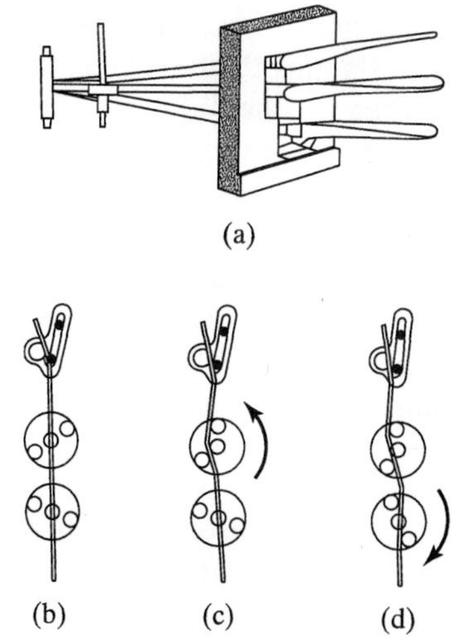

Figure 12-6 Tuning mechanism used in the double-pedal harp. When the pedal is in the central position, shown in (a), the tuning discs are in the positions shown in (c), and the strings are tuned to a C major scale. Pushing a pedal up and down rotates the disks on all strings having the same letter name. A lowered string is shown in (b), and a raised string is shown in (d). (From *The History of Musical Instruments* by Curt Sachs. Copyright 1940 by W. W. Norton & Company, Inc., renewed © 1968 by Irene Sachs. Used by permission of W. W. Norton & Company, Inc.)

The pizzicato technique for a violin-family instrument must be somewhat different from that for the plucked strings because of the mechanism by which the bridge couples the oscillations to the body. A violin player is taught to pluck with a finger motion parallel to the front plate, not outward, away from the plate. This creates a string vibration parallel to the plate, which can then couple to the front plate by the same rocking motion of the bridge. Plucking the violin perpendicular to the front plate will not produce this rocking motion of the bridge, and will therefore yield a weaker plucked sound. An interesting extreme case is the "Bartok" pizzicato, named after the twentieth-century composer Bela Bartok, who popularized its use. The string is pulled far away from the fingerboard and released, slapping the fingerboard with a sharp crack.

The harp is the only true plucked stringed instrument regularly found in the modern orchestra. The big sound obtained by a modern harp arises from vibrations in the frame, rather than a sounding board or plates, as in the other stringed instruments. Whereas the medieval harp was limited to a diatonic scale with a range of about two octaves, the modern orchestral harp can play chromatically over a range of several octaves. To obtain sharp and flat notes, seven double-action foot pedals are used, often with the performer changing pedals rapidly while playing. Each pedal controls all the strings with a given letter name and has three fixed positions (sharp, natural, and flat), which control stops that move against the string, as shown in Figure 12-6.

Virtually all plucked instruments except the harp are fretted; the positions of the frets are set to produce the correct intonation of the fretted notes. Two effects are present in determining the frequency of stopped notes: the position of the frets and the slight increase in tension of the strings when they are pushed down against the frets.

The "rule of 18" has been used since medieval times in determining fret positions. According to this procedure, each successive half-step is obtained by positioning a fret at one-eighteenth of the remaining length of the string. This produces a ratio of frequencies of approximately 1.0588 for notes a half-step apart, a reasonable approximation to the equal-tempered half-step, which is 1.0595. However, because of the difference between strings, the frequency ratios do not scale identically for all strings, and playing using higher frets will often introduce successively greater intonation errors between the strings.

Attempts were made to obtain proper intonation, including some irregular temperaments, on some lutes and other early plucked strings by slanting the frets and other techniques. The additional tension created in the string when it is held down against a fret also affects the intonation. Intonation problems seem to appear to some extent in all fretted instruments, the bowed viols as well as the plucked instruments. The best that can be done is to situate the frets in their best compromise positions.

The most significant recent development in plucked stringed instruments is the use of electronic pickups, particularly with the guitar. The acoustic guitar simply uses the normal wood and air resonances of the instrument to amplify the sound of the string. In an electric guitar, a transducer (an electronic pickup) or perhaps several such pickups mounted on the wood plates of the guitar, convert the sound vibration into an electronic signal. This signal is then fed into a preamplifier and power amplifier, and then to a speaker system. The electric guitar has become immensely popular in recent years, especially in show music, jazz, and rock and roll. A great variety in sizes, shapes, and basic designs for electric guitars has appeared recently, some of which bear little resemblance to the traditional acoustic guitar.

SUMMARY

(***Section 12.1***) The earliest stringed instruments were single-string plucked instruments such as the **monochord**, which attained a place in Greek philosophy. Early stringed instruments such as the medieval harp and the lyre took on mystical meanings. Hammering and bowing developed as techniques for stringed instruments with the introduction of such instruments as the **hurdy-gurdy** and the **dulcimer**. During the twelfth century, families of unfretted bowed strings, such as the **rebec** and **fiedel**, and fretted instruments, such as the **lute** and the **viola da gamba** families, developed. The only record of many of these early stringed instruments is in preserved paintings of the periods. During the Renaissance the **violin family** became popular, and the **guitar** developed in nearly its contemporary form. The **clavichord** and the **harpsichord**, precursors to the piano, were also popular. Of the myriad of stringed instruments used during the Renaissance, only a few remain in popular use today. (***Section 12.2***) Stringed instruments behave acoustically in similar ways: a **bridge** transfers the vibration of a string, either bowed or plucked, to the wooden plates of the instrument, which couple the vibrations to the air through a series of **Chladni plate resonances**. (***Section 12.3***) In a violin, the **bow** uses horsehair coated with resin to scrape the string, creating a triangular displacement that moves back and forth along the string. The **soundpost** aids in transferring the vibration from the **belly** to the **back** of the instrument; the **bass bar** reinforces the belly of the instrument so that pressure from the bridge will not damage the wooden belly, and it also causes the back to vibrate more coherently. A Helmholtz type **air resonance** enhances lower-frequency notes. A violinmaker uses **tap tones** to set the wood resonances to the proper frequencies, and makes **loudness curves** to locate resonances that affect the tone of the instrument. In general, the number of harmonics depends on the position at which the string is bowed: **sul tasto** bowing produces fewer harmonics and a more mellow tone, whereas **sul ponticello** bowing produces more harmonics. A number of factors affect the tone of a violin, including the type and cut of the wood, the glue and varnish, the strings and bow, and playing history. Acoustical details of the violin have changed significantly over the centuries as the pitch level has risen. **Natural harmonics** and **artificial harmonics** are used to achieve special effects. Other members of the violin family have similar properties but have some problems resulting from changes in the wood and air resonances with respect to the played notes. One such problem is the cello **wolf tone**. (***Section 12.4***) Members of the violin family must have their strings plucked parallel to the wood plate, for the most efficient transfer of sound to the plates; because of the different geometry of the guitar, its strings must be plucked perpendicular to the plate. The **harp** is the only orchestral plucked stringed instrument. Other common plucked instruments, such as the guitar, possess **frets**, which are positioned on the fingerboard of the instrument according to the **rule of 18**.

QUESTIONS

1. List and describe the resonances that play an important role in the violin. What happens when these resonances occur at the wrong frequencies?

2. Calculate the sizes of "full-size" cellos and basses by assuming that the size of the instrument should scale up from the violin as the wavelength of the lowest note. Compare this with a normal cello and bass. Roughly, how much would a hand have to move to play an octave above the open string on one string of these "full-size" instruments?

3. Obtain a cello, preferably an inexpensive one. Play notes around the note $F_2^\sharp$ on the lowest (C) string, and try to find the wolf tone. Why is it easier to produce the wolf tone on an inexpensive cello?

4. **a.** Compare the sound produced by a guitar when it is plucked with motion parallel to the front plate and with motion perpendicular to the front plate—that is, away from the plate,

b. Compare the pizzicato sound obtained on a violin using these two types of motion. Why is plucking parallel to the body of the instrument better than plucking perpendicular to the body?

5. Remove the sound post from an inexpensive violin; then play the violin. Compare the sounds before and after the sound post is removed. Explain the difference you hear.

6. Listen to a recording of a baroque-style violin. Compare this sound with that of a contemporary violin.

7. Play some natural and artificial harmonics on a violin. Describe the sound quality and explain its origin. Describe possible sources of these differences.

8. Obtain a cello and try to create the seagull effect. Describe in detail what is happening.

9. Examine some folk instruments, such as the psaltery, banjo, ukulele, and dulcimer. Describe the mechanisms by which string vibrations are transformed into sound waves in the air.

10. Use a pair of cords to suspend a bow above a horizontal violin so that the bow is perpendicular to one of the strings and gently touches one of them at a point. Hold or fasten the violin in its horizontal position and bow the string with a second bow. Observe any motion of the suspended bow for various positions along the string. What conclusions can be drawn from this experiment concerning bowed violin strings? Where is the motion the greatest? How does the position of the bow used to bow the string affect the motion of the string?

11. Put cotton in the *f*-holes of a violin or cover the *f*-holes with masking tape (completely or partially) to investigate the role of air resonance in the production of violin tone. How do your modifications affect volume and tone quality at different frequencies (different notes)? Note particularly changes near the frequency of the air resonance.

PROBLEMS

1. Suppose that you play a natural harmonic on your violin by touching gently the A string (440 Hz) at exactly 1/10 of its length. What note will you be playing, and what is its frequency?

2. Suppose that you play an artificial harmonic on your violin by stopping the A string at one-quarter of its length while simultaneously touching it gently at one quarter of the small (vibrating) section. What note will you be playing, and what is its frequency?

3. Suppose that you wish to play a beatless double stop (two notes at once on different strings) on your violin, listening for a difference tone to tune the interval. If you play the open A string and the F_4 below that A (finger stopped on the D string) forming a major third, what will the exact frequency of the note F_4 need to be in order to make the interval beatless? What will be the frequency of the difference tone, and what will be its letter name?

4. Suppose that you play a double stop consisting of the note F_5, one half-step above the open E string, and the note C_5, a minor third above the open A string. What is the musical interval between these notes? What are the frequencies of these notes in the equal-tempered scale? What will be the frequency of the note C_5 after it has been adjusted to make the interval beatless? What are the frequency and the letter name of the difference tone?

REFERENCES

BACKUS, JOHN. *The Acoustical Foundations of Music.* 2nd ed. New York: W. W. Norton & Co., Inc., 1977. See his reference list.

BENADE, ARTHUR H. *Fundamentals of Musical Acoustics.* New York: Oxford University Press, Inc., 1976. See his reference list.

FIRTH, IAN M., and J. MICHAEL BUCHANAN. "The Wolf in the Cello." *Journal of the Acoustical Society of America,* 53, no. 2 (1973) A somewhat detailed description of the physical mechanism by which the cello wolf tone is created.

FORD, CHARLES, ed. *Making Musical Instruments: Strings and Keyboard.* New York: Pantheon Books, Inc., 1979. Contains much of the detailed information necessary to appreciate the construction and workings of stringed instruments. It contains excellent and complete information on how to build many instruments.

HUTCHINS, CARLEEN, M. "The Physics of Violins." *Scientific American,* November 1962. An early survey article by one of the important researchers in the field of the acoustics of stringed instruments.

PICKERING, NORMAN. "A Study of Bow Hair and Rosin," *Journal of the Violin Society of America* 8, No. 1, p. 46 (1984). Important study of the violin bowing mechanism.

SCHELLING, JOHN C. "The Bowed String and the Player," *Journal of the Acoustical Society of America,* 53, no. 3 (1973). Deals with several aspects of the physics of bowed stringed instruments.

———. "The Physics of the Bowed String." *Scientific American,* January 1974. Deals primarily with the mechanism by which the bow imparts energy to the string of a bowed stringed instrument.

WHEELER, THOMAS H. Rev. ed. *The Guitar Book: A Handbook for Electric and Acoustic Guitarists.* New York: Harper & Row, Publishers, Inc., 1978. Contains a wealth of information on all kinds of acoustical and electric guitars, including some history.

Chapter 13 The Piano

In this chapter, we shall cover the history of the development of the modern piano, which involves other stringed keyboard instruments, such as the harpsichord and clavichord. We shall discuss the construction and operation of the modern piano and relate this to basic physical principles, some of which have been treated earlier, such as Mersenne's laws, although others may be new to the reader. Performance technique will be discussed when appropriate.

13.1 History of the Piano

The modern piano, or pianoforte, has historical roots stretching back to antiquity, with the discovery that a stretched string emits a tone when plucked. Although the origins of the classical lyre are unknown, several inscriptions suggest that the ancient Assyrians had some form of this primitive stringed instrument. The Greeks also had simple lyres for which the sound box was a hollow tortoise shell; they also had citharas, which consisted of a wooden soundbox, two curved arms, and a crosspiece that held taut the several strings connected to the sound box. The number of strings on citharas was as high as 11 or more by the fifth century B.C. Both lyres and citharas were plucked with a hand-held implement called a plectrum.

About 582 B.C., Pythagoras constructed a monochord, an instrument that consisted of a single string stretched across a resonator box. It had a movable wooden bridge that could divide the string into two sections, each of definite length and thus definite pitch. Pythagoras used his monochord to investigate the relationship between the pitch of the fundamental and the length of the string and discovered that sections of string whose lengths were related by integers emitted tones that sounded pleasant together.

The psaltery, mentioned in the Bible, consisted of a number of strings stretched across a resonator box. The performer could pluck several strings at a time. Keyboards,

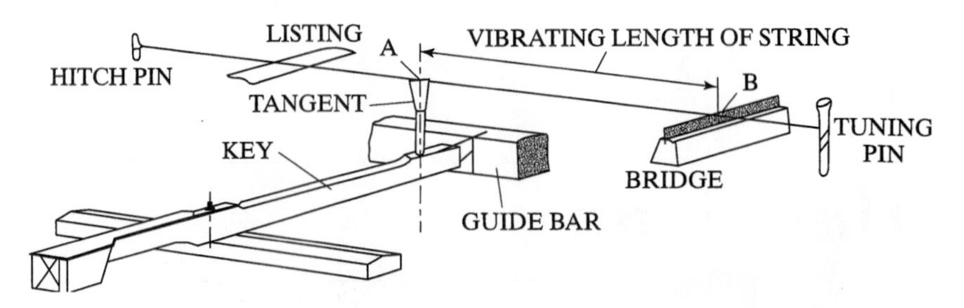

Figure 13-1 Clavichord mechanism. As the key is pressed, the tangent strikes the string and sets the length of string at the right vibrating. (Reprinted by permission of the publisher from *The Harvard Dictionary of Music*, Fourth Edition, edited by Don Michael Randel, p. 187, Cambridge, Mass.: The Belknap Press of Harvard University Press, Copyright © 1986, 2003 by the President and Fellows of Harvard College.)

essential in further development toward the modern piano, were first used with pipe organs in about the second century B.C.

Around the beginning of the eleventh century a three-stringed instrument, the hurdy-gurdy or organistrum arose, in which one of the strings could be pressed by metal tangents at different positions, producing different pitches; the other two strings were drones. Church musicians of the period used monochords to help teach singers chants. These monochords were then grouped together on the same sound box to form polychords. Near the beginning of the fifteenth century, polychords were fitted with keys that controlled tangents similar to those of the organistrum. This was the first clavichord.

In the clavichord there were several tangents per string, each tangent striking a different point along the string. One end of the string was damped with a cloth strip. When a key was depressed, its tangent struck the string, set it vibrating, and forced a node at the point of contact. The undamped part vibrated at a frequency determined by its length, while the damped part did not vibrate. When the key was released, the tangent left the string and the tone stopped because of the cloth damping. Figure 13-1 shows the action of a clavichord.

Although the tone of the clavichord is soft, there is a fair dynamic range available. As the key is struck more strongly, more energy is imparted to the string, and it sounds louder. An unusual type of vibrato is obtained when the key pressure is changed while its note is sounding; the tension and thus the pitch of the string change slightly. Because the positions of the tangents along the string are fixed, pitch relationships cannot be changed. For that reason, temperament changes required to perform in harmonically distant keys cannot be made. By the end of the seventeenth century, however, clavichords with only one tangent per string were being made, which avoided the temperament limitation.

Parallel to the development of the clavichord was the progression from the psaltery to the harpsichord. The triangular psaltery was fitted with keys, and by the beginning of the sixteenth century an elaborate action had developed. In harpsichords, each string vibrates in its entirety and is excited by one key. As a key is depressed, a vertical jack moves up and a crow-quill plectrum plucks the string, as illustrated in Figure 13-2. As the key is released, the jack descends until the angled, lower side of the

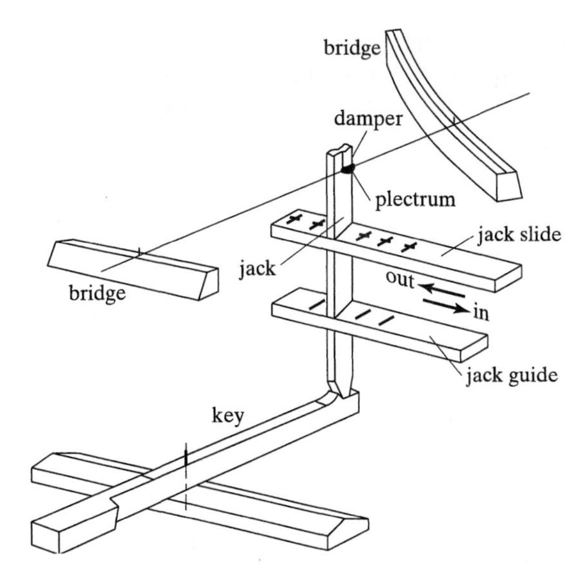

Figure 13-2 Harpsichord mechanism. As the key is pressed, the jack rises and the quill (plectrum) plucks the string. When the key is released, the jack descends and the quill tips to the side so the string is not plucked again. (Reprinted by permission of the publisher from *The Harvard Dictionary of Music*, Fourth Edition, edited by Don Michael Randel, p. 386, Cambridge, Mass.: The Belknap Press of Harvard University Press, Copyright © 1986, 2003 by the President and Fellows of Harvard College.)

plectrum touches the string. The jack then tips sideways, slightly, and the jack can descend farther. Mounted on the jack above the plectrum is a cloth damper, which then damps the string. As with the clavichord, the harpsichord string can vibrate only as long as the key is depressed.

Although generally louder than the clavichord, the harpsichord has one important drawback: Its dynamic range is limited. Striking a key harder does not result in a significantly louder sound, because the string moves about the same distance to the side as it is plucked. A partial solution was to add one, two, or more "choirs" of strings, tuned in unison and/or octaves, such that one key activated corresponding strings simultaneously, producing a louder sound. A jack slide could be hand-shifted to disengage the plectra of any choir of strings. As a jack slide is moved to the left, in Figure 13-2, the plectrum is disengaged, and the string will not sound when the key is struck. Because only one, two, or three of the choirs can be activated at any one time, the harpsichord dynamics are terraced.

In 1709, Bartolommeo Cristofori, a harpsichord maker in Florence, Italy, produced the first hammer action to replace the quilled-jack harpsichord action. Because his hammers were covered with leather, the tone quality differed from the harpsichord and, most importantly, it allowed a large, even (nonterraced) dynamic range. He called his new instrument *gravicembalo col piano e forte*[1] (soft and loud), expressing the new dynamic spectrum available. This was the origin of the "pianoforte." Cristofori's later, improved action contained many of the essential features found in the actions of today.

Because considerable force could be transmitted with the hammer action, strings that were tighter, stronger, and longer could be used to increase the dynamics. Also, multiple strings for each key were used for the same reason.

[1]Gravicembalo is a modification of *clavicembalo*, the Italian word for harpsichord; *gravi* means "lower."

These changes required a longer and more solid piano body, and in 1822 Babcock in Boston introduced a cast-metal frame, which permitted very high string tension. Modern grand pianos withstand enormous force owing to this tension, a total of some 60,000 pounds, the weight of 20 small cars!

13.2 Construction and Action of the Piano

Figure 13-3 shows an exploded view of a modern grand piano. The body or case is made of wood veneer, but recently other materials, such as clear plastic, have been tried. The large flat soundboard is made of spruce planks joined together, is bellied slightly upward, and fits snugly into the case. The pin block is made of very strong, laminated maple and contains about 230 steel tuning pins, or pegs, to which the ends of the metal strings are attached. Both the pin block and the soundboard sit under the heavy cast iron frame, with the pin block positioned such that the enormous tension of the strings is transmitted to the frame. The strings are attached to tuning pins that pass through holes in the frame. The strings pass over the bridge, around pegs set at the far end of the frame, then back to the pin block. Some manufacturers use felt strips to damp this unstruck portion of the string between the bridge and pegs. Other manufacturers leave it undamped, and this may help to produce a "brighter" sound.

The keyboard consists of 88 keys, 36 black and 52 white, which when depressed initiate a complicated motion of the action that sends a hammer to strike the unison of strings. Each key has its own action, such as the one shown in Figure 13-4, except the actions of the upper 21 keys of many pianos do not have dampers. Also, the hammers for high notes are smaller and lighter than those for low notes. This permits better *coupling* between the hammer and string, increasing ease of play: a heavy hammer is more effective in exciting a heavy string, a lighter hammer is more effective in exciting a lighter string.

As the performer depresses the key, the key pivots about a point just beyond its midpoint (the section nearest the player is longer), sending the far end upward, like a seesaw. This upward motion leads to two subsequent motions: (1) the underlever and damper head are raised, and (2) the hammer is thrown or pushed upward to strike the strings. As the key is released, the damper returns to the string and the sound stops.

Another important feature of modern piano action is that it allows the player to repeat notes without fully removing the finger from the key. With proper adjustment of the action, or *regulation*, as the key is released, first the hammer will reset, then the damper drops onto the strings. This feature produces two important results: (1) Notes can be repeated extremely rapidly on a well-regulated piano, much more so than on the harpsichord, for example. Many composers of the nineteenth century took advantage of this and wrote music exploiting this new technical capability. (2) A note can be repeated while it is still sounding, without using the pedal. This can be useful in some sustained passages.

Consider a note played and held for several seconds. No matter what performance technique is used to strike the key, the string only "knows" how fast the hammer strikes it, and this speed determines the loudness of the sound. For a given key, many performance techniques (such as playing from the finger, wrist, elbow, or shoulders)

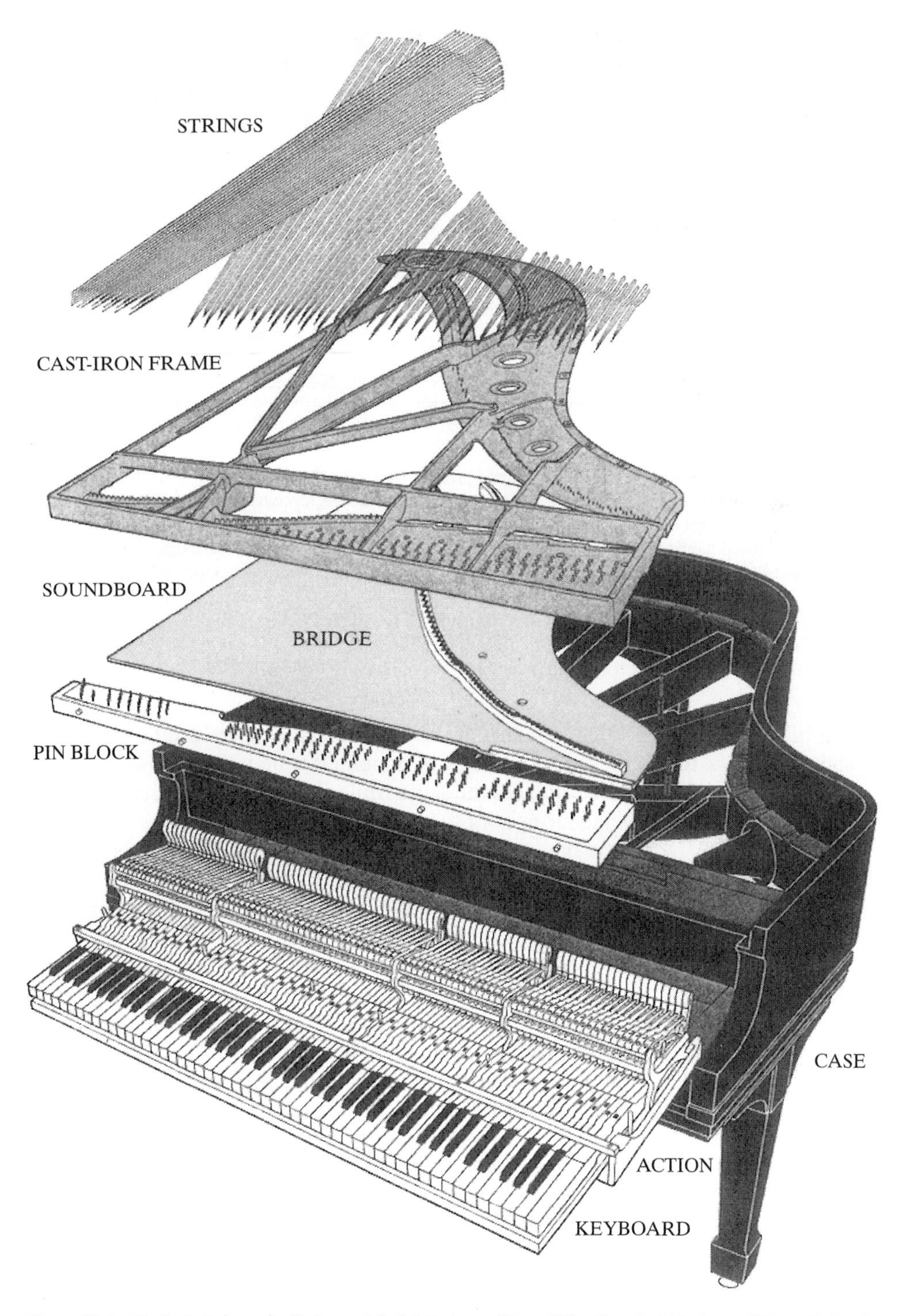

Figure 13-3 Exploded view of a Steinway Model B piano. (From "The Coupled Motions of Piano Strings" by Gabriel Weinreich, *Scientific American*, January 1979. Graphics by Dan Todd; used with permission.)

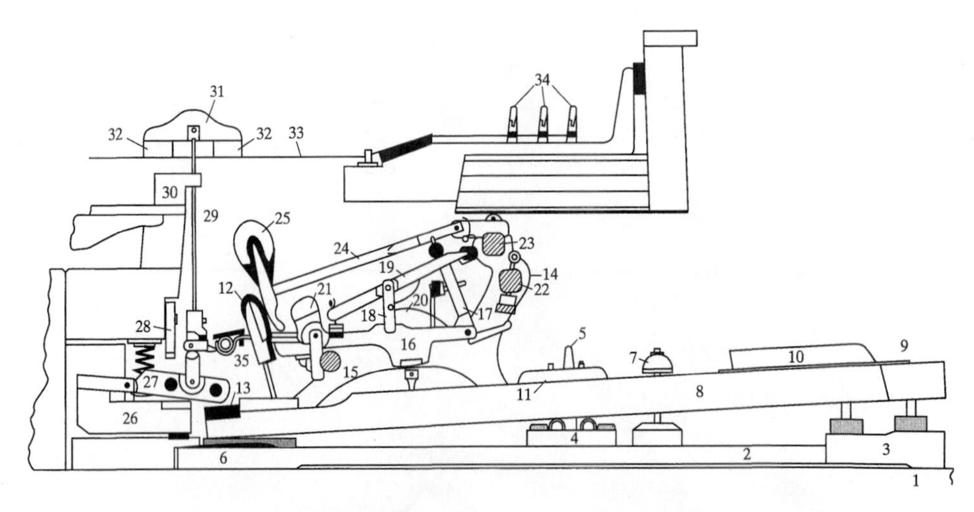

Figure 13-4 Cross section of a modern grand piano action: 1. keybed; 2. keyframe; 3. front rail; 4. balance rail; 5. balance rail stud; 6. back rail; 7. key-stop rail; 8. white key; 9. key covering; 10. black key; 11. key button; 12. backcheck; 13. underlever key cushion; 14. action hanger; 15. support rail; 16. support; 17. fly; 18. support top flange; 19. balancer; 20. repetition spring; 21. hammer rest; 22. regulating rail; 23. hammer rail; 24. hammershank; 25. hammer; 26. underlever frame; 27. underlever; 28. damper stop rail; 29. damper wire; 30. damper guide rail; 31. damper head; 32. damper felts; 33. string; 34. tuning pin; 35. sostenuto rod. (Reprinted by permission of the publisher from *The Harvard Dictionary of Music*, Fourth Edition, edited by Don Michael Randel, p. 653, Cambridge, Mass.: The Belknap Press of Harvard University Press, Copyright © 1986, 2003 by the President and Fellows of Harvard College.)

can be used to get the same loudness. If different techniques are used but produce the same loudness, the sounds of the string will nonetheless be indistinguishable. In short, there is no way to keep the loudness the same while changing warmth, crispness, or any other attribute of the piano string sound.

The attack technique may have more subtle effects, however, effects that are discernible by a well-trained ear. The noise of the action can differ between notes of equal loudness, depending on performance technique, and the slapping of a finger against the key can sometimes be heard. Although the string motion is only determined by the final speed of the hammer, the noise at the attack need not be strictly determined by this final speed.

This is not to say that the attributes mentioned cannot be obtained, or that the pianist should not learn different performance techniques. Varying the loudness can change the sound quality, as we shall see in the next section, and different techniques are appropriate for different loudnesses. Of course, the duration that the key is depressed can be controlled, as this will result in different effects, and different performance techniques are appropriate for producing notes of different duration. Thus one can have staccato (short) notes and legato (long) notes of the same loudness, but these differences result from differences in the duration of the note and cannot be attributed to the attack per se.

Because the strings in the upright piano are vertical and those in a grand piano horizontal, the corresponding actions and sounds differ. For example, hammers in grand pianos fall back because of gravity, while a more complicated system is used in uprights.

The sound due to the initial impact of hammer with strings is thrown back, usually toward a wall, for uprights. In the grand piano, however, this sound is projected up and out, reaching the listener before any reflections occur. This leads to a crisper sound.

Because most effort went into perfecting the horizontal piano, and the upright was introduced primarily for cost and size considerations, we shall restrict our attention to the grand piano in the following section.

13.3 Piano Strings and Sound Production

The piano strings are the heart of the piano, and their motions are the most important factor in determining the sound quality of the piano. Their motions are complicated and, in many respects, deviate from the "ideal" discussed in previous chapters. Some of the most important aspects of piano string motion have only been properly understood recently, and some aspects may still not be completely understood.

Mersenne's laws can be used to understand the most obvious features of piano strings. The first is that the strings of high notes are shorter than those of lower notes, as described by Mersenne's first law. If length were the only difference between the strings, however, and the shortest were, say 6 inches, the longest would have to be about 76 feet long! Instead, the mass per unit length is changed, in accordance with Mersenne's third law, to make the pitch of the lower strings low while keeping the length reasonable. This extra weight is added in the form of multiple wrappings of the strings. Lower strings are wrapped in copper wire, and the very lowest strings have two such wrappings. The tension does not vary significantly from string to string, because this might lead to uneven stress on the frame and perhaps buckling.

Inharmonicities play a very important role in the sound of the piano. Because there is some inherent stiffness in the strings, the overtones produced do not have the simple, integral relationship that we have discussed throughout the book. This inherent stiffness makes the overtones too high, as if the oscillations were in a more tightly stretched string. The higher overtones result from modes in which the string has many loops, and thus a great deal of bending, and these are successively higher in pitch than the simple harmonics of $2f$, $3f$, and so on. In a piano string, the sixteenth overtone may be as much as a half-tone higher, and the fiftieth overtone may be as much as seven half-steps higher than in the ideal case. Wrapping the strings in the manner discussed is done to minimize stiffness. A thicker string, like a stiff rod, would raise the overtones unacceptably. Electronically synthesized "piano" sounds that are made of integrally related overtones do not sound as full or as rich as when the overtones are inharmonic; the tone does not sound quite like that of a piano.

This inharmonicity affects piano tuning. If the notes an octave apart were tuned such that their fundamentals had a ratio of frequencies of precisely 2, then the slightly raised second overtone of the lower note would beat with the fundamental of the upper note when played simultaneously. To avoid the beating, notes an octave apart are *stretched*; that is, they are tuned to have a ratio of frequencies slightly greater than 2.

The interaction of the hammer with the strings affects the sound as well. If any string such as a guitar string is plucked or struck near its end, many higher harmonics will be excited, because the string has a sharp bend in it. Guitarists sometimes pluck strings near the bridge to produce this type of sound for variety. To produce uniform tone along the keyboard, then, the hammers must strike their respective strings at the same ratio along their lengths. What should this ratio be? If the hammer were to strike the string at its midpoint, low-frequency overtones would sound; if the hammer were to strike near the end, higher overtones would predominate. Because the sound is considered more beautiful and rich when many high and low overtones are present, the position is chosen such that many of the overtones sound. Hammers are set to strike strings at a point ranging between one-seventh and one-eighth of their lengths, positions found to yield rich tones when struck. If the hammers were very sharp, a node might be produced at the striking point, as in the clavichord. Because the hammer is not a point, but fairly soft and curved, a node is not produced there, and many overtones sound.

The sharpness or hardness of the hammer also affects the sound quality by exciting many high overtones. If the felt on the hammer is worn and hardened, the string will have a sharp cusp in it when struck. It will thus have a large number of higher overtones with appreciable amplitudes. This produces the "tinny" sound of an old, worn barroom piano. This tinny quality can be reduced by pricking the felt on the hammers with a needle, loosening and softening the felt.

Up to now, we have been discussing single strings, but many important aspects of the piano sound result from the fact that most keys activate several strings tuned very nearly in unison. The first such aspect is loudness. The highest 68 keys have triplets of strings, and the lowest keys have singlets or doublets. This arrangement helps to keep the loudness uniform from note to note, because a small, single upper string is not very loud when sounded alone.

The multiple strings have other, more complicated effects involving the *prompt* and *sustained* sounds. The loudness of a single long note drops quickly for the first few seconds. After that, the decay is much more gradual, lasting as long as 30 seconds or more. The existence of the early (prompt) and later (sustained) sounds can be explained using several different mechanisms.

The first of these involves polarization. Recall from Chap. 2 that a transverse wave, such as a wave on a string, can be polarized. For instance, the motion of the string can be up and down, or it can be side to side (or combinations of both). When the hammer strikes a string, most of the energy goes into forcing the string to vibrate in this up-down (vertical) polarization, but because there are slight irregularities in the hammer surface and striking angle, the string vibrates in its side-side (horizontal) polarization as well, although not as much. The sound resulting from the vertical polarization dies down very quickly as the vertical motion of the string is transmitted very effectively to the soundboard, where it is radiated as sound. The horizontal polarization does not radiate its energy as effectively, because the string pushes horizontally on the soundboard; thus the sound produced by horizontal string motion will last a long time. Even though the sound, because of vertical polarization, is initially louder (the prompt sound), there comes a time, about 5 seconds after the string is struck, that the sound due to the horizontal polarization is louder (the sustained sound). This partial explanation of the origin of the prompt and sustained sounds is also applicable to the guitar, as we saw in Section 12.4.

The full explanation must take into account the multiple strings of a single key. Consider a key that has a doublet of strings. Just after hammer impact, the strings are moving up and down in phase (we will neglect the horizontal motion for the time being), but because of irregularities in the hammer, as mentioned earlier, one string (call it A) will have a slightly higher amplitude than the other (B). Because they are initially moving in phase, they push and pull hard on the bridge, and sound energy is radiated from the soundboard. Later, as the amplitude of the strings decreases, there comes a point at which the amplitude of B is essentially 0, although A is still oscillating. The motion of each string couples to the motion of the other through the bridge, and A begins to drive B; this is an example of driven oscillation. Thus B begins to oscillate, although now it oscillates out of phase with respect to A. A continues to decrease its amplitude and B to increase its amplitude until they are approximately equal. At this time, energy is radiated much more slowly, as the bridge does not move much; as A pulls the bridge up, B pulls it down, and vice versa. There is, essentially, no net vertical force on the bridge. This is another cause for the sustained sound. If this were a perfect system, no energy would be lost through the support, only through the strings pushing the air directly. This basic mechanism also occurs with three strings, but it is more complicated and we shall not discuss it in detail.

For some piano keys, these complex interactions may occur more readily with certain overtones than others. The result is that different overtones will decay at different rates. This can be heard by striking and holding the note C_3. The sound of the third overtone, at G_4, decays more slowly than that of the fundamental, as can be easily heard.

We can now understand the function of the left piano pedal, the *una corda*, or soft pedal. During soft passages, the sustained part of a held note may be too soft, softer than the background noise in the concert hall, while the prompt sound is audible. The problem then is to make the sustained part of the note louder without making the prompt sound louder. The una corda pedal solves this problem. When the pedal is pressed, the entire keyboard and action are shifted slightly to the side, so a hammer that would normally strike three strings only strikes two. The two struck strings begin moving in phase and immediately start to drive the unstruck string to move out of phase. Thus the sustained sound will occur before the amplitudes of the struck strings have dropped very much. Consequently, the sustained sound is louder relative to the prompt sound when the una corda pedal is pressed than when it is not pressed.

On an upright piano or spinet, the *soft pedal* replaces the una corda pedal. Rather than slide the keyboard and action, the soft pedal modifies the action so that the hammer will not hit the strings as hard as when the pedal is not depressed.

The middle pedal, or *sustain pedal*, on a grand piano is missing from most uprights and some grand pianos. The pedal sustains only keys that are being held at the time the pedal is pushed. For a few grand pianos and most uprights that have a middle pedal, it simply acts as a damper pedal for the notes in the low range, say below about C_3. This is useful for some types of music in which chords must sound while both hands play other notes.

The last mechanism for producing the sustained tone involves several concepts and results from the complicated motion of deliberately mistuned strings. First, let us recall some aspects of driven oscillations and resonance, as discussed in Chap. 3. When the driver's frequency is slightly different from the natural, or resonant, frequency of the resonator, only a little energy passes from the driver to the resonator. (Recall the

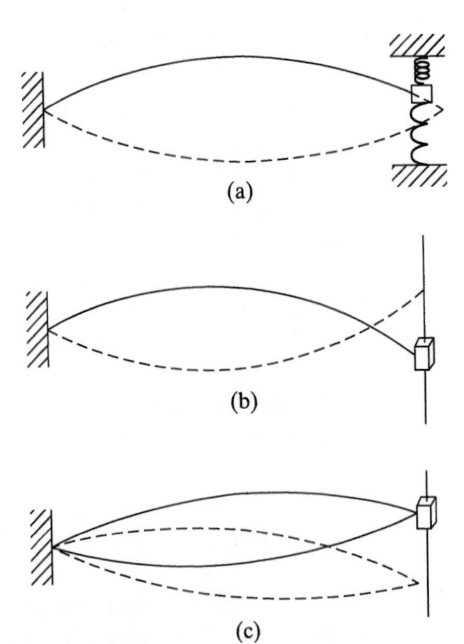

Figure 13-5 String oscillating with one fixed end (at left) and the other end attached to (a) a springy support, (b) a massive support, and (c) a resistive support. (From "The Coupled Motions of Piano Strings" by Gabriel Weinreich, *Scientific American*, January 1979. Graphics by Dan Todd; used with permission.)

experiments with pendula.) But at the resonant frequency, a great deal of energy is passed from the driver to the resonator, and the driver leads the resonator by 90° in phase.

Consider a vibrating string attached to an idealized frictionless "springy" support on one end, as shown in Figure 13-5(a). For such a springy support, the motion of the string and support are in phase. The string loses no energy, because after one complete oscillation the system is back in its original configuration. The string vibrates as if it had a node out beyond the support, and thus has a lower frequency than if the support were the usual fixed type described previously.

Now consider the idealized, frictionless "massive" support shown in Figure 13-5(b). This consists of a very heavy object free to slide without friction up and down a rod. Here the motion of the string and the support will be out of phase, and, as before, no energy will be lost by the string to the support. After one complete oscillation, the system is back in its original configuration. Because the node is shifted inward in the massive-support case, the frequency of the string oscillation is higher than if there were a fixed support.

A resistive support is different. It does have friction, so after each oscillation the string has lost some energy to the heating of the support by friction. The node, however, is right at the support, so that the string's frequency is the same as in the case of a rigid support, as shown in Figure 13-5(c). The motion of the support is 90° behind that of the force of the strings, a point to which we shall soon refer.

These support cases are similar to the driven oscillations mentioned in earlier chapters and will help explain the interaction of the piano string with its bridge. Real bridges have friction and are, for the most part, resistive supports.

Imagine two piano strings, slightly mistuned, that are momentarily oscillating out of phase with each other. As we saw earlier, the bridge will not move appreciably, as the forces of the strings cancel. Then, as the string with the slightly higher frequency (call it A) oscillates, it gains in phase over the other string (B). The strings are no longer out of phase, and a net force results on the bridge. In fact, this force differs in phase by 90° compared with either string—ahead of one and behind the other. The resistive bridge responds by oscillating 90° out of phase with respect to the force, as we saw previously; it is moving in phase with the higher-frequency string (A) and out of phase with B. Thus A "feels" a springy support, and its frequency is lowered; B "feels" a massive support, and its frequency is raised. The two strings reach the same intermediate frequency and sound in perfect unison. We saw how two ideal, identically tuned strings oscillating out of phase would not lose energy to their support. Here, with the slightly mistuned strings, energy can be slowly transmitted to the support because the bridge moves.

Consider now the mistuned strings just after they are struck by a hammer; they begin oscillating in phase. As A gains in phase, the resistive support cannot be 90° behind both A and B, because the two strings have different phases. Thus the initial energy loss is reduced in this case compared to the perfect-unison case. Also, as one string "feels" springy support and the other string a massive support, the frequencies of the strings are made identical. Thus the mistuned strings sound at the same pitch, and because they never lost much initial energy (as they would have had they been perfectly in tune), more of the energy goes into the sustained sound.

Of the three mechanisms for the production of the sustained sound—horizontal polarization, the out-of-phase motion of two strings, and the interaction of mistuned strings—the first two depend on subtleties of the piano itself and may vary from note to note. The mistunings, on the other hand, are set by the piano tuner and can compensate for irregularities resulting from the other mechanisms. That the mistunings set by an expert piano tuner seem random, while the tone characteristics, including the prompt and sustained sounds, are quite uniform from note to note, substantiates this view. If the same piano is deliberately detuned, the same expert tuner will, in general, retune it by making the same subtle mistunings to achieve uniform tone.

One final mechanism is significant in the production of good piano tone: the coupling of the vibration from the sounding board to the air. Consider a tuning fork. When you strike and hold the tuning fork its sound is very soft, often hardly audible at a distance of a few feet. However, if the base of the tuning fork is held against a piece of wood, like a chair seat or table top, the sound is dramatically increased, because the vibrations from the tuning fork are transmitted to the wood, which is coupled efficiently to the air. The entire wooden surface oscillates in the manner of a loudspeaker, producing motion of the air adjacent to the surface. The ability of a piece of vibrating wood like a sounding board to couple its vibrations to the air goes up dramatically with size, accounting for the tremendous volume of sound that can be produced by a large grand piano. This is another illustration of impedance matching.

13.4 Recent Innovations for the Piano

For over a century, both musicians and patrons of piano music have tried to devise a method by which piano music could be recorded and played in a natural and expressive way. Early attempts at recording and playback on real pianos include the old piano-roll technique. The piano roll, a strip of paper with a series of holes and slots, passes by an electromechanical device that senses the holes in the roll of paper, indicating when to play the notes and how long to hold them. This technique produced a large number of individual musical pieces on rolls, which are to this day of great interest to collectors. However, these recordings were limited, in that they provided very little expression. Later, of course, recordings were available that played back the music through an audio system using loudspeakers.

The timely development of digital electronics and computers, along with great improvements in the reliability of electromechanical transducers, has led to the development of new systems of recording a piano performance and playing it back using an actual piano. A transducer is placed on each key, indicating when and how hard the key is struck, the only two factors that affect the sound of the piano. This information is recorded digitally in the memory of a computer and transferred to a hard disk; in this way an entire musical performance can be recorded. The transducer is also able to carry out the reverse process: given the information about when and how hard the key was originally pressed, it can play the key just as it was played in the original performance. Using an array of electromechanical transducers attached to each key of a piano, a computer, and software, the complete performance of a piece of piano music can be recorded and replayed on the same piano or on any piano on which a similar device is installed. One example of such a device is the *Disklavier*.

Another example is the Bösendorfer 290-SE Computer-Based Piano Performance Reproduction System; Figure 13-6 is a photograph of the Bösendorfer system

Figure 13-6 The Bösendorfer 290-SE Computer-Based Piano Performance Reproduction System at the International Piano Archives at the University of Maryland, College Park.

installed at the International Piano Archives at the University of Maryland, College Park. The system includes the array of electromechanical transducers, which is in the large box hanging under the piano, and the computer, which is on a desk behind the piano bench in the picture. An interesting feature of the 9′6″ Imperial Bösendorfer Grand piano in the photograph is that it has 97 keys. The nine extra keys (in black at the low end of the keyboard) extend the range of the keyboard one octave below that of the standard 88-key piano, providing extra resonance for bass notes and allowing this piano to perform music specially written for a Bösendorfer or other piano that includes those extra keys. These extra keys are occasionally useful when performing transcriptions of orchestral music, which might include such low notes.

After recording the music on the 290-SE System, it can be edited using the computer keyboard to remove any performance problems, or to modify or enhance any aspect of the performance: changing notes, adjusting volume, or varying timing. When the recording is played, the keys on the piano move just as if they were being struck, and the music sounds exactly the way it would if the performer were actually playing the piano. With this system, for the first time in history it is possible to construct a literally perfect performance—exactly the way the performer intends it to be. In addition, the performance is on an actual piano, not a recording, so the sound is as good as that of the original piano performance. This device can then be used to play the piano for a normal recording session or can serve as one piano in a live dual-piano performance by the same person. These recordings can be distributed on floppy disks or over the Internet.

Most of us have seen guitars and even violins played "left handed"—that is, with the strings reversed and the functions of the left and right hands interchanged. The construction of the piano has also been very strongly influenced by the fact that most people are right handed. For example, higher notes are on the right of the keyboard, and the more technically difficult parts are generally assigned to the right hand. In recent years there has been an increasing realization that left-handed people are at a distinct disadvantage in playing instruments such as the piano. Some contemporary piano historians believe that the left- or right-handedness of composers significantly affected their compositions. In the Renaissance period most woodwind instruments were constructed so that either hand could be positioned on top or the bottom finger holes, avoiding this disadvantage. When technically complex chromatic music made fancy machinery necessary on these instruments, they became "right handed."

A piano asymmetric to the standard "right handed" piano would allow a left-handed person to play the piano *exactly the same in every respect* except that high and low notes are reversed. The music is the same and the fingerings are the same, except that the hands are reversed! A commercially available electronic device, called the "Keyboard Mirror®", can be used with an electronic keyboard and a MIDI system to reverse the keys of a modern electronic keyboard so that it can be played left handed. After his experience with the Keyboard Mirror, Christopher Seed, a left-handed English professional pianist and piano teacher, produced a left-handed piano, the first of its kind in the world. Mr. Seed found that it is relatively easy to interchange hands after the basic piano technique had been mastered. This development even allowed him to expand his repertoire to include certain music that was particularly suited for right-handed pianists. Making it possible for the people in the world who are left handed to participate fully in piano performance might be the most important development in the acoustic piano of the late twentieth century.

In conclusion, we have seen how the keyboard instruments have evolved throughout history and how many physical principles and processes are involved in the modern piano. It is hoped that the student has a greater appreciation for the piano and recognizes it as truly "grand."

SUMMARY

(*Section 13.1*) The ancestry of the piano can be traced back over 2,000 years to such antique instruments as the monochord and the psaltery. The immediate ancestors of the modern pianoforte are the **clavichord** and the **harpsichord**. In 1709 Cristofori invented the *gravicembalo col piano e forte*, or **pianoforte**, increasing the expressive and the technical capabilities of the instrument. (*Section 13.2*) The modern grand piano, expanded to 88 keys, has a steel frame with a large, heavy wooden **sounding board** and heavy strings under great tension. The **action** of each key allows a great dynamic range and expression, and provides a mechanism for very rapid repetition. Important parts of the action include the **key**, the **hammer**, and the **damper**. (*Section 13.3*) **Mersenne's laws** govern the size and length of the strings, and the tension in all strings is kept approximately constant so forces on the frame are uniform. The **stiffness** of the lower strings leads to **inharmonicity** in their overtones and to tuning adjustments, such as **stretched octaves**. The shape and hardness of the **hammer** is selected to provide a full harmonic spectrum. **Polarization** of the string vibrations creates prompt sound, whereas **coupling** between strings for the same note is responsible for the delayed sound. The **una corda** pedal allows only two of the three strings for each note to be struck, reducing the prompt sound but not the delayed sound. The **sustain** pedal lifts the dampers to allow the strings to vibrate freely. (*Section 13.4*) Recent innovations for the piano include electromechanical reproduction systems, such as the Bösendorfer 290-SE Computer-Based Piano Performance Reproduction System, and the left-handed piano.

QUESTIONS

1. Trace, roughly, the production of a piano tone from the performer's finger motion to the final emergence of the sound.

2. Play and hold a piano note. Listen for the prompt and sustained sounds. Describe three mechanisms by which these sounds are produced. Which of these mechanisms can be controlled by a piano tuner? Which cannot?

3. Define inharmonicity. What makes piano notes inharmonic? What effects does this have on tuning and tone quality?

4. Listen to harpsichords and clavichords and compare their volume, range, and tone quality with each other and with that of a grand piano. Discuss the physical reasons for these differences.

5. Obtain recordings of piano music played on historical pianos, beginning from about 1750. Analyze the sound for each type of music being heard. Investigate some of the physical features of these old pianos that gave rise to these early sounds.

6. Listen and compare the effect of the una corda pedal on the tone of a grand piano with the effect of the soft pedal on an upright piano or a spinet. Try to observe the source of these differences in actual pianos. Explain the origin of the differences.

7. Investigate the strings of a grand piano and notice how the entire keyboard is made up of ranges of similar strings and/or groups of strings. Write down how the strings are constructed for the entire piano (e.g., groups of 1, 2, or 3 per note; wirewound or simple strings; etc.).

8. Describe what is meant by the "touch" of a performer on the piano. Discuss why it is impossible for a performer to independently control both the touch and the loudness.

9. Strike a tuning fork and touch it to the soundboard of a piano to see how the soundboard amplifies the sound. Describe how this happens. Does the amplification depend on where you touch the soundboard? Is the amount of amplification dependent on the frequency of the tuning fork?

PROBLEMS

1. The lowest note on the piano is A = 27.5 Hz, and the highest note is C = 4186 Hz. What are the wavelengths of these two frequencies in air at room temperature?

2. The lowest note on the piano is A = 27.5 Hz and its length on a Kohler & Campbell baby grand piano is about 1.00 m. What is the speed of sound for that string?

3. The highest note on the piano is C = 4186 Hz and its length on a Kohler & Campbell baby grand piano is about 5.0 cm. What is the speed of sound for that string?

4. The shortest string (C = 4186 Hz) on a Kohler & Campbell baby grand piano is about 5.0 cm long. If the tension and the mass per unit length of the strings were to be kept the same, how long would the longest string (A = 27.5 Hz) be? Comment.

5. The longest string (A = 27.5 Hz) on a Kohler & Campbell baby grand piano is about 1.00 m long. If the tension and the mass per unit length of the strings were to be kept the same, how long would the shortest string (C = 4186 Hz) be? Comment.

6. The shortest string (C = 4186 Hz) on a Kohler & Campbell baby grand piano is about 5.0 cm long, and the longest string (A = 27.5 Hz) is about 1.00 m long. In order to keep the sounding board from warping it is desirable to keep the tension in all strings the same. To achieve the requirement of uniform tension for all strings, what is the ratio of mass per unit length for these two strings? What would be the ratio of their diameters if they were made of the same material? Examine the strings in a baby grand piano and comment on whether your calculations seem reasonable.

7. Suppose that you wished to construct a piano with all of the strings of the same length and tension, changing only the mass per unit length. If the mass per unit length of the string with the lowest frequency were W g/cm, what would be the mass per unit length of the string with the highest frequency?

8. Suppose that you wish to redesign a Kohler & Campbell baby grand piano to make the highest string, now length 5.0 cm, the same length as the lowest string, 1.0 m. If the diameter of the smallest string on a regular grand piano is now about 0.75 mm, what would it be after your redesign? Comment on the practicality of a string of this diameter. (A human hair is about 0.05 mm in diameter.)

REFERENCES

ASKENFELT, ANDERS (Editor). *Five Lectures on the Acoustics of the Piano*. Stockholm, Sweden, 1990. Compilation of lectures given by Harold A. Conklin, Jr., Anders Askenfelt & Erik Jamsson, Donald E. Hall, Gabriel Weinreich, and Klaus Sogram. Also available on the World Wide Web, including a nice set of audio examples of various piano sounds, at the URL: **http://www.speech.kth.se/music/5_lectures/contents.html**

BENADE, ARTHUR H. *Fundamentals of Musical Acoustics*. New York: Oxford University Press, Inc., 1976. Slightly more advanced than the present text.

BLACKHAM, E. DONNELL. "The Physics of the Piano." *Scientific American*, December 1965. Stresses the role of inharmonicity in piano tone and tone perception.

Bösendorfer Pianos of Vienna, A Division of Kimball International, Inc. P.O. Box 460, Jasper, Ind. 47546. (812) 482-1600. Information on the Bösendorfer extended piano and the computer-based piano performance reproduction system.

HART, HARRY C., MELVILLE W. FULLER, and WALTER S. LUSBY. "A Precision Study of Piano Touch and Tone." *Journal of the Acoustical Society of America*, October 1934, pp. 80–94. Describes experiments and results showing that the tone of a piano cannot be altered without changing the loudness.

HOLLIS, HELEN R. *The Piano: A Pictorial Account of Its Ancestry and Development*. London: David and Charles Publishers, 1975. A musician's introduction to the history of the piano.

RANDEL, DON MICHAEL, ed. *The Harvard Dictionary of Music*. 4th ed. Cambridge, Mass.: Harvard University Press, 1986, 2003. A classic dictionary of music and instruments.

WEINREICH, GABRIEL. "The Coupled Motions of Piano Strings." *Scientific American*, January 1979, pp. 118–127. Contains a clear description of the complicated interactions of the strings of a single piano note.

WHITE, WILLIAM B. *Theory and Practice of Piano Construction*. New York: Dover Publications, Inc., 1906, reprinted 1975. A readable account of how technical innovation and experimentation have affected the piano throughout history.

WOLFENDEN, SAMUEL. *Treatise on the Art of Pianoforte Construction*. Surrey, England: Unwin Brothers Ltd., 1977. A good starting point for the student who wants to understand the workings, including actions, of a piano.

Chapter 14

Percussion Instruments

Of all the instrument families, the percussion family has the greatest variety of sound, in both dynamics and sound quality. Such sounds range from the delicate "ting" of a triangle, through the rich tone of a marimba, the crack of a snare drum, the clear tone of a bell lyra, the majestic peal of church bells, to the thundering roll of timpani and the crash of cymbals and gong.

In this chapter, we shall describe the physics of two important classes of percussion instruments, the idiophones (e.g., xylophone, vibraphone, chimes, bells, and cymbals) and the membranophones (e.g., timpani, drums, and tabla). Chordophones, such as the piano, and aerophones, such as the whistle, are classified with percussion instruments but will not be treated here. We shall examine the features that characterize the sound of most percussion instruments: significant inharmonicities, a large number of overtones, and the unique attack and decay transients. These characteristics arise by different mechanisms in different instruments.

14.1 Bar Instruments

The bell lyra, shown in Figure 14-1, is acoustically one of the simplest bar percussion instruments, a subgroup of the idiophones, or struck instruments lacking membranes. In marching bands, a bell lyra is usually held with one hand at the top, the other end being supported at the performer's waist. A hand mallet in the performer's free hand is used to strike the bars, which are laid out similarly to a piano keyboard. In concert situations, the bell lyra is sometimes laid on a table and played like the more familiar glockenspiel.

A bell lyra bar, when struck, vibrates transversely at a frequency determined by its dimensions, the properties of the metal, and the location of the suspension points. Such a bar is a length of rectangular metal held to supporting felt pads by a loose rivet.

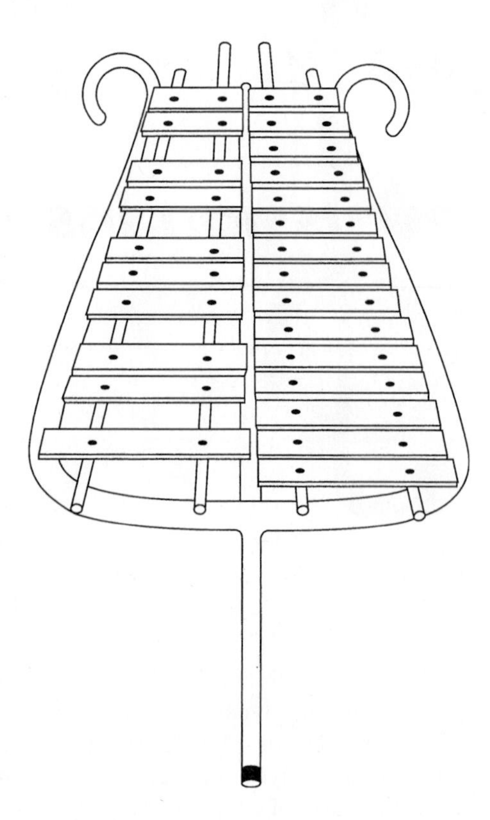

Figure 14-1 The bell lyra.

As with strings, there are multiple modes of oscillation, although here the bar ends are free to vibrate.

Figure 14-2 shows a side view of the first four modes of oscillation of a bell lyra bar and their relative frequencies. Although the bars have slight arches cut underneath, we shall ignore their effects for the moment. Note particularly that the frequencies are not harmonically related. This is because of the considerable inherent stiffness in the bar, an extreme example of the effect also seen in piano strings. These modes can also exist for oscillations across the width of the bar, as if Figure 14-2 represented a view of the narrow end of the bar. Because the width of the bar is less than its length, corresponding oscillations across the width will, in general, be higher in frequency than those along the length. In general, none of these modes is harmonically related to the fundamental or to each other.

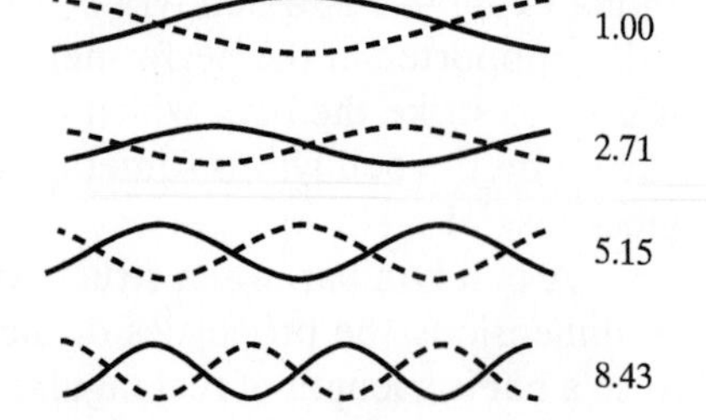

Figure 14-2 The first four transverse modes in a metal bar. (After Rossing, "Acoustics of Percussion Instruments I," *The Physics Teacher*, December 1976)

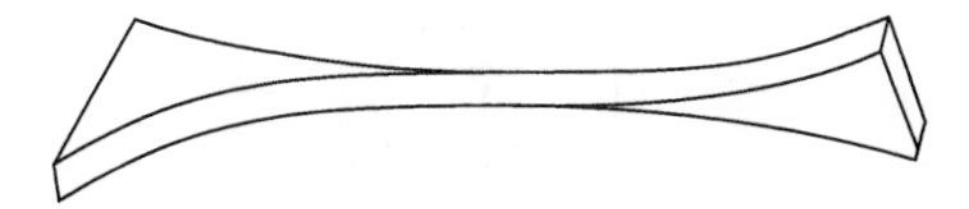

Figure 14-3 The simplest torsional mode in a metal bar (greatly exaggerated).

There are other bar modes. The bar can oscillate in a torsional or twisting mode, much as the Tacoma Narrows Bridge did before it collapsed. The simplest such mode has diagonally opposite corners moving up while the other diagonal pair moves down, and vice versa, as shown in Figure 14-3. Other, higher torsional modes have more twists; as a general rule, none is harmonically related to the fundamental.

Although the bar is not prevented from oscillating in longitudinal modes, these are not excited significantly by striking the top of the bar with a mallet. Also, the longitudinal modes have frequencies much higher than the transverse modes and do not significantly affect the bar's sound.

When the bar is struck, many transverse modes are excited. The higher modes decay quickly, whereas the fundamental sounds much longer, because the bar is supported near the nodes of the fundamental mode. Thus, after an initial sharp impact sound, the bar sounds quite clear and pure for several seconds before becoming inaudible.

The marimba, shown in Figure 14-4, is another bar instrument, but differs from the bell lyra in several important ways. The bars are made of rosewood (or keylon, an artificial substance less affected by changes in temperature and humidity) and are supported by strings that pass through two horizontal holes near the nodes of the fundamental. Each bar has an arch cut out of its bottom, and a tubular resonator sits vertically beneath each bar. These differences influence the sound quality. Because wood does not have the high stiffness and resonance characteristics that metal does, marimba notes die away faster than do bell lyra notes. The vibration of artificial marimba bars does not decay as fast as does that of wood bars.

The arch on the underside of the bars, shown in Figure 14-5, is important for several reasons. The deep arch in each low-frequency bar helps keep the size,

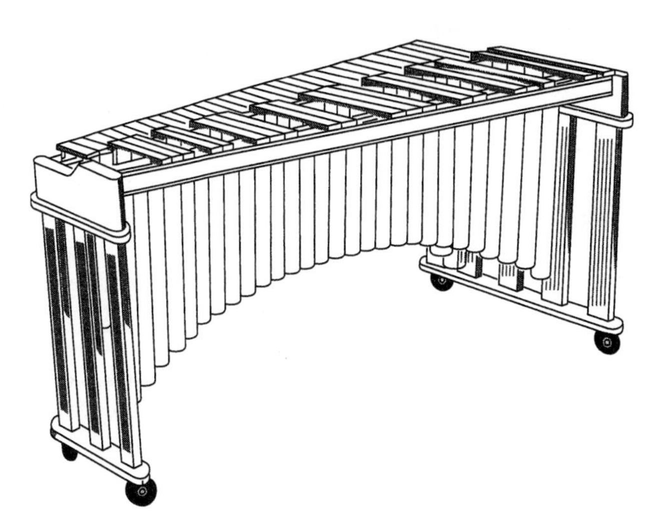

Figure 14-4 The marimba. Note the vertical resonator tubes under the bars. (Reprinted by permission of the publisher from *The Harvard Dictionary of Music*, Fourth Edition, edited by Don Michael Randel, p 645(3), Cambridge, Mass.: The Belknap Press of Harvard University Press, Copyright © 1986, 2003 by the President and Fellows of Harvard College.)

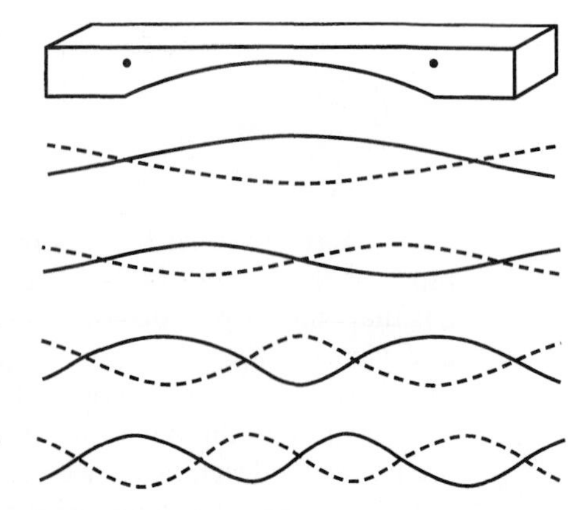

Figure 14-5 The first four transverse modes in a marimba bar. Note that the wavelength of each mode is greater toward the end of the bar than at the center. (After Rossing, "Acoustics of Percussion Instruments I," *The Physics Teacher*, December 1976)

weight, and cost of the marimba to acceptable limits. If the arches were not present, each bar would have to be twice as long as the bar one octave above it, which would result in very long bars for the low notes. Deep arches in the lower bars decrease the wave velocity in the bar. In this way, the arched low bars can be shorter than the unarched bars while still having the proper frequencies. In any bar the arch affects the modes of oscillation. Because the wave speed is lower at the thin center of the bar than at the thick ends, the wavelength of a standing wave of a certain frequency is shorter in the middle of the bar than at the ends. This can be seen by comparing the modes of a marimba bar (see Figure 14-5) with the modes of a bell lyra bar (see Figure 14-2).

Another important effect of the arch concerns tuning and tone quality: The first two overtones are approximately two octaves and three octaves plus a minor third above the fundamental.

Finally, each resonator is closed at the bottom and has a length chosen so as to resonate at the fundamental of the corresponding bar. The resonators affect the sound in two ways: (1) Because energy in the vibrating bar is transferred efficiently to the air, the loudness is increased and the sound decays more rapidly. (2) The resonator reinforces other frequencies much less than the fundamental, because they are not at the resonant frequencies of the tube. A marimba without resonators sounds richer in overtones, almost like a milk bottle being struck by a spoon, and the tone is softer and lasts longer.

Many principles applicable to the marimba are also useful in understanding the xylophone, shown in Figure 14-6. Xylophone bars are narrower than those of the marimba, and because they are also thicker, they have a higher wave velocity and thus a higher frequency for a bar of the same length. The arch cut in a xylophone bar is shallower than that in a marimba bar, because the first overtone is tuned to three times the fundamental frequency. Both the fundamental and this first overtone will be reinforced by the closed xylophone resonator tubes. Oscillations in the thick bars die away quickly; this and the presence of the reinforced first overtone give the xylophone its sharp, bright sound.

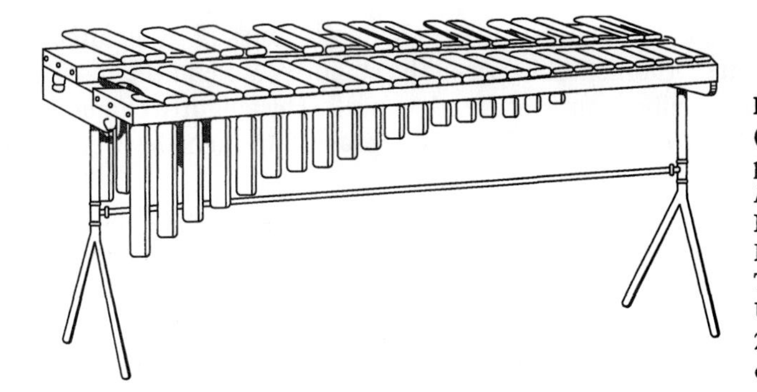

Figure 14-6 The xylophone. (Reprinted by permission of the publisher from *The Harvard Dictionary of Music*, Fourth Edition, edited by Don Michael Randel, p 645(2), Cambridge, Mass.: The Belknap Press of Harvard University Press, Copyright © 1986, 2003 by the President and Fellows of Harvard College.)

The vibraphone (or vibraharp), shown in Figure 14-7, has deeply arched aluminum bars in which the first overtone is two octaves above the fundamental. Because the decay time for undamped vibraphone bars is very long, a horizontal damper rod running under the bars, controlled by a foot pedal, is used when short notes are desired.

A unique effect obtainable with the vibraphone involves motor-driven butterfly valves that open and close the top end of each of the resonators. Each resonator tube has its own butterfly valve, and all the valves open and close together. When the valves are open, the tubes resonate and the sound is loud; when closed, the sound is soft. The rate of this opening and closing can be controlled by the performer to obtain fast vibe, medium vibe, or slow vibe, or the motor can be turned off to avoid the tremolo effect. Although the valve action primarily affects the volume of the sound, the pitch varies slightly as well, producing vibrato. Because the effective length of a tube is affected by

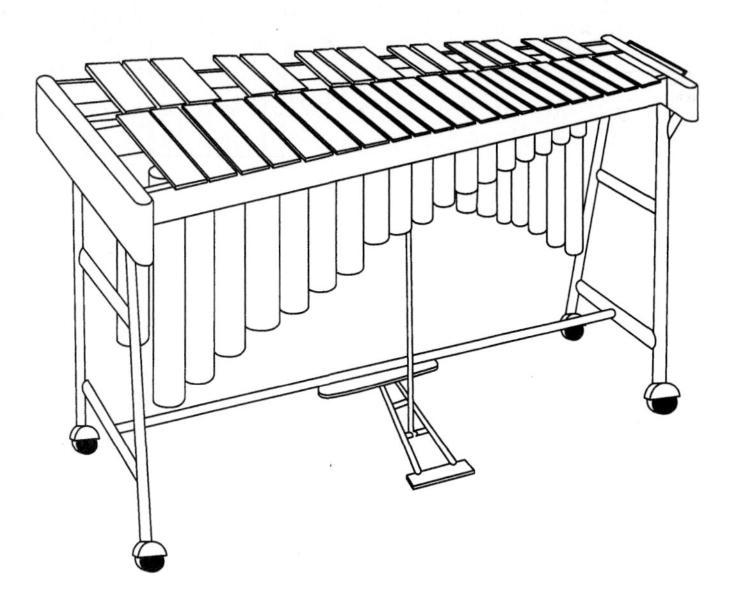

Figure 14-7 The vibraphone.

the width of its opening, the resonant frequency changes as the size of the tube opening changes. By a complicated interaction between the standing wave in the resonator and the vibrating bar, the bar vibration is affected, as well. This slightly changes the frequency of the emitted wave, producing a slight vibrato, although the tremolo effect predominates.

The long decay time and excellent resonance characteristics of the vibraphone permit an effect not attainable with the preceding instruments, that of *bending* a note. If a hard mallet is pressed on a node of an oscillating vibraphone bar, that is, over a support point, there will be little damping. As the mallet is slid toward the center of the bar, however, damping will occur, but, more importantly, the pitch will become lower. The mallet head acts to increase the effective weight of the bar and, by a law analogous to Mersenne's third, the pitch will drop. This technique is sometimes used in jazz and contemporary classical music.

14.2 Chimes and Triangles

Orchestral chimes (see Figure 14-8) consist of a set of hollow vertical pipes suspended at the top by thin ropes. The tubes range in length and wall thickness to obtain, usually, a range of about one and one-half octaves, as shown in Figure 14-8. Each pipe is struck with hammers on the edge of the top in a motion that is both downward and sideward.

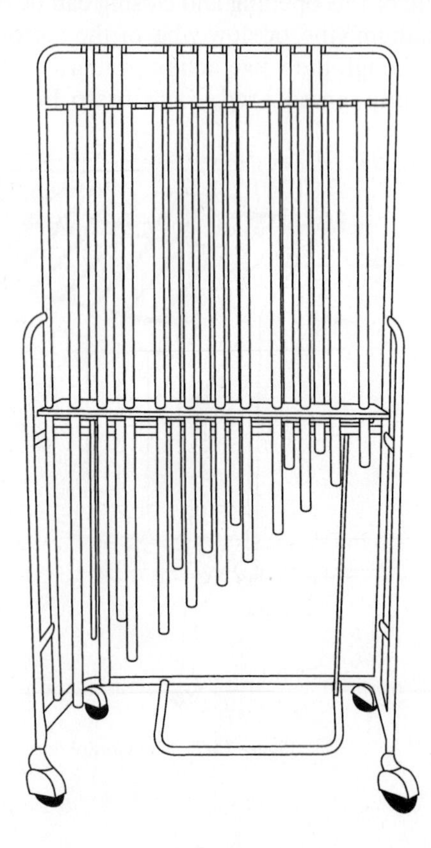

Figure 14-8 Orchestral chimes.

The three lowest modes of such a tube have frequencies with ratios of approximately 2:3:4. The ear then perceives a note an octave below the fundamental—a missing fundamental. Other overtones deviate from this approximate linearity and help give the unique sound quality of the chime. Chime tubes are often fitted with metal plugs in their ends. The plugs damp out many higher overtones, lower the frequencies of the low overtones, and reduce wear on the tube.

Nonuniformities in the thickness on the different sides of a chime tube can detune the overtones of corresponding transverse modes. If the frequencies of these modes are close but not equal, beating will result. This problem can be alleviated by squeezing the tube in a vise to make it more symmetric around its perimeter. The frequencies become equal, and the beats are therefore eliminated.

Triangles are steel bars bent into a triangle with one open corner; they are struck with metal rods called beaters. The bends do not affect the tone of the bar significantly, and thus the tone of the triangle is very similar to that of a straight bar of equal length.

When the triangle is struck in the standard way, most of the vibration occurs within the plane of the triangle, although some motion perpendicular to this plane also occurs. Because the triangle bar is thick and solid, there is a great deal of stiffness and thus inharmonicity. The initial "click" of metal-on-metal is followed by a tone made up of inharmonic overtones that then die down gradually, as there is little friction with the support. The higher overtones die down more rapidly, leaving the fairly clear tone of the lower overtones.

14.3 Membranophones

Whereas strings and air columns are essentially one-dimensional media, surfaces, such as drumheads, are two dimensional. The sound-production mechanism in membranophones begins with the oscillation of such membranes. The overtone structure in a one-dimensional medium is particularly simple in the ideal case; the corresponding overtone structure in two-dimensional media is not so simple. In particular, the modes are more complicated, and their frequencies are not integrally related.

In a string we can describe the modes by a single integer, the harmonic number. Because a drumhead is two dimensional, however, we need two integers to describe each of its modes precisely. The two integers needed are the number of nodal diameters and the number of nodal circles. These numbers are shown, along with the modes, in Figure 14-9. Figure 14-9(a) shows the lowest frequency mode. The center of the membrane moves up and down, almost like your chest as you inhale and exhale. Figure 14-9(b) shows a higher mode. As the center moves up, the outer ring of the membrane moves down, and vice versa. Just as there are nodes at the ends of a fixed string, there is a nodal ring around the edge of the membrane where it is held fixed to the edge of the drum. On Figure 14-9(b), however, there is another nodal line, a ring around the center. Points along this ring do not move up or down, but remain fixed. Figure 14-9(c) shows a mode that is not symmetric around the center point; as the left side moves up, the right side moves down, and vice versa. Here the nodal line is a straight line running perpendicularly through the center of the membrane. Rather than draw the complicated figures in Figure 14-9, we can just draw the nodal lines with + and − symbols

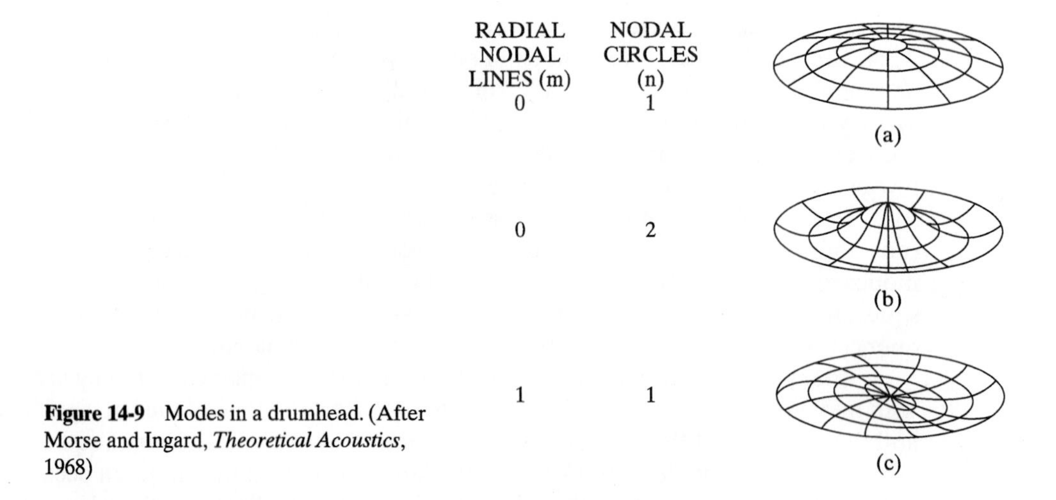

Figure 14-9 Modes in a drumhead. (After Morse and Ingard, *Theoretical Acoustics*, 1968)

to represent the relative motion of the different parts of the membrane at any instant, as in Figure 14-10.

Because the rim of the head is held fixed, all modes will have a nodal ring around the outside. Half a period later, the parts of the membrane will be in the opposite phase, a + becomes a −, and vice versa. Also shown in Figure 14-10 are the relative frequencies of the modes. The frequency of the lowest mode, the fundamental, has been set equal to 1.

Figure 14-10 depicts an ideal membrane, one that can support tension but does not have any inherent stiffness or complex interaction with the air. The figure is also valid for low-amplitude oscillations in real membranes, where these effects are not important.

Just as the tension in a string affects its frequency, the tension in a drumhead affects the pitch produced: the higher the tension, the higher the frequency. A tauter head has a higher wave speed, and for a given wavelength oscillation, a higher frequency.

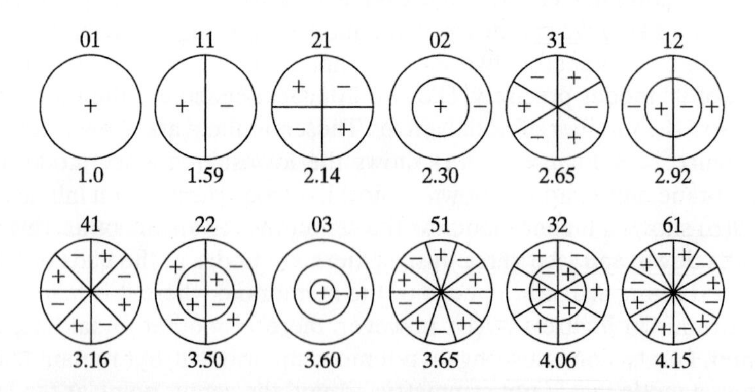

Figure 14-10 Schematic showing modes in a drumhead, the number of radial nodal lines *m*, and the number of circular nodal lines *n*. Also shown are the relative frequencies. (After Rossing, "Acoustics of Percussion Instruments II," *The Physics Teacher*, May 1977)

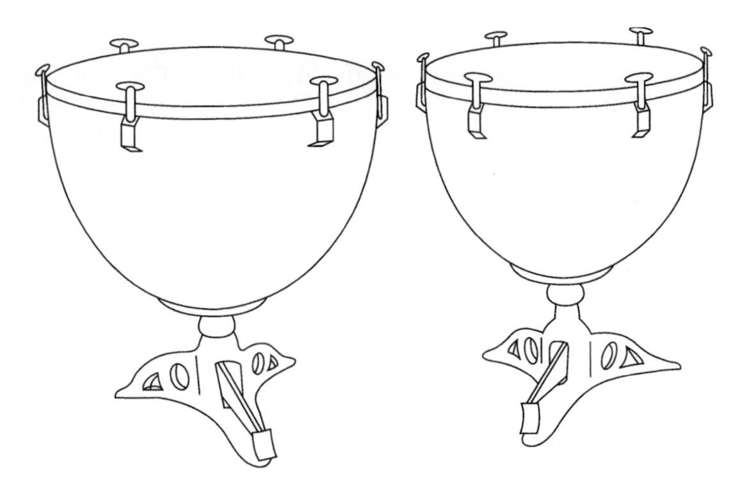

Figure 14-11 The timpani, or kettle drums.

Two important features of drumhead oscillations should be noticed. The frequencies of the modes are not integrally related, and the modes that have more closely spaced nodal lines have, in general, higher frequencies.

Timpani

Timpani, or kettle drums, shown in Figure 14-11, are the most important percussion instrument in the standard orchestra; they appear in the music of composers from before the baroque era to the present. They are in the drum tradition that traces back to antiquity; the ancient Persians had bowl-shaped drums that resemble modern timpani. Timpani are played in sets of two to six or more drums, each sounding at a different pitch. Each drum consists of a skin or plastic head stretched over a copper or fiberglass bowl. Tuning pegs around the rim of the bowl hold the head tight and can be adjusted so that the head is stretched evenly in all directions. The tension on the heads, and thus the pitch of the drum, is controlled by a foot pedal that pulls the head uniformly tighter about the rim.

Timpani are usually struck with wood sticks whose wood heads are covered with cloth or felt, or occasionally left bare. The head is usually struck about one-quarter of the diameter from the rim. Just as a string can vibrate in many modes simultaneously, the timpani head vibrates in many of the modes shown in Figure 14-10 simultaneously. We saw that ideal strings had a particularly simple overtone structure, and nonideal strings had a complicated (inharmonic) overtone structure. For timpani heads this situation is reversed; the ideal head supports modes that are inharmonic, but the nonideal head can support modes that are nearly harmonic. An actual timpani differs from the ideal case in two ways: there is inherent stiffness in the drumhead, and there is complicated coupling, or interaction, between the head and the air.

The pitch of the drum is determined by the 11 (one-one) mode. The lowest-frequency mode, the 01 mode, dies down very quickly as the large surface of the head pushes coherently against the air. If the drum is struck in the center (and some contemporary music requires this), the 01, 02, 03, and so on modes are excited, while the 11 is not. The resulting sound is a dull "thud" that dies away very quickly.

The stiffness of the head raises the frequencies of the higher modes, whereas the complex interaction of the head with the air lowers the frequencies of the low modes. The latter effect dominates and tends to make the overtones nearly harmonic. The pitch heard is determined primarily by the 11 mode, which is lowered slightly because of the complex interaction between the vibrating head and the air. The louder the note, the more this interaction affects the pitch. Careful timpanists take into consideration the loudness of the note when tuning. For instance, in the second movement of Beethoven's ninth Symphony there is an important three-note phrase played by the timpani alone, first loud then very soft. Some timpanists make subtle tuning adjustments—lowering the tension slightly for the soft notes—so that the phrases sound at the same pitch. The effect is noticeable in the decay of a single note, too. A loud timpani note rises slightly in pitch as the amplitude of the head motion decreases and the interaction of the head with the air diminishes.

Omitting the quickly damped 01 mode, the next lowest modes have frequencies with ratios of approximately 2:3:4:5 (modes 11, 21, 31, and 41), a harmonic series built on a missing fundamental. Some timpanists hear the pitch of the timpani as that which is written; others hearing the same tone claim to hear a note an octave lower. The latter hear a missing fundamental, whereas the former do not. This may be because of the large amplitude of the 11 mode and the low amplitudes and short duration of the higher modes, along with subtle hearing variations between performers. This issue has not been settled.

Tom-toms, snare drums, and bass drums

Tom-toms are drums that consist of a cylindrical shell (like a tin can with the ends missing) with either one head and an open bottom or two heads, one over the top opening and one over the bottom. Complicated interaction of the motion of the struck head and the motion of the head on the bottom tends to give such two-headed drums an indefinite pitch. The depth of the drum (length of the cylindrical shell) can affect the tone quality of the sounds produced. In general, a deep tom-tom will have a longer standing wave between its heads, and thus a lower tone, than a shallow tom-tom.

The snare drum resembles a shallow two-headed tom-tom with one important difference: lengths of gut or wire stretched along the outside of the bottom head. These snares vibrate against the head when the drum is played and give the crisp sound that is characteristic of the snare drum. The shallow body also aids this end. A lever is usually present on snare drums to release the snares so that the drum can sound like a tom-tom (when desired).

The bass drum consists of two large heads stretched over a large cylindrical shell. The drum is usually used to add *weight* to the sound of the lower sections of the band or orchestra. Each head has a loosely defined pitch center, but because they are tuned to close but different pitches, the sound of the drum does not have a well-defined pitch center. Thus the bass drum can be used unchanged for music in any key.

Large beaters are used to permit good coupling and louder sound. Often, short, loud bass drum notes are desired; these can be obtained by striking on or near the center. This excites the 01 mode, which decays rapidly. Bass drums in drum sets have a beater that is operated by a foot pedal; the beater strikes the head with a direct stroke near the center.

Tabla

The preceding drums all have heads of uniform thickness. Some Asian Indian drums, however, have weighted heads; that is, a circular portion of the head is heavier than the rest, as shown in Figure 14-12. Examples of these Indian drums, collectively called *tabla*, are often played in duet with the sitar. Figure 14-12 shows these two drums; the right-hand drum is called the dayan (or tabla) and the left-hand drum is called the bayan. The dayan is usually made of wood with a calfskin head held taut by calfskin straps; the larger bayan is bowl shaped and made of metal. The weighting, or loading, of the membranes is achieved through a mixture of soot, iron rust, and gum that is allowed to dry on the head. The circular area so weighted is called the *ak*. The first four dayan modes are very nearly harmonic because of this loading, which creates the clear ringing tone of this drum. One stroke, called *tin*, is produced by lightly touching the head with the fourth or the little finger while simultaneously slapping the head a quarter of the way around. This forces the equivalent 11 mode to predominate, yielding a high-frequency tone.

The bayan has a deeper, more "tubby" sound, because of its larger size and lower head tension. The bayan is unusual, in that its head tension is often adjusted during the playing of a single note. The heel of the left hand rests on the head, about midway between the center and the edge opposite the ak. To produce the stroke *ga*, the fingertips strike the head on the other side of the ak; then the heel of the hand is quickly pressed into the head, and sometimes slid along the head. This raises the pitch of the struck tone, and a tubby, "shlurping" sound results.

The pitch ranges of the percussion instruments previously described are shown in Table 14-1.

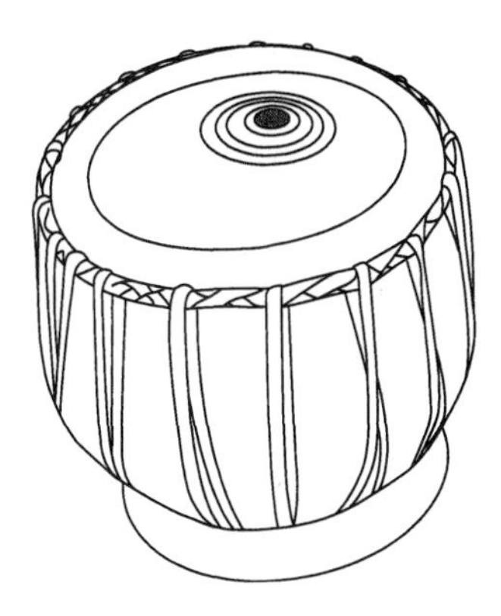

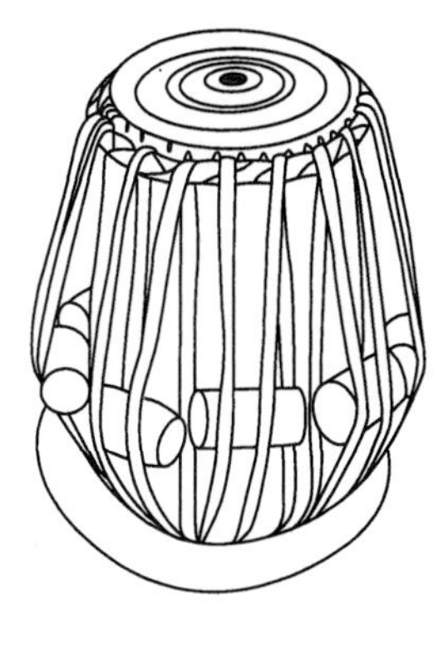

Figure 14-12 Tabla (performer's view), Bayan (left) and dayan (right).

Table 14-1 Approximate Ranges of Pitched Percussion Instruments

		Written	Sounds
Idiophones	Bell lyra	A_3–A_5	2 octaves higher
	Glockenspiel	G_3–C_6	2 octaves higher
	Xylophone	F_4–C_7	1 octave higher
	Marimba	C_3–C_7	As written
	Vibraphone	F_4–F_6	As written
	Chimes	C_4–F_5	As written
Membranophones	Timpani 30 in.[a]	$D_2^{\flat}$–A_2	As written
	28 in.	F_2–C_3	See text
	25 in.	$B_2^{\flat}$–F_3	See text
	23 in.	D_3–A_3	See text
	Tabla: bayan	C_3–C_4	As written
	dayan	$F_3^{\sharp}$–G_4	As written

[a] Diameter of head.

14.4 Gongs, Tam-Tams, Cymbals, and Bells

Gongs and tam-tams are circular-shaped metal plates that are struck with large, soft beaters. Gongs (and their Indonesian relative, the gamelan) generally have a domed central area, thick metal, and a general pitch center. Tam-tams, on the other hand, are generally flat, made of thin metal, and have a poorly defined pitch.

When a tam-tam is struck near the middle, low frequencies sound first, and later the higher modes sound. The attack is thus gradual and consists of a rushing of the initial sound. Because the tam-tam does not speak immediately after it is struck, it is tricky for the percussionist to make it sound at the desired moment. To avoid this problem, percussionists "prime" the tam-tam by lightly tapping it with the beater. Low-frequency, low-amplitude, inaudible vibrations result, which nonetheless allow the tam-tam to speak sooner after it is struck than if it were not primed.

Cymbals are circular metal plates, slightly arched with a more strongly arched cup, or *dome*, at the center. When supported at the center, cymbals can be struck with soft or hard sticks; soft sticks produce a prolonged "whooshing" sound as many modes are excited. When the cymbal is struck on the edge with a hard wood stick, many higher modes are excited, and a brighter sound results.

Crash cymbals are a pair of hand-held cymbals struck sharply against one another. This technique generates many high-frequency overtones, especially during the attack. To project the full sound from the cymbals, the performer holds their faces toward the audience immediately after striking. Because the cymbal plates vibrate perpendicularly to the face, most of the sound projects out. If crash cymbals are played in an anechoic chamber (very absorbent walls), no sound will be heard along directions perpendicular to the face of the cymbals.

Diffraction and interference also affect the tone quality of the cymbal. The out-of-phase waves emitted from the back of the cymbal can diffract around and interfere destructively with the waves emitted by the front; the sound will thus be reduced. Because low frequencies diffract more than the higher frequencies, the low frequencies are not as loud. This effect was described in our speaker-with-baffle experiment in Chapter 2.

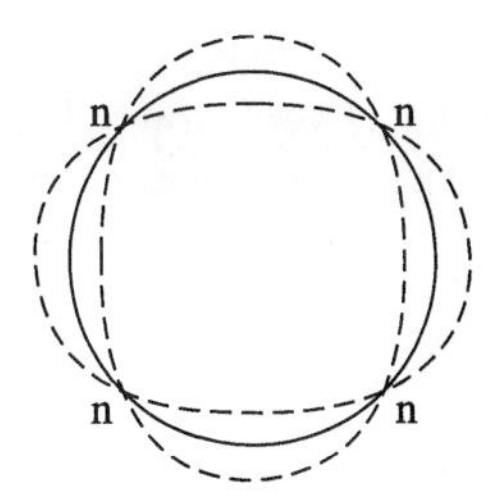

Figure 14-13 Fundamental mode in a bell, viewed from above.

The vibrations of bells are similar to those in a thick metal plate, with radial nodal lines and circular nodal lines. These modes are similar to those in a beaker or glass flicked with a fingernail. As with plates and drumheads, the lower-frequency modes have fewer nodal lines, whereas the higher modes have more nodal lines, more closely spaced. Unlike a drumhead, the bell edge is free to vibrate, and thus there can be antinodes there.

The lowest mode that sounds with appreciable amplitude has four meridian nodal lines and produces the "hum" tone. In this mode, opposite sides move inward and outward together while the motion one-fourth of the way around the bell is out of phase with this motion, as can be seen in Figure 14-13. The strike tone is adjusted to be nearly an octave above the hum tone. Both have four nodal meridians, but the strike tone has a circular nodal line partway up the bell, whereas the hum tone does not. The pitch of the bell is associated with the striking tone, whereas the hum tone is usually a vague, low-pitched sound.

English bells, designed for sequential, or *change*, ringing, are not as carefully tuned as Flemish bells or carillon bells, which must sound pleasant (and beatless) together. Because the thickness of the bell wall is tailored to adjust the overtones, careful study of the overtones of the bells has suggested different thickness profiles. Bells are often tuned in mean-tone temperament, so that notes a major third apart played simultaneously will be beatless.

Metal casting and tuning of bells is slow and expensive, so other techniques have been used to produce bell and carillon tones. The most common employs metal bars or rods that are shaped to give the desired overtones. An electrical pickup converts the motion into electrical signals, which are then amplified and played through a loudspeaker.

SUMMARY

Percussion instruments, including **idiophones** and **membranophones**, have the greatest variety of sounds of any modern orchestral family. (***Section 14.1***) The simplest of bar instruments is the **bell lyra**, a keyboard-like array of rectangular bars with soft supports struck with a hand mallet. Because of its rigidity, higher modes are inharmonic, but they are relatively soft and short-lived. **Marimba** bars are made of wood or a similar material, so their tone is of short duration, and they have an **arch** on the bottom to tune them; a closed resonant tube mounted below the bar reinforces the fundamental frequency. The **xylophone** uses a thicker and narrower metal bar with a shallower arch, so its tone is also short in duration; a closed tube below each bar of the xylophone resonates both the fundamental and the third harmonic. The **vibraphone** uses deeply arched aluminum bars to achieve tones that decay very slowly, with a damper to shorten the notes. A set of butterfly valves, one at the top of each resonator tube, is used to obtain a tremolo effect. Because vibraphone tones decay very slowly and can be adjusted in frequency by sliding the head of the mallet along the bar, **bending** of the notes is often used in jazz bands. (***Section 14.2***) **Orchestral chimes** are a set of vertically suspended tubes struck near the top. Their three lowest modes have frequencies in the ratio of 2:3:4, so the perceived tone is an octave below the lowest mode. When fitted with metal plugs in their ends, their frequencies are lowered. A triangle is a steel rod bent into the shape of a triangle, suspended by a string and struck with a metal beater. The overtones are inharmonic; after the triangle is struck, the higher harmonics die down, and the tone becomes more pure. (***Section 14.3***) Membranophones are two dimensional, and are even more inharmonic than chimes and bells. Each mode requires a two-number description: one number for circles and one for diameters. **Timpani** (kettle drums) have tunable membranes stretched over large bowls, and are found in sets in the orchestra. The timpani head is struck by a mallet at one-quarter of the radius of the head to excite an asymmetric mode that is enhanced by the air oscillating back and forth in the bowl. The lowest mode is generally not used, but the frequencies of the next four modes are in the ratio of about 2:3:4:5, forming the perceived pitch. **Tom-toms** are cylindrical drums that may have a membrane at either one or both ends, and have indefinite pitch; the length of the drum determines its pitch level. The **snare drum** is a short-banded, two-headed tom-tom with lengths of wire contacting the lower head to produce crisp noise. The bass drum has two large heads tuned to slightly different frequencies, giving it a low undefined pitch. A set of Asian Indian drums called **tabla** use a weighted area, the **ak**, on the membrane to achieve a nearly harmonic set of modes, thus producing a more tonal sound. They are played in pairs, the **dayan** and the **bayan**, with the **sitar**. (***Section 14.4***) **Gongs** are thick circular metal plates with a general pitch center. **Tam-tams** are thin metal plates with a poorly defined pitch. After a tam-tam is struck, the lower-frequency modes sound quickly and the higher-frequency modes are delayed; it is necessary to prime the tam-tam to get it to sound more quickly. **Cymbals** are slightly arched circular plates with a greater arch near the center, and can produce a variety of sounds, depending on how they are excited. Crash cymbals are two cymbals struck together. **Bells** vibrate in a two-dimensional pattern similar to that of the vibrations of drumheads, with four vertical nodal lines. The **strike tone** has an additional circular nodal line, but the **hum tone** does not. Bells are often tuned in mean-tone temperament so that they will sound good together when played in thirds.

QUESTIONS

1. **a.** Compare and contrast the oscillations in bars with those in strings. Be sure to discuss harmonicity, nodes, and modes.

b. What types of modes can exist in a bar that cannot exist in a string?

2. Describe tone production in a triangle.

3. **a.** Give two reasons for the arches cut in the bottom of the bars of certain tuned percussion instruments, such as the xylophone and marimba.

b. How do resonators affect the tone of these instruments?

4. a. How do the butterfly valves affect the tone of the vibraphone?
   b. Using a vibraphone, compare the loudness and duration of a tone produced when the valves are open with a tone produced when the valves are closed.

5. a. Why do drums have bodies or shells?
   b. Why do single-headed tom-toms, for instance, need a cylindrical shell?

6. How is the indefinite pitch achieved in a bass drum?

7. a. How is the pitch of a bayan changed during a single note?
   b. How does the ak affect the tone of the tabla?

8. Stretch a thin rubber or similar membrane uniformly over a large ring and hold it in place tightly at the circumference. Excite the membrane using a sine wave oscillator and a loudspeaker placed under the *center* of the membrane. Using a stroboscope, slow down the motion of the membrane so that you can clearly see the motion of each mode. Make appropriate sketches showing each of the modes that you excite.

9. Stretch a thin rubber or similar membrane uniformly over a large ring and hold it in place tightly at the circumference. Excite the membrane using a sine wave oscillator and a loudspeaker placed under the *center* of the membrane. List the frequencies of the 01, 02, and 03 resonances and calculate the ratio of each to that of the fundamental mode. Compare the list that you obtain with the frequencies of the modes in Figure 14-10.

10. Stretch a thin rubber or similar membrane uniformly over a large ring and hold it in place tightly at the circumference. Excite the membrane using a sine wave oscillator and a loudspeaker placed under the *center* of the membrane to find the fundamental (01) mode. Then excite the membrane at about half the radius of the membrane from its center to determine the frequencies of the asymmetric resonances. List the frequencies of as many resonances as you can find, identify them and calculate the ratio of each to that of the fundamental mode. Compare the list that you obtain with the frequencies of the asymmetric modes in Figure 14-10.

11. Stretch a thin rubber or similar membrane uniformly over a large ring and hold it in place tightly at the circumference. Use heavy tape to form a weighted region about halfway between the center and the edge of the membrane. Excite the membrane as described in Questions 9 and 10, and list the frequencies of the modes. Can you find a position for the weighted region that will cause the frequencies of the modes to line up at octave intervals, or in an overtone series?

12. Obtain some thin aluminum tubing and construct a one-octave diatonic set of tubular chimes. Suspend each tube from a horizontal bar by a string passing through two small holes on opposite sides of the top of each tube. Find the best place to strike each chime with a small rubber hammer. Play "Twinkle, twinkle little star" or another simple tune on your chimes. Describe the sound, and the difficulties in obtaining a musical tone.

13. Visit the studio of a serious percussionist to see a variety of percussion instruments. Examine them to see their similarities and differences, and categorize them into the families described in the text.

PROBLEMS

1. Consider two uniform rectangular bars with no arch (unlike the bars for the idiophones discussed here), resting on soft supports at about one-quarter of their length from each end. Suppose that one has twice the length of the other. Will the frequency of the longer one be greater than, less than, or equal to the frequency of the shorter one?

2. Consider two uniform rectangular bars with no arch (unlike the bars for the idiophones discussed here), resting on soft supports at about one-quarter of their length from each end. Suppose that one has twice the width of the other. Will the frequency of the wider one be greater than, less than, or equal to the frequency of the narrower one?

3. Consider two uniform rectangular bars with no arch (unlike the bars for the idiophones discussed here), resting on soft supports at about one-quarter of their length from each end. Suppose that one has twice the thickness of the other. Will the frequency of the thicker one be greater than, less than, or equal to the frequency of the thinner one?

REFERENCES

BLADES, JAMES. *Percussion Instruments and Their History*. New York: Praeger Publishers, Inc., 1970. A clear presentation of the history of most important percussion instrument groups.

BRINDLE, REGINALD S. *Contemporary Percussion*. New York: Oxford University Press, Inc., 1978. Gives examples of how composers use different percussion instruments for different effects. It is particularly suited for a composer who does not know very much about percussion instruments.

FLETCHER, N. H., and T. D. ROSSING. *The Physics of Musical Instruments*. New York: Springer-Verlag, 1990 (hardcover), 1993 (softcover). An excellent, up-to-date book containing a wealth of detailed information about virtually all types of musical instruments.

MORSE, P. M. *Theoretical Acoustics*. New York: McGraw-Hill Book Co., 1968. A very advanced book that requires considerable advanced mathematics and is not directed toward music or musical instruments per se.

RAMAKRISHNA, B. S., and M. M. SONDHI. "Vibrations of Indian Drums Regarded as Composite Membranes." *Journal of the Acoustical Society of America* 26 (1954). A very advanced article; however, students can appreciate the Chladni figures made using tabla heads.

RANDEL, DON MICHAEL, ed. *The Harvard Dictionary of Music*. 4th ed. Cambridge, Mass.: Harvard University Press, 1986, 2003. Classic dictionary of music and instruments.

ROBERTSON, DONALD. *Tabla: A Rhythmic Introduction to Indian Music*. New York: Peer International, 1968. A short, clear introduction to the instruments and their parts, as well as a primer for tabla performance.

ROSSING, THOMAS D. "Acoustics of Percussion Instruments I." *Physics Teacher*, December 1976, pp. 546–555.

———. "Acoustics of Percussion Instruments II." *Physics Teacher*, May 1977, pp. 278–288. An excellent two-part introduction to the physical principles and their consequences when applied to standard and some Eastern percussion instruments.

ROSSING, THOMAS D. *Science of Percussion Instruments*. River Edge, N.J.: World Scientific, 2000. An excellent book containing a wealth of information at a technical level that can be understood by most musicians.

Appendix A

Elementary Music Theory

This Appendix contains the basic music theory assumed in the discussions of the overtone series. We shall begin with the correspondence of musical notation with keys on the piano (or organ or music synthesizer) keyboard. This will allow us to identify simple musical intervals and establish the sequence of notes in the overtone series.

Figure A-1 shows the correspondence of the notes of the piano keyboard with notes on the ten horizontal lines of a musical staff system. Notes on the piano keyboard are labeled (for example, C_4) according to the convention suggested by the U.S. Standards Association. The notes are placed on the staff as heavy filled or open ovals either on a line or in a space between two adjacent lines; the staff system most commonly used consists of a *bass clef* (lower five lines) and a *treble clef* (upper five lines), as shown. The treble clef symbol indicates that the note G_4 is represented by the next to bottom line on that staff, whereas the base clef symbol indicates that the note F_3 is the next to top line on that staff. Between the two clefs are three notes, as shown in the figure. The line between the two clefs is C_4, called middle C because it is just about at the middle of the normal piano keyboard. To represent notes above or below the staff, additional lines called *ledger lines* are used.

The white keys on the piano keyboard are labeled successively by the letters A through G, which then repeat. Keys that occupy the same position relative to the groups of either two or three black keys have the same letter name. For example, the letter C labels any key just below a group of two black keys. The subscripts describe the particular C (of approximately ten within the range of hearing). The black keys can be labeled in two ways. For example, the black key between F and G can be called $F^\sharp$ ($^\sharp$ reads "sharp"), meaning it is just above F, or it can be called $G^\flat$ ($^\flat$ reads "flat") meaning it is just below G. Because there is no black key between B and C, $B^\sharp$ is the same as C and $C^\flat$ is the same as B; similarly, $E^\sharp$ is F and $F^\flat$ is E.

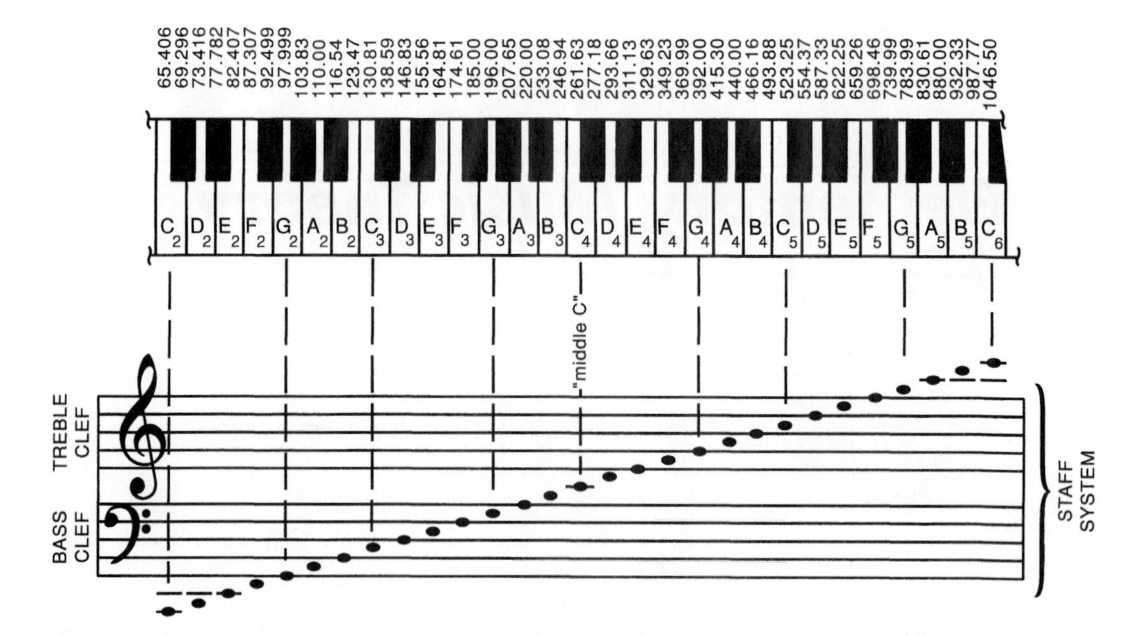

Figure A-1 Correspondence between notes on the musical staff system and the keys on the middle of the piano keyboard. Shown on each key is its U. S. Standards Association identification; above each key is its frequency.

Two notes that are adjacent on the piano keyboard (either both white or one black and one white) are said to be a musical interval of one half step apart; for example, the interval between E and F or between G and G♯ is one half-step. A whole step is equal to two half steps, as in the interval between F and G or between E and F♯.

In this book (except for Chap. 9) use of the overtone series and the musical intervals between the notes of the overtone series will be limited to the white keys only. Table A-1 lists the musical intervals forming the overtone series, along with the number of half steps in each interval. For example, one octave, which is the musical interval between any two nearest notes with the same letter name, consists of 12 half-steps. There is an interval of a perfect fifth between any two white keys where there are 7 half steps between the two notes, such as F to C or G to D. B to F is not a perfect fifth, because it contains only 6 half steps. Similarly, G to C and A to D are perfect fourths, because they contain 5 half steps. G to B and F to A are major thirds, separated by 4 half steps, and B to D and E to G are minor thirds, containing 3 half steps.

Knowledge of scales gives us another way to determine the interval between two notes. The name given to the interval is the number of the upper note in the scale beginning on the lower note. An interval of a major third contains the first and third notes of a major scale—for example C to E or G to B. An interval of a minor third contains the first and third notes of the minor scale—for example E to G or B to D. A perfect fourth contains the first and fourth notes of the scale, and a perfect fifth contains the first and fifth notes of the scale. A perfect fourth can be obtained from either a major or a minor scale, as can a perfect fifth.

Table A-1 Musical Intervals, the Number of Half-Steps Included in Each, and the Frequency Ratio Between the Two Notes

Name of interval	Number of half-steps	Approximate frequency ratios
Octave (8va)	12	2:1
Minor seventh	10	7:4
Perfect fifth	7	3:2
Perfect fourth	5	4:3
Major third	4	5:4
Minor third	3	6:5
Major second (whole step)	2	—

For a given musical interval, the frequencies of the two notes involved will always be in a certain ratio; these approximate frequency ratios are also shown in Table A-1. The frequency ratio between two notes one octave apart is exactly 2:1. That is, if the frequency of A_4 is 440 Hz (vibrations per second), then the frequency of the next highest A, which is A_5, is 880 Hz, a factor of 2 higher. The frequency of A_3 is 220 Hz, a factor of 2 lower. Thus the frequency ratio between any two adjacent "A"s is 2:1. For any musical interval of a perfect fifth, the frequency ratio between the notes is approximately 3 to 2; that is, if the lower note has a frequency of 600 Hz, then the note a perfect fifth higher will have a frequency of approximately 900 Hz. Two notes a major third apart have a frequency ratio of approximately 5:4, so if the lower frequency is 600 Hz, then the upper frequency would be about 750 Hz. It is important to remember that the particular ratios hold for the intervals wherever the interval is taken on the keyboard.

In the equal-tempered scale, the octave is the only interval that has an exact integral frequency ratio, exactly 2:1. Other intervals in this temperament have frequency ratios that are only approximately integral. In the equal-tempered scale the frequency ratio between two notes one half-step apart is always 1.0595. This is the twelfth root of 2 (to four decimal places), and use of this factor guarantees that octaves, which contain 12 half steps, have the ratio of frequencies of exactly 2:1. A true perfect fifth above $A_3 = 220$ Hz would be $220 \times \frac{3}{2} = 330$ Hz, but the frequency of E_4 is actually 329.63 Hz, a difference of 0.37 Hz.

The sequence of notes called the overtone series, as illustrated in Figure A-2, is of particular interest, because the frequency ratios between the notes are the integers

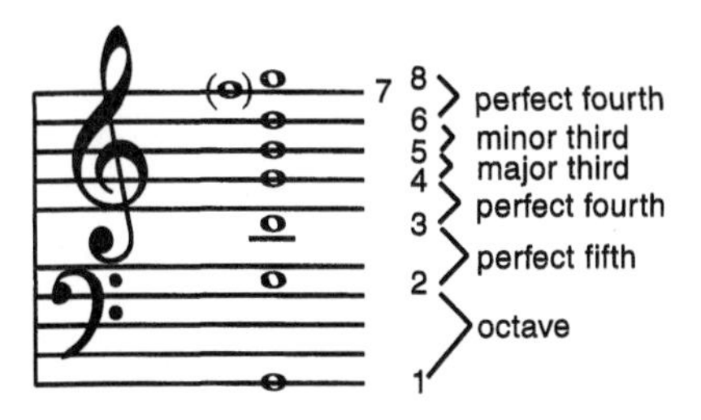

Figure A-2 The first eight notes of the overtone series built on G_2.

Table A-2 Frequencies of Exact Harmonics of the Note G_2 Compared with Frequencies of the Nearest Notes of the Equal-Tempered Scale

Note	Harmonic number	Exact frequency (Hz)	Piano frequency (Hz)	Difference frequency (Hz)
G_2	1	98.0	98.0	0.0
G_3	2	196.0	196.0	0.0
D_4	3	294.0	293.7	0.3
G_4	4	392.0	392.0	0.0
B_4	5	490.0	493.9	3.9
D_5	6	588.0	587.3	0.7
F_5	7	686.0	698.5	13.5
G_5	8	784.0	784.0	0.0

1, 2, ... as tabulated in Table A-2. Starting with G_2, the note G_3 one octave higher has a frequency two times the frequency of G_2. D_4 has a frequency of 3:2 times the frequency of G_3, or three times G_2. G_4 has a frequency of 4:3 times the frequency of D_4, or four times the frequency of G_2. Table A-2 shows these frequencies up through the eighth harmonic for exact frequency multiples and the actual frequencies of piano notes of the equal-tempered scale. For the octaves, the frequencies of the integral multiples are exact; for the third, fifth, and sixth notes, the frequencies of the exact multiples are very close to the frequencies of the notes on the piano. Because the frequency ratio of 7:1 is only very roughly approximated by $\left(\sqrt[12]{2}\right)^{34}$ (34 half-steps in the equal-tempered scale), we put the seventh harmonic in parentheses when it is written on the staff, as in Figure A-2. The modern "musical" ear, accustomed to the equal-tempered scale and the sound of musical intervals using the note that best approximates the true seventh harmonic, generally finds intervals involving the true seventh harmonic foreign and somewhat "out of tune."

The notes of the overtone series are called harmonics of the fundamental, or lowest note, and are labeled by their harmonic numbers, which are the integers giving the frequency ratio. In the case discussed here, G_2 is the fundamental or first harmonic and, for example, the fifth harmonic would be B_4.

It is possible to build such an overtone series on any note of the piano, using the intervals shown in Figure A-2. However, for our purposes, we shall use only the overtone series built on the note G_2 as presented here to avoid the use of sharp or flat notes.

Appendix B

Terms and Units

Quantity	Symbol	Common units
distance	x, y, z	meter (m); km, cm, mm, mi, ft
time	t	second (s)
velocity or speed	v	m/s, cm/s, mm/ms
speed of sound	S	345 m/s or 34,500 cm/s
acceleration	a	m/s^2, cm/s^2
acceleration of gravity	g	9.8 m/s^2 or 980 cm/s^2
mass	m	gram (g), kg, mg
force	F	Newton (N), pounds (lb)
density	ρ(rho)	g/cm^3, kg/m^3
pressure	P	Pascal (1 Pa = 1 N/m^2), $lb/in.^2$
period	T	s, ms
frequency	f	Hertz (Hz), kHz, MHz
wavelength	λ(lambda)	m, cm
phase	φ(phi)	degree (°)
tension	F	Newton
weight	W	Newton, pound
mass per unit length	W	kg/m
harmonic number	N	—
intensity	I	W/m^2
sound intensity level	SIL	decibel (dB)
loudness level	LL	phon
electrical voltage	V	volt (V)
electrical current	I	ampere (A)
electrical resistance	R	ohm (Ω, omega)
electrical power	P	watt (W)
sound absorption	—	sabin

Appendix C

Prefixes with Common Units

Quantity	Symbol	Value	Examples
pico-	p	10^{-12}	1 pA = 10^{-12} A
nano-	n	10^{-9}	1 nm = 10^{-9} m
micro-	μ(mu)	10^{-6}	1 μA = 10^{-6} A
milli-	m	10^{-3}	1 mW = 10^{-3} W
centi-	c	10^{-2}	1 cm = 10^{-2} m
deci-	d	10^{-1}	1 dB = 10^{-1} bel
kilo-	k	10^{3}	1 km = 10^{3} m
mega-	M	10^{6}	1 MV = 10^{6} V
giga-	G	10^{9}	1 GHz = 10^{9} Hz

Glossary

ac *Alternating Current*; electrical current oscillating in direction.

acceleration The change in velocity per unit time.

acoustic suspension An airtight loudspeaker enclosure, without an opening to the outside of the box.

action The moving parts in a piano that allow a finger on a piano key to control the sound.

addition of waves Combining two waves by adding their values point for point throughout their duration.

air resonance Helmholtz-type resonance of the body of a stringed instrument such as the violin or guitar.

ak The weighted area of the tabla drumhead.

aliasing The introduction of spurious frequencies during the digital-to-analog process.

AM See *amplitude modulation*.

amplitude A measure of how large a vibration is; can be a distance, pressure, voltage, or other unit.

amplitude modulation A transmitting process in which a modulating signal varies the amplitude of the carrier wave, or signal.

analog synthesizer An early type of musical synthesizer that used analog audio signals.

analog-to-digital converters Electronic devices for converting analog signals to digital signals.

antinode The point of maximum vibration along a standing wave.

anvil The second of the three "ossicles," the bone chain that converts vibrations of the ear drum into vibrations of the cochlear fluid.

arch A curved segment under a tuned percussion bar that is removed to obtain the proper frequency response.

artificial harmonics A violin effect obtained by playing a stopped note and then touching the vibrating string gently at a nodal point for the desired harmonic of that note.
audio spectrogram A graph showing the spectrum of a sound as a function of time.
auditory nerve The nerve that carries electrical impulses from the cochlea to the brain.
back The plate opposite the string side of a violin or other stringed instrument.
bagpipe An early instrument that used an air bag as source of wind.
balance The distribution of sound between the two speakers of a stereo system.
balanced modulation A type of modulation in which the offset level is the same as the signal zero level.
baritone A low-frequency brass instrument with a wide conical bore.
baroque trumpet A valveless trumpet typically played in its upper range to obtain a complete scale.
base The wide end of the cochlea where it connects to the middle ear.
basilar membrane The membrane between the two scala of the inner ear adjacent to the frequency-sensitive organs.
bass A descriptive term referring to low frequencies.
bass The lowest-voiced member of the orchestral string family.
bass bar A wooden structure that gives strength to the violin belly and enhances the vibration of the plate.
bass reflex See *tuned-port.*
bassoon A low-frequency double-reed instrument with a very narrow conical bore.
bayan The larger of the two drums in the *tabla* set.
beats A pattern of alternating constructive and destructive interference between two pure tones of similar amplitude and frequency.
bell An increase in the diameter of the bore at the end of musical instruments.
bell-in-vacuum experiment An experiment in which a bell sounds in a container as the air is pumped out.
bell lyra A tuned percussion instrument that uses simple rectangular bars.
bellows A collapsible device used to pump air into a musical instrument, such as a small organ.
bells Traditional-shaped metal instruments generally found in diatonic sets.
belly The front plate on a violin or other stringed instrument.
bending An effect involving changing the pitch of a note during its duration, often used in jazz.
Bernoulli effect The reduction of pressure in rapidly moving air streams.
bias A high-frequency ($\sim$100 kHz) magnetic field applied to magnetic recording tape to reduce the effects of hysteresis.
binaural Involving both ears together; unobservable by a single ear.
binaural beats Beats in which separate signals are sent to the two ears.
bird flageolet A small fipple flute, similar to the sopranino recorder, used to teach songs to birds during the Renaissance.
blend A property of room acoustics in which sound from various performers mixes well.
Böhm fingering system A set of rings and keys used to simplify and systematize the fingering of many woodwind instruments.
bore The inner diameter of the tube of woodwind and brass instruments.

bow An implement used to scrape the strings of instruments of the violin family to produce the sound; uses hair from a horse's tail.
bridge A construction feature that supports one end of the active string for stringed instruments; transfers energy to the wooden plates.
brilliance An acoustical property of rooms with low-frequency T_R below average.
burst errors Errors in reading CD data resulting from unreadable bits on the disc.
capped reeds Early woodwind instruments, which in playing the performer blows on an air enclosure rather than with lips contacting a reed.
cardioid A microphone with a heart-shaped directional response.
CD *Compact Disc*; a medium for digital audio storage.
central auditory system The final part of the hearing system: the auditory nerve and the auditory cortex of the brain.
Chladni plate resonances Resonances in the plates of stringed instruments, similar to two-dimensional standing waves in metal plates.
chorus effect An aspect of tone quality resulting from many identical instruments playing the same music.
circle of fifths A series of musical key signatures obtained by changing the key signature up or down by successive fifths until the original key is reached.
clarinet A family of single-reed instruments with cylindrical bores that function like an acoustical closed tube.
clarity An acoustical property of rooms in which the amount of reflected sound is below average.
clavichord An early keyboard instrument on which a string is struck by a *tangent* that also sets the pitch.
closed-temperament A keyboard tuning technique that allows use of all key signatures.
closed tube A tube with one open end and one closed end.
closed unequal temperament A keyboard tuning technique that allows use of all key signatures but excludes equal temperament.
cochlea The coiled part of the inner ear in which the mechanical vibration is converted into a series of electrical impulses.
cochlear implant An electronic device implanted in the cochlea that stimulates nerves in the cochlea to send signals to the brain representing audible sounds.
colored noise Noise with a pitch center containing increased amplitudes of vibrations.
combination tones Frequencies created by the nonlinearity of the ear mechanism when two or more notes are sounded simultaneously.
comma of Pythagoras The frequency difference between any note and the "same" note obtained after a complete circle of true fifths from that note.
complex waves Waves consisting of two or more overtones.
compression Reduction of the number of bits of digital data needed for a musical wave or other signal.
condenser microphone An electronic device that outputs the varying voltage across two charged plates with variable spacing.
constructive interference Interference in which the component waves are in phase, so that the intensity increases.
control oscillator A low-frequency oscillator used to control vibrato in an analog synthesizer.

Cooper scale flute A modern flute with tone hole spacing modified from the Böhm original design to compensate for pitch rise from A = 435 Hz to 440 Hz.
cornet A high-frequency brass instrument with a conical bore.
cornetto An early hybrid wind instrument blown like a trumpet but fingered like a woodwind.
coupling The transfer of energy between two oscillators with similar frequencies.
coupling resonance The total transfer of energy between two oscillators with the same frequency.
critical band The region of the basilar membrane that is excited by a pure tone.
crossover network An electrical circuit in a system with two or more loudspeakers that directs various frequencies of sound into the proper speaker.
current The net motion of electrons in an electrical circuit.
cymbals Percussion instruments consisting of thin circular metal plates, usually sounded by crashing two of them together or by striking one with a stick or metal whisk.
damper A soft rectangular block that rests on piano strings to prevent their vibration.
damping A force that absorbs energy from an oscillator, causing vibration to decrease in amplitude.
dayan The smaller of the two drums in the *tabla* set.
dc *Direct Current*; electrical current that always moves in the same direction.
decibel The unit of *Sound Intensity Level*, a logarithmic sound scale.
demodulate The process of extracting the modulator signal from a modulated carrier.
destructive interference Interference in which the component waves are out of phase, causing the intensity to decrease.
difference tones Combination tones where $f_{comb} = nf_1 - mf_2$.
diffraction The spreading out or bending around corners of a wave.
digital A method of transmission that uses stepwise representation, as contrasted with analog, or smooth-waveform representation.
digital audiotape (DAT) Audiotape that employs the digital representation of sound.
digital compression The process of reducing the amount of data that represents a signal.
digital synthesizer A musical synthesizer that uses digital waveforms.
digital-to-analog converters A device that converts a digitized wave into an analog (smooth) wave.
digitizers An electronic device that converts an analog wave into a digital (stepwise) form.
distance A measure of how far two points are from each other.
Dolby noise reduction A technique for removing high-frequency tape hiss from audiotape.
Doppler effect The change in the perceived frequency of a sound source resulting from relative motion between the source and the observer.
double-sideband modulation See *balanced modulation*.
double-valued reverberation time Room reverberation that has two values.
driven resonance Condition in which a periodic force acts on a vibrating system so as to increase its amplitude.
dulcimer An ancient plucked or hammered stringed instrument, sometimes used in folk music.

dynamic loudspeaker An electronic device that uses force on current in a magnetic field to produce sound.

dynamic microphone An electronic device that uses Faraday's law of magnetic induction to generate a signal.

dynamic range The full range of frequency or amplitude response—e.g., for ear or CD.

eardrum The membrane separating the outer from the middle ear that vibrates when a sound wave impinges on it.

echo The reflection of sound from a wall or other obstacle.

edge tone The sound generated when air flow strikes a sharp edge.

electroacoustic music Music composed of sounds from both electronic and acoustical sources.

electrostatic microphone An electronic device that uses the difference in voltage between two charged plates to generate a signal.

embouchure hole The hole on a flute into which the player blows.

ensemble An acoustical property of a room that enhances the ability of performers to hear each other.

envelope generator An analog-synthesizer function that controls the overall intensity of a wave as a function of time.

equal loudness curves Curves displaying the intensity of pure tones at various frequencies that are perceived at the same loudness level, measured in phons.

equal-temperament Keyboard tuning in which the same interval in all key signatures has the same frequency ratio.

equalization The process of undoing pre-emphasis on high frequencies that is necessary to reduce high-frequency tape hiss.

euphonium A low-voiced brass instrument with a large conical bore.

external noise Noise that invades a concert hall from the outside.

Faraday's law The principle that a change of magnetic field in a coil of wire *induces* an electrical voltage.

fiedel A medieval stringed instrument, precursor to the viola da gamba and violin families.

filter An electronic device used in an audio system or analog synthesizer to remove undesired frequencies.

Finale A computer program used in the MIDI system to process music.

finger holes Holes on woodwind instruments covered by fingers to change the sounded note.

fipple flute A recorder or similar instrument that uses a windway and an edge tone to produce sound.

flue pipe An organ pipe that uses an edge tone for sound production.

flügelhorn An alto member of the conical brass family.

flute A woodwind played by directing the player's breath onto the edge of an embouchure hole.

FM See *frequency modulation.*

focusing of sound An acoustics problem in which a concave wall shape focuses sound to one particular location in the room.

force A push or a pull that causes something to accelerate, changing its velocity.

formants An emphasized frequency band in the sound spectrum, creating the timbre for that wave.

Fourier analysis (spectrum analysis) A mathematical procedure by which the overtones are extracted from a complex wave.

Fourier spectrum A set of frequency components of a complex wave.

Fourier synthesis The addition of frequency components to create a complex wave.

French horn A narrow-bore conical brass instrument, often played with a hand in the bell to adjust tone quality.

frequency The number of vibrations per second, measured in Hertz.

frequency modulation A transmission process in which the modulating signal varies the frequency of the carrier.

frequency response The relative efficiency of electronic equipment as a function of frequency.

frets Narrow humps across the fingerboard of the guitar and some other instruments, used to ensure the accurate pitching of notes.

fullness An acoustical property of a room that produces an above-average amount of reflected sound.

fundamental tracking The ability of the ear to hear fundamentals when a series of waves consisting of higher harmonics is sounded.

gating The process of using one signal to turn another signal on and off.

gong A round metal instrument with a thick physical center and a general pitch center.

guitar A six-stringed plucked instrument with a nearly flat belly and back.

hair cells Frequency-sensitive nerve endings along the basilar membrane.

hammer The part of the piano mechanism that strikes the string to produce vibrations.

hammer A stick with a enlarged end used to strike some percussion instruments, such as chimes.

hammer The first of the three ossicles, in the middle ear, a bone chain that converts vibrations of the eardrum into vibrations of the cochlear fluid.

hand stopping The technique of using a hand in the bell of a French horn to affect the tone.

harmonics Overtones that are integral multiples of the frequency of the fundamental.

harp A plucked stringed instrument often used in symphony orchestras.

harpsichord An early keyboard instrument on which plucking activated the strings.

helicotrema An opening at the apex of the cochlea that allows vibrations to return to the round window at the base, where they are damped.

high level See *line level.*

horn chords Chords produced on low brass instruments when one note is played and another is sung, producing a third.

hum tone A low frequency that sounds below the strike tone of a bell.

hurdy-gurdy A medieval stringed instrument that uses a rotating rosined wheel to scrape the strings.

Huygens's principle A physical principle that states that every point on a wave front acts as a source of spherical waves.

hysteresis The failure of the magnetization of a magnetic tape to return to its original amorphic orientation when the external magnetic field is removed.

idiophones Instruments that consist of elastic materials capable of producing sound, such as bars, chimes, bells, and cymbals.
impedance The actual resistance to the flow of electricity in an alternating-current system.
impedance match The condition in which input and output circuits are matched in order to transfer signals most efficiently.
infinite baffle A large wall, effectively infinite in size, containing a hole in which a speaker is mounted.
infrasound Sound waves with frequencies lower than humans can hear.
inharmonicity A condition in which overtones are not harmonics—i.e., not multiples of the fundamental frequency.
inner ear The cochlea and semicircular canals.
intensity The strength of a signal, proportional to the square of its amplitude.
interference The addition of two or more waves to obtain their sum; may be *constructive* or *destructive*.
interval The musical distance, or separation, between two notes.
intimacy A feature of room acoustics, in which the first reflection reaches the listener within 20 ms of the direct sound.
intonation Production of musical tones with regard to accuracy of pitch.
inverse square law The principle that sound attenuates in proportion to the square of the distance from the source.
just noticeable difference The minimum frequency change of a single tone that can be heard by the normal ear.
just temperament A tuning procedure based on beatless major or minor chords.
kettle drums Timpani.
key A set of notes used in a particular scale.
keyboard An electronic instrument having a piano-type keyboard.
kilogram A unit of mass, equal to 1000 grams.
knife edge The term for a sharp edge toward which an air stream is directed by the fipple to produce sound in a fipple flute.
larynx The cavity of the vocal system above the vocal folds and below the glottis.
limit of frequency discrimination The minimum frequency difference between two tones sounding simultaneously that can be distinguished by the average ear.
line level The signal-voltage level used to transmit waves between audio components.
Lissajous figures Patterns created when a point executes two orthogonal harmonic oscillations.
liveness The property of an auditorium when T_R is longer than average for similar rooms.
logarithmic frequency response A type of nonlinear frequency response of the ear.
logarithmic intensity response A type of nonlinear intensity response of the ear.
longitudinal wave A wave in which the medium vibrates parallel to the direction of propagation.
loop One full repetition of the spatial pattern in a standing wave.
lossy compression Compression in which some of the signal information is lost—e.g., MP3.
loudness The psychophysical response to the intensity of a wave.

loudness Preamplifier control that boosts bass response relative to treble for low intensity levels.

loudness curves Graphs of intensity as a function of frequency for notes played on the violin, used to check the behavior of wood and air resonances.

lute A plucked, many-stringed instrument popular during the Renaissance and baroque periods.

magnetic domain A small collection of atoms in a magnetic material that magnetize as a unit.

magnetic induction See *Faraday's law*.

manuals Keyboards on a pipe organ.

marimba A percussion instrument that produces sound using wooden bars with resonator tubes.

masking The phenomenon of one tone covering up another, softer tone.

mass A measure of the amount of matter, expressed in units of grams and kilograms; does not depend on gravity.

mean-tone temperament A tuning technique in which thirds are made beatless.

membranophones Percussion instruments, such as timpani and snare drum, that use vibrating membranes to produce sound.

Mersenne's laws Laws that define the fundamental frequency of a stretched wire in terms of its length, tension, and mass per unit length.

meter A unit of length or distance in the metric system, often expressed in meters.

metric units A system of measurement, common worldwide, based on seconds, meters, and grams.

middle ear The section of ear between the eardrum and cochlea, containing the *ossicles*.

MIDI *Musical Instrument Digital Interface*; system for communication in electronic music.

midrange speaker The loudspeaker in a three-way system that handles the middle frequencies.

mode One of the many patterns of vibration in a vibrating system.

monochord An ancient musical instrument that used a single string.

mouth The cavity behind the lips; very important acoustically in singing.

mouthpiece The part of the woodwind or brass instrument that the player blows on or into.

MP3 A compression technique for digital music storage.

multiphonics See *horn chords*.

nasal cavity The cavity within the nose, important in sound production.

natural frequency The frequency at which an object vibrates when excited and left to vibrate unimpeded.

natural harmonics A violin performance technique in which the string is touched very lightly to create a high harmonic of its fundamental note.

node A location along a standing wave at which there is no motion of the medium.

noise Sound consisting of many nonintegrally related frequencies, typically not possessing a well-defined pitch.

noise generator An electronic instrument that produces noise.

nonlinearities Physical responses of a system or component that are not linearly proportional to the excitation.

oboe A treble double-reed woodwind instrument with a narrow conical bore.

Ohm's law of hearing The principle that the timbre of a complex wave depends only on the amplitudes of its harmonics, and not on the phases.

Ohm's law of circuits Voltage = current × resistance in an electrical circuit.

omnidirectional Equally sensitive in all directions.

open temperament Tuning scheme in which some key signatures are usable but others are not.

open tube A tube with both ends open.

orchestral chimes A percussion instrument that uses hollow tubes suspended from one end and struck by a mallet or hammer.

organ A keyboard instrument that produces sound using wind from a bellows or pump.

organ of Corti The central part of the cochlea, in which mechanical vibration is converted into electrical signals.

ossicles The three bones in the auditory chain that convert vibrations of the eardrum into vibrations of the fluid in the inner ear.

outer ear The part of the ear that lies outside of the eardrum, including the external pinna.

oval window The membrane through which the stirrup transfers its vibrations to the cochlea.

overtone A mode of oscillation of any system with a frequency higher than the fundamental.

overtone series A series of frequencies that are whole-number multiples of the fundamental frequency.

parabolic The shape of a surface used to focus waves, such as sound and light.

period The time required for a periodic wave to repeat itself.

periodic motion Motion that repeats itself.

periodic wave A wave that repeats continuously as time passes.

periodicity pitch The fundamental frequency perceived by the brain when only harmonics are physically present in the sound.

peripheral auditory system The outer, middle, and inner ears.

permanent hearing loss Hearing loss that will not become better and cannot be medically corrected.

pharynx The section of the vocal tract in which the oral cavity and the nasal cavity meet.

phase A measure of relative time or distance between corresponding points on two identical waves.

phase distortion The change of phase of Fourier components of complex waves resulting from electronic effects in audio components.

pianoforte The full name for the original piano, meaning "soft and loud."

pilot signal A 19-kHz signal on FM radio telling the receiver that the broadcast is in stereo.

pink noise Noise attenuating at 3 dB per octave; equal energy in each octave interval.

pinna The visible part of the outer ear.

pipe organ An organ that uses standing waves in pipes as source of sound.

piston valve A brass-instrument valve that uses piston motion to open a valve tube.

pitch The psychological response to the physical characteristic of frequency.

pitch level The rising pitch standard (e.g., from A = 415 Hz to A = 440 Hz) over several centuries.
place theory of hearing A leading theory of hearing that posits a correlation between frequency and the position of the response along the basilar membrane.
polarization A term defining the direction of vibration of transverse waves.
polarization The direction of vibration of strings in a piano.
power Energy per unit time.
preamplifier An audio-system component used to organize and standardize all source inputs.
pre-emphasis Amplification of high-frequency sound components on magnetic tape to raise their level above that of high-frequency tape hiss.
presbycusis Loss of hearing with age resulting from decreasing flexibility of the mechanism in the organ of Corti.
pressure Force per unit area that a liquid or gas exerts.
projection A one-dimensional view of two-dimensional motion, as seen from in the plane of motion.
psychokinesis The mythical claim that brain waves can move physical objects.
pulse-width modulation Modulation that changes the width of a flat-top pulse wave over the duration of the sound.
Pythagorean scale A musical scale based on beatless fifths.
rank A set of pipes in a pipe organ having the same tone quality.
rebec A medieval bowed stringed instrument.
recording Teaching a bird to sing a song by playing the song repeatedly on a *bird flageolet*.
reed pipe An organ pipe that uses a vibrating reed to produce sound.
reflection The bouncing of a wave backward as it strikes a new medium.
refraction The bending of a wave as it moves into a different medium.
regime of oscillation A group of overtone frequencies that combine to stabilize a note on a wind instrument.
resistance An electrical property that limits the current flow when a voltage is applied to a circuit.
resonance A large response of a vibrating system to certain frequencies (see *coupling resonance* or *driven resonance*).
resonance curve A graph of the response of a resonant system as a function of excitation frequency.
reverberation time (T_R) The time required for the sound in a room to decrease by 60 dB after the sound source ceases.
ring modulation See *balanced modulation*.
rotary valve A brass-instrument valve that uses rotary motion to open a valve tube.
round window The flexible area on the cochlea base that absorbs the residual wave energy of the sound wave that has passed over the basilar membrane.
rule of 18 A general rule used to determine stringed-instrument fret positions.
sabin A unit of absorption in architectural acoustics; one square foot of perfectly absorbing material.
sackbut A Renaissance-era narrow-bore trombone.

saxophone A single-reed, wide-bore conical woodwind instrument invented in the mid-nineteenth century.

scala tympani The return path for sound in the cochlea, which runs from the helicotrema to the round window.

scala vestibuli The path taken by sound in the cochlea as it excites the auditory nerve endings.

second The metric unit of time.

second-order beats Subtle changes in timbre between pure tones one octave apart.

shadows Regions behind obstacles in concert hall where sound waves do not reach at full strength.

sharpening The ability of the brain to sense a very wide cochlear excitation as a single frequency.

shock wave A large disturbance with a single, very narrow wave front.

shotgun A type of microphone that senses sound only in a very narrow range of directions.

simple addition Point-by-point addition of two or more waves.

simple harmonic motion Motion with a single frequency of vibration.

singing formant A group of harmonics with a frequency of 2,500–3,000 Hz that helps to project a male voice.

sitar A large Indian instrument with many strings; often played with tabla.

slide A tube fitted inside another tube, used to adjust the length of the sackbut or trombone.

snare drum A short drum with two heads that has wires stretched across the bottom head.

sonic boom A shock wave created by any motion faster than the speed of sound.

sound intensity level A logarithmic scale used to describe sound intensity, in units of decibels.

sounding board The wooden plate of a stringed instrument that aids in transferring string motion to the air.

soundpost A wooden dowel that holds apart the back and belly of violin-family instruments and transfers vibrations from the belly to the back.

sousaphone A deep bass brass instrument, wrapped around on itself so it can be held while marching.

spatial localization The binaural ability of the ears to locate a sound in three-dimensional space.

spectrum analysis See *Fourier analysis*.

standing waves Wave motion that appears to be stationary.

stiffness Resistance to bending of strings, such as in the piano, leading to inharmonicity of the overtones.

stirrup The third of the three ossicles, the bone chain that transfers vibrations from the eardrum to the cochlear fluid.

stopped An organ flue pipe that is closed at one end.

stretched octave A slight increase in frequency ratio between octaves to accommodate the inharmonicity of piano strings.

strike tone The pitch at which a bell note is heard (contrast with *hum tone*).

subcarrier The 38-kHz signal of an FM radio signal on which the second stereo channel is broadcast using amplitude modulation.
sul ponticello Bowing a violin or other stringed instrument *near the bridge*.
sul tasto Bowing a violin or other stringed instrument *near the fingerboard*.
sum tones Combination tones in which $f_{comb} = nf_1 + mf_2$.
superposition The principle that many waves can exist at the same time in the same place without changing each other.
sustain An analog-synthesizer envelope-generator function that determines the amplitude while the key is being held.
T_R See *reverberation time*.
tabla A set of Indian drums often played together with the sitar.
tam-tam A percussion instrument made of a flat metal plate that is struck with a beater and with poorly defined pitch.
tap tones Tones generated by tapping the back and belly of violin-family instruments.
tape hiss High-frequency noise on magnetic tapes resulting from random magnetic-domain orientation.
temperament A scheme by which the notes within one octave of a keyboard instrument are tuned.
temporary threshold shifts A reduction in hearing sensitivity that will naturally return to normal.
texture An acoustical property of a room in which the reverberated sound does not decrease monotonically after the sound ceases.
timbre See *tone quality*.
time A measure of how fast things happen, in units of seconds.
timpani (kettle drums) Large, bowl-shaped percussion instruments struck with mallet or stick, used in sets in the contemporary orchestra.
tinnitus Ringing in the ear.
tom-tom A cylindrical drum that may have either one or two heads.
tone quality The sound differences among various waves.
transient A rapidly changing wave shape at the beginning of a musical note.
transient response The response of an electronic audio system to rapidly changing signals.
transposing instrument One that plays a sounded note that is different from the written note.
transverse wave One in which the vibration of the medium is perpendicular to the direction of propagation.
treble Referring to high frequencies.
tremolo Periodic changes in amplitude of a musical note.
trombone A low-frequency, cylindrical brass instrument that uses a slide to change its length.
trumpet A high-frequency, brass instrument with a cylindrical bore.
tuba A very low-voiced, conical brass instrument.
tuned port A loudspeaker cabinet with an opening shaped so that the enclosure is resonant at low frequencies.
tweeter A small loudspeaker that broadcasts high-frequency sound.
ultrasound Sound composed of frequencies above the audible range (>20 kHz).

una corda The "soft" pedal on a grand piano.
uniform circular motion Motion in a circle with a constant speed.
valve The moving part on a brass instrument that is used to change the length of the tube.
velocity A vector change in position with time, including both magnitude and direction.
vibraphone A tuned percussion instrument that uses aluminum bars with resonators within which butterfly valves have been positioned to create tremolo.
vibrato Periodic changes in frequency of a musical note.
viola da gamba A six-stringed, fretted precursor of the violin family.
violin family A group of similar stringed instruments: violin, viola, cello, and string bass.
vocal formants An emphasized groups of harmonics creating various vowel and other vocal sounds.
voicing The process of making all of the notes on an instrument sound similar in timbre.
voltage An electrical force that causes electrons to move in circuits, creating electrical current.
voltage-controlled amplifier An analog-synthesizer unit that modulates the amplitude of the note controlled by the envelope generator.
voltage-controlled oscillator An analog-synthesizer wave source, frequency controlled by keys.
volume The intensity of an audio signal.
warmth An acoustical property of rooms for which T_R at low frequencies is above average for similar rooms.
wave A change in air pressure or signal voltage as a function of time.
wave shape (or wave form) A voltage or pressure variation over time, associated with a wave.
wavelength The distance in space over which a periodic wave repeats itself.
white noise Sound composed of equal-intensity frequencies over the entire audible spectrum.
windway A slot in a fipple flute that directs the player's breath onto a sharp edge to produce edge tones.
wolf tone A growling sound resulting from beating between a note and the wood resonance in a cello.
woofer A large loudspeaker used for lower frequencies.
xylophone A tuned percussion instrument that uses thick bars with a third harmonic reinforced by a closed-tube resonator.

Index

Absorbers, retractable architectural, 228, 230–231
Absorption, 226–227
air, 226
audience, 227
coefficients, 227
materials and surfaces, 227
ultrasonic. *See* Ultrasound, absorption
Acceleration, 2
sensing, 148
Acoustic criteria, auditorium, 216–221
Acoustic power, 184, 232
Acoustic problems, auditorium, 221–225
Acoustic suspension speaker. *See* Loudspeaker, acoustic suspension
Adam's apple, 167
Addition of waves. *See* Wave addition
ADSR. *See* Synthesizer, analog, envelope generator
Aerophone, 351
Air pressure, 3
Air resonance, 110
string instrument, 322
violin, 326
Ak, 361
Aliasing, 209
Alternating current (ac), 184
Altissimo register, clarinet, 281
Amati, 243
AM-FM tuner. *See* Tuner, AM-FM
Amniotic fluid, 61
Ampere (A), 183
Amplification, in auditorium, 229
Amplifier, power. *See* Power amplifier
Amplifier, voltage controlled. *See* Voltage-controlled amplifier (VCA)
Amplitude (A), 5
of driven harmonic motion, 11–12
Amplitude modulation (AM), 122
See also Modulation, amplitude
AM radio. *See* Radio, AM
Analog synthesizer. *See* Synthesizer, analog
Analog-to-digital converter (ADC), 138, 205
Analysis of tone quality. *See* Timbre
Anechoic chamber, 227
Antinodal line, 43
Antinodes, standing wave. *See* Standing wave, antinodes
Anvil. *See* Ossicles
Apex of cochlea. *See* Cochlea
Arch, bar instrument, 353–355
Armadillo shell, 319
Arteriosclerosis, diagnosis, 61
Artificial harmonics, violin. *See* Violin, harmonics, artificial
A-scan, ultrasonic, 60
Atmospheric layer location, 37
Atmospheric pressure, 3
Atmospheric refraction. *See* Refraction, atmospheric
Attack. *See* Envelope generator
Attack transients. *See* Transient
woodwind. *See* Woodwind instruments, attack transients in
Audience absorption, 226–227
Audience noise, 221
Audiocassettes. *See* Tape recorder
Audiogram, 164
Audio spectrogram, 170–176
birds, 176
computer voice, 175
formant visualization, 169–170, 173–174
pouring water into tube, 113–115
teaching of hearing impaired, 175
vocal sounds, 174–175
vocal sounds, schematic, 173
Audio system, 184
See also Reproduction systems, audio
components in, 185–187
noise in, 187
positioning in room, 232
power, 187
power transfer in, 187
reflection of signals in, 187
signal level in, 187
Auditorium acoustics, design, 227–231
See also Room acoustics
Auditorium, open air, 227–228
Auditory canal, 146, 153
Auditory nerve, 148
Auditory system, peripheral, 145–148
Aural harmonics, 157
Avery Fisher Hall, 227

Babcock, 338
Bach, Johann Sebastian (1685–1750), 239, 298
Bach fugues, 161
Baffle, 46
Bagpipe, 260–261
Balance, 148
Balance control, 18, 196
Balanced modulation, 123–124
See also Modulation, balanced
Bandora. *See* String instruments, pandora
Bandpass filter. *See* Filters, bandbass
Band-reject filter. *See* Filters, notch
Bandshell, 228
Banjo, 318
Baritone, 313

Bar modes, 351–356
inherent stiffness, effect of, 352
longitudinal, 353
torsional, 353
transverse, 352
Bartok pizzicato, 331
Base of cochlea. *See* Cochlea
Basilar membrane, 147, 148–151
damage, 164
distance versus frequency, 148
Bass bar, violin, 325
Bass control, 197
Bass drum, 360
Bass notes, hearing, 159
Bassoon, 286–288
baroque, 261
bore, 286
fingering system, 287
intonation problems, 287
octave keys, 286
photograph, 287
pitch variation, 286
range, 275
voicing defects, 287
Bass viol, 320, 329
Bathroom resonance. *See* Resonance, architectural
Bats, 59
Bayan, 361
Beats, 50–51, 149,
balanced modulation, equivalency, 124
beat rate of, 51
between two waves, 51
binaural, 162
and brain alpha state, 140
chime, 3
critical band experiment, 149
envelope of, 51
first-order, 50
frequency of, 51
quality, 159
second-order, 159
tuning of intervals, 245
wolf note, 246
Beethoven, Ludwig van (1770–1827), 243
Bel, 153
Bell, Alexander Graham (1846–1922), 153
Bell in vacuum experiment, 24
Bell lyra, 351–353
bar, 351
modes, 352
Bells, 86, 363
See also Percussion, tuned
Bell tuning, organ pipe, 291
Bending notes:
saxophone, 284
vibraphone, 356
Bernoulli, Daniel (1700–1782), 166
Bernoulli force, 166
Bernoulli principle, 167
brasses, 304
organ, 290
voice, 167
woodwinds, 266
Bias signal. *See* Tape, bias
Binary numbering, 203–204
coding on CD, 204
computer, 203
Binaural beats, 162
comparison with monaural beats, 162
difficulty in matching pitches, 162
Binaural effects, 161–163
sound localization, 162
Bird flageolet, 272
Blend, 220
Block, organ pipe, 289
Block diagram, in musical synthesizer, 129, 131, 133
Blood flow, ultrasonic measurement, 61
Bob. *See* Pendulum(s)
Bocal, 286
Böhm, Theobald (1794–1881), 261
Böhm fingering system, 261
Böhm flute, tuning error, 277
Boiling process, 62
Bore:
brass, 308
conical, 267
cylindrical, 267
exponential horn, 266
woodwind, 266–267
Bose QuietComfort headphones, 45
Bösendorfer grand piano, 346
Bottle band, 110
Bow, position on string, 323
Bow mechanism, 323
pressure, 323–324
Bowed instruments, theory, 322–325
Brain lesions, ultrasonic treatment, 62
Brain waves, 13
Brass instruments:
baritone, 299
bell, 304–305
bore, 304, 308
bugle, keyed, 300
contemporary, 299–300, 308–314
cornet, 299
picture, 300
cornetto, 297–298
euphonium,
picture, 301
families, 299
flügelhorn, 299
picture, 300
range, 303
French horn,
hand stopping, 299
picture, 301
range, 303
frequencies, 305
frequency ratios, 306
history, 297–303
keys, 300
lip tension, 307
mouthpiece, 304
mute, 308
overtone series in, 305
ranges, 303
sackbut, 298
slide, 298
relative frequencies, 306
tuning, 307
sound production, 303–308
sousaphone, 299
photograph, 302
range, 303
straight tube, 303–304
trombone, 298
bore, 308, 310
horn chords, 312
pedal tones, 311
picture, 301
positions, 311
range, 303
trumpet, 298, 308–310
baroque, 299
bore, 308
frequencies, open, 309
picture, 300
range, 303
regime of oscillation, 309
resonance curve, 309–310
tuning slide, 309
tuba, 299
picture, 301
range, 303
valves, relative frequencies, 306
valves, 299
effect of, 306–307
piston, 300, 302
rotary, 300, 302
tuning compromises, 306–307
variety, 297
Breaking, ocean waves, 39
Bridge (instrument):
bowed strings, 322, 325
piano, 343
plucked strings, 330
violin, 325, 326
Bridge (traffic):
marching soldier resonance, 88
Tacoma Narrows, 13, 88
Brilliance, 220
B-scan, ultrasonic, 60
Bugle, 300
Burst errors, 208
Butterfly valves, vibraphone, 355

Cancer, ultrasonic treatment, 62
Canopy, reflecting, 219
Cap, organ pipe, 289
Capacitor, 188
Capped-reed instruments, 261
See also Krummhorn; Kortholt; Rauschpfeife
Carillon bells, 363
Carrier, radio frequency, 122, 125
Carry of voices, 40
temperature inversion effect, 40
wind effect, 40
Casting, of bells, 363
Cathedral dome, standing waves. *See* Standing wave, dome of cathedral
Cavitation, 62
CD. *See* Compact disc
Cello:
range, 322
wolf note in, 329
Chalumeau, 262, 280
Chalumeau register, clarinet, 281
Chamfering, 271
Change ringing, 363
Chanter, bagpipe, 261
Charango, 319
Chimes, 86, 356–357
frequencies of modes, 357
missing fundamental, 357
Chimney draw, 166
Chladni, Ernst Florens Friedrich (1756–1827), 87
Chladni figures, 87
Chladni plates, 87
Choirs of strings, harpsichord, 337
Chordophone, 351
Chorus effect, 197
Chromatic scale, valved brass, 306
Circle, SHM reference, 6
Circle of fifths, 238
Circuit, electrical, 183
hydraulic model, 182

Cithara, 335
Cittern, 320
Clarinet, 280–283
 Böhm, 282
 bore, 280
  flare, 282
 cross section, 281
 family, 261
 fingering, 281
  forked, 282
 harmonics, 101, 280
 material, effect of, 272
 mouthpiece, and oboe, 286
 overblowing, 280
 photographs, 283
 ranges, 275
 reed, 265
 register hole, 281
 register key, 281
 registers, 281
 spectrum, 102, 280
 standing wave, 266
 throat tones, 271, 281
 transposing, 263, 282
 tuning errors, 282
 wave form, 102
Clarion register, clarinet, 281
Clarity, 219
Clavicembalo, 337
Clavichord, 320, 336
 action, 336
 tangent, 336
 vibrato, 336
Cleaning, ultrasonic. *See* Ultrasonic cleaning
Clef, treble and bass, 367
Closed temperament. *See* Temperament, closed
Closed tube, 85, 280
 brass instrument, 303–304
 clarinet, 85
  modes in, 280
 harmonics in, 84–85
 standing wave. *See* Standing wave, closed tube
 vocal system, 168
Clouds, architectural, 222
Coal, location using sonic reflections, 37
Coarseness, perceived, 149
Cochlea, 147,
Cochlear fluid, wave in, 147, 158
Cochlear implants, 165–166
 components, 165
 frequency range, 166
 usage, 165
Coherent light, 33
Colored noise. *See* Noise, colored
Combination of waves. *See* Modulation; Wave addition
Combination tones, 157–159
Comma of Didymus, 247, 249
Comma of Pythagoras, 244
Compact disc (CD), 202–210
 comparison with LP discs, 202–203
 distortion, 203, 209
 dynamic range, 203,207
 error correction, 208–209
 features, 203
 introduction to consumer, 203
 memory requirements, 211
 noise, 203
 optical system, reading, 207–208
 oversampling, 209
 pits on surface, 204–205, 208
 recording devices, 209
 rotational speed, 206
 sampling frequency, 209
 surface, microphotograph, 204–205
 tracking, 207–208
Compatibility, FM stereo, 199
Complex waves:
 analysis, 98–104
 effects of amplitude, 100–101
 effects of higher harmonics, 100
 harmonic content. *See* Fourier analysis
 period, 94, 97
 pitch heard, 97
 standard, 94–96, 100
 synthesis, 94–96
  effects of amplitude, 100–101
  effects of phase, 92–94, 100
 timbre, 100
Compression, 26
Compression, dynamic range, 210–211
Computer, 136–141, 203
 binary numbers in, 203
Computer-based piano reproduction system, 346–347
Computer software, 137–138
 Finale, 137
 Sibelius, 137–138
Computerized Speech Laboratory, 175
Condenser microphone. *See* Microphone, condenser
Conical bore, woodwinds. *See* Bore, conical
Consonants, synthesis of. *See* Synthesizer, voice
Constant linear velocity, CD, 206
Continuant, 171
Contrabassoon, 287–288
 doubling bass viols, 288
 range, 275
Contrast media, 62
Control oscillator, low-frequency, 131
Control voltage, 128
Cooper, Albert, 278
Cooper scale flute, 278
Cork, flute, 276–277
Cornet. *See* Cornetto; Brass, cornet
Cornetto, 297–298
Corti, organ of. *See* Organ of Corti
Cosine motion. *See* Simple harmonic motion (SHM)
Coupled vibrations, 12
Cristofori, Bartolomeo (1655–1731), 337
Critical band, 149
 arm analog experiment, 150–151
 experiment, 149
 flutes, demonstration, 149
 subjectivity of , 151
 widening with age, 164
 width, 149
Crossing musical lines, synthesis of, 134
Cross Interleave Reed-Solomon Code (CIRC), 208
Crossover network, 194
Current, electrical, 183
Cycles per second. *See* Hertz (Hz)
Cylindrical bore. *See* Bore, cylindrical
Cylindrical tube, modes, 84–86
Cymbals, 362
 diffraction and interference in, 362
Damage, to ear mechanism, 164–165
Damped oscillation, 11
Damper, piano action, 338
Damper pedal, piano, 338
Damped SHM, 11
Damping, 11
Dayan, 361
dB. *See* Decibel
Deadening of listening room, 233
Decay. *See* Synthesizer, analog, envelope generator
Decay transients. *See* Transient
Deceleration, 2
Decibel, 153
  and equal loudness curves, 156
Dekelboum, Elsie and Marvin, Concert Hall, 229–231
Delay of sound, 217
Denner, J. C. (1655–1707), 280
Density, 3
  linear, 3
Depth finder, 37
Diaphragm vibrato. *See* Vibrato, diaphragm
Didymus, comma of, 247, 249
Difference tones, 158–159
  example, 158
  in small loudspeakers, 159
  use in tuning, 158
Diffraction, 45–46
  architectural, 222
  around corners, 45
  compact disc pits, reading, 207
  cymbal sound, 362
  directional speakers, 46
  effect of wavelength, 45
  effect on loudspeakers, 46
  grating, 207
  Huygens's principle description, 31, 45
  ripple tank, 45
  sound localization effect, 163
  ultrasonic. *See* Ultrasound, diffraction
  voices, 45
Diffraction and sound localization. *See* Sound localization, diffraction
Digital audio tape (DAT), 138, 210
Digital sampler, 135–136
Digital sound reproduction, 203–206
Digital synthesizer. *See* Synthesizer, digital
Digital-to-analog converter (DAC), 133, 205
Digital Video Disc (DVD), 204
Digitization, of picture, 210
Diplacusis, 162
Direct current (dc), 183
Directional microphone. *See* Microphone, directional
Direct sound, 217
Disease and hearing loss. *See* Hearing loss, disease
Disklavier, 346
Distance, 1
Distortion:
  phase. *See* Phase distortion
  power amplifier, 197
  preamplifier, 197
  tape, 201
Dog whistle, 59
Dolby noise reduction, 202
Dolcian, 261
Dolphins, 37
Domains, magnetic. *See* Magnetic domains

Domes, standing waves in, 87
Donald Duck sound. *See* Formants, helium effect
Doppler, Christian Johann (1803–1853), 51
Doppler effect, 51–57
boat in water, 51
critical band research, 151
frequency and loudness, 57
and galaxies, 51
graphs, 53–57
light, 51
moving observer, 54
moving source, 52–54
ripple tank example, 52
sound, 57
Doppler shifted roar, 59
Double action foot pedals, harp, 331
Double horn, 313
Double reeds. *See* Woodwind instruments, double reed
Double-sideband modulation. *See* Modulation, balanced
Double-valued reverberation time. *See* Reverberation time, double-valued
Draw of chimney, 166
Driven oscillation, 11, 14
in piano, 344
Drones, bagpipe, 261
Drum, 360–361
Drum-and-bugle corps, 300
Drumhead, 357–359
ideal, 357
modes of, 358
tension, 358
weighted. *See* Tabla
Dulcimer, 320
Duplex scanning, ultrasonic, 61
Dust motion. *See* Kundt's tube
Dynaco, 138
Dynamic expression, 265, 274
Dynamic microphone. *See* Microphone, dynamic
Dynamic range:
compact disc, 207
compression, 210
ear mechanism, 145
MP3, 211

Ear, 145–166
anatomy, 145–148
sensitivity, 152
Ear canal. *See* Auditory canal
resonances in, 153
Eardrum, 3, 146
motion of, 146
Ear response, non-linearity of, 148–149, 153
frequency response, 145
intensity response, 145, 151–153, 155
Earth, structure, 64Earthquake, 63
Echo, 34
Echoes, architectural, 222
Echolocation, 59
Edge tone, 264
Eighteen, rule of, 331
Eight-to-fourteen modulation (EFM), 208
Elastic limit, of ear mechanism, 147
Electrical action, organ, 288
Electrical current, 183
Electrical pickup:
carillon bells, 363
guitar, 321, 332
Electroacoustical Music Laboratory, 138
Electromagnetic (EM) wave. *See* Wave, electromagnetic
Electromotive force. *See* Voltage
Electronic amplification, for rooms. *See* Room acoustics, electronic amplification
Electronic music performance, 140–141
Electronic tuners, 255–256
Electrostatic microphone. *See* Microphone, condenser
Elephant, pressure of footprints, 3
Ellipse
See also Lissajous figures
focusing, 36, 221
Embouchure hole. *See* Flute, embouchure hole
EM waves. *See* Wave, electromagnetic
Endoscopic ultrasonic disintegration, 62
Energy, in waves, 23
Energy conservation, effect on organs, 292
English bells, 363
English horn, 286
formants, 286
range, 275
Enharmonic equivalent, 245
Ensemble, 221
Envelope:
beats. *See* Beats, envelope of
generator. *See* Synthesizer, analog, envelope generator
keyboard signal. *See* Synthesizer, analog, keyboard
Equalization:
FM radio, 199
tape, 202
Equalizer, room, 233
Equal loudness curves, 152
Equal temperament. *See* Temperament, equal
Equal tempered scale, 369
See also Temperament, equal
Equilibrium position, 4
Erase head. *See* Tape head
Euphonium, 313
Eustachian tube, 3, 147
Exponential horn. *See* Bore, exponential horn
Extender, string bass, 329
External noise, architectural, 224
Eye, frequency response, 145

Falsetto voice, 166
Faraday, Michael (1791–1867), 188
Faraday's law, 188
Feedback, and transients. *See* Transient, feedback
in air column, 113, 265
Ferrichrome, 200
Fetal heartbeat. *See* Ultrasonic fetal stethoscope
Fetus, hearing loss in, 164
sonogram of, 61
f-holes, 325
Fidelity, 185
MP3, 210–211
Fiedel, 319
Fifth:
perfect, 77, 245, 369
wolf, 244
Fifths, circle of. *See* Circle of fifths
Filtering, musical sounds, 196
Filters:
bandpass, 131–132
Q, 132
CD player, use in, 209
equalizer, 233
fixed frequency, 132
high-pass, 131–132, 196
low-pass, 131–132, 196
notch, 132
preamplifier, 196
tracking, 132
transition frequency, 131
Finale. *See* Computer software
Fingerboard, 325, 328
Finger damping, 324
Finger holes, 267–270
Fingering, fork, 270–271
clarinet, 281–282
recorder, 270
Fingering chart. *See* Recorder, baroque, fingering
Fipple, recorder. *See* Recorder, fipple
Fipple flute, 262
First, David, 140
First-in-first-out buffer, 206
Fish, location by ultrasound, 37
Fixed end. *See* Longitudinal wave, reflection, spring and air correspondence; Transverse pulse, reflection of
Flageolet, bird, 272
Flap, organ pipe. *See* Organ pipe, tuning
Flat pulse, 127
Flemish bells, 363
Fletcher-Munson curves, 152, 156, 211
See also Equal loudness curves
Flow diagram. *See* Block diagram
Flue pipe. *See* Organ, pipe, pipes, flue
Flügelhorn, 300, 313
Flute, 276–280
alto, 279
baroque, 251, 276
bass, 279
Böhm, 277
bore, 276
Cooper scale, 278
cork, 277
cross section, 276
dynamic expression, 265
embouchure hole, 276
fingering, 276
headjoint, 277
intonation defects, 265, 277
material, effect of, 271
modern, 277
nineteenth century development, 276
photographs, 279
ranges, 275
Renaissance, 261
sound production, 265
tabor, 262
tone hole spacing, 278
tuning error, 277–278
Flute, fipple. *See* Recorder
Flutter echo, 222
FM radio:
muting, 199
noise in. *See* Radio, FM
FM stereo, 201
Focal point. *See* Focus
Focus, architectural, 221

Focus:
- ellipse, 36
- parabolic reflector, 35

Focusing:
- by air temperature, 39–40
- architectural, 221
- by wind, 40

Foot pedals, harp, 331
Force, 3
- linear restoring, 4–5
- magnetic, on current, 191
- units of, 3

Forced harmonic motion, 12
Forced oscillations, 11
Forearm, JND and limit of discrimination experiment. *See* Touch analogy to JND and limit of discrimination
Fork fingering, 270
- clarinet, 281
- recorder, 270, 273

Formants, 106,
- English horn, 286
- graphs, 169, 170
- helium effect, 171
- singing, 170
- sulfur hexafluoride effect, 171
- table, 169
- vocal, 106, 167–171
  - male-female difference, 169
- vowel sounds, 168

Fourier, Jean Baptiste (1768–1830), 92
Fourier analysis, 100–103
Fourier analyzer, 100
Fourier components, 92
- synthesis of, 92–97
- in tape playback, 201

Fourier spectrum, 98–104
- blowing across microphone, 108
- helium, effect of, 171
- instruments, 101–103
  - clarinet, 101, 102
    - square wave similarity, 101
  - krummhorn, 101, 103
  - recorder, 101
  - violin, 102
- noise, 107
- open tube, blown, 112–113
- phase information lacking, 100
- projection of voice, 170
- pure tones, 98
- singing, 170
- standard waves, 99–100
- timbre, effect on, 100
- limits of, 104
- sulfur hexafluoride, effect of, 171
- Tibetan monks, 170
- two-component waves, 98–99
- voice, 167–171
- vowels, 168–169
- water, pouring, 114–115
- wave shape, relation to, 100

Fourier synthesis, 92–97
- *See also* Wave addition
- amplitude effect, 92
- complex wave, 92
- missing fundamental, 97
- phase effect, 94
- standard waves, 94–96
  - amplitudes of harmonics, 96
  - pulse train, 96
  - triangular wave, 94
  - sawtooth wave, 95
  - square wave, 95
- timbre, relation, 94
- two components, 92–94

Fourier's theorem, 97
Fourth, perfect, 76, 369
French horn, 312–313
- *See also* Brass instruments, French horn
- double, 313
- hand stopping, 299
- range, 303
- rotary valves, 300–302

Frequency (*f*), 5
Frequency discrimination, limit of, 150
- experiment, 150–151

Frequency just noticeable difference, 150
- critical band relationship, 150

Frequency modulation (FM), 124–127
- *See also* Modulation, frequency
- use in JND experiment, 150

Frequency ratio:
- human hearing, 149
- intervals, 76–77

Frets, 331
Fullness, 219
Fundamental, 76
Fundamental frequency, of stretched string, 77–78
Fundamental tracking, 157

Gamba, 319
Gamelon, 362
Gamma rays. *See* Wave, electromagnetic
Gating, 121
- level, 121

Gee Haw Whimmey Diddle, 17
Glare, 33
Glass, breaking by sound resonance, 87
Glockenspiel, 351
Glue, violin, 328
Gong, 362
Gram, 2
Grand piano. *See* Piano
Graphical wave addition. *See* Wave addition, graphical
Gravicembalo col piano e forte, 337
Gravitational field, 148
Gravity, force of, 3
- in piano action, 340

Great manual, on organ, 288
Greek open air theaters. *See* Auditorium, open air
Guitar, 320–321, 332
Gurgling water, 114
Gut strings. *See* String, gut

Hair cells, 147
- damage, 164

Hammer. *See* Ossicles
Hammer, piano. *See* Piano, hammer
Handel, George Frederick (1685–1759), 243
Hand stopping, French horn, 299
Harmonic Choir, 171, 181
Harmonic Content, envelope generator control of, 127
Harmonic flavor, of keys, 241
Harmonic motion. *See* Simple harmonic motion (SHM)
Harmonic number (*N*), 75
Harmonic relationships, 238
Harmonic singing, 171
Harmonics:
- artificial, 330
- aural, 157–158
- closed tubes, 84
- emphasized. *See* Formants
- natural, 330
- open tubes, 84
- "seagull" on cello, 330
- stretched strings, 75

Harmonics and timbre, 101–104
Harp, 331
Harpsichord, 320, 336
- dynamics, 337

Headjoint, flute. *See* Flute, headjoint
Headphones, 45, 195,
- Bose, noise cancellation, 45
- hearing damage from, 195
- stereo separation, 195
- transient response, 195

Hearing, in animals, 59
Hearing, threshold. *See* Threshold of hearing
Hearing aid, 164
Hearing loss, 163–165
- disease, 164
- drugs, 164
- exposure to sounds, 163–164
- restoration, 164
- temporary threshold shift, 164

Heart valve defects, 61
Helicotrema, 147
Helium, and formants. *See* Formants, helium effect
Helmholtz, Hermann von (1821–1894), 109, 159
Helmholtz resonator, 109–110
- air resonance, violin, 322
- bottle band, 110
- loudspeakers, 110, 194
- original use, 109
- resonance curve, 110
- string instruments, 326, 329

Hertz, Heinrich (1857–1894), 6
Hertz (Hz), 6
High-pass filter. *See* Filters, high-pass
Highway noise barrier, 46
- effect of diffraction, 46

Hiss:
- FM radio, 107
- tape, 202

Holographic interferograms, of violin plates, 327
Home listening room, 231–233
- equalizers, 233
- furnishings, 231
- good listening area, 232
- materials, 231
- placement of speakers, 233

Hooey stick, 17
Hooke, Robert (1635–1702), 5
Hooke's law, 5
Horn chords, 312
Howling, wind. *See* Wind, howling
Human ear:
- amplitude response. *See* Human ear, intensity response
- frequency response, 145
- intensity response, 145,151
- nonlinearity, 148, 153
- physiology of, 146
- pressure response. *See* Human ear, intensity response
- range of, 145

Hum tone, of bell, 363
Hurdy-gurdy. *See* String instruments, organistrum
Huun-Huur-Tu, 141
Huygens, Christian, (1629–1695), 30
Huygens's principle, 29–31
  and diffraction. *See* Diffraction, Huygens's principle description
  and reflection. *See* Reflection, Huygens's wavelets
  and wave propagation, 31
Huygens's wavelets, 30, 34, 38, 45
Hydraulic circuit, 182
Hyperbolic reflector, 36
Hysteresis, 201

Idiophones, 351
Impedance, 184
  input, 186
  matching, 184–185, 304
  in brasses, 304, 313
  power amplifier to speaker, 198
  power transfer, 198
  microphone, 186
  output, 186
Impotence, male, 61
Incidence, angle of, 34
Induction. *See* Magnetic induction
Infinity Kappa 8.1 speaker system, 194–195
Infrasound, 63
  animal sensitivity, 64
  earthquakes, 63
Inharmonicity, 105–106
  bar instruments, 352
  chime, 356–357
  effect on piano string tuning, 341
  membranophones, 357–358
  piano strings, 105, 341
  synthesizer ring modulator, 133
  tuned percussion, 105
Inner ear, 146–148
  Ultrasonic treatment, 62
Input impedance. *See* Impedance, input
Instant people, 227
Instruments, musical, new, 138–139
Intensity, decrease with distance, 32–33
  sound localization use, 162
  sound wave, units, 152–153
  See also Inverse square law
Intensity just noticeable difference, 155, 160
Interference, 41–45
  constructive, 41
  cymbal sound,
  destructive, 41
  loudspeaker waves, 44
  Moire pattern model, 41–44
  ripple tank, 43
  speakers (Young's experiment), 44
    effect of separation and wavelength, 44
  stereo speaker experiment, 44
  two point sources, 43
Interferogram. *See* Holographic interferograms, of violin plates
Intermodulation distortion (IMD), 197
Interval, musical, 76–77, 368–369
Intimacy, 218, 222
Intravascular ultrasound (IVUS), 62
Inverse square law, 29, 32–33
  three dimensions, 33
  two dimensions, 32
Iron oxide, 200

Jack, harpsichord, 336–337
Jet. *See* Sonic boom
JND. *See* Frequency just noticeable difference; Intensity just noticeable difference
John F. Kennedy Center for the Performing Arts, 224
Jug band, 110
Just scale. *See* Temperament, just

Kay Elemetrics, 175
Keyboard, 133–136
  analog synthesizer, 128
  historical, 320, 336
  touch-sensitive, electronic. *See* Synthesizer, digital
Keyboard Mirror, 347
Key, piano. *See* Piano, key
Keys, 237–238
Kilogram. *See* Gram
Kirnberger, Phillip (1721–1783), 241, 251
Kirnberger temperament. *See* Temperament, Kirnberger
Korg:
  Triton Pro Digital Performer, 138
  Auto Chromatic Tuner AT-12, 255
Kortholt, 260
Kronos String Quartet, 140–141
Krummhorn:
  history, 261
  photograph, 103
  wave shape and spectrum, 103
Kundt, August Adolph (1839–1894), 80
Kundt's tube, 80
  striations, 80
Kurzweil, 138
  K2600 workstation, 138

Lambda (λ). *See* Wavelength
Larynx, 167
Laser:
  light properties, 33
  sound-and-light show, 16
Laser diode, 207
Ledger lines, 367
Legato, piano technique, 338–340
Length, 1
  stretched string (L), 77
  units of, 1
Limit of frequency discrimination, 150
Linearity, audio electronics, 187
Linear density. *See* Density, linear
Linear restoring force, 4
Lip, 264
  organ pipe, 288
Lissajous, Jules Antoine (1822–1880), 15
Lissajous art, 14, 16
Lissajous figures, 15
  phase, 15
  reference graph, 15
  speed of sound determination, 18, 20
Listening area, stereophonic sound, 232
Listening rooms, 231–233
Liveness, 218
  home listening rooms, 231
Localization, of sound source
  See also Binaural effects; Sound localization
  high frequency, 162–163
  low frequency, 162
Logarithms, 153–155
  useful properties, 153

Longitudinal standing wave, 79–86
  antinodes, 79
  brass instruments, 86
  clarinet, 85–86
  closed tube, 84
  conical tube, 85
  cylindrical tube, 85
  frequencies and wavelengths, 84–85
  harmonic number, 84, 85
  nodes, 79
  open tube, 84
  pressure and velocity correspondence, 79
  slinky spring, 79
Longitudinal wave, 23–24
  phase change on reflection, 80–81
    experiment to illustrate, 82–83
  pulse reflection, 80–81
    closed end, 83
    open end, 83
  reflection, spring and air correspondence, 80–81
  transverse representation, 26, 79
Loss of voice, in wind, 40
Loudness (sensation), 156
  and amplitude, 8
Loudness control, 196–197
Loudness curve, violin, 326–328
Loudness level, 156
Loud sounds, ear damage from, 163
Loudspeaker, 191–195
  acoustic suspension, 193
  baffle, effect of, 46
  bass reflex. *See* Loudspeaker, tuned port
  configuration, in room, 232
  crossover network, 194
  cross-section view, 192
  efficiency, 193, 194
  electromagnetic, 191
  electrostatic, 191
  enclosures, 192
  equalization, 233
  fidelity, 195
  frequency response, 193–194
  impedance, 186, 198
  infinite baffle, 192
  Infinity Kappa 8.1, 194
  low-frequency response, 194
  midbass, 194
  midrange, 194
  nonlinearity, 191
  phase distortion in, 195
  physical principle, 191–192
  placement, in room, 232
  polarity effects, 195
  power levels in, 186, 194
  resonances, 193
  "smart," 194
  system, 194
  transient response, 192
  tuned port, 193–194
    resonance, 194
  tweeter, 194
  two-way system, 194
  unmounted, 46, 192
  woofer, 194
Low-frequency control oscillator. *See* Control oscillator, low frequency
Low-pass filter. *See* Filters, low-pass
Lungs, 167
Lute, 320
L-wave, 63
Lyre, 335

Mach wedge. *See* Sonic boom; Mach wedge
Magnet and coil, experiment, 188
Magnetic domains, 200
Magnetic induction, 188, 201
  Phase distortion, 201
Magnetic tape, 200–202
  cassettes, 199–200
  materials, 202
  playback of, 201
Magnetization, of tape, 200
Major third, 76–77, 247, 249, 369
Manual, organ, 288
Many sources, selection by ear, 163
Marching band, ensemble problems, 221
  example of refraction, 37–38
Marimba, 353–354
  bars, 353
  resonator, 353
Maryland, University of, MIDI lab, 138
Masked tone, 161
Masking, 160–161
  Bach fugues, in, 161
  level, 161
  perception of timbre, 161
  rejecting unwanted stimuli, 161
  relation to critical band, 161
  tone, 161
Mass, 2
  on spring, 5
  units of, 3
Massive support, piano string, 344
Matching impedance. *See* Impedance matching
Mean-tone tuning. *See* Temperament, mean-tone, quarter-comma
  of bells, 363
Medium, in sound transmission, 24–25
Megaphone, 36
Membrane. *See* Drumhead
Membranophone, 351, 357–362
  modes, description, 357–359
Memling, Hans, 319
Mersenne, Marin (1588–1648), 77
Mersenne's laws, 77–78
  equation, 78
  piano, 78
  violin and cello, 78
  vibraphone, 356
Meter, 1
Microphone, 188–191
  carbon, 189
  cardioid, 189
  condenser, 188, 190
  crystal, 189
  damping, 190
  directional, 189
  dynamic, 188
  electret condenser, 189
  electrostatic. *See* Microphone, electret condenser
  frequency response, 190
  impedance of, 186
  moving coil, 188
  moving magnet, 188
  omnidirectional, 189
  overshooting, 190
  parabolic, 36, 190
  phase distortion in, 188
  pressure, response to, 191
  ribbon, 189
  shotgun, 190
  transient response, 190–191
  wind screen, 191
Middle ear, 146–147
  damage, 147
  muscular control, 147
MIDI, 133, 136–140
  interface box, 136
  Keyboard Mirror, 347
  teaching lab, 138
Minerals, location using sonic reflections, 64
Mirror, keyboard, 347
Missing fundamental, 157
  chimes, 357
  timpani, 360
Mixer, 138–139
Mode, 76
Modulation:
  amplitude, 122–123
    AM radio, 122
    100 percent, 123
    tremolo, 122
  balanced, 123–124
    beats, 124
  frequency, 124–127
    FM radio, 122
    vibrato, 125
  pulse width, 127
    "electronic" sound, 127
  unbalanced, 123
Modulator, 122
Moire pattern, 41
Monks. *See* Formants; Tibetan monks
Monochord, 335
Motion, 2, 4–8
Motion Picture Experts Group (MPEG), 210
Motion sickness, 64
Motorboat wake. *See* Sonic boom
Motor drives, tape, 201
MOTU, 138
Mouth, 167, 170
Mouthpiece, brass. *See* Brass instruments, mouthpiece
Mozart, Wolfgang Amadeus (1756–1791), 243
Mozart symphonies, clarity, 219
MP3, 210–211
  compression techniques, 211
  legality, 211
  memory requirements, 211
  production procedure, 211
MPEG, 210
Muffled voices, 45
Multiphonics, 312
Multiple reflection, 217
Multiplexer, 205
Muscle vibration, 152
Musical Instrument Digital Interface. *See* MIDI
Musical interval. *See* Interval, musical
Musicalische Temperatur, 239
Musical reproduction rooms, 231–233
Musical temperament. *See* Temperament
Music theory, elementary, 367–370
Muting, FM, 199

Nasal cavity, 167, 169
Natural frequency, 12
Natural gas, location using sonic reflections, 37
Natural harmonics, violin. *See* Violin, harmonics, natural
Neidhardt, Johan George (1685–1739), 240, 251
Neidhardt temperament. *See* Temperament, Neidhardt
Neural impulses, 148
Newton, Isaac (1642–1727), 3
Newton (unit of force), 3
Nodal lines, 43, 87, 350–351
  bell, 363
  Chladni figures, 87
  violin, 87
Nodal ring, in drumhead, 350–351
Node, standing wave. *See* Standing wave, nodes
Noise, 107–109
  air handlers, 224
  architectural, external, 224
  broadband, 107
  colored, 108, 132
  environmental, 224
  external, 224
  filtered, 107
  generator, 132
  gurgling water, 107
  highway, 46
  pink, 109
  use, 109, 233
  reduction. *See* Dolby noise reduction
  sound of, 107
  spectrum, 107–108
  wave form, 108
  white, 107, 132
  wind, 108, 132
Noise cancellation, active, 45
Nonintegral overtones: 105–106, 341, 351
  percussion instruments, 133, 352, 363
  piano, 105, 341
Nonionizing radiation, 60
Normal, 34
Nose, 167
Notch filter. *See* Filters, notch
Notched stick with rotor, 17
Nuclear bomb testing, 63
Nun's fiddle. *See* Stringed instruments, tromba marina

Oboe, 285–287
  acoustical problems, 285
  baroque, 261
  bore, 285
  clarinet mouthpiece, 286
  fingering system, 285
  photograph, 287
  range, 275
  timbre, 286
Occult, 13
Ocean waves, breaking of, 39
  refraction, 39
Octave, 76–77, 148
  response of ear, 148
Octave, tuning, stretching in pianos, 341
Octave hole, 268
Offset level, AM, 122
Ohm, Georg Simon (1787–1854), 159, 183
Ohm (Ω), 183
Ohm's law of electricity, 182–184
  hydraulic analog, 182
Ohm's law of hearing, 94, 159–160, 184, 195, 201
  limitations of, 159
Oil, location using sonic reflections, 37
Omnidirectional microphone. *See* Microphone, omnidirectional
100 percent modulation. *See* Modulation, amplitude, 100 percent
Open string, sound, 324

Open temperament. *See* Temperament, open
Open tube, 84
harmonics in, 84
sound production in, 113
standing waves. *See* Standing wave, open tube
Oral cavity, 167
Orchestral chimes. *See* Chimes
Organ, pipe, 288–292
electrical action, 288
fundamental tracking in, 157
great manual, 288
hydraulic, 260
keyboard, 288
manuals, 288
pipes:
flue, 288–289
reed, 290
portative, 260
ranks, 288
reed pipe, 288
stops, 288
swell manual, 288
tracker, 260, 288
tuning, 291–292
temperature effect, 292
variety of sound, 292
wind chest, 289
Organistrum. *See* String instruments, organistrum
Organ of Corti, 147, 164
Ornamentation, musical, 240, 250
cornetto, 297–298
Oscillation:
damped, 11
driven, 11–12
forced, 12
torsional, 87–88
Oscillator, voltage-controlled, 128
Oscilloscope, 9
dual-trace, 28
Ossicles, 147
Ouija windmill, 17
Outdoor auditoriums, 224, 227
Outer ear, 146
Output impedance. *See* Impedance, output
Oval window, 147
Over-modulation. *See* Modulation, amplitude
Oversampling, 209
Overtone series, 71–77, 368–370
comparison with equal-tempered scale notes, 370
musical intervals, 76, 369
musical notes in, 76, 369
piano notes, frequency comparison, 77, 370
Overtones:
decay of, in piano, 343
decay of, in triangle, 357
frequencies, in stretched string, 75
frequencies, open and closed tubes, 84–86
nonintegrally related, 105–106, 341, 351

Pain, threshold of. *See* Threshold of pain
Palestrina, 242
Pandora, 320
Parabolic microphone. *See* Microphone, parabolic
Parabolic mirror, 35
Parabolic reflector, 36
auditorium, 221
Paraboloids, 35, 221
Parabolic taper, flute headjoint, 276
Parallel walls, effect, of, 222
Parkinson's disease, ultrasonic treatment, 62
Pascal (Pa), 3
Passive room equalizer, 233
Paths of sound in room, 217
Pedal, damper. *See* Damper pedal, piano
soft. *See* Soft pedal
sustain. *See* Sustain pedal
una corda. *See* Una corda pedal
Pendulum(s), 6
coupled, 12–13
Percussion, 351–366
bar instruments, 351–356
tuned, 86, 351–359
ranges, 362
synthesis of, 133
variety of sound, 351
Period (T), 5
driven harmonic motion, 11–12
wave, 26
Periodicity pitch, 156–157
Periodic motion, 4
Periodic wave, 26
Peripheral auditory system, 145–148
cross section, 146
Perlon, violin strings, 327
Pharynx, 167
Phase (phi or ϕ):
control, by computer, 194
relative, 160
sound localization using, 162
stereo effect, 162
Phase change, on reflection:
longitudinal wave, 80–81
transverse wave, 71
Phase distortion, 160
See also Distortion, phase
in CD player, 209
loudspeakers, 191, 194
in tape playback, 201
Phon scale, 156
Piano:
action, 338–341
attack technique, 340
cast iron frame, 338
coupling, between strings, 343–345
coupling, hammer and string, 338
damper, 338
dynamics, 337
exploded view, 339
extended keyboard, 347
frame, stress on, 338
hammer, 338
coupling to strings, 341
effect of hardness, 341
effect of position, 341
history, 335–338
inharmonicities, 341
key, 340, 367
keyboard, note of, 338
left-handed, 347
pedals:
damper, 338
soft, 343
sustain, 343
una corda, 343
performance technique, 338, 340
pin block, 338
prompt sound, 342
regulation, 338
soundboard,
stretched octave, 341
strings, 341–345
coupling between strings, 343–345
coupling to soundboard, 345
Mersenne's laws, 341
motion, 3, 341
multiple, 342
polarization, 342
stiffness, 341
tension in, 341
wrapping of, 341
sustained sound, 342
synthesis of sound, 341
timbre, 341
tone decay, 343
transients, 342
tuning, 256, 341, 343–345
upright, 340
Pianoforte. *See* Piano
Piano Performance Reproduction System, 346–347
Picardy third, 77
Piccolo, 279
range, 275
Piezoelectric ultrasonic transducer, 60
Pilot signal, FM radio, 199
Pin block, piano. *See* Piano, pin block
Pink noise. *See* Noise, pink
Pinna, 1, 146
moving, 146
sound localization use, 163
Pipe organ. *See* Organ, pipe
Pitch, and frequency, 8
Pitch level, 242–244,
adjustment for instruments, 243–244
Bach, 243
baroque, 237
Beethoven, 243
consequences of further increase, 244
effect on brasses, 244
effect on string instruments, 321
effect on woodwinds, 243
fifteenth century, 242
historical data, 242
historical development, 242–244
information from organ pipes, 244
motivation to increase, 243
Mozart, 243
nineteenth century, 243
physics standard, 237
Renaissance, variation, 242
rise in pitch:
during instrument warmup, 86
effect on flute, 277–278
over centuries, 242
standard, 237
twentieth century, 243
U. S. standard, 237
Pits, on CD surface, 205, 208
Pizzicato, 331
"Bartok," 331
plucked instruments, 330–331
violin, 331
Place theory of hearing, 94, 148–151
Plane of polarization. *See* Polarization, plane of
Playback head. *See* Tape head; Playback head
Plectrum, harpsichord, 337
Plosive consonants, 174
Plucked string instruments, 330–332
prompt and sustained components, 330

synthesis of sound, 130–131
transients, 330
Polarization:
light, 33
piano strings, 342–343
plane of, 33
transverse waves. *See* Transverse wave, polarization
Polaroid sunglasses, 33
Polychord, 336
Portative organ, 260
Position, 1
Power:
acoustical, 184, 186, 198
electrical, 184
transfer of, 184, 186
Power amplifier, 198
linearity, 198
output impedance, 198
Practice rooms, music, 224
Preamplifier, 187, 196–197
controls, 196
distortion, 197
linearity, 197
Precedence effect, 163
in auditoriums, 229
Preemphasis:
FM radio, 199
tape, 202
Prefixes, metric, 372
Presbycusis, 164
Pressure, 3
atmospheric, 3
examples, 3
units of, 3
variation of, 3
Pressure antinodes. *See* Longitudinal standing wave, pressure and velocity correspondence
Pressure nodes. *See* Longitudinal standing wave, pressure and velocity correspondence
Pressure variation:
painful, 152
smallest audible, 152
weather, 152
Priming, tam-tam, 362
Prism, 107
Projection, vocal, 170
Prompt sound, in piano, 342
Psaltery, 335–336
Psychoacoustic vibration transducer, 13
Psychoacoustic waves, 13
Psychokinesis, 13, 17
Pulse train 9
See also Standard waves
analysis, 100
synthesis, 96
Pulse width modulation (PWM), 127
See also Modulation, pulse width
Pure motion. *See* Simple harmonic motion (SHM)
Pure wave, 4
P-wave, 63
Pythagoras, 335
comma of, 244
Pythagorean temperament. *See* Temperament, Pythagorean

Q of filter. *See* Filters, bandpass, Q
Quality beats. *See* Beats, quality
Quarter-comma mean tone temperament. *See* Temperament, mean-tone, quarter-comma
Quartz, 60
Quiet region. *See* Shadows, acoustical
Quincke's tube, 41

Rackett, 261
Radar waves. *See* Wave, electromagnetic
Radio:
AM, 122, 198
fidelity of, 198
FM, 127, 199
FM frequency excursion, 127
FM muting, 199
FM noise, 199
FM pilot signal, 199
FM stereo, 199
FM subcarrier, 199
Radio waves. *See* Wave, electromagnetic
Ramp wave. *See* Sawtooth wave
Range of human hearing. *See* Human ear, frequency response
Rank, organ, 288
Rarefaction, 26
Rauschpfeife, 261
Ray, 38
Rebec, 319
Recorder, 272–275
baroque, 262, 272
fingering, 270
bore, 273
dynamic expression, 274
family, 275
fipple, 272
fork fingering, 273
holes, 273–274
intonation, 274
octave hole, 273
range, 275
Renaissance, 262, 272
scaling with pitch, 274
spectrum, 101
timbre, 101
tuning compromises, 274
voicing, 273
wave form, 101
Recorder, tape. *See* Tape recorder
Recording (birds), 272
Reed:
clarinet. *See* Clarinet, reed
organ, 265
resonances in, 265
woodwind instruments, 265
Reed organ pipe. *See* Organ pipe, pipes, reed
Reed tongue. *See* Organ pipe, reed
Reel-to-reel tape, 203
Reference graphs. *See* Lissajous figures, reference graph
Reflected sound, 217
Reflection, 34–37
angle of, 34
atmospheric layers, 37
boundary, 36, 37
ellipsoid, 36
geological layers, 37
Huygens's wavelets, 34
law of, 34
multiple, 217
parabolic, 35
signals in audio system, 185
transmission and, 36
ultrasound, 61
wind instruments, 37
Reflection of longitudinal pulse. *See* Longitudinal wave, pulse reflection
Reflection of transverse pulse. *See* Transverse pulse, reflection of
Reflective walls, 217
Reflector:
ellipsoidal, 36
hyperbolic, 36
parabolic, 35
Refraction, 37–41
atmospheric, 39–40
change of direction of wavefront, 37–39
Huygens's wavelet description, 38–39
marching band model, 38
ocean waves, 39
breaking, 39
ripple tank description, 39
temperature inversion, 39, 40
thunder, 41
maximum range, 41
wave approaching beach, 39
wind, 40
Register key, clarinet, 85
Regulation, of piano action, 338
Release. *See* Synthesizer, analog, envelope generator
Reproduction systems, audio, 185–187
Resistance, electrical, 183
Resistive support, piano string, 344
Resistor, 183
Resonance:
See also Tacoma Narrows bridge
air, violin, 110
architectural, 222
bridge, soldiers marching, 88
coupled, 12–13
driven, 12
English horn, 286
forced oscillation, 12, 14, 88
Helmholtz. *See* Helmholtz resonator
manipulated in vocal tract, 168, 170–171
nuclear, reactor tubes, 88
pendulum, 11
piano, 343–344
room, 223
seashell, 152
speaker, 193–194
speaker enclosures, 194, 224
wooden plates, of string instruments, 326, 327
Resonance curve, 110–112
application, 112
closed tube, 111
difference from Fourier spectrum, 111
experiment, 110
Helmholtz resonator, 110
loudspeaker, 194
open tube, 111
Resonator:
marimba, 353
vibraphone, 355
xylophone, 354
Reverberation time:
adjustment, 227, 230
appropriate, 218
calculation, 225–226
churches, 218
conference rooms, 218
control of, 225–227
definition, 217
double-valued, 225
formula, 225
lecture rooms, 218

music reproduction rooms, 218
recording studios, 218
Riley, Terry, 140
Ringing in ears, 164
Ring modulation. *See* Modulation, balanced
Ripple tank, 30
diffraction. *See* Diffraction, ripple tank
Doppler effect. *See* Doppler effect, ripple tank example
interference. *See* Interference, ripple tank
waves in, 30
Rise in pitch. *See* Pitch level, rise in pitch
Roland RD-700 Expandable Keyboard, 136
Room acoustics:
criteria, good acoustics, 216–221
electronic amplification, 229
problems in design, 221–225
Room resonance, 222–224
Rope wave. *See* Wave, rope
Rotary head digital audio tape (RDAT). *See* Digital audio tape (DAT)
Rotary valves, 302
on French horn, 313
Roughness, two-tone:
experiment, 149
obtained with recorders, 149
Round window, 147
Rubella, and hearing loss, 164
Rule of 18, 331
intonation, 332

Sabin, 225
Sabine, Wallace C. (1868–1919), 225
Sackbut, 298
Sample and hold, 128
Sampler, digital, 13, 136
Sampling frequency, 205, 209
Sawtooth wave
See also Standard waves
analysis, 100
synthesis, 95
Sax, Antoine Joseph (Adolphe) (1814–1894), 282
Saxophone, 282–285
Böhm fingering system, 283
bore, 282, 284
cross section, 284
expression, 285
family, 261, 262, 285
jazz, use in, 262
octave holes, 283–284
overblowing, 283
photographs, 285
ranges, 275
Scala timpani, 147
Scala vestibuli, 147
Scales, musical, 238, 368–369
Schlick, Arnolt (born *c.* 1460), 240, 250
Schumann resonances, 140
Scope. *See* Oscilloscope
Seagull effect, cello, 330
Searchlight, 33
Seashell resonance. *See* Resonance, seashell
Second-order beats. *See* Beats, second-order
Seed, Christopher, 347
Seismic shocks, use of, 64
Semicircular canals, 148
Serpent, 298
Seventh, minor, 76, 369
Shadows, acoustical, 222
Shallot, organ pipe, 290
Sharpening, 149, 150,
Shawm, 261
Shell, architectural, 228
SHM. *See* Simple harmonic motion (SHM)
Shock wave. *See* Sonic boom
Shore, John (*c.*1662–1752), 243
Sibelius. *See* Computer software
Sibilant, 171
Sigma-Delta oversampling, 209
SIL. *See* Sound intensity level (SIL)
Simple addition. *See* Addition of waves
Simple harmonic motion (SHM), 4
amplitude, 5
application to sound, 8
conditions for, 4
damped, 11
driven (forced), 11
frequency, 5
mass on spring, 4
pendulum, 6
period, 5
phase, 7, 8
position in, 5, 7
and uniform circular motion, 6
velocity in, 7
Sine wave, 4
See also Standing wave
analysis, 98
timbre, 101
Singing formant. *See* Formants, singing
Sinusoidal motion. *See* Simple harmonic motion (SHM)
Sitar, 361
Slide, brass. *See* Brass instruments, slide
Slinky, standing waves in. *See* Longitudinal Standing wave, slinky spring
Slinky spring, 81
Slope, auditorium seating, 228
Snare drum, 360
Snares, 360
Soft pedal, 343
Soldiers on bridge. *See* Resonance, bridge, soldiers marching
Sonogram, fetus, 61
Sonar, 37, 60
bats, 37, 59
depth finder, 37, 60
dolphins, 37
locating fish, 37, 60
velocity of submarine, 37
Sone scale, 156
Sonic boom, 58–59
circular wave construction, 58
cowboy's whip, 59
electrical spark, 59
lightening, 59
Mach wedge, 58
nature of, 58
ripple tank, 58
supersonic plane, 58
thunder, 41, 59
Sonic ranging, 37
Sonoluminescence, 62
Soprano range, 169–170
Sound, and SHM, 8
speed of, 27
experiment, 27–29
temperature effect, 40
Soundboard, piano. *See* Piano, soundboard
Sound intensity level (SIL), 153–154
addition of sources, 154–155
reference level, 154
reverberation time, 217
scale (SIL), 153
usefulness, 153, 155
Sound localization, 162–163
diffraction, 163
high frequency, 162–163
low frequency, 162
phase ambiguity, 162
pinna, effect of, 163
precedence effect, 163
Sound post, 325
Sound production, 112–113
feedback mechanism, 113
musical, 107–115
wind instruments, 264–267
Sound spectrogram. *See* Audio spectrogram
Sousa, John Philip (1854–1932), 312
Speaker
See also Loudspeaker
baffle. *See* Diffraction, effect on loudspeakers
efficiency of, 193–194
power transfer to, 198
Speaker and baffle experiment, 46
Speaking, and wind, 40
Spectra, Fourier. *See* Fourier spectrum
Spectrogram, audio. *See* Audio spectrogram
Spectrum analyzer. *See* Fourier analyzer
Speed, 2
unit of, 2
Sphere, radius of and inverse square law, 33
Spiegel der Orgelmacher und Organister, 240
Spirograph, 14
Spring, force, 4–5
See also Hooke's law
Springy support, piano string, 344
Square wave
See also Standard waves
analysis, 100
clarinet-similarity, 101
synthesis, 95
Staccato, piano, 340
Staff, musical, 367
Standard waves:
analog synthesizer use, 129
Fourier spectra of, 99–100
synthesis of, 94–97
Standing wave:
addition to form, 71
antinodes, 71, 94
architectural, 87
bassoon, 286
bell end, 305
bore shape, 85
bowed, 323–324
brass instrument, 85
Chladin plates, 87
circular drumhead, 86
closed tube, 84
development by reflection, 72
dome of cathedral, 87
end configurations, 71
fipple, 85
formation in strings, 68–71
frequency of, 75, 78, 84
fundamental, 75, 84
human vocal tract, 168
impossible, 74
longitudinal. *See* Longitudinal standing wave
loop, 71, 84
metal plates, 87
nodes, 71, 84
open tube, 84
overtone series in stretched string, 75
phase change upon reflection, 72

picture, 70
reed end, 85
stretched string, 68–71
Tacoma Narrows bridge, 87
torsional, 87–88
transverse. *See* Transverse standing wave
two-dimensional, 86–89
wavelength of, 75
wave speed, 75
wineglass, 87
woodwind instruments, 85–86
Steinway, 339
Stereo, FM radio, 199
Stereo headphones. *See* Headphones
Stereo speaker interference. *See* Interference, stereo speaker experiment
Stethoscope:
phase sensitivity experiment, 163
ultrasonic fetal. *See* Ultrasonic fetal stethoscope
Stirrup. *See* Ossicles
Stops, organ, 288
Stradivari, Antonio (*c.*1644–1737), 328, 243
Stradivarius violin, 328
Stretching octaves, 105, 341
Striations, in Kundt's tube, 80
Strike tone, of bell, 363
String:
end effects, 71–72
gut, 327
length, effect, 77
mass per unit length, effect, 77
mistuned, in piano, 343–345
overtone series in, 71–77
perlon, 327
resonances in, 71–77
tension, effect, 77
wire-wound, 327, 341
String bass, 320, 329
extender, 329
range, 322
Stringed instruments, 318–334
air resonance, 322, 326
banjo, 318
Baroque innovations, 320
bass viol. *See* String bass
bowed, 319
bowing, theory, 322–325
sound production, 322–325
charango, 319
cittern, 320
dulcimer, 318
fiedel, 319
fretted, bowed, 324
gamba, 319
guitar, 320, 321
hammered, 319
harp, 331
history, 318–321
keyboard, 320
clavichord, 320
harpsichord, 320
hurdy-gurdy, 319
lute, 319
lyre, 319
medieval, 319
monochord, 318
organistrum, 319, 336
pandora, 320
plucked, 320, 330–332
bridge on, 322, 330
decay of transients, 330
polychord, 336
psaltery, 335
rebec, 319, 322
scaling, 329
standing waves in, 75–77
string bass, 320
string vibration, 323
finger damping, 324
harmonics in, 323–324
timbre, 323, 324
tromba marina, 319, 326
ud, 319
ukulele, 318
vielle, 319
vihuela, 320
viola da braccio, 320
viola da gamba, 320
viola d'amore, 320
violin
See also Violin
Subjective fundamental, 157
Subjective listening response, and audio component choice, 197
Subjective loudness, 156
Subjectivity, of audio experiments, 151
Sul ponticello, 324
Sul tasto, 324
Sulfur hexafluoride, 171
Sum tones, 158
masking of, 161
Sun Rings, 140
Superposition:
principle of, 29, 31–32
pulses, 47
waves, 48
Surgery:
ear, 62
trackless, 62
ultrasonic, 61–62
Sustain. *See* Synthesizer, analog, envelope generator
Sustained sound, in piano, 342
Sustain pedal, 343
S-wave, 63
Swell manual, organ, 288
Swing, and driven SHM, 11
Sympathetic vibrations, 13
Synchronization of instruments, 221
Synclavier, 135
Synthesis, Fourier. *See* Fourier synthesis
Synthesizer, analog, 127–133
audio spectrogram of "wow" sound, 175
block diagram, 127, 129, 131, 133
envelope generator, 130
ADSR, 130
attack, 130
decay, 130
release, 130
sustain, 130
filters, 131–132
gating, 121
keyboard, 128
pitch control, 128
trigger control, 128, 130
limitations, 134
low-frequency control oscillator, 131
modulation in, 131, 133
noise generator, 132
sample-and-hold, 128
signals, 128
audio, 128
control, 128
two-note production, 134
voltage-controlled amplifier (VCA), 129
voltage-controlled oscillator (VCO), 128–129
Synthesizer, digital, 133–136
digital wave form, 134
features of, 134–136
FM, 134
MIDI use, 133
multiple voice lines, 134
musical temperament, 135
performance with, 134
piano substitute, 135
sampling, 135
temperaments, use of, 256
touch-sensitive keys, 135
Synthesizer, FM, 134
Synthesizer, use, 135
Synthesizer, voice, 174–175

Tabla, 361
resonances in, 361
Tabor flute, 262
Tacoma Narrows Bridge, 13, 87, 88
Tam-tam, modes, 362
Tangent, clavichord, 336
Tape:
bias, 201
hiss, 202
magnetization, 200
See also Magnetic tape
Tape head, 200
Tape recorder, 199–202
Tap tone, 87, 326
See also Violin, tap tone
Tartini, Giuseppe (1692–1770), 158
Tectorial membrane, 147
Telescope, 36
Temperament:
choice, 237
closed, 239–240
compromises, 238
criteria, 239–240
effect on musical performance, 240
equal, 240, 252–255
development, 240
frequencies of notes, 241, 368
compared with overtone series, 253, 370
frequency ratios, 241, 252
reservations, historical, 254
setting, 254
sound of scales, 253
flavor, 241
future, 254
history, 237–241
"ideal," 238
just, 247–249
chords, correct, 247
major, frequencies of notes, 241
major, frequency ratios, 241, 248, 249
minor, 248
setting, 247
keyboard instruments, 237
Kirnberger, 240, 251
mean-tone, quarter-comma, 249–250
advantage, 249
comma of Didymus, 249
frequencies of notes, 241
frequency ratios, 241
setting, 250
tuning of bells, 363
wolf note, 250
See also Temperament, quarter, mean tone
Neidhardt, 240, 251

open, 239, 250
Pythagorean, 244–247
comma:
of Didymus, 247
of Pythagoreas, 244
frequencies of notes, 241
frequency ratios, 241, 246
setting, 246
third, 247
quarter-comma mean tone. *See* Temperament, mean-tone, quarter-comma
Renaissance, 239
setting, 237
unequal, closed, 240, 250–252
Valotti, 240, 251
"well-tempered," 239
Werckmeister Correct Temperament No. 1, 251
frequencies of notes, 241
frequency ratios, 241
setting, 251–252
Werckmeister III. *See* Werckmeister Correct Temperament No. 1
Temperature inversion, 39
sound refraction during, 39–40
Temporary threshold shift, 163
Terms and units, acoustical, 371
Texture, 220
Thermoacoustic cooling, 63
Third, major (interval), 76, 369
temperaments, 247, 249–250, 253
Threshold of hearing, 151
Threshold of pain, 151
Throat tone, clarinet, 271
Thunder. *See* Sonic boom, thunder
refraction. *See* Refraction, thunder
Thyroid cartilage, 167
Tibetan monks, 170
Tierce de Picardie. *See* Picardy third
Timbre, 9
analysis, 104–107
chorus effect, 107
complex tones, amplitudes of harmonics, 94
effects of harmonic amplitudes and phases, 93–94
formants, 106
harmonics, effect of, 104
inharmonicities, 105–106
piano tone, 105
stretched strings, 105
tabla, 105
transients in, 105
tuned percussion, 105
masking, effect of, 161
transients, attack and decay, 104–105
experiment, 104
research on, 104
vibrato, 106
and wave shape, 9
Time, 2
Timpani, 86, 359–360
head, modes in, 359–360
missing fundamental, 360
pitch, 360
See also Percussion, tuned
Tin, 361
Tinnitus, 164
Tom-tom, 360
Tone, musical, production, 113–114
Tone control, preamplifier, 196–197
Tone-hole cutoff frequency, 270
Tone quality. *See* Timbre
Tongue, 167, 168
Tonic, 238
Torsional oscillation, 87–88
Touch analogy to JND and limit of discrimination, 150–151
Tracker action, organ, 260
Tracking filter, 132
Trackless surgery, 62
Transducer, 60
guitar, 332
Transient:
attack, 265
auditory response to, 104
computers and, 104, 135–136
feedback, 265
percussion instruments, 351
piano, 104
plucked strings, 104
response, microphone, 190
response, power amplifier, 198
response, speaker, 194
timbre. *See* Timbre, transients, attack and decay
Transition frequency of filter. *See* Filters, transition frequency
Transposing instruments, 263
Transverse pulse, reflection of, 71
Transverse standing wave, 68–71
See also Standing waves
antinode, 70, 71
formation of, 68
loop, 70, 71
motion in, 70, 71
node, 70, 71
Transverse wave, 23, 24
graph, 25
period, 25
piano strings, 342
polarization, 29, 33
representation, 26
speed of, 26
wavelength, 25–26
Treble control, 197
Tremolo, 122, 125, 131
vibraphone, 355
Triangle, 357
Triangular wave
See also Standard waves
analysis, 100
synthesis, 94
timbre of, 100
Tromba marina, 319, 326
Trombone, 301, 310–312
See also Brass instruments, trombone
bass, 301
equal-tempered scale, 310
horn chords, 312
notes of, 311
pedal tone, 311
range, 303
slide positions, 311
Trumpet, 300, 308–310
See also Brass instruments, trumpet
chromatic scale, 309
frequencies, open, 305
range, 303
tuning slide, 307
valve, tuning, 305–307
Tuba, 301, 314
Tuned percussion. *See* Percussion, tuned
Tuned-port speaker. *See* Loudspeaker, tuned port
Tuner, AM-FM, 186, 198–199
Tuner, electronic, 255
"Tuning" a room, 233
Tuning bars, 13, 86
Tuning fork, invention, 243
Tuvan throat singers, 141
Tweeter, 194
Two-way speaker system. *See* Loudspeaker, two-way system

Ud, 319
Ukulele, 320
Ultrasonic cleaning, 63
Ultrasonic dental drilling, 62
Ultrasonic dentistry, 62
Ultrasonic duplex scanning, 61
Ultrasonic fetal stethoscope, 61
Ultrasonic frequencies, 59
Ultrasonic lithotripter, 62
Ultrasonic microscope, 62
Ultrasonic machining, 63
Ultrasonic soldering, 63
Ultrasonic sonogram, 61
Ultrasonic wave, 59
Ultrasonic weld inspection, 63
Ultrasonics, 59–63
diagnostic, 60–61
diffraction effects, 60
industrial applications, 63
medical applications, 60–62
nondestructive testing, 63
relation to sound, 59
relation to X-rays, 60
scientific research, 62
"seeing" with, 62
treatment, 61–62
Ultrasound:
absorption, 60
diffraction, 60
heating by, 62
Ultraviolet wave. *See* Wave, electromagnetic
Una corda pedal, 343
Unbalanced modulation. *See* Modulation, unbalanced
Uniform circular motion, and SHM, 6
Units, acoustical terms, 371
U. S. Standard pitch. *See* Pitch level, U. S. Standard

Valotti, Francesco Antonio (1697–1780), 240, 251
Valotti temperament. *See* Temperament, Valotti
Valve:
alternatives, 306
brass, 299
piston, 300, 302
rotary, 300, 302
tuning errors, 307
Valve tuning compromise, 307
Varnish, violin, 322 VCO. *See* Voltage-controlled oscillator
Velocity, 2
See also Speed
Velocity (anti)nodes. *See* Longitudinal standing wave, pressure and velocity correspondence
VGA. *See* Voltage-controlled amplifier
Vibraharp. *See* Vibraphone
Vibraphone, 355–356
bars, 355
butterfly valves, 355
decay time, 356
tremolo, 355

Vibrato, 106, 122, 131
clavichord, 336
diaphragm, 106
violin, 324
Vielle, 320
Vihuela, 320
Viola, 329
range, 322
Viola da braccio, 320
Viola da gamba, 319
range, 322
Viola d'amore, 319
Violin, 325–330
air resonance, 326
bass bar, 325
bridge, 325–326
double stops, 158
family, 320, 325–330, 329
f-hole, 325
fingerboard, 329
glue, 328
harmonics, 102, 330
artificial, 330
natural, 330
holographic interferograms, wooden plates, 327
loudness curve, 326–327, 328
modifications, since baroque, 329
pitch rise, effect of, 321, 329
pizzicato, 330, 331
range, 322
regular playing, effect of, 328
scaling, 327
solo and group. *See* Chorus effect
sound post, 325
spectrum of, 102
strings, 327
tension in, 328
tap tone, 87, 326
varnish, 328
wave form, 102
wood, 327
wood resonance, 326
Violoncello. *See* Cello
VirSyn CUBE Software Synthesizer, 136–137
Vocal folds, 166
air rushing through, 166
tension, 167
Vocal formants. *See* Formants, vocal
Vocal sounds, Fourier spectra of, 169, 172
sound spectrogram, 170, 174–175
Vocal tract:
anatomy, 166–167
resonances, 167
Voice, human. *See* Human voice
Voice programming, 174
Voice recognition, 174
use by hearing impaired, 175
Voice synthesis, 174
Voice spectrogram. *See* Audio spectrogram
Voicing, 269
Volt (V), 183
Voltage, 183
Voltage-controlled amplifier (VCA), 129
Voltage-controlled oscillator (VCO), 128
Volume control, 196
logarithmic scale, 196
Volume unit (VU) meter, 202
Vowels, 168–170
sound spectrograms, 171, 173–174
spectral analysis, 169–170

Warmth, 219
Water, poured-in-cylinder experiment, 113–114
Wave:
circular, 30
definition, 23
electromagnetic, 24, 25
energy transfer, 23
front, 30–31
infrasonic, 63–64
intensity of, 33
longitudinal, 21
See also Longitudinal wave
periodic, 9–11
propagation of, 30–31
properties of, 29
fundamental, 29
general, 34
pulse train. *See* Pulse train
radio. *See* Radio waves
rope, 25
sawtooth. *See* Sawtooth wave
sound, 27
slinky spring sound model, 27
speed, 27–29
spherical, 33
spring, 27
square. *See* Square wave
standing. *See* Standing wave
transverse. *See* transverse wave
triangular. *See* Triangular wave
ultrasonic, 59–63
Wave addition, 47–50, 70–71, 120
different amplitudes, 93
different periodicity, 92–93
different phase, 93
graphical, 49
with missing fundamental, 97
pulses, 47
with two harmonics, 92–94, 97
See also Complex waves; Fourier synthesis
Wave form, 9
digital, 133–134
musical tone, 10
noise, 10
open tube, blown, 113
pouring water, 114
pulse train, 9
pure tone, 9
sawtooth, 9
sine, 9
singing voice, 10
square wave, 9
standard waves, 9
triangular, 9
Warming up. *See* Pitch level, rise in pitch, during instrument warmup
Warmth, 219
Wave, periodic, 9
circular, 30
plane, 30
Wavelength (λ), 26
Weight, 3
"Well-tempered," 239
Well-Tempered Clavier, 239, 240
Werckmeister, Andreas (1645–1706), 239, 251
Werckmeister temperaments. *See* Temperament, Werckmeister Correct Temperament No. 1

Whip, 59
Whispering chamber, 36
Whisper key, bassoon, 286
White light, 107
White noise, 107
See also Noise, white
filtered, 107
Wind:
electronic generation of, 132
howling, 132
refraction effect, 40
Wind chest, organ, 289
Wind screen, 191
Wine glass, singing, 87
Wire, stretched. *See* Standing wave, stretched string
Wire, wrapping of, piano, 341
Wire-wound strings, violin, 327
Wolf fifth, 244
Wolf note, cello, 326
Wood prime resonance, violin, 326
Wood resonance, violin, 326
Woodwind instruments
See also Clarinet; Flute; Recorder
antinode, location, 268
attack transients in, 265
baroque, 269
bore, 262, 266–267
capped-reed, 261
chamfering holes, 271
diameter and tone, 266
modification, 266, 271
shape and overtones, 266
shapes, 266
double reed, 266
edge tone, 264
families, 262, 263
finger holes, 266–271
fork fingering, 270
history, 260–264
hole diameter, 269
hole effective size, 271
hole spacing, 269
hole volume, 271
intonation problems, 269
material, effect of, 271
noise source, 265
octave hole, 268
open holes, effect on tuning, 271
pitch errors, 269
reed, 265–266
sound production and, 265
Renaissance, 269
tone hole cutoff frequency, 270
tone quality, 264–272
elements, 264
transposing, 263
voicing, 269
Woofer, 194

X-rays. *See* Wave, electromagnetic
Xylophone, 354
bars, 86, 354
resonators, 354

Yamaha WX5 Wind MIDI Controller, 139